Carbon Nanostructures

More information about this series at http://www.springer.com/series/8633

Gil Gonçalves · Paula Marques
Mercedes Vila
Editors

Graphene-based Materials in Health and Environment

New Paradigms

Springer

Editors
Gil Gonçalves
Nanoengineering Research Division, Department of Mechanical Engineering
Centre for Mechanical Technology and Automation
University of Aveiro
Aveiro
Portugal

Mercedes Vila
Nanoengineering Research Division, Department of Mechanical Engineering
Centre for Mechanical Technology and Automation
University of Aveiro
Aveiro
Portugal

Paula Marques
Nanoengineering Research Division, Department of Mechanical Engineering
Centre for Mechanical Technology and Automation
University of Aveiro
Aveiro
Portugal

ISSN 2191-3005 ISSN 2191-3013 (electronic)
Carbon Nanostructures
ISBN 978-3-319-45637-9 ISBN 978-3-319-45639-3 (eBook)
DOI 10.1007/978-3-319-45639-3

Library of Congress Control Number: 2016949578

Printed on acid-free paper

This Springer imprint is published by Springer Nature
The registered company is Springer International Publishing AG
The registered company address is: Gewerbestrasse 11, 6330 Cham, Switzerland

Contents

Part II Medicine: Biosensors

Part III Medicine: Tissue Engineering

Part I
Medicine: Nanomedicine

Potential and Challenges of Graphene in Medicine

Marta Skoda, Ilona Dudek and Dariusz Szukiewicz

Abstract Graphene due to its excellent properties has attracted the great attention in the area of nanomedicine. Due to the high surface area and capability of biofunctionalization graphene provides an efficient platform for drug and gene delivery. Many studies indicate that graphene is an attractive tool for cancer diagnosis and therapy, allowing the improvement of already existing techniques by providing more precision and effectiveness in cancer treatment and also by reducing secondary side effects. Furthermore, graphene is able to induce tissue-specific inductive capabilities which are desirable in tissue engineering and its high biocompatiblity makes it very suitable for the growth and maintenence of adherent cells. In vitro studies show that graphene promotes stem cell growth and differentiation which makes it a valuable nanomaterial in regenerative medicine. However, because of the variety of different forms of graphene and different methods of synthesis, the existing findings regarding graphene toxicity and biological interactions are ambiguous and sometimes even contradictory. The inconsistency of available data and the lack of sufficient information make it hard to fully assess the suitability of graphene as a biomaterial or nanocarrier. Indeed, more systematic and standarized research procedures in graphene production are required. In this chapter we will focus on the possible applications of graphene-based materials in numerous areas of medicine such as cancer therapy, drug and gene delivery, tissue engineering and bioimaging.

M. Skoda (✉) · I. Dudek · D. Szukiewicz
Department of General and Experimental Pathology, Centre for Preclinical Research and Technology (CEPT), Medical University of Warsaw, Pawinskiego 3C, 02-106 Warsaw, Poland
e-mail: skoda.marta@gmail.com

I. Dudek
e-mail: ilodudek@gmail.com

D. Szukiewicz
e-mail: dariusz.szukiewicz@wum.edu.pl

G. Gonçalves et al. (eds.), *Graphene-based Materials in Health and Environment*, Carbon Nanostructures, DOI 10.1007/978-3-319-45639-3_1

1 Introduction

Nanotechnology provides an essential improvement in many areas of science ranging from molecular physics to medicine. The medical applications of nano-technology provide a wide range of new strategies useful in diagnosis, treatment, and disease prevention. The discovery of graphene, a two-dimensional crystal composed of carbon atoms arranged in a single-layer structure, triggered a multidisciplinary research in many areas of science. Because of its unique properties such as chemical inertness, electrical conductivity, high flexibility, and ability to adapt to different surfaces, graphene has enabled a wide application potential in biomedicine [1]. However, many of currently existing systems still lack stability, specificity, or induce cytotoxicity which reduces their efficacy. In the last few years, an unlimited potential application of graphene was extensively explored. Graphene family nanostructures provide new application opportunities in medicine (Fig. 1). Indeed, they have been extensively explored in tissue engineering, drug and gene delivery nanoplatforms, and as bioimaging agents due to their unique structure and ability to interact with living cells and tissues. Numerous studies have proved graphene as a promising biomaterial with the potential to replace and improve existing ones. However, some concerns remain relatively to the potential toxic effects of graphene-based materials. The biocompatibility studies performed until now showed very divergent results, which do not allow to obtain clear evidences regarding the safe use of graphene derivatives in the human body.

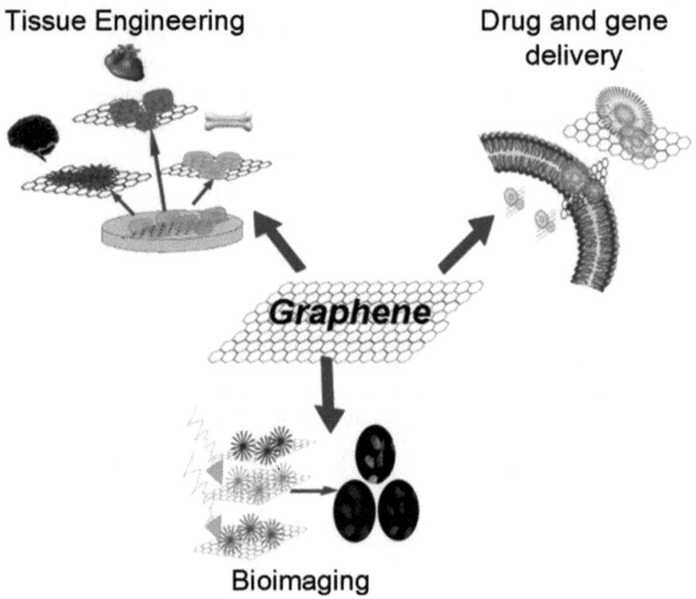

Fig. 1 Possible applications of graphene-based nanomaterials in medicine

1.1 Graphene Nanomaterials Biocompatibility

Graphene is composed of sp^2-hybridized carbon atoms arranged in monolayer with large flat surface (2630 m^2/g) [2]. The planar structure of graphene is susceptible to participate in many reactions and surface modifications allowing attachment of various chemical compounds in high concentrations via chemical bonding, electrostatic interactions, or physisorption. However, graphene functionalization required for specific applications slightly changes its properties. The bending rigidity of single-layered graphene equals 1.44 eV (2.31×10^{-19} N m) which resembles the values of lipid cell bilayer (1–2 eV) and makes graphene suitable for many biological applications [3]. Most of the nanovector systems enter into the cells by endocytosis and release the drug into the endosomal compartment. The physical interactions of graphene with the cell membranes are one of the major causes the cytotoxicity of graphene family materials [4]. Due to the 2D structure and specific surface area, graphene can interact with different organic molecules and even pass through the cell membranes without damaging it. Inside the cells, one graphene derivative, graphene oxide (GO), can be surrounded by endosome-like structures or can accumulate in the cytoplasm which suggest a combination of active and passive mechanisms involved in the GO uptake [5]. However, the precise uptake mechanisms are not fully recognized and precise questions concerning nonspecific interactions of cell membrane with nanoparticles still remain to be answered. The Langmuir monolayer technique used to understand the relation between GO and lipid models showed that they are governed by electrostatic interactions. GO can be incorporated only into the monolayer of positively charged lipids, but not into negatively and neutrally charged phospholipids present in the cell membranes. GO enters the cells through the membrane rather than by electrostatic interactions [6].

Graphene influences the processes occurring on the cell surface including cellular transport or polarization. Hydrophobic forms of pristine graphene interact with the cell membrane lipids, while the other forms may bond to the cell receptors and modify the cell metabolism, inhibit nutrient supply, and induce oxidative stress or cell death [7]. The interactions of GO with the cells may lead to excessive ROS generation, which is the first step in the mechanisms of aging or cancerogenesis. More dangerous are direct interactions of graphene with cell genetic material based on DNA intercalations indicating mutagenic potential [8, 9]. Reduced graphene oxide (rGO) has high affinity to the cell membranes and the irregular and sharp edges may affect their integrity and activate mitochondrial pathways causing apoptosis [10].

The cytotoxicity of graphene depends on its size, shape, or surface coating which determines the interactions with the cell membranes, intracellular uptake, and clearance pathways. In a comparative study, large, small, and very small GO sheets differed in the behavior toward the examined cells. The smaller particles are more cytotoxic than bigger ones and cause apoptosis by the direct cellular membrane damage (Fig. 2) [5]. GO, as an amphiphilic molecule, contains both free surface π

electrons from unmodified areas of graphene capable for noncovalent functionalization and negatively charged groups which provide colloidal stability allowing weak interactions, hydrogen bonding, and other various reactions [11]. These characteristics can be exploited while designing high loading graphene nanocarriers. The surface functionalization of nanomaterials is crucial for their biocompatibility and controlled behavior. Electrostatic interactions between modified carbon surface and cell phospholipid bilayer may facilitate the transmembrane drug delivery [11].

Graphene biocompatibility and solubility can be improved by several surface modifications. However, additional coatings change the graphene properties and influence the interactions with the cells. One of the most frequently used polymers for graphene surface functionalization is poly(ethylene glycol) (PEG). PEG coating provides good dispersibility, enhances photothermal effect, helps GO to avoid macrophage recognition, and does not cause hemolysis [12]. The comparison of GO samples engineered with PEG, poly(ether imide) (PEI), and bovine serum albumin (BSA) demonstrated different mechanisms of macrophage endocytosis. Positively charged GO-PEI favored membrane binding and endocytosis, amphiphilic BSA restricted cell interactions, while negatively charged GO-PEG complexes hindered endocytosis almost completely. Additionally, GO-PEI with a high positive zeta potential appeared to be the most harmful to macrophages, whereas PEGylated GO with more neutral potential had a negligible impact on the cell viability [13]. Polyethylene glycol-grafted GO had no influence on mice survival after intravenous administration and improved the delivery of tested drugs (Fig. 3) [14].

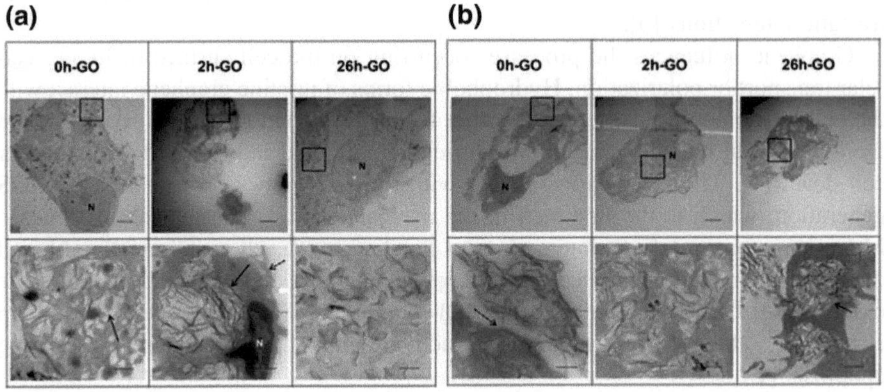

Fig. 2 TEM images of human monocyte-derived macrophages (hMDM) (**a**) and primary murine intraperitoneal macrophages (mIPM) (**b**) incubated with 0 h-GO, 2 h-GO, and 26 h-GO (50 μg ml^{-1}) for 24 h. Pictures at the *bottom* consist of the enlargement of the *dotted areas* identified in the respective images at the top. *Black dotted arrows* indicate cellular membrane; *black arrows* indicate vesicles; *N* nucleus. *Scale bars* correspond to 2 μm (*top line*) and 500 nm (*bottom line*), respectively. Reproduced from Russier et al. [5] with permissions, Copyright 2013, Royal Society of Chemistry

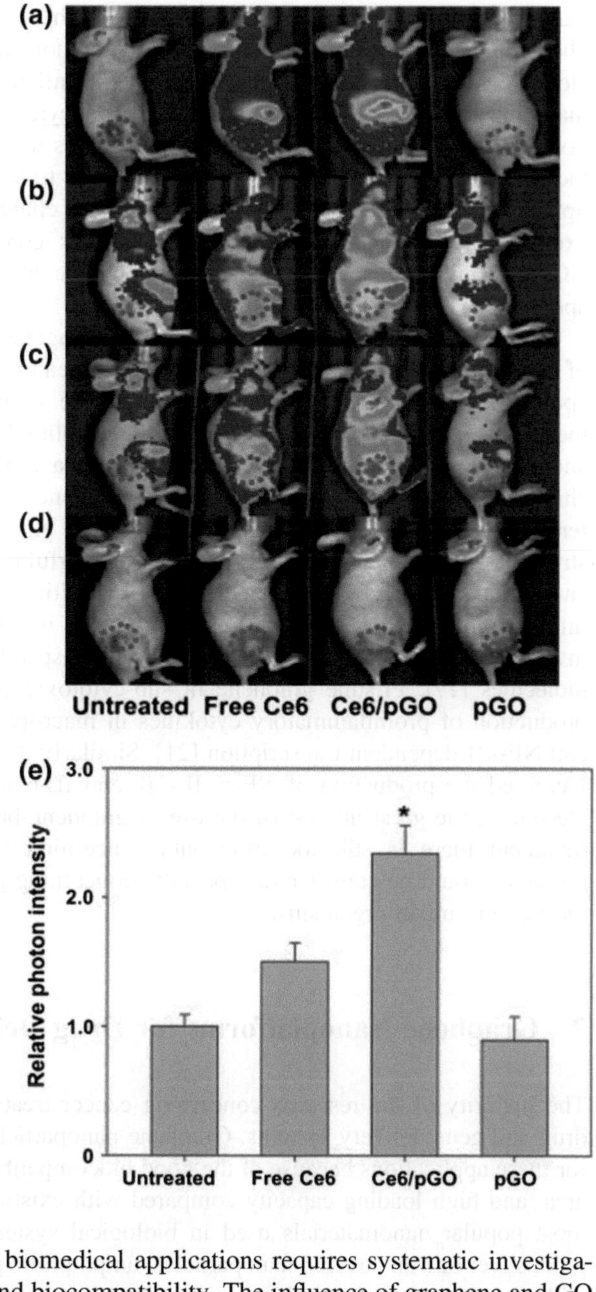

Fig. 3 In vivo biodistribution of pGO nanophysisorplexes. SCC7-bearing mice were systemically treated with pGO, free Ce6, Ce6/pGO, or with Ce6/Dox/pGO (Ce6 10 mg/kg and Dox 5 mg/kg). After 1 h (**a**), 24 h (**b**), and 48 h (**c**), the in vivo distributions of Ce6 fluorescence were visualized using a molecular imaging system. **d** Optical images are provided for location of tumors. **e** Relative photon counts of tumor sites in comparison with untreated group were quantitated at 48 post-dose by in vivo imaging system. *Significantly higher ($p < 0.05$) compared to the other groups (assessed by ANOVA and the Student–Newman–Keuls test). Reproduced from Miao et al. [14] with permissions, Copyright 2013, Elsevier Limited

The use of graphene for biomedical applications requires systematic investigations regarding its toxicity and biocompatibility. The influence of graphene and GO on the immune system depends not only on the shape, chemical composition, and method of synthesis but also on the route of administration (e.g., intravenous, nasal) [15]. The interactions with blood are especially important while introducing graphene intravenously in case of cancer treatment or in the contact with blood and material surface. The research showed that both pristine and functionalized graphene possess anticoagulation properties and do not cause red blood cell hemolysis. However, they induced the production of proinflammatory cytokines in murine macrophages without the influence on lymphocyte proliferation. Graphene functionalization reduced toxicity of pristine samples and allowed the entrance inside the cells [16]. Pristine graphene flakes stay close to the cell membranes, hinder the uptake of nutrients, and interact with surface channels and molecules causing excessive ROS generation. Another study has confirmed high cytotoxicity of GO-PEI conjugates concerned with lymphocyte membrane damage resulting in apoptosis [17].

Toxicity of graphene is desirable when used against cancer cells but not in case of surrounding healthy tissue. Pristine graphene can induce cytotoxicity and trigger apoptosis in murine macrophages through the reduction of the mitochondrial membrane potential and intracellular ROS generation [18]. However, both graphene and GO can effectively inhibit cancer cell metastasis by the influence on mitochondrial respiration [19]. Additionally, graphene can directly interact with different genes encoding important proteins and enzymes. The difference in the structure of rGO and GO makes rGO more powerful to penetrate cell compartments and directly interacts with the nuclear DNA resulting in genotoxic effects [20]. The influence of graphene on the immune system is still at a very early stage of investigation together with the activation of specific immunological response molecules [17]. Pristine graphene in sub-cytotoxic concentrations stimulates the production of proinflammatory cytokines in macrophages through TLR-mediated and NF-κB-dependent transcription [21]. Similarly, GO alone or with PVP coating increased the production of NF-α, IL-1β, and IL-6 in dendritic cells (DCs) [22]. Despite of the great interest of the use of graphene-based nanomaterials in cancer treatment there is still too little data concerning graphene immunity. Further research should be considered, especially concerning graphene biodegradation and the fate in human organisms.

2 Graphene Nanoplatforms for Drug Delivery

The majority of the research concerning cancer treatment (73 %) focuses on the drug and gene delivery systems. Graphene nanoparticles are the perfect candidates for these applications because of the good biocompatibility, extremely large surface area, and high loading capacity compared with existing systems [23]. Among the most popular nanomaterials used in biological systems there are relatively inert gold, silver or iron covered nanoparticles, amphiphilic polymers and nanoliposomes with the central hydrophobic region, dendrimers, carbon nanotubes, fullerenes, carbon bullets, and recently graphene nanomaterials with the flat surface prone for the modification [24]. In the last few years an unlimited potential of GO was extensively explored which can be confirmed by a growing number of publications regarding its recent applications in drug delivery (Fig. 4).

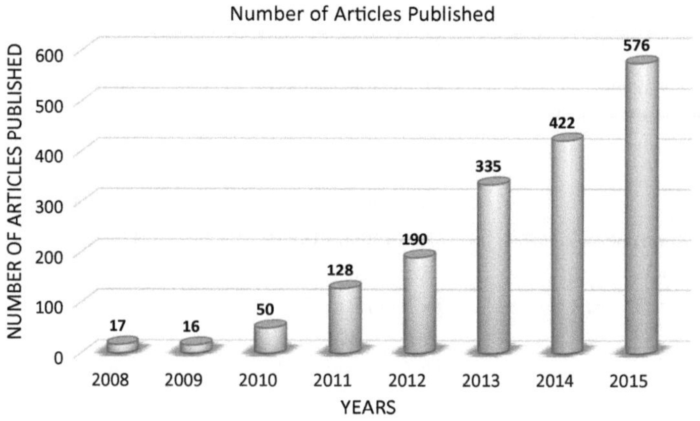

Fig. 4 The growing number of scientific publication regarding GO as a drug delivery system. Data based on ScienceDirect statistics

Nanoscience gives the possibility to create multifunctional drugs able to recognize and exterminate cancer cells. Specially designed nanoparticle systems loaded or bound to anti-cancer drugs can enhance the delivery and intracellular accumulation of desired substances in selected organs or tissues. The drugs can be protected from the enzyme degradation and released from nanoplatforms in a controlled manner, under appropriate pH or temperature [23]. For the first time in 2005, the U.S. Food and Drug Administration (FDA) approved Abraxane® as a novel nanosystem solution for more effective treatment option for patients with metastatic breast cancer. It is a combination of paclitaxel and albumin particles which facilitate intratumor drug accumulation and allow to overcome the solvent-related problems. The FDA is currently reviewing 130 similar nanoparticles with various drugs for more precise biodistribution and cancer treatment [25]. Different graphene-based materials have been designed in order to eliminate undesirable side effects such as toxicity or immunogenicity, to achieve maximum efficacy and to improve drug pharmacokinetics and pharmacodynamics [26, 27]. Additionally, graphene-based nanomaterials functionalized with known biopolymers can be successfully loaded with several drugs and used for even more precise targeted cancer therapies (Tables 1, 2). Poorly soluble substances can be conjugated with graphene and its derivatives to increase their solubility and stability without losing their efficiency.

Among all the graphene family members, GO possesses the highest drug loading capacity and a number of advantages. GO can be successfully bound with doxorubicin (DOX), widely used anti-cancer drug, as an intracellular cytosolic delivery system [28]. The administration of DOX immobilized on GO surface through π-π interactions bypasses cellular barriers and allows the direct drug release in appropriate acidic pH around the cancer [29]. Additionally, the proper surface coating of

Table 1 Graphene-based in vitro models for drug delivery

Nanohybrid	Drug	Target cells
PF127-GN (pluronic F127-graphene nanosheets)	Doxorubicin (DOX)	MCF-7 [105]
FA-CD-GO (folic acid-modified β-cyclodextrin)	Doxorubicin (DOX)	HeLa [106]
NGO-HDex	Doxorubicin (DOX)	MCF-7 [107]
HA-GO (hyaluronic acid)	Doxorubicin (DOX)	HepG2 [108]
PEG-BPEI-rGO	Doxorubicin (DOX)	PC-3 [109]
GO@Ag (Asn-Gly-Arg peptide motif)	Doxorubicin (DOX)	MCF-7 [110]
GO-DTPA-Gd (diethylenetriaminepentaacetic acid-gadolinium)	Doxorubicin (DOX)	HepG2 [26]
MGMSPI (magnetic graphene-mesoporous silica)	Doxorubicin (DOX)	U251 [64]
PEG-GO/CuS	Doxorubicin (DOX)	HeLa [111]
CGN (rGO-chitosan)	Doxorubicin (DOX)	TRAMP-C1 [112]
GO-SiO_2 microspheres	Doxorubicin (DOX)	HepG2 [63]
NGO-PEI-Ag	Doxorubicin (DOX)	HeLa [113]
GON-Cy-ALG-PEG	Doxorubicin (DOX)	HepG2 [114]
MSP-BA-GOF (mesoporous silica-broic acid-folate GO)	Doxorubicin (DOX)	HeLa [115]
RGD-RC-GO (cyclic RGD-modified chitosan)	Doxorubicin (DOX)	Bel-7402, SMMC-7721, HepG2 [64]
rNGO@mSiO_2@pNIPAM-*co*-pAAm (mesoporous silica shell)	Doxorubicin (DOX)	HeLa [116]
GIAN (graphene-isolated-Au-nanocrystal)	Doxorubicin (DOX)	MCF-7 [117]
MGMSPI (mesoporous silica coating)	Doxorubicin (DOX)	U251 [64]
CHA-rGO (cholesteryl hyaluronic acid)	Doxorubicin (DOX)	HeLa [118]
nGO-PEG	Doxorubicin (DOX)	EMT6 [32]
FA-NGO-PVP	Doxorubicin (DOX)	HeLa, A549 [119]
FA-nGO (folic acid)	Doxorubicin (DOX), Camptothecin (CPT)	MCF-7 [120]
rGO-PEI-FA	Elsinochrome A (EA)/ Doxorubicin (DOX)	CBRH7919 [121]
GO	Adriamycin (ADR)	MCF-7 [122]
PPG (polyethylenimine/poly (sodium 4-styrenesulfonates))/GO)	Adriamycin (ADR)	MCF-7 [123]

(continued)

Table 1 (continued)

Nanohybrid	Drug	Target cells
GO-CD-HAADA (hyaluronated adamantine)	Camptothecin (CPT)	MDA-MB-231 [124]
GO-CS (chitosan)	Camptothecin (CPT)	HepG2 [30]
FA-PEG-Pep-GO	Camptothecin (CPT), Curcumin (CUR), Evodiamine (EVO), Silbin (SIL)	HeLa, A549, HUVEC [125]
GO-PEG	Paclitaxel (PTX)	A549, MCF-7 [126]
GO-IO	Anastrozole (ANS)	MCF-7 [127]
Gn	Gambogic acid (GA)	MCF-7, Panc-1 [128]
GO-F38 (pluronic F38), GO-T80 (Tween 80), GO-MD (maltodextrin)	Ellagic acid (EA)	MCF-7, HT29 [129]
GO, DGO, GQD	Curcumin (CUR)	HCT 116 [130]
G@C-folate	Curcumin, Paclitaxol, Camptothecin, Doxorubicin	HeLa [131]
GR-PY^+-Chol	Tamoxifen Citrate (TmC)	A549, HeLa, HepG2, NIH3T3, MCF-7, MDA-MB231 [132]
Fe_3O_4-rGO	β-lapachone (β-lap)	MCF-7 [133]
Fe_3O_4-GN (MGNs)	5-fluorouracil (5-FU)	HepG2 [134]
GO-PAA (polyacrylic acid)	BCNU (1,3-bis (2-chloroethyl)-1-nitroso urea)	GL261 [135]

GO with chitosan, a naturally occurring polysaccharide, helps to eliminate GO hemolytic activity. Chitosan-GO complexes possess high loading capacity for water-insoluble anticancer drug, camptothecin (CPT) [30]. To fully utilize the properties of GO and avoid possible limitations some other efficient methods have been used. GO was successfully functionalized with various highly biocompatible substances such as PEG, BSA, dextran (DEX), or poly(amidoamine) (PAMAM) dendrimer [12]. PEG has been approved by the FDA for the human use as a non-toxic polymer easily eliminated from organism. PEG in conjugation with GO provides good dispersibility and prevents the macrophage uptake [12]. For the first time, in 2008 Liu and coworkers synthesized PEGylated GO conjugated with water-insoluble aromatic molecule SN38 which is a camptothecin (CPT) analog. They presented an ultrasmall SN38 drug molecule complex with graphene as a potential colon cancer treatment [31]. Zhang et al. demonstrated an effective drug delivery system composed of polyethylenimine (PEI)-functionalized GO (PEI-GO)

Table 2 Graphene-based in vivo models for drug delivery

Nanohybrid	Drug	Target cells
HA-GO (hyaluronic acid)	Doxorubicin (DOX)	H22 hepatic cancer cell-bearing mice [108]
GO, DGO, GQD	Curcumin (CUR)	HCT tumor-bearing mice [130]
GO@Ag-NGR (Asn-Gly-Arg peptide motif)	Doxorubicin (DOX)	S180 tumor-bearing mice [110]
PEG-GO/CuS	Doxorubicin (DOX)	HeLa cell-bearing mice [38]
GO	Cisplatin (CDDP)	CT26 colon tumor-bearing mice [136]
nGO-PEG	Doxorubicin (DOX)	EMT6 cell-bearing mice [32]
HA-GO/pluronic	Mitoxantrone (MIT)	MCF-7 cell-bearing mice [137]
FA-PEG-Pep-GO	Camptothecin (CPT), Curcumin (CUR), Evodiamine (EVO), Silbin (SIL)	HeLa tumor-bearing mice [125]
Tf-PEG-GO (transferrin)	Doxorubicin (DOX)	C6 glioma-bearing rats [23]
rGO-CHA (cholesteryl hyaluronic acid)	Doxorubicin (DOX)	KB tumor-bearing mice [14]

with high effectiveness in chemotherapy [32]. However, studies on surface-engineered graphene nanomaterials revealed that GO-PEI conjugates tend to interact with mitochondria and change the membrane potential followed by intracellular ROS generation and cytochrome C release [33]. On the other hand, Kim and colleagues successfully synthesized biocompatible nanotemplate composed of reduced GO covalently conjugated with both branched polyethylenimine (BPEI) and PEG for the direct release of DOX into the cancer cells by endosomal disruption triggered by photothermal stimulation [34] (Fig. 5). Doxorubicin hydrochloride (DXR) was also applied on the nanohybrid formed by a chemical deposition of Fe_3O_4 particles on GO surface [35]. Soluble GO-Fe_3O_4-DXR complexes accumulate in acidic environment, move in the magnetic field, and hence they can serve as a candidate for pH-triggered targeted therapy [35, 36]. An outstanding property of gold nanoparticles has been used for graphene functionalization which resulted in drug delivery platform with near-infrared thermal properties useful in cancer treatment [37]. Similarly, the combination of PEG-GO nanohybrid with CuS particles grafted on its surface exhibited pH-dependent drug release enhanced additionally by NIR laser irradiation [38].

Graphene-based materials have been extensible explored as a bioplatform for multiple drug conjugations, including doxorubicin (DOX) camptothecin, rituxan,

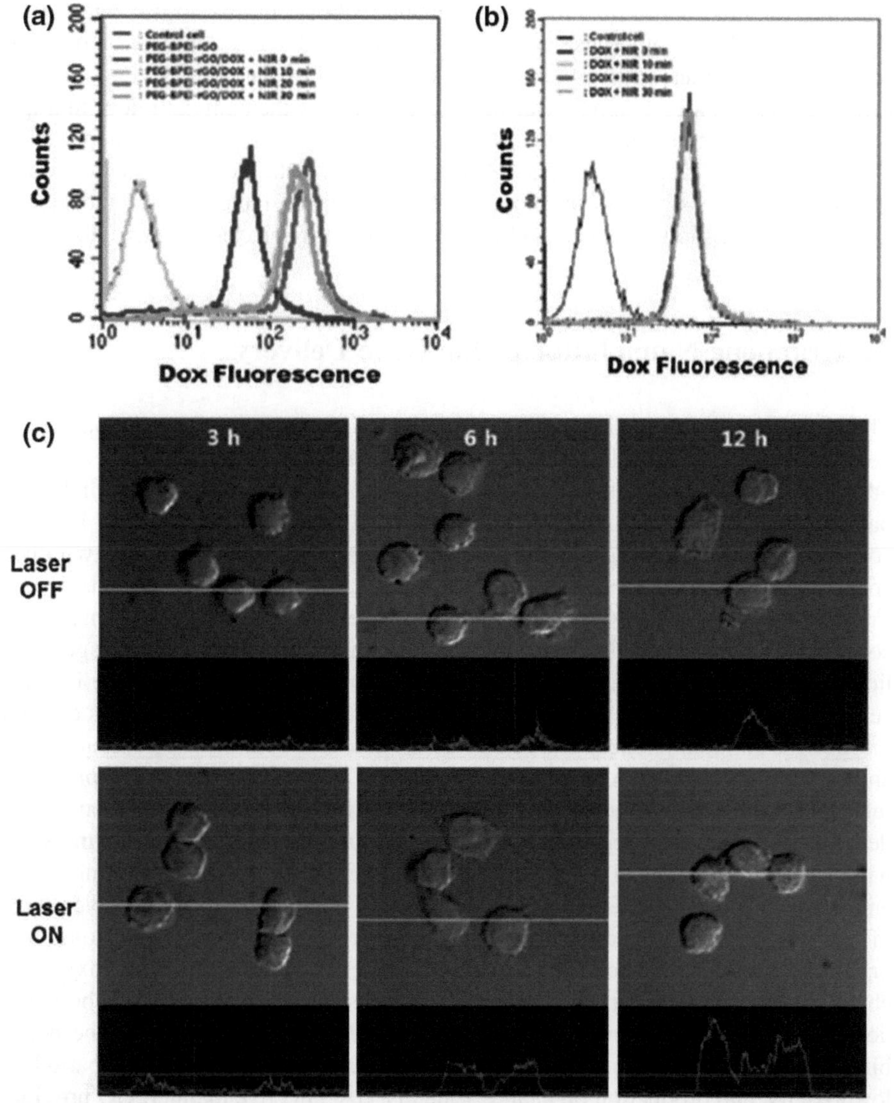

Fig. 5 NIR-responsive DOX release from PEG-BPEI-rGO in vitro. Flow cytometry histogram profile of recovered DOX fluorescence from PEG-BPEI-rGO/DOX complexes (**a**) and free DOX fluorescence (**b**) in PC-3 cells upon different NIR irradiation times. **c** Confocal microscopic images in PC-3 cells treated with PEG-BPEI-rGO/DOX complexes in the dark (*top*) and under NIR irradiation (*bottom*) after various incubation times (3, 6, 12 h). Nuclei were stained with DAPI (blue), DOX has specific light emission (*red*). Reproduced from Kim et al. [34] with permissions, Copyright 2013, American Chemical Society

cisplatin, adriamycin, paclitaxel, and many more [24]. Tables 1 and 2 present some already existing graphene-based nanosystems used both in in vitro and in vivo models which prove their efficacy in cancer treatment. Graphene nanohybrids allow to reduce the dose needed for the treatment and make it possible to use some of the drugs that would not otherwise be applied independently. The conjugation of drugs with graphene enhances their cellular uptake and inhibits the proliferation of cancer cells [12]. The unique graphene surface not only allows the conjugation of multiple drugs but also of different proteins, RNA, DNA, antibiotics, or other nanoparticles. The possibility of graphene functionalization together with the reduced cytotoxicity offers endless possibilities in biomedical applications including diagnosis and treatment of infectious diseases and cardiovascular or neurodegenerative disorders.

3 Graphene Nanoplatforms for Gene Delivery

The evolution of nanomedicine and molecular biology has led to the development of new alternative and novel therapies associated with the targeted drug delivery or other therapeutic molecules including genes locally administered at the site of disease pathology. Gene therapy can be used not only to treat cancer but also in other conditions which results from the presence of defective genes. There are two main approaches that can be utilized in gene therapy, transduction, and transfection. Transduction mediated by viral vectors is highly efficient but presents some safety concerns. Non-viral transfection involves plasmids carrying the transgene or liposome-entrapped plasmids, which facilitates the entry of the transgene into the cell. Nucleic acids due to their hydrophilic properties and size cannot pass directly through the cell membranes. The role of gene vectors is to allow the cellular uptake and protect DNA or RNA from nucleases [39]. To overcome the safety problems related to viral vectors the chemical approaches for the gene delivery have been developed. Currently used non-viral gene vectors include lipids, polymers, or nanomaterials (gold and silica nanoparticles), which interact with the plasmid DNA and form nanocarriers able to pass through the cell membranes [40]. However, most of these systems present several drawbacks concerning aggregation, low solubility, or immunogenicity. To increase the transfer efficiency and minimize the toxicity of chemical reagents many modifications of these vectors have been established. The development of nanotechnology and growing interest in graphene gave the possibility to apply it as an efficient delivery system. Despite of good biocompatibility, low toxicity, and production on a large scale in a cost effective manner, GO provides an effective immobilizing agent for chemical drugs and nucleic acids [23]. Lu et al. demonstrated fast absorption of single-stranded DNA (ssDNA) on GO at room temperature but low binding affinity toward double-stranded DNA (dsDNA) molecules [41]. The graphene-based nanomaterials, thanks to the large surface area, high loading capacity, and the possibility of surface modifications, allow to create an efficient nanocarriers widely used for gene therapy. Zang and coworkers successfully and for the first time delivered to cancer cells small interfering RNA (siRNA) silencing Bcl-2 protein using polyethylenimine PEI-GO as a nanocarrier [32]. Formation of covalent, amide bonding between GO and PEI improved solubility and stability of GO in saline solution and changed the surface charge for more positive to facilitate the cellular uptake. Attachment of negatively charged RNA or DNA was also possible by positive zeta potential of PEI-GO complex. This may suggest that properly functionalized GO is able to interact with negatively charged biological membranes and directly deliver nucleic acids into the cells [32]. After RNA/DNA dissociation graphene is released from the cell via exocytosis (Fig. 6).

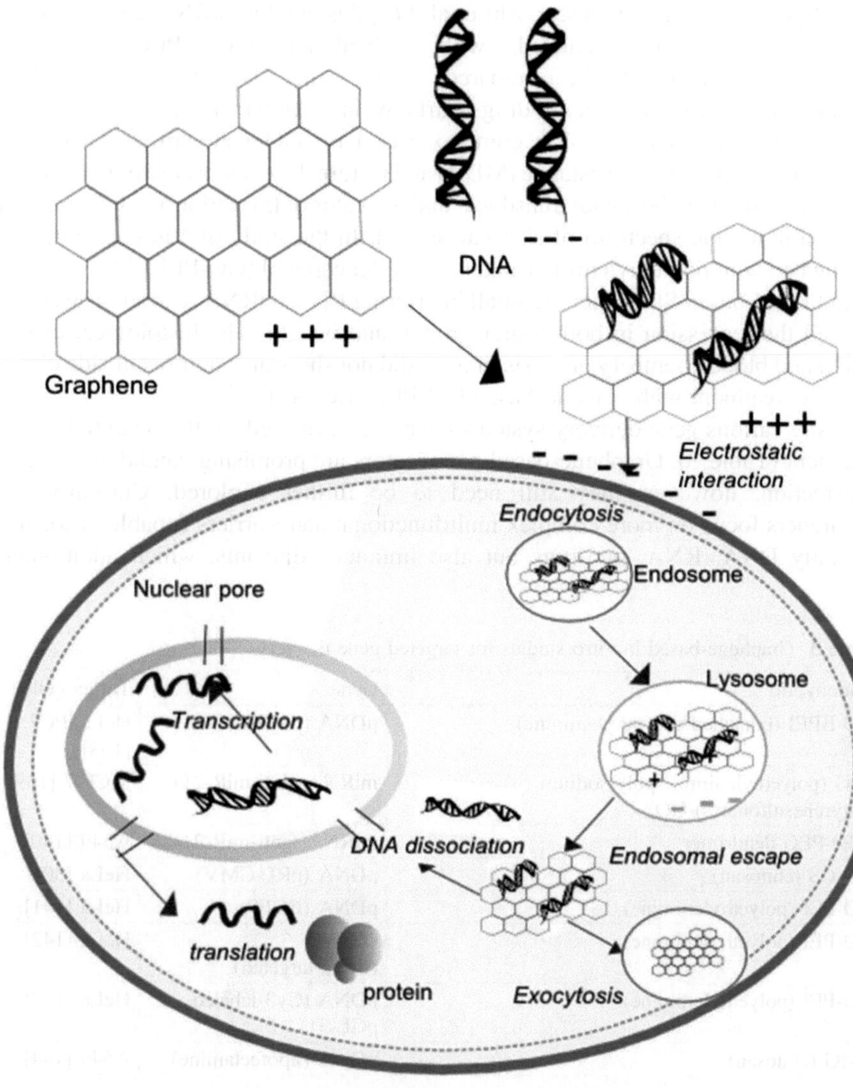

Fig. 6 Schematic representation of graphene-based gene delivery process. Functionalized graphene surface is able to interact with negatively charged nucleic acids. Graphene protects DNA from degradation and facilitates endocytosis. After DNA dissociation graphene is removed through exocytosis. The dissociation of graphene-DNA complexes is the key step for the transfection efficiency

Single-stranded, short microribonucleic acids (miRNAs) present both in the nucleus and the cytoplasm are responsible for the regulation of several gene expressions by direct interactions with their mRNA. Additionally, miRNAs can function directly on the DNA influencing transcription and epigenetic alterations and thus may serve as a potential therapeutic target. Deregulations in miRNA expression are often present in most of the breast cancer cells and frequently correlate with the drug resistance in patients with this type of cancer. Zhi et al. [42] designed a miRNA complex composed of GO functionalized with polyethylenimine (PEI)/poly(sodium 4-styrenesulfonates) (PSS), prepared using the layer-by-layer method. Administration of anticancer drug adriamycin (ADR) along with targeted silencing-miRNA (anti-miR-21) complex based on GO nanoparticles may help overcome the multidrug resistance (MDR) and restore the chemosensitivity of cancer drugs. Mutations in the signal transducer and activator of transcription Stat3 promote survival in a wide spectrum of skin cancer cell. In the study of Yin et al. GO was conjugated with polyethylenimine and polyethylene glycol (GO-PEI-PEG) and used as a plasmid-based Stat3-specific small interfering RNA (siRNA) carrier. The results showed the regression in both tumor growth and weight. The histological examination and blood chemistry analysis in mice did not show any significant side effects after the treatment with plasmid-based GO-PEI-PEG [43].

Many various gene delivery systems have been explored for the potential cancer treatment (Table 3). Graphene-based nanovectors are promising candidates in gene transfection; however, they still need to be further explored. Currently, the researchers focus on more complex multifunctional nanocarriers capable of loading not only DNA, RNA, or drugs but also immune adjuvants, which additionally

Table 3 Graphene-based in vitro studies for targeted gene delivery

Nanohybrid	Gene	Target cells
GO-BPEI (branched polyethylenimine)	pDNA (pCMV-Luc)	HeLa, PC3 [138]
PPG (polyethylenimine-poly(sodium 4-styrenesulfonates)-GO)	miRNA (anti-miR-21)	MCF-7 [139]
nGO-PEG-dendrimer	miRNA (anti-miR-21)	A549 [140]
GO-CS (chitosan)	pDNA (pRL-CMV)	HeLa [30]
GO-PEI (polyethylenimine)	pDNA (EGFP)	HeLa [141]
GO-PEI (polyethylenimine)	siRNA (Bcl-2-targeted)	HeLa [142]
GO-PEI (polyethylenimine)	pDNA (Cy3-labeled pGL-3)	HeLa [143]
CMG (chitosan)	pDNA (lipofectamine)	A549 [144]

stimulate immune response in the host organisms [44]. Such nanovectors can provide synergistic therapeutic effects triggered by the appropriate pH or photothermal irradiation.

4 Graphene-Based Materials for Photothermal and Photodynamic Therapy

4.1 Photothermal Therapy (PTT)

Electromagnetic radiation, mostly near-infrared (NIR), has been widely used for the treatment of cancer and various medical conditions. Photothermal therapy (PTT) is based on the use of photosensitizing agents which generate heat after light absorption and destroy the surrounding cells. The major challenge in such approach is to place photosensitizer selectively to avoid nonspecific reactions in healthy cells. Materials such as gold nanoparticles and carbon nanotubes (CNTs) have recently been explored extensively because of the high optical absorbance of NIR [45]. Robinson et al. prepared highly effective alternate photothermal agent from reduced GO nanosheets uncovalently bound with PEG for the potential use in photothermal therapy [46]. The studies using PEGylated graphene nanosheets in mice revealed high tumor accumulation, strong NIR optical absorption and efficient tumor destruction with no obvious cytotoxicity toward the healthy cells [47]. Sheng et al. used rGO to create a photothermal platform without any surface modifications. The researchers developed an effective nano-rGO theranostic agent in one-step uncomplicated method using bovine BSA as reductant and stabilizer of the synthesis [48] (Fig. 7). The same method was used for the construction of BSA-functionalized rGO loaded with anticancer drug DOX. The combination of high drug loading ability of

Fig. 7 Infrared thermographic maps of mice in different conditions and histological staining of the excised tumors (*bar* 50 μm). (The NIR laser, 808 nm, 0.6 W/cm^2). Reproduced from Kim et al. [48] with permissions, Copyright 2013, Elsevier

graphene with excellent light absorption resulted in twofold greater release of DOX from the DOX-BSA-rGO nanosheets when the NIR irradiation was used [49]. Laser NIR heating can also influence the cell membrane permeability allowing more efficient cellular uptake of GO nanocarriers [50]. However, the greatest improvement in photothermal ablation effect on tumor cells was achieved by the combination of gold nanoparticles with GO functionalized with an NIR dye-labeled matrix metalloproteinase-14 (MMP-14) peptide substrate (CP). The researchers presented an efficient, image-guided enhanced photothermal therapy method. Synthesized nanocomplexes exhibited efficient tumor ablation with additional fluorescence signal triggered by MMP-14 endopeptidase present on the cancer cell surface [51]. Graphene-based dual therapies for the synergistic combination of PTT and chemotherapy significantly reduce the possibility of side effects in comparison to the conventional chemotherapy and bring novel opportunities to the treatment of cancer and other diseases. Most of the studies combining graphene and PPT in cancer treatment focus on breast, cervical, lung, and brain cancers [45].

4.2 Photodynamic Therapy (PDT)

The number of research concerning the use of graphene in photodynamic therapy (PDT) is relatively low. In comparison to chemo- and radiotherapy, PDT is more specific, shows less side effects, and gives the possibility to avoid repeated doses of the drug. It is based on photosensitizers which after suitable irradiation trigger production of reactive oxygen species (ROS) inducing cytotoxicity. Most of the photosensitizers are based on porphyrin molecules. However, alone they possess some limitations including prolonged cutaneous photosensitivity, low selectivity, and poor water solubility. To eliminate these drawbacks such photosensitizing agents can be efficiently loaded on different carriers including polymers, liposomes, gold, or magnetic nanoparticles and recently also on the surface of graphene [52]. In most PDT therapies using graphene as a nanoplatform, PEG, polyvinylpyrrolidone (PVP), and other coatings are used to improve graphene biocompatibility and to ensure the stability in biological fluids. Huang and coworkers proposed an uptake mechanism of graphene-based nanocarrier for Chlorin e6 (Ce6), the second-generation photosensitizer, for the targeted photodynamic therapy against gastric tumor cells. According to them the process begins with the complex endocytosis directed by the conjugation of folic acid molecules present on the surface of GO with the specific folate cell receptors. The change in pH during endosome into lysosome conversion allows the Ce6 release and appropriate photodynamic activation after single-wavelength laser irradiation [52]. The combination of PDT and PTT was proposed by Sahu et al. The researchers used noncovalently functionalized GO with pluronic block copolymer complexed through electrostatic interaction with methylene blue, a positively charged photosensitizer. The release of photosensitizer was obtained at acidic environment of the extracellular tissues of tumor and both photothermal and photodynamic therapy were initiated after NIR light irradiation [53]. Another type of combined anticancer approach was obtained by simultaneous loading of chemotherapy drug (SN-38) and photosensitizer (Hypocrellin A) on the surface of GO using noncovalent interactions [54]. Such strategy highly improved anticancer activity of both antiproliferative agents giving the opportunity for more efficient cancer treatment. Tumor cells thanks to the overexpression of many biomarkers on their surface can be precisely localized by different molecules which are especially useful in cancer therapy and diagnosis. However, targeting of the key cellular organelles responsible for the apoptosis would be even more effective therapeutic approach. The development of such a nanosystem capable of accumulation in tumor mitochondria was recently achieved [55] (Fig. 8). The researchers constructed a nanodrug based on GO

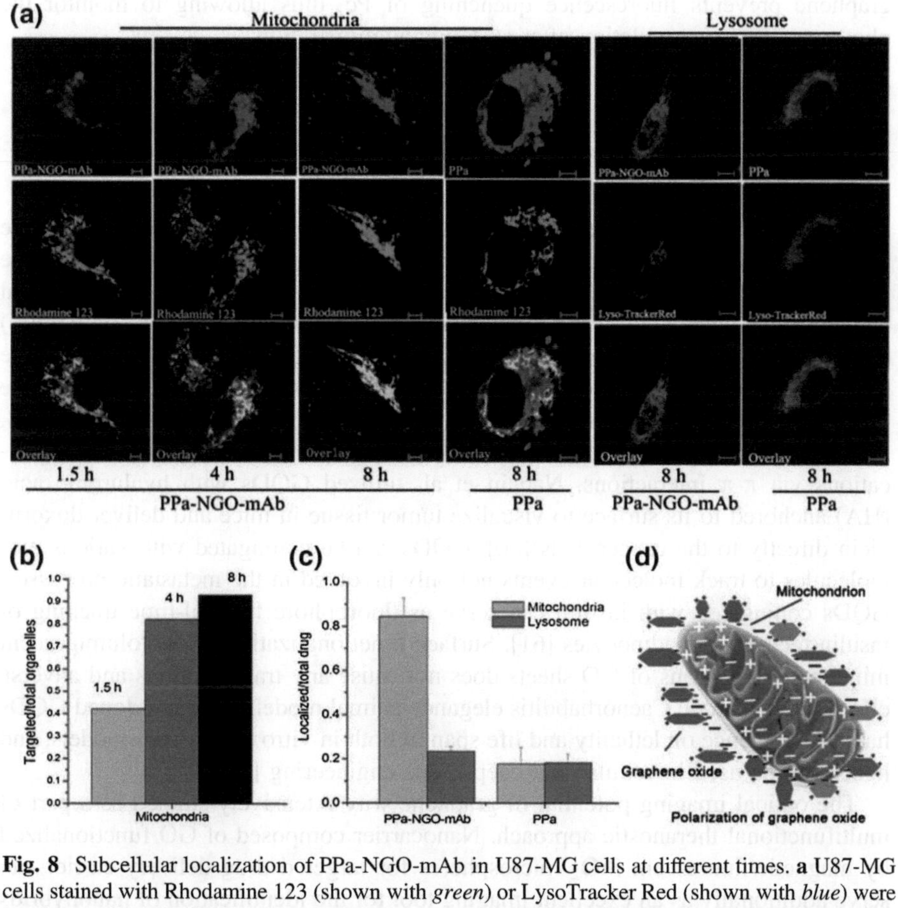

Fig. 8 Subcellular localization of PPa-NGO-mAb in U87-MG cells at different times. **a** U87-MG cells stained with Rhodamine 123 (shown with *green*) or LysoTracker Red (shown with *blue*) were incubated with PPa-NGO-mAb or only PPa for various time durations. The confocal images show the subcellular localization of PPa-NGO-mAb and PPa, using the mitochondria and lysosome-specific probes, respectively. *Scale bar* 10 μm. The histograms show the ratio of targeted mitochondria with different incubating times (**b**) and the ratio of drug localization at the mitochondria and lysosome (**c**). **d** The possible localization mechanism of NGO induced by the mitochondrion. Reproduced from Wei et al. [55] with permissions, Copyright 2016, Royal Society of Chemistry. (Color figure online)

platform coated with polyethylene glycol and conjugated with pyropheophorbide-a (PPa) to induce phototoxicity and integrin αvβ3 monoclonal antibody (mAb) for tumor targeting. They created a system which is activated not until it reaches its final destination which diminishes side effects and enhances the treatment outcome. To cope with the needs of constantly developing medicine novel dual single-light-induced phototherapy with imaging capability was proposed for the therapy of unresected ovarian cancer. Taratula et al. [56] used luteinizing hormone-releasing hormone (LHRH) to target tumor cells and phthalocyanine (Pc) as a photosensitizer loaded on polypropylenimine dendrimers (PPIG4). Graphene nanosheets conjugated with poly(ethylene glycol) served as a nanoplatform for both Pc and LHRH. Low-power single-wavelength (690 nm) NIR irradiation was used for ROS production and heat generation exhibiting pronounced antitumor effect. In addition, graphene prevents fluorescence quenching of Pc, thus allowing to monitor the photosensitizer accumulation after systemic administration.

5 Graphene Nanostructured Materials for Bioimaging

Multifunctional nanoparticles for cancer treatment and imaging are recently in the area of interest. In the context of imaging, the use of graphene is highly profitable because of excellent photostability and possibility to avoid toxic dyes and chemical agents [57]. Graphene derivatives such as GO and graphene quantum dots (GQDs) exhibit intrinsic photo-luminescence emitted after UV excitation and exhibit stable strong fluorescence along with electrical and thermal conductivity. GQDs can be used in in vivo and in vitro drug delivery monitoring or photodynamic therapy [58, 59]. Optical properties of GQDs can be additionally enhanced by surface modifications via π-π interactions. Nahain et al. utilized GQDs with hyaluronic acid (HA) anchored to its surface to visualize tumor tissue in mice and deliver doxorubicin directly to the cancer cells [60]. GQDs can be conjugated with various biomolecules to track molecular events not only involved in the metastatic processes. GQDs conjugated with insulin can serve as fluorophore for real-time tracking of insulin receptors in adipocytes [61]. Surface functionalization of photoluminescent miniaturized versions of GO sheets does not cause any translocations and adverse effects on nematode Caenorhabditis elegance animal model. Nitrogen-doped GQDs have no influence on lethality and life span in both in vitro and in vivo models, and hence can be used in cellular and deep-tissue engineering [62].

The optical imaging potential of graphene was extensively studied as a part of multifunctional theranostic approach. Nanocarrier composed of GO functionalized by magnetic/fluorescent SiO_2 microspheres for targeted drug delivery carrier can serve additionally as an excellent imaging tool for the identification of nanohybrids presence inside the tumor cells [63]. Magnetic graphene-based mesoporous silica nanosheets (MGMS) modified by a targeting peptide were used as a nanoplatform to directed delivery of anticancer drugs with the possibility of magnetic resonance imaging (MRI) and monitoring of therapeutic progress [64]. Positron emission

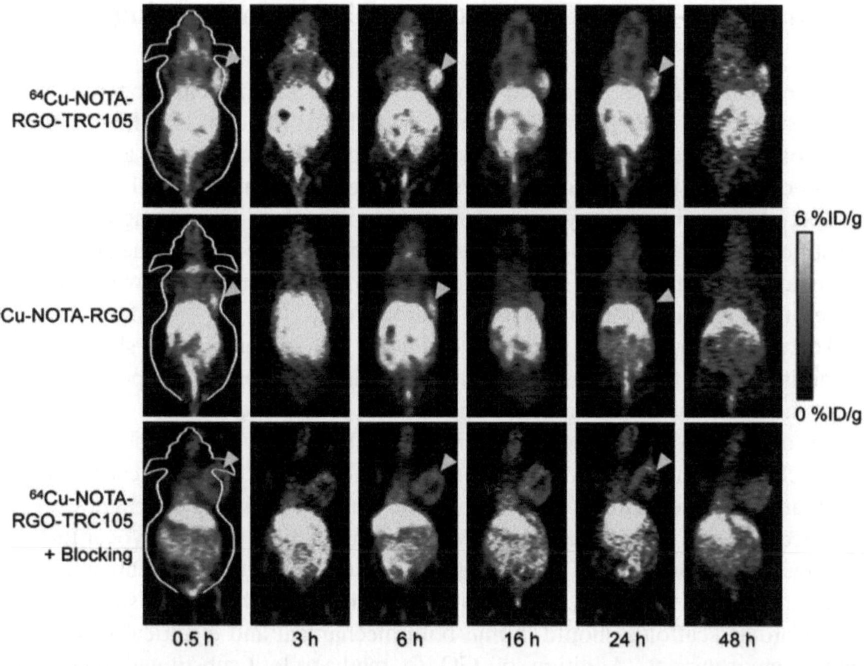

Fig. 9 Serial coronal PET images of 4T1 tumor-bearing mice at different time points post-injection of 64Cu-NOTA-RGO-TRC105, 64Cu-NOTA-RGO, or 64Cu-NOTA-RGO-TRC105 after a pre-injected blocking dose of TRC105. Tumors are indicated by *arrowheads*. Reproduced from Shi et al. [65] with permissions, Copyright 2013, Elsevier

tomography (PET) imaging, widely used technique in oncology, detects gamma rays emitted indirectly by a positron-emitting radionuclide. Such a radioactive isotope could be incorporated in GO conjugate and used as the PET label for imaging and pharmacokinetics assessment of tumor-targeted drug in living organisms [65] (Fig. 9). Optical properties of graphene can be used in imaging techniques for precise visualization of therapeutic processes but also at the sub-cellular level to identify cancer even at the very early stage [57]. Graphene-based materials can serve in both optical and non-optical imaging studies. Raman spectra of graphene derived from vibrational excitation of wide range of molecules are used to characterize graphene structure but also as a valuable bioimaging approach. The Raman spectroscopy signal can be additionally enhanced using metal nanoparticles (Au, Ag) due to the dipole and plasmon resonance (SER) effects [66, 67]. Surface-enhanced Raman scattering (SERS) is a perfect analytical tool for the precise detection at the single-molecule level. Huang et al. employed SERS spectroscopy for the study of the cellular uptake of GO-Au particles [68]. They showed that SERS spectroscopy can be very useful for the understanding of GO-cell interactions, especially while studying GO-based targeting for the drug delivery applications.

6 Graphene-Based Templates for Tissue Engineering

The chemical inertness, electrical conductivity, and mechanical properties of graphene such as high flexibility, elasticity, and ability to adapt to different surfaces can provide a specialized environment for cell culture. Graphene is a nanomaterial whose chemical, physical, or mechanical properties permit the active tissue integration of desirable cell types and tissue components. Tissue-engineered bone constructs based on graphene has emerged high interest because of the possibility to augment bone regeneration and replace already existing materials which still lack some of the characteristic of an ideal scaffold [69].

The preliminary studies on the interactions with adherent cells show that graphene is suitable for their growth and maintenance suggesting the potential use of graphene as a biomaterial in tissue engineering [70]. Cardiac tissue engineering is based on ability of stem cells to repair and regenerate injured myocardium. Mesenchymal stem cells (MSCs) promote angiogenesis, attenuate cardiac remodeling, and improve cardiac regeneration. The study of Park and colleagues revealed that incorporation of rGO flakes into MSC spheroids used for the myocardial repair improves the expression of cardiac-specific markers resulting in enhanced therapeutic efficacy of rGO-MSC hybrids in comparison to either hMSCs or rGO alone [71]. Cardiac scaffolds should mimic both mechanical and electrical properties of natural environment. Addition of GO to methacryloyl-substituted recombinant human tropoelastin (MeTro), highly elastic and conductive biomaterial for cardiac tissue engineering, improved greatly its physical properties. Enhanced electrical signal propagation doubled the number of cardiomyocytes after one day of culture and increased the expression of specific cardiac biomarkers [72]. Different approaches have been used in an attempt to improve the biocompatibility of stents including antimitotic coatings or regrowth endothelium coverage (stent grafts). An ideal stent material should be inert, hemo-, and biocompatibile and not cause toxicity and injury in a biological system. Recently used materials have been tested as biocompatible implant surfaces, but they still lack some properties and do not meet all the criteria for an ideal material for stent development. The investigations of graphene as a surface coating for nitinol stents showed no significant in vitro toxicity for endothelial and smooth muscle cell lines which confirmed its biocompatibility and additionally improved durability and impermeability of nitinol alloys [73]. Stainless steel 316L is one of the most commonly used materials in implant devices. It exhibits limited resistance to corrosion and wear which eventually leads to the release of potentially harmful metallic ions contributing to medical complications such as thrombus formation or cell apoptosis. Coating SS316L with graphene improved its mechanical and anticorrosive properties. Cytotoxicity studies with human endothelial cells (huvec) on ss316L coated with rGO showed no adverse effects against the cell morphology or spreading [74].

Because of excellent physical and chemical properties graphene was also implemented as a substrate for the growth of neuronal cells and human osteoblasts. The use of stem cell technologies is especially attractive for the treatment of neurodegenerative disorders because neurons are unable to undergo mitosis. Human neural stem cells (hNSCs) after growth on graphene films transferred to glass coverslips and covered with laminin showed proper adhesion, better attachment, and enhanced neuronal differentiation. Graphene worked as an excellent growth substrate even for the long-term period and induced differentiation of hNSCs toward neurons rather than glial cells [75]. The report concerning the cellular neurogenesis on pristine graphene-covered glass showed successful and significantly greater differentiation of human neuroblastoma cells SH-SY5Y relative to glass control even with the absence of the soluble neurogenic factor. Additional studies within the same experiment suggested focal adhesion (FAK) and p38 MPAK signaling involvement in the process of cellular neurogenesis on graphene [76]. Graphene substrates influence the activity of neuronal networks and stimulate both growth and development of primary motor neurons. This can be due to the activation of calmodulin kinase pathways achieved by the depolarizing effect of graphene on voltage-gated Ca^{2+} channels which increases calcium concentration inside the cells [77]. 3D graphene foam (3D-GF) was found to support the neural stem cells (NSCs) growth, proliferation, and differentiation toward neurons [78]. Neuronal differentiation may be enhanced in such 3D porous structures in comparison to 2D graphene-based materials because of their anisotropic microstructure. Revolutionary three-dimension (3D) printing technology was used to fabricate graphene scaffold by layer-by-layer method using liquid ink composed of graphene flakes and biocompatible and biodegradable poly (lactide-co-glycolide) (PLG). The researchers revealed that PLG, a hyperelastic polyester used in biomaterials research, together with GO has neuronal-inducing capabilities. 3DG scaffold served as conducting microenvironment and favored hMSCs differentiation without additional neurogenic stimuli showing the great potential in nerve tissue engineering [79].

The biosafety is the critical concern in the research of implantable medical devices. In hard tissue engineering it is essential to develop high-performance and biologically active scaffolds to repair and regenerate the tissue properly. The scaffold should provide mechanical support and be able to control the proliferation and differentiation of the cells. The unique properties of graphene, good biocompatibility, and tissue-specific inductive capabilities are responsible for its great potential in tissue engineering [69]. Graphene as a biomaterial should be safe and not cause any inappropriate host response. Bone tissue engineering starts with the osteoprogenitor cells which form the extracellular matrix and provide the tissue remodeling. Typical bone scaffolds are made of porous biodegradable materials which support the cell growth and regeneration of damaged tissue. Kalbacova et al. studied the growth of human mesenchymal stem cells (hMSCs) and osteoblasts on single-layer graphene on glass substrates (SiO_2/Si). The research resulted in increased cell proliferation of cells compared to SiO_2/Si substrate alone together with the higher rate of differentiation of hMSCs into osteoblasts [80]. Similar studies of Nayak et al. showed that graphene not only does not affect cell morphology and viability but also does not disturb the natural growth and differentiation of hMSCs into osteoblasts. They also proved that hMSCs differentiate into bone cells in a comparable manner to samples after growth factor administration [81].

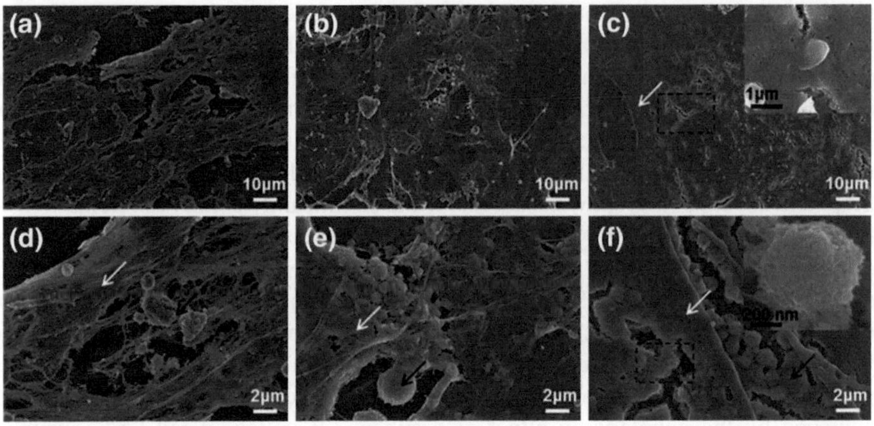

Fig. 10 SEM images of MC3T3-E1 cells after 14 days of culture in osteogenic medium on glass (**a**, **d**), GO (**b**, **e**), and GO-Gel (**c**, **f**). *White arrows* indicate cells and *black arrows* indicate the mineral. The *inset* shows the magnified image from the boxed portion of the micrograph. Reproduced from Liu et al. [83] with permissions, Copyright 2014, Royal Society of Chemistry

Human mesenchymal stem cells are so likely to differentiate into osteoblasts on graphene substrates due to the accumulation of osteogenic inducers, rough surface morphology, and electrical conductivity. Additionally, strength and high elastic modulus of graphene-based materials ($\sim$1-2,4 TPa) can be a driving force to induce spontaneous osteogenic differentiation [82]. The major goal for the successful tissue engineering is the development of material able to act as a natural carrier or template for the cell growth which would facilitate additionally tissue formation processes. Hydroxyapatite (HA), currently used for bone repair because of close similarity to bone minerals, was reinforced by GO-gelatin composite. Detailed characteristics of GO-Gel hybrid interactions with osteoblast cells showed improved cell adhesion, proliferation, and differentiation together with enhanced HA mineralization [83] (Fig. 10). GO nanoparticles can improve mechanical properties of PCL (polycaprolactone) composites causing better proliferation of preosteoblasts on planar 2D substrates and higher differentiation and mineralization rate in 3D composite scaffolds [84]. The great potential of graphene to promote osteogenic differentiation was confirmed in both 2D and 3D graphene models.

Bone-related protein expression was upregulated in periodontal ligament stem cells (PDLSCs) grown on the graphene substrates without the use of osteogenic stimuli [85]. GO shows great osseopromotive properties not only when incorporated into 3D hydroxyapatite structures [86, 87] but also as a coating of collagen scaffolds [88]. The researchers made a comparative study to show the differences between bioactivity of GO and rGO collagen-modified scaffolds directly in rats connective tissue. The results were different for both types of graphene suggesting that rGO is more effective for osteogenic differentiation. However, both samples were tissue compatible while inflammatory cells were rarely present around the probes [89]. Incorporation of GO into the chitosan scaffold provided an increase in osteoblast attachment, growth, and proliferation [90] as well as reinforcement of tensile strength in silicone constructs covered with GO [91]. Graphene was also successfully incorporated into the nano-58S bioactive glass scaffold. Despite of the great reinforcement of the mechanical properties of the composite, it was favorable for the growth and colonization of human bone cells [92]. Therefore, graphene can serve as a multifunctional material, which can assembly bone-like structures, enhance stem cells differentiation into bone cells, and promote mineralization [93]. Orthopedic implants used to replace a missing bone or joint are usually made of stainless steel or titanium alloys covered with plastic to act as an artificial cartilage. Among major complications present after implantation are poor osseointegration resulting in the lack of bone tissue integration, wear, and infections [69]. Zancanela et al. showed that the presence of GO in the osteoblast cell culture increased the viability of the cells on the plastic surface and mineralization rate of titanium substrates [94]. Additionally, some studies have shown that GO possesses antibacterial properties [95]. Graphene-based nanomaterials have an enormous potential in bone regeneration not only because of the unique properties but also the possibility of surface functionalization and ability to tailor various structures [96]. In contrast to bone tissue, cartilage is unable of spontaneous regeneration because of the presence of only 1 % of chondrocytes and the lack of vasculature. Therapies using MSCs are especially useful in cartilage tissue regeneration. Graphene cell composites were also proposed as a platform for cartilage tissue engineering. The presence of GO accelerated cellular chondrogenic differentiation and improved mechanical strength and conductivity of the cartilage scaffolds [97–99]. Similarly, GO can serve as a skin or adipose tissue engineering platform [100, 101]. 3D graphene foam loaded with bone marrow-derived MSCs improved wound healing in animal models. Despite of good biocompatibility 3D graphene scaffold promoted growth and proliferation of MSCs and improved wound healing through enhanced vascularization and reduced scarring [100]. Adipose-derived stem cells (hASCs) showed strong affinity to GO films and enhanced differentiation rate toward different types of cells including osteoblasts, adipocytes, and epithelial cells depending on the culture medium composition [101]. All these results confirm graphene potential as a coating and reinforcement material for tissue regeneration or as a scaffold enhancing cell adhesion, proliferation, and desired cellular differentiation which may find various applications in tissue engineering.

7 Conclusions

The unique properties of graphene, good biocompatibility, and tissue-specific inductive capabilities are responsible for its great potential for medical applications. However, existing data regarding the adherent cell behavior on the pristine graphene surface do not include the precise control over the quality, number of layers or substrate coatings. Because of the variety of different forms of graphene and methods of synthesis existing findings regarding graphene toxicity and biological interactions are ambiguous and sometimes even contradictory [102]. Most of the experiments, including biological response and safety tests, have been conducted on graphene derivatives such as GO or rGO. This is mainly because of the easier methods of synthesis and administration and also because GO and rGO are more hydrophilic than pristine graphene. Additional oxygen functional groups increase their solubility and allow them to remain stably dispersed in water. These polar groups reduce the thermal stability of graphene but facilitate interactions with other molecules and cell components. The research concerning nanoparticles used as adjuvants showed that positively charged nanoparticles are more likely to induce inflammatory responses in comparison to neutral or negatively charged variants [103]. Pristine graphene has exceptional chemophysical properties, yet it is strongly hydrophobic. Stable solutions of pristine graphene in polar solvents can be obtained only with the addition of proper surfactants. On the other hand, the hydrophobicity of large-area graphene synthesized by chemical vapor deposition (CVD) seems to support cell adhesion, growth and differentiation [104].

More comprehensive investigation is needed regarding the interactions of pristine graphene with human cells not only in the form of graphene flakes suspension but as a surface coating material. So far no report has been closely associated with the biosafety of single-layer pristine graphene and the limited information is available regarding the cellular events, including cell viability, adhesion, and spreading, which occur when mammalian cells interface with pure form of graphene. There are relatively a low number of researches concerning graphene hemocompatibility and the influence of graphene on the immune system is still at a very early stage of investigation. However, graphene due to its excellent physicochemical properties and good biocompatibility is still a perfect candidate innovative therapeutic strategy. The researches on the biomedical use of graphene and GO that have been being carried out until now are promising and show high medical potential of graphene nanomaterials. However, the insufficient knowledge about two-dimensional carbon structure needs further explanations especially regarding the biodistribution and foreign body response at the implant-tissue interface. Before any clinical use also the long-term biocompatibility of graphene should be carefully addressed. Utilizing graphene as a nanomaterial in biomedical applications requires more thorough and standardized research procedures due to various sizes, shapes, and production methods.

References

1. Singh Z (2016) Applications and toxicity of graphene family nanomaterials and their composites. Nanotechnol Sci Appl 16(9):15–28
2. Bolotin K, Sikes K, Jiang Z (2008) Ultrahigh electron mobility in suspended graphen. State Commun 146:351–355

3. Wei Y, Wang B, Wu J, Yang R, Dunn ML (2013) Bending rigidity and Gaussian bending stiffness of single-layered graphene. Nano Lett 13(1):26–30
4. Seabra AB, Paula AJ, De Lima R, Alves OL, Duran N (2014) Nanotoxicity of graphene and graphene oxide. Chem Res Toxicol 27(2):159–168
5. Russier J, Treossi E, Scarsi A, Perrozzi F, Dumortier H, Ottaviano L, Meneghetti M, Palermo V, Bianco A (2013) Evidencing the mask effect of graphene oxide: a comparative study on primary human and murine phagocytic cells. Nanoscale 5(22):11234–11247
6. Li S, Stein AJ, Kruger A, Leblanc RM (2013) Head groups of lipids govern the interaction and orientation between graphene oxide and lipids. J Phys Chem C 117:16150–16158
7. Jaworski S, Sawosz E, Grodzik M, Winnicka A, Prasek M, Wierzbicki M, Chwalibog A (2013) In vitro evaluation of the effects of graphene platelets on glioblastoma multiforme cells. Int J Nanomed 8:413–420
8. De Marzi L et al (2014) Flake sizedependent cyto and genotoxic evaluation of graphene oxide on in vitro A549, CaCo2 and vero cell lines. J Biol Regul Homeost Agents 28(2):281–289
9. Liu Y et al (2013) Graphene oxide can induce in vitro and in vivo mutagenesis. Sci Rep 3, article 3469
10. Hinzmann M et al (2014) Nanoparticles containing allotropes of carbon have genotoxic effects on glioblastoma multiforme cells. Int J Nanomed 15(9):2409–2417
11. Goenka S, Sant V, Sant S (2014) Graphene-based nanomaterials for drug delivery and tissue engineering. J Control Release 10(173):75–88
12. Kiew SF, Kiew LV, Lee HB, Imae T, Chung LY (2016) Assessing biocompatibility of graphene oxide-based nanocarriers: a review. J Control Release 28(226):217–228
13. Luo N, Ni D, Yue H, Wei W, Ma G (2015) Surface-engineered graphene navigate divergent biological outcomes toward macrophages. ACS Appl Mater Interfaces 7:5239–5247
14. Miao W, Shim G, Lee S, Lee S, Choe YS, Oh YK (2013) Safety and tumor tissue accumulation of pegylated graphene oxide nanosheets for co-delivery of anticancer drug and photosensitizer. Biomaterials 34(13):3402–3410
15. Ding Z, Zhang Z, Ma H, Chen Y (2014) In vitro hemocompatibility and toxic mechanism of graphene oxide on human peripheral blood T lymphocytes and serum albumin. ACS Appl Mater Interfaces 6(22):19797–19807
16. Sasidharan et al (2012) Hemocompatibility and macrophage response of pristine and functionalized graphene. Small 8(8):1251–1263
17. Dudek I, Skoda M, Jarosz A, Szukiewicz D (2015) The molecular influence of graphene and graphene oxide on the immune system under in vitro and in vivo conditions. Arch Immunol Ther Exp (Warsz) 2015:1–21
18. Li Y, Liu Y, Fu Y et al (2012) The triggering of apoptosis in macrophages by pristine graphene through the MAPK and TGF-beta signaling pathways. Biomaterials 33(2):402–411
19. Zhou H et al (2014) The inhibition of migration and invasion of cancer cells by graphene via the impairment of mitochondrial respiration. Biomaterials 35(5):1597–1607
20. Hinzmann V et al (2014) Nanoparticles containing allotropes of carbon have genotoxic effects on glioblastoma multiforme cells. Int J Nanomed 15(9):2409–2417
21. Zhou H, Zhao K, Wei L et al (2012) The interactions between pristine graphene and macrophages and the production of cytokines/chemokines via TLR- and NF-kB-related signaling pathways. Biomaterials 33:6933–6942
22. Zhi X, Fang H, Bao C, Shen G, Zhang J, Wang K, Guo S, Wan T, Cui D (2013) The immunotoxicity of graphene oxides and the effect of PVP-coating. Biomaterials 34 (21):5254–5261
23. Liu J, Cui L, Losic D (2013) Graphene and graphene oxide as new nanocarriers for drug delivery applications. Acta Biomater 9(12):9243–9257
24. Orecchioni M, Cabizza R, Bianco A, Delogu LG (2015) Graphene as cancer theranostic tool: progress and future challenges. Theranostics 5(7):710–723
25. Smidt KF (2007) Nanofronties: visions for the future of nanotechnology. Project on emerging Nanotechnologies, PEN 6. Available at http://www.nanotechproject.org/file.download/files/PEN6.NanoFrontiers.pdf

26. Zhang G, Zeng X, Ping L (2013) Nanomaterials in cancer-therapy drug delivery system. J Biomed Nanotechnol 9:741–750
27. Zeng S, Baillargeat D, Ho H, Yong K (2014) Nanomaterials enhanced surface plasmon resonance for biological and chemical sensing applications. Chem Soc Rev 43:3426–3452
28. Depan D, Shah J, Misra RDK (2011) Controlled release of drug from folate-decorated and graphene mediated drug delivery system: synthesis, loading efficiency, and drug release response. Mater Sci Eng C 31:1305–1312
29. Yang XY, Zhang XY, Liu ZF, Ma YF, Huang Y, Chen Y (2008) High-efficiency loading and controlled release of doxorubicin hydrochloride on graphene oxide. J Phys Chem C 112:17554–17558
30. Bao H, Pan Y, Ping Y, Sahoo NG, Wu T, Li L, Li J, Gan LH (2011) Chitosan-functionalized graphene oxide as a nanocarrier for drug and gene delivery. Small 7(11):1569–1578
31. Liu Z, Robinson JT, Sun X, Dai H (2008) PEGylated nanographene oxide for delivery of water-insoluble cancer drugs. J Am Chem Soc 130(33):10876–10877
32. Zhang L, Lu Z, Zhao Q, Huang J, Shen H, Zhang Z (2011) Enhanced chemotherapy efficacy by sequential delivery of siRNA and anticancer drugs using PEI-grafted graphene oxide. Small 7(4):417–545
33. Luo N, Ni D, Yue H, Wei W, Ma G (2015) Surface-engineered graphene navigate divergent biological outcomes toward macrophages. ACS Appl Mater Interfaces 7(9):5239–5247
34. Kim H, Lee D, Kim J, Kim TI, Kim WJ (2013) Photothermally triggered cytosolic drug delivery via endosome disruption using a functionalized reduced graphene oxide. ACS Nano 7:6735–6746
35. Yang X, Zhang X, Ma Y, Huang Y, Wang Y, Chen Y (2009) Superparamagnetic graphene oxide-Fe(3) O(4) nanoparticles hybrid for controlled targeted drug carriers. J Mater Chem 19 (18):2710–2714
36. Yang X, Wang Y, Huang X, Ma Y, Huang Y, Yang R, Duan H, Chen Y (2011) Multi-functionalized graphene oxide based anticancer drug-carrier with dualtargeting function and pH-sensitivity. J Mater Chem 21(10):3448–3454
37. Wang C, Li J, Amatore C, Chen Y, Jiang H, Wang XM (2011) Gold nanoclusters and graphene nanocomposites for drug delivery and imaging of cancer cells. Angew Chem Int Ed Engl 50(49):11644–11648
38. Bai J, Liu Y, Jiang X (2014) Multifunctional PEG-GO/CuS nanocomposites for near-infrared chemo-photothermal therapy. Biomaterials 35(22):5805–5813
39. Jin L, Zeng X, Liu M, Deng Y, He N (2014) Current progress in gene delivery technology based on chemical methods and nano-carriers. Theranostics 4(3):240–255
40. Chen B, Liu M, Zhang LM, Huang J, Yao J, Zhang Z (2011) Polyethylenimine-functionalized graphene oxide as an efficient gene delivery vector. J Mater Chem 21:7736–7741
41. Lu CH, Yang HH, Zhu CL, Chen X, Chen GN (2009) A graphene platform for sensing biomolecules. Angew Chem Int Ed Engl 48(26):4785–4787
42. Zhi F, Dong H, Jia X, Guo W, Lu H, Yang Y, Ju H, Zhang X, Hu Y (2013) Functionalized graphene oxide mediated adriamycin delivery and miR-21 gene silencing to overcome tumor multidrug resistance in vitro. PLoS ONE 8(3):e60034
43. Yin D, Li Y, Lin H, Guo B, Du Y, Li X, Jia H, Zhao X, Tang J, Zhang L (2013) Functional graphene oxide as a plasmid-based Stat3 siRNA carrier inhibits mouse malignant melanoma growth in vivo. Nanotechnology 24(10):105102
44. Tao Y, Ju E, Ren J, Qu X (2014) Immunostimulatory oligonucleotides-loaded cationic graphene oxide with photothermally enhanced immunogenicity for photothermal/immune cancer therapy. Biomaterials 35(37):9963–9971
45. Yang K, Feng L, Shi X, Liu Z (2013) Nano-graphene in biomedicine: theranostic applications. Chem Soc Rev 42(2):530–547
46. Robinson JT, Tabakman SM, Liang Y, Wang H, Casalongue HS, Vinh D, Dai H (2011) Ultrasmall reduced graphene oxide with high near-infrared absorbance for photothermal therapy. J Am Chem Soc 133(17):6825–6831

47. Yang K, Zhang S, Zhang G, Sun X, Lee ST, Liu Z (2010) Graphene in mice: ultrahigh in vivo tumor uptake and efficient photothermal therapy. Nano Lett 10:3318–3323
48. Sheng Z et al (2013) Protein-assisted fabrication of nano-reduced graphene oxide for combined in vivo photoacoustic imaging and photothermal therapy. Biomaterials 34 (21):5236–5243
49. Cheon YA, Bae JH, Chung BG (2016) Reduced graphene oxide nanosheet for chemo-photothermal therapy. Langmuir 32(11):2731–2736
50. Feng L, Yang X, Shi X, Tan X, Peng R, Wang J, Liu Z (2013) Polyethylene glycol and polyethylenimine dual-functionalized nano-graphene oxide for photothermally enhanced gene delivery. Small 9:1989–1997
51. Gao S, Zhang L, Wang G, Yang K, Chen M, Tian R, Ma Q, Zhu L (2016) Hybrid graphene/Au activatable theranostic agent for multimodalities imaging guided enhanced photothermal therapy. Biomaterials 79:36–45
52. Huang P, Xu C, Lin J, Wang C, Wang X, Zhang C, Zhou X, Guo S, Cui D (2011) Folic acid-conjugated graphene oxide loaded with photosensitizers for targeting photodynamic therapy. Theranostics 13(1):240–250
53. Sahu A, Choi WI, Lee JH, Tae G (2013) Graphene oxide mediated delivery of methylene blue for combined photodynamic and photothermal therapy. Biomaterials 34:6239–6248
54. Zhou L, Zhou L, Wei S, Ge X, Zhou J, Jiang H, Li F, Shen J (2014) Combination of chemotherapy and photodynamic therapy using graphene oxide as drug delivery system. J Photochem Photobiol, B 5(135):7–16
55. Wei Y, Zhou F, Zhang D, Chen Q, Xing D (2016) A graphene oxide based smart drug delivery system for tumor mitochondria-targeting photodynamic therapy. Nanoscale 8 (6):3530–3538
56. Taratula O, Patel M, Schumann C, Naleway MA, Pang AJ, He H (2015) Phthalocyanine-loaded graphene nanoplatform for imaging-guided combinatorial phototherapy. Int J Nanomed 24(10):2347–2362
57. Li JL, Tang B, Yuan B, Sun L, Wang XG (2013) A review of optical imaging and therapy using nanosized graphene and graphene oxide. Biomaterials 34:9519–9534
58. Yoo JM, Kang JH, Hong BH (2015) Graphene based nanomaterials for versatile imaging studies. Chem Soc Rev 44(14):4835–4852
59. Markovic ZM, Ristic BZ, Arsikinb KM, Klisic DG, Harhaji-Trajkovic LM, Todorovic-Markovic BM, Kepic DP, Kravic-Stevovic TK, Jovanovic SP, Milenkovic MM, Milivojevic DD, Bumbasirevic VZ, Dramicanina MD, Trajkovic VS (2012) Graphene quantum dots as autophagy-inducing photodynamic agents. Biomaterials 33(29):7084–7092
60. Nahain AA, Lee JE, In I, Lee H, Lee KD, Jeong JH, Park SY (2013) Target delivery and cell imaging using hyaluronic acid-functionalized graphene quantum dots. Mol Pharm 10 (10):3736–3744
61. Zheng XT, Than A, Ananthanaraya A, Kim DH, Chen P (2013) Graphene quantum dots as universal fluorophores and their use in revealing regulated trafficking of insulin receptors in adipocytes. ACS Nano 7(7):6278–6286
62. Zhao Y, Liu Q, Shakoor S, Gong JR, Wang D (2015) Transgenerational safety of nitrogen-doped graphene quantum dots and the underlying cellular mechanism in Caenorhabditis elegans. Toxicol Res 4:270–280
63. Gao Y, Zou X, Zhao JX, Li Y, Su X (2013) Graphene oxide-based magnetic fluorescent hybrids for drug delivery and cellular imaging. Colloids Surf, B 112:128–133
64. Wang Y, Huang R, Liang G, Zhang Z, Zhang P, Yu S et al (2014) MRI-visualized, dual-targeting, combined tumor therapy using magnetic graphene-based mesoporous silica. Small 10:109–116
65. Shi S, Yang K, Hong H, Valdovinos HF, Nayak TR, Zhang Y et al (2013) Tumor vasculature targeting and imaging in living mice with reduced graphene oxide. Biomaterials 34:3002–3009

66. Schedin F, Lidorikis E, Lombardo A, Kravets V, Geim A, Grigorenko A, Novoselov K, Ferrari A (2010) Surface-enhanced Raman spectroscopy of graphene. ACS Nano 4 (10):5617–5626
67. Sun S, Wu P (2011) Competitive surface-enhanced Raman scattering effects in noble metal nanoparticle-decorated graphene sheets. Phys Chem Chem Phys 13(47):21116–21120
68. Huang J, Zong C, Shen H, Liu M, Chen B, Ren B, Zhang Z (2012) Mechanism of cellular uptake of graphene oxide studied by surface-enhanced Raman spectroscopy. Small 8 (16):2577–2584
69. Shadjou N, Hasanzadeh M (2016) Graphene and its nanostructure derivatives for use in bone tissue engineering: recent advances. J Biomed Mater Res, Part A 104(5):1250–1275
70. Li N, Cheng Y, Song Q, Jiang Z, Tang M, Cheng G (2014) Graphene meets biology. Chin Sci Bull 59(13):1341–1354
71. Park J et al (2015) Graphene potentiates the myocardial repair efficacy of mesenchymal stem cells by stimulating the expression of angiogenic growth factors and gap junction protein. Adv Funct Mater 25(17):2590–2600
72. Annabi N et al (2016) Highly elastic and conductive human-based protein hybrid hydrogels. Adv Mater 28(1):40–49
73. Podila R, Moore T, Alexis F, Rao AM (2013) Graphene coatings for enhanced hemo-compatibility of nitinol stents. RSC Adv 3:1660–1665
74. Cardenas L, MacLeod J, Lipton-Duffin J, Seifu DG, Popescu F, Siaj M, Mantovani D, Rosei F (2014) Reduced graphene oxide growth on 316L stainless steel for medical applications. Nanoscale 6(15):8664–8670
75. Park SY, Park J, Sim SH, Sung MG, Kim KS, Hong BH, Hong S (2011) Enhanced differentiation of human neural stem cells into neurons on graphene. Adv Mater 23(36): H263–H267
76. Lee JS, Lipatov A, Ha L, Shekhirev M, Andalib MN, Sinitskii A, Lim JY (2015) Graphene substrate for inducing neurite outgrowth. Biochem Biophys Res Commun 460 (2):267–273
77. Feng ZQ, Wang T, Zhao B, Li J, Jin L (2015) Soft graphene nanofibers designed for the acceleration of nerve growth and development. Adv Mater 27(41):6462–6468
78. Li N, Zhang Q, Gao S, Song Q, Huang R, Wang L, Liu L, Dai J, Tang M, Cheng G (2013) Three-dimensional graphene foam as a biocompatible and conductive scaffold for neural stem cells. Sci Rep 3:1604
79. Jakus AE, Secor EB, Rutz AL, Jordan SW, Hersam MC, Shah RN (2015) Three-dimensional printing of high-content graphene scaffolds for electronic and bio-medical applications. ACS Nano 9(4):4636–4648
80. Kalbacova M, Broz A, Kong J, Kalbac M (2010) Graphene substrates promote adherence of human osteoblasts and mesenchymal stromal cells. Carbon 48(15):4323–4329
81. Nayak et al (2011) Graphene for controlled and accelerated osteogenic differentiation of human mesenchymal stem cells. ACS Nano 5(6):4670–4678
82. Lee JU, Yoon D, Cheong H (2012) Estimation of Young's modulus of graphene by Raman spectroscopy. Nano Lett 12(9):4444–4448
83. Liu H, Cheng J, Chen F, Bai D, Shao C, Wang J, Xi P, Zeng Z (2014) Gelatin functionalized graphene oxide for mineralization of hydroxyapatite: biomimetic and in vitro evaluation. Nanoscale 6(10):5315–5322
84. Kumar S, Azam MD, Raj S, Kolanthai E, Vasu KS, Sood AK, Chatterjee K (2015) 3D scaffold alters cellular response to graphene in a polymer composite for orthopedic applications. J Biomed Mater Res B Appl Biomater 104(4):732–749
85. Xie H, Cao T, Gomes JV, Neto AHCN, Rosa V (2015) Two and three-dimensional graphene substrates to magnify osteogenic differentiation of periodontal ligament stem cells. Carbon 93:266–275
86. Nair M, Nancy D, Krishnan AG, Anjusree GS, Vadukumpully S, Nair SV (2015) Graphene oxide nanoflakes incorporated gelatin-hydroxyapatite scaffolds enhance osteogenic differentiation of human mesenchymal stem cells. Nanotechnology 26(16):161001

87. Liu L, Guo Y, Chen X, Li R, Li Z, Wang L, Wan Z, Li J, Hao Q, Li H, Zhang X (2012) Three-dimensional dynamic culture of pre-osteoblasts seeded in HA-CS/Col/nHAP composite scaffolds and treated with α-ZAL. Acta Biochim Biophys Sin (Shanghai) 44(8):669–677
88. Nishida E, Miyaji H, Kato A, Takita H, Iwanaga T, Momose T, Ogawa K, Murakami S, Sugaya T, Kawanami M (2016) Graphene oxide scaffold accelerates cellular proliferative response and alveolar bone healing of tooth extraction socket. Int J Nanomed 11:2265–2277
89. Kanayama I, Miyaji H, Takita H, Nishida E, Tsuji M, Fugetsu B, Sun L, Inoue K, Ibara A, Akasaka T, Sugaya T, Kawanami M (2014) Comparative study of bioactivity of collagen scaffolds coated with graphene oxide and reduced graphene oxide. Int J Nanomed 11 (9):3363–3373
90. Depan D, Girase B, Shah JS, Misra RD (2011) Structure-process-property relationship of the polar graphene oxide-mediated cellular response and stimulated growth of osteoblasts on hybrid chitosan network structure nanocomposite scaffolds. Acta Biomater 7(9):3432–3445
91. Girase B, Shah JS, Misra RDK (2012) Cellular mechanics of modulated osteoblasts functions in graphene oxide reinforced elastomers. Adv Eng Mater 14:B101–B111
92. Gao Ch, Liu T, Shuai C, Peng S (2014) Enhancement mechanisms of graphene in nano-58S bioactive glass scaffold: mechanical and biological performance. Sci Rep 4, Art No. 4712
93. Lee WG et al (2011) Origin of enhanced stem cell growth and differentiation on graphene and graphene oxide. ACS Nano 5(9):7332–7341
94. Zancanela DC et al (2016) Graphene oxide and titanium: synergistic effects on the biomineralization ability of osteoblast cultures. J Mater Sci Mater Med 27(4):71
95. Tian T, Shi X, Cheng L, Luo Y, Dong Z et al (2014) Graphene-based nanocomposite as an effective, multifunctional, and recyclable antibacterial agent. ACS Appl Mater Interfaces 6:8542–8548
96. Wang Y, Li Z, Wang J, Li J, Lin Y (2011) Graphene and graphene oxide: biofunctionalization and applications in biotechnology. Trends Biotechnol 29(5):205–212
97. Lee WC, Lim CH, Kenry, Su C, Loh KP, Lim CT (2015) Cell-assembled graphene biocomposite for enhanced chondrogenic differentiation. Small 11(8):963–969
98. Yoon HH, Bhang SH, Kim T, Yu T, Hyeon T, Kim BS (2014) Dual roles of graphene oxide in chondrogenic differentiation of adult stem cells: cell-adhesion substrate and growth factor-delivery carrier. Adv Funct Mater 24:6455–6464
99. Liao L, Qu Y, Chu B, Zhang X, Qian Z (2015) Biodegradable CSMA/PECA/graphene porous hybrid scaffold for cartilage tissue engineering. Sci Rep 5:9879
100. Li Z, Wang H, Yang B, Sun Y, Huo R (2015) Three-dimensional graphene foams loaded with bone marrow derived mesenchymal stem cells promote skin wound healing with reduced scarring. Mater Sci Eng, C 27:181–188
101. Kim J et al (2013) Bioactive effects of graphene oxide cell culture substratum on structure and function of human adipose-derived stem cells. J Biomed Mater Res A 101:3520–3530
102. Skoda M, Dudek D, Jarosz A, Szukiewicz D (2014) Graphene: one material, many possibilities—application difficulties in biological systems. J Nanomater 14, Art No. 890246
103. Dobrovolskaia MA, Germolec DR, Weaver JL (2009) Evaluation of nanoparticle immunotoxicity. Nat Nanotechnol 4:411–414
104. Pinto AM, Concalves IC, Magalhaes FD (2013) Graphene-based materials biocompatibility: a review. Colloids Surf Biointerfaces 111:188–202
105. Hu H, Yu J, Li Y, Zhao J, Dong H (2012) Engineering of a novel pluronic F127/graphene nanohybrid for pH responsive drug delivery. J Biomed Mater Res A 100(1):141–148
106. Yang Y, Zhang Y-M, Chen Y, Zhao D, Chen J-T, Liu y (2012) Construction of a graphene oxide based noncovalent multiple nanosupramolecular assembly as a scaffold for drug delivery. Chem. Eur J 18(14):4208–4215
107. Jin R, Ji X, Yang Y, Wang H and Cao A (2013) Self-assembled graphene–dextran nanohybrid for killing drug-resistant cancer cells. ACS Appl Mater Interfaces 5(15):7181–7189
108. Song E, Han W, Li C, Cheng D, Li L, Liu L, Zhu G, Song Y, Tan W (2014) Hyaluronic acid-decorated graphene oxide nanohybrids as nanocarriers for targeted and pH-responsive anticancer drug delivery. ACS Appl Mater Interfaces 6(15):11882–11890

109. Kim H, Lee D, Kim J, Kim TI, Kim WJ (2013) Photothermally triggered cytosolic drug delivery via endosome disruption using a functionalized reduced graphene oxide ACS Nano 7(8):6735–6746
110. Shi J, Zhamg J, Ma R, Gao J, Liu Y, Zhang Ch, Zhang Z (2014) A tumor-targeting near-infrared laser-triggered drug delivery system based on GO@Ag nanoparticles for chemo-photothermal therapy and X-ray imaging. Biomaterials 35(22):5847–5861
111. Bai J, Liu Y, Jiang X (2014) Multifunctional PEG-GO/CuS nanocomposites for near-infrared chemo-photothermal therapy. Biomaterials 35(22):5805-5813
112. Wang C, Mallela J, Garapati US, Ravi S, Chinnasamy V, Girard Y, Howell M, Mohapatra S (2013) A chitosan-modified graphene nanogel for noninvasive controlled drug release. Nanomedicine 9(7):903–911
113. Chen H, Wang Z, Zong S, Wu L, Chen P, Zhu D, Wang C, Xu S, Cui Y (2014) SERS-fluorescence monitored drug release of a redox-responsive nanocarrier based on graphene oxide in tumor cells. ACS Appl Mater Interfaces 6(20):17526–17533
114. Zhao X, Liu L, Li X, Zeng J, Jia X, Liu P (2014) Biocompatible graphene oxide nanoparticle-based drug delivery platform for tumor microenvironment-responsive triggered release of doxorubicin. Langmuir 30(34):10419–10429
115. He D, He X, Wang K, Zou Z, Yang X and Li X (2014) Remote-controlled drug release from graphene oxide-capped mesoporous silica to cancer cells by photoinduced pH-jump activation. Langmuir 30(24):7182–7189
116. Wan H et al (2014) Facile fabrication of a near-infrared responsive nanocarrier for spatiotemporally controlled chemo-photothermal synergistic cancer therapy. Nanoscale 6:8743–8753
117. Bian X et al (2014) Fabrication of graphene-isolated-Au-nanocrystal nanostructures for multimodal cell imaging and photothermal-enhanced chemotherapy. Sci Rep 4:6093
118. Miao W, Shim G, Kang CM, Lee S, Choe YS, Choi HG, Oh YK (2013) Cholesteryl hyaluronic acid-coated, reduced graphene oxide nanosheets for anti-cancer drug delivery. Biomaterials 34(37):9638–9647
119. Qin XC, Guo ZY, Liu ZM, Zhang W, Wan MM, Yang BW (2013) Folic acid-conjugated graphene oxide for cancer targeted chemo-photothermal therapy. J Photochem Photobiol B 120:156–162
120. Zhang L, Xia J, Zhao Q, Liu L, Zhang Z (2010) Functional graphene oxide as a nanocarrier for controlled loading and targeted delivery of mixed anticancer drugs. Small 6(4):537–544
121. Wei G, Yan M, Dong R, Wang D, Zhou X, Chen J, Hao J (2012) Covalent modification of reduced graphene oxide by means of diazonium chemistry and use as a drug-delivery system. Chemistry 18(46):14708–14716
122. Wu J, Wang YS, Yang XY, Liu YY, Yang JR, Yang R, Zhang N (2012) Graphene oxide used as a carrier for adriamycin can reverse drug resistance in breast cancer cells. Nanotechnology 23(35):355101
123. Zhi F, Dong H, Jia X, Guo W, Lu H, Yang Y, Ju H, Zhang X, Hu Y (2013) Functionalized graphene oxide mediated adriamycin delivery and miR-21 gene silencing to overcome tumor multidrug resistance in vitro. PLoS One 8(3):e60034
124. Zhang YM, Cao Y, Yang Y, Chen JT, Liu Y (2014) A small-sized graphene oxide supramolecular assembly for targeted delivery of camptothecin. Chem Commun (Camb) 50 (86):13066–13069
125. Tian J, Luo Y, Huang L, Feng Y, Ju H, Yu BY (2016) Pegylated folate and peptide-decorated graphene oxide nanovehicle for in vivo targeted delivery of anticancer drugs and therapeutic self-monitoring. Biosens Bioelectron 80:519–524
126. Xu Z, Wang S, Li Y, Wang M, Shi P, Huang X (2014) Covalent functionalization of graphene oxide with biocompatible poly (ethylene glycol) for delivery of paclitaxel. ACS Appl Mater Interfaces 6(19):17268–17276
127. Chaudhari NS et al (2014) Graphene oxide based magnetic nanocomposites for efficient treatment of breast cancer. Mater Sci Eng 37:278–285

128. Saeed LM et al (2014) Single-walled carbon nanotube and graphene nanodelivery of gambogic acid increases its cytotoxicity in breast and pancreatic cancer cells. J Appl Toxicol 34(11):1188–1199
129. Kakran M, Sahoo NG, Bao H, Pan Y, Li L (2011) Functionalized graphene oxide as nanocarrier for loading and delivery of ellagic Acid. Curr Med Chem 18(29):4503–4512
130. Some S et al (2014) Cancer therapy using ultrahigh hydrophobic drug-loaded graphene derivatives. Scientif Rep 4, Art No. 6314
131. Maity AR, Chakraborty A, Mondal A, Jana NR (2014) Carbohydrate coated, folate functionalized colloidal graphene as a nanocarrier for both hydrophobic and hydrophilic drugs. Nanoscale 6(5):2752–2758
132. Misra SK, Kondaiah P, Bhattacharya S, Rao CNR (2012) Graphene as a nanocarrier for tamoxifen induces apoptosis in transformed cancer cell lines of different origins. Small 8:131–143
133. Zheng XT, Li CM (2012) Restoring basal planes of graphene oxides for highly efficient loading and delivery of β-lapachone. Mol Pharm 9(3):615–621
134. Fan X, Jiao G, Zhao W, Jin P, Li X (2013) Magnetic Fe3O4-graphene composites as targeted drug nanocarriers for pH-activated release. Nanoscale 5(3):1143–1152
135. Lu Y-J et al (2012) Improving thermal stability and efficacy of BCNU in treating glioma cells using PAA-functionalized graphene oxide. Int J Nanomed 7: 1737–1747
136. Chen GY et al (2015) Graphene oxide as a chemosensitizer: diverted autophagic flux, enhanced nuclear import, elevated necrosis and improved antitumor effects. Biomaterials 40:12–22
137. Hou L et al (2016) Multifunctional hyaluronic acid modified graphene oxide loaded with mitoxantrone for overcoming drug resistance in cancer. Nanotechnology 27(1):015701
138. Kim H, Namgung R, Singha K, Oh IK, Kim WJ (2011) Graphene oxide-polyethylenimine nanoconstruct as a gene delivery vector and bioimaging tool. Bioconjug Chem 22 (12):2558–2567
139. Zhi F, Dong H, Jia X, Guo W, Lu H, Yang Y et al (2013) Functionalized graphene oxide mediated adriamycin delivery and miR-21 gene silencing to overcome tumor multidrug resistance in vitro. PLoS ONE 8(3):e60034
140. Wang F, Zhang B, Zhou L, Shi Y, Li Z, Xia Y, Tian J (2016) Imaging dendrimer-grafted graphene oxide mediated anti-miR-21 delivery with an activatable luciferase reporter. ACS Appl Mater Interfaces 8(14):9014–9021
141. Feng L, Zhang S and Liu Z (2011) Graphene based gene transfection. Nanoscale 3:1252–1257
142. Zhang L, Lu Z, Zhao Q, Huang J, Shen H, Zhang Z (2011) Enhanced chemotherapy efficacy by sequential delivery of siRNA and anticancer drugs using PEI-grafted graphene oxide. Small 7(4):460–464
143. Chen B, Liu M, Zhang L, Huang J, Yao J, Zhang Z (2011) Polyethylenimine-functionalized graphene oxide as an efficient gene delivery vector. J Mater Chem 21:7736–7741
144. Wang Ch et al (2013) Multifunctional chitosan magnetic-graphene (CMG) nanoparticles: a theranostic platform for tumor-targeted co-delivery of drugs, genes and MRI contrast agents. J Mater Chem B 1:4396–4405

Graphene-Based Materials in Biosensing, Bioimaging, and Therapeutics

Sivaramapanicker Sreejith, Hrishikesh Joshi and Yanli Zhao

Abstract Biomedical research has become extremely important in these days due to its direct impact on human health. The quest for the development of sophisticated materials for sensitive sensing, selective imaging and effective therapeutics has led to the creation of a unique class of materials known as graphene-based materials (GBMs). GBMs can be broadly classified into three groups: graphene-based nanocomposites, graphene quantum dots, and graphene-wrapped hybrids. These materials possess remarkable electrical, physical, and chemical properties, which can be exploited to develop efficient sensors, probes, and drugs. In this chapter, a detailed account about the synthetic strategies of these materials along with the mechanisms governing their performance in biosensing, bioimaging, and therapeutics is presented. The chapter highlights the suitability of GBMs in non-conventional and emerging techniques such as nonlinear photonics and photoacoustic imaging. The GBMs can also be employed to fabricate synergistic materials that are capable of simultaneous imaging and therapeutic actions. Therefore, the GBMs provide a promising platform for cutting-edge developments in the field of biomedical research.

Keywords Bioimaging · Biosensing · Graphene · Hybrid materials · Therapeutics

S. Sreejith · H. Joshi · Y. Zhao (✉)
Division of Chemistry and Biological Chemistry, School of Physical and Mathematical Sciences, Nanyang Technological University, 21 Nanyang Link, Singapore 637371, Singapore
e-mail: zhaoyanli@ntu.edu.sg

Y. Zhao
School of Materials Science and Engineering, Nanyang Technological University, 60 Nanyang Avenue, Singapore 639798, Singapore

G. Gonçalves et al. (eds.), *Graphene-based Materials in Health and Environment*, Carbon Nanostructures, DOI 10.1007/978-3-319-45639-3_2

1 Introduction

Biosensing, bioimaging, and therapeutics are three important aspects of biomedical research. The major focus of this area lies in the development of probes for selective detection and treatment of diseases. Therefore, this research area has grabbed a lot of attention due to its immense applicability in terms of analyte detection, photonics, and drug delivery. The synthetic diversity that enables the fabrication of inorganic, organic and inorganic–organic hybrid materials serves as a unique platform for targeted applications. As a result, a wide range of materials have been developed for biomedical applications, including mesoporous silica [1, 2], zeolites [3], graphene [4], quantum dots [5], organic probes [6, 7], polymer composites [8], and supramolecular self-assemblies [9]. However, introducing biocompatibility and selectivity are two important challenges that accompany the material synthesis pursuits. Amongst the aforementioned list, materials encoded with graphene are extensively employed in biology-based research due to their inherent biocompatibility, structural flexibility, and easy synthesis. These properties have led to the development of several classes of graphene-based materials (GBMs) such as graphene nanocomposites, graphene quantum dots (GQDs) and graphene-wrapped hybrids, which serve as effective biosensors, targeted imaging probes, and drug carriers.

The three directions of biomedical research are discretely classified on the basis of their functionality. For instance, biosensing deals with the qualitative/quantitative recognition of a specific type of analytes by characterizing spectroscopic, electrochemical, or magneto-chemical behavior of the systems. Due to immense advancements in technology, a wide range of characterization techniques can be used to enhance the biosensing capabilities and develop highly efficient, stable and commercially viable sensors. Most importantly, biosensing facilitates the detection of biomolecules (e.g., enzymes, proteins and nucleic acids) and bioessential chemical analytes (e.g., metals and sugars) that help in monitoring the biochemical processes.

Bioimaging can be considered as the extrapolation of biosensing toward the detection of the specific type of biological components for diagnostic purposes. Along with high specificity to target molecules, materials used for bioimaging have to be biocompatible and nontoxic. As bioimaging is concerned with the visual detection of components, the most commonly employed detection technique is optical spectroscopy. Increasingly, a wide range of techniques have been used for bioimaging such as fluorescence spectroscopy, photoacoustic imaging (PAI), positron emitting tomography (PET), and nonlinear optical spectroscopy that provide a large scope for the development of new imaging probes. Bioimaging studies are generally performed in two types of environments, i.e., in vitro and in vivo, where the efficacy of the developed imaging probes could be determined [10].

Therapeutics is a field that deals with drug delivery and treatment of infected biological components (e.g., cells, tissues, and bones). The materials used for therapeutics have to be highly selective and nontoxic to minimize the adverse effects on the host cells [11]. Each direction of biomedical research demands a unique set of properties from the materials, while some requirements related to the physiochemical behavior are quite similar. In this chapter, we conduct a comprehensive review to explicate the extensive applicability of GMBs in all directions of the biomedical field, i.e., biosensing, bioimaging, and therapeutics.

2 Graphene

Graphene is a single layer two-dimensional (2D) nanomaterial composed of π conjugated benzene rings, which has high mechanical flexibility and electrical conductivity [12]. Graphene possesses a delocalized π electron cloud that enables it to interact non-covalently with other chemical species. Graphene has typically three important forms, i.e., graphene, graphene oxide (GO), and reduced graphene oxide (rGO). The structures are shown in Fig. 1. Interestingly, GO and rGO are highly susceptible to chemical modification and coupling reactions due to the presence of different kinds of functional groups.

Typically, graphene can be synthesized by several top-down approaches that can exfoliate graphite by chemical or physical means. The most popular synthetic route to synthesize GO and rGO is the Hummer's method. A chemical disintegration of graphite sheets is used to provide thin graphene sheets that are then subjected to oxidation and reduction reactions to form GO and rGO, respectively [13]. Over the years, Hummer's method has undergone several modifications, which is collectively known as the modified Hummer's method [14, 15]. Another popular approach to synthesize graphene is the physical exfoliation of graphite. This approach was devised by Geim and Novoselov in 2004, which was a cutting-edge development in the field of graphene science [16]. Graphene can also be

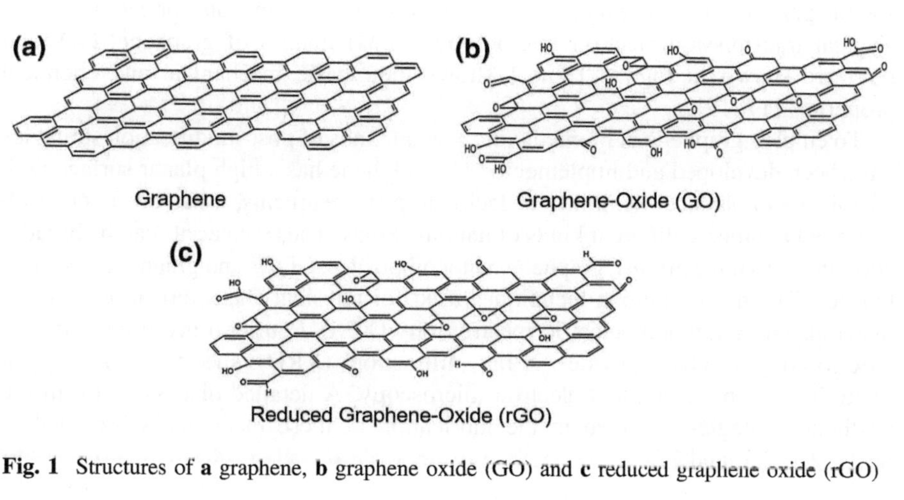

Fig. 1 Structures of **a** graphene, **b** graphene oxide (GO) and **c** reduced graphene oxide (rGO)

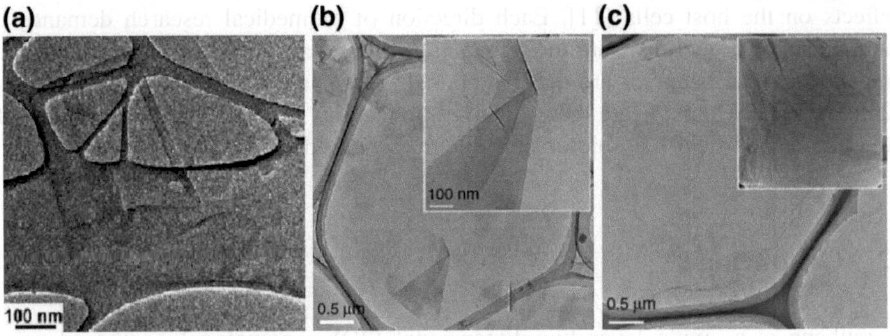

Fig. 2 TEM image of **a** graphene. Reprinted with permission from [18]. © 2011 American Chemical Society. TEM images of **b** GO and **c** rGO. Reprinted with permission from [19]. © 2010 Nature Publishing Group

Table 1 Zeta potential, tensile strength, powder X-ray diffraction (XRD) peak and *d* spacing values of graphene, GO, and rGO

Materials	Zeta potential (mV)[a]	Tensile strength (GPa)	XRD peak (2θ)[b]	*d* spacing (nm)	Ref.
Graphene	−42.0	130	25.0[c]	~0.350	[17, 18]
GO	−50.3	0.150	10.2	0.875	[19]
rGO	−34.2	0.100	26.0	0.350	[19]

[a]The values are reported at pH = 10 as zeta potential is dependent on pH
[b]Wide angle XRD peak
[c]Single layer graphene shows a broad peak in wide angle XRD spectrum

synthesized in the gas phase by chemical vapor deposition (CVD) on nickel (Ni) and copper (Cu) substrates [12], Recently, graphene was synthesized by the use of spark plasma sintering process without the use of substrates or catalysts [17]. Typical transmission electron microscopy (TEM) images of graphene, GO, and rGO are shown in Fig. 2. Table 1 shows their basic mechanical and chemical properties [17–19].

To employ graphene in biomedicine, several kinds of post-modification strategies have been developed and implemented. As graphene has a high planar surface with chemical and physical flexibility to facilitate post-treatments, it has been very well exploited to prepare different kinds of nanomaterials. These materials can be broadly classified into three groups, graphene nanocomposites, GQDs and graphene-wrapped hybrids. The most common techniques used for physical characterization of these materials are X-ray photoelectric spectroscopy (XPS), Fourier transformed infrared spectroscopy (FTIR), powder X-ray diffraction (XRD) spectroscopy, Raman scattering spectroscopy, and electron microscopy. A detailed discussion of all the synthetic strategies involved in the fabrication of these materials is provided in Sects. 2.1, 2.2, and 2.3.

2.1 Graphene Nanocomposites

Graphene nanocomposites are fairly simple to synthesize and can be used in a wide range of biomedical applications. GO and rGO possess a lot of functional groups such as carbonyl, hydroxyl, epoxy, and carboxyl groups to couple with biomolecules, chemical receptors, imaging probes and drugs for biomedical applications [12]. The basic strategy for synthesizing graphene-based nanocomposites is shown in Fig. 3. Some of the typical graphene-based nanocomposites synthesized from this strategy are poly(diallyldimethylammonium chloride)-capped gold nanoparticle functionalized graphene [20], polyethylenimine-graphene nanocomposite [21], rGO-AuPd alloy [22], and Ru–ZnO graphene nanocomposites [23].

The major advantage of such nanocomposites is that the inherent properties of graphene are preserved along with the properties of constituent materials. Therefore, these materials provide a way for introducing additional properties by encoding different kinds of substrates for targeted applications. On the other hand, most of the nanocomposites have an open structure that is often detrimental to bioimaging applications. This gives rise to a new class of graphene-based materials called graphene-wrapped hybrid, which is discussed in Sect. 2.3.

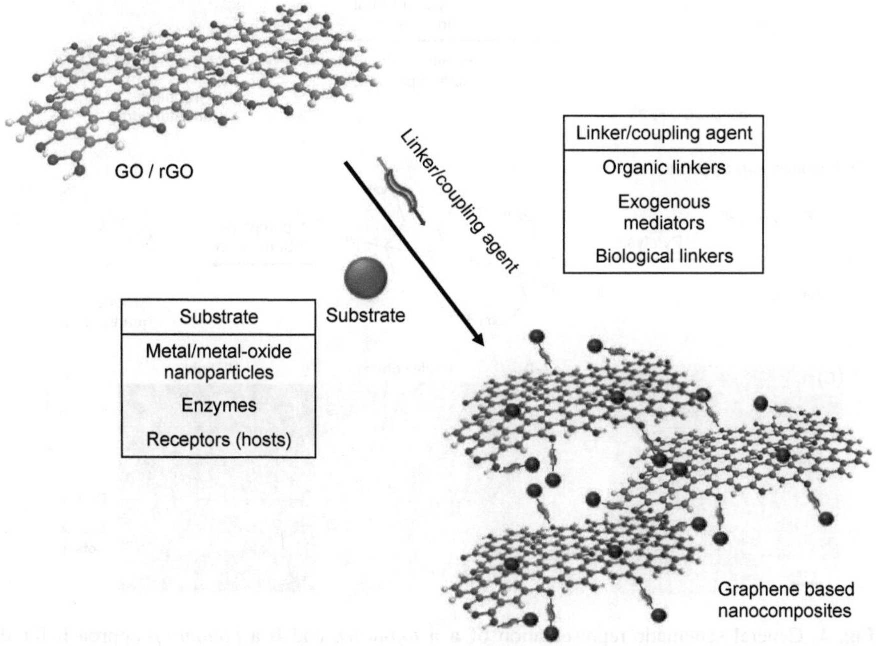

Fig. 3 Schematic representation of the synthetic strategies for graphene-based nanocomposites

2.2 Graphene Quantum Dots

GQDs are nano-dimension fragments of graphene with remarkable photophysical properties owing to the quantum confinement effects and edge effects. These materials can be synthesized by simple top-down approaches involving a hydrothermal treatment or cutting process. The synthesis schemes may be modified slightly to suit the application. The underline mechanism that breaks down graphene sheets into nano-sized dots (quantum dots) remains universal to all the top-down approaches [24]. In this hydrothermal approach, graphene sheets are subjected to high pressure and temperature, causing them to disintegrate into nanosized dots. On the other hand, there are some bottom-up approaches for the synthesis of GQDs such as cage-opening reactions of fullerene and solution chemistry [25]. A schematic diagram showing the two types of processes for the formation of GQDs is illustrated in Fig. 4a, b. Though the bottom-up approach provides a better morphological control, it is more intricate as compared to the top-down approaches [26].

Interestingly, GQDs can also be found naturally in coal. The extraction process gives about 20 % yield, and the obtained GQDs are soluble and fluorescent in

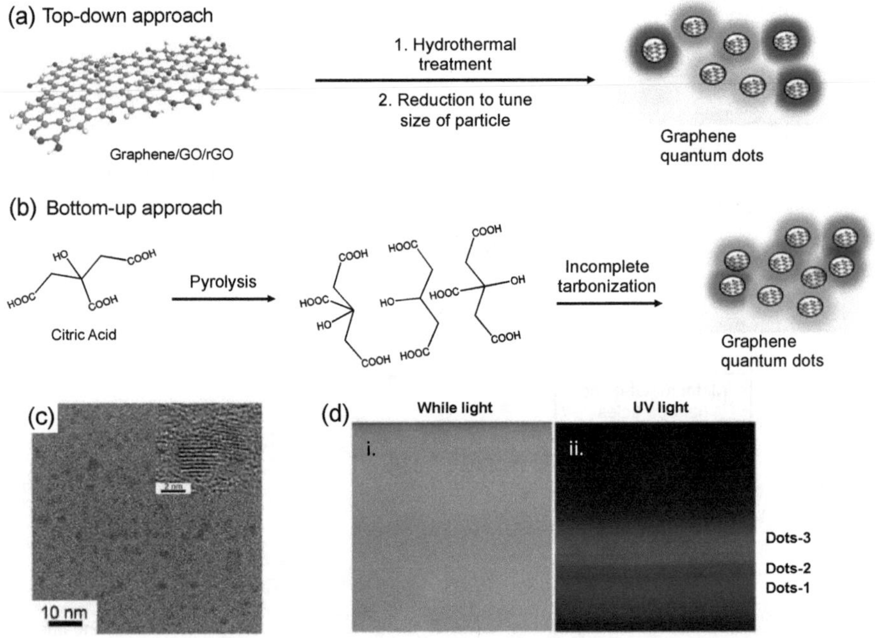

Fig. 4 General schematic representation of **a** a *top-down* and **b** a *bottom-up* approach for the synthesis of GQDs. **c** Typical TEM image of GQDs. *Inset* HRTEM of GQDs showing the fringe pattern. Reprinted with permission from [25]. © 2012 American Chemical Society. **d** Agarose gel (0.75 %) of the as-prepared GQDs in aqueous solution. The images were taken under (*i*) white and (*ii*) UV light (302 nm). Reprinted with permission from [26]. © 2016 American Chemical Society

aqueous solutions [27]. The TEM image of GQDs prepared by hydrothermal reaction is shown in Fig. 4c with an inset graph showing its fringe pattern. The dots are uniformly spherical with a relatively broad size distribution. Moreover, the photophysical property of GQDs varies with its particle size, which is inversely proportional to band gap. This synthetic approach provides a range of distinctly fluorescent particles (Fig. 4d). Owing to extensive electronic and morphological flexibility, GQDs are widely employed in bioimaging applications.

2.3 *Graphene-Wrapped Hybrids*

Graphene-wrapped hybrids are a new kind of GBMs that were developed to overcome limitations of open structured graphene nanocomposites. These were essentially synthesized to isolate biological and chemical species. They can be considered as a special case of graphene-based polymers where the graphene sheet is used as an encapsulation material rather than a two-dimensional support. A combination of mesoporous silica and graphene serves as an excellent host for the encapsulation of dyes/probes required in imaging. To synthesize these hybrids, each component of the hybrids is synthesized independently and then subjected to conjugation reactions. Recently, we reported the synthesis of GO-wrapped squaraine-loaded mesoporous silica for bioimaging using above strategy [28]. The squaraine dye was loaded into the mesoporous silica prior to its coupling reaction with graphene as shown in Fig. 5a. This encapsulated dye showed high applicability in fluorescence bioimaging.

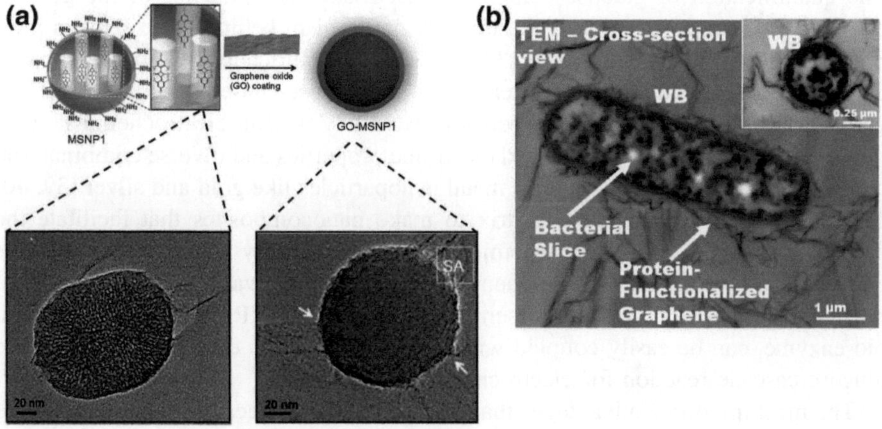

Fig. 5 a Scheme for the preparation of graphene-wrapped mesoporous silica loaded with squaraine dye. *Inset* TEM images of unwrapped and wrapped hybrids. Adapted with permission from [28]. © 2012 American Chemical Society. **b** TEM image of protein-functionalized graphene wrapped bacteria. Reprinted with permission from [30]. © 2011 American Chemical Society

Another interesting wrapped hybrid is the wrapped biological hybrid. Akhavan et al. [29] demonstrated such an encapsulation by a simple photo-reduction reaction between the graphene sheets and melatonin-bacterial suspension. Similarly, Mohanty et al. [30] reported the encapsulation of bacteria by obtaining a wet phase TEM image of Gram-positive bacteria, Bacillus subtilis, as shown in Fig. 5b. While graphene-wrapped hybrids have not been studied extensively, the most common usage of these materials is in the field of imaging.

3 Biosensing

Biosensing particularly deals with the detection of biologically abundant analytes such as ions, amino acids, small biomolecules, RNA (ribonucleic acid), DNA (deoxyribonucleic acid), and proteins. The most commonly used techniques for biosensing are electrochemical and Förster resonance energy transfer (FRET)-based detections. This is because these techniques are very efficient and sensitive. However, techniques like fluorescence spectroscopy, surface plasmon resonance (SPR) and surface enhanced Raman scattering (SERS) have also shown promising results in biosensors. The analytes under investigation in biosensing are generally essential components that can be used to monitor the changes within a biological system. Although biosensing has been established since the late 1900s [31–33], the initial studies on the use of graphene in biosensing were reported in 2008. These reports showed the fabrication of graphene-based transistors [34] and electrodes [35] for selective detection of dopamine and DNA, respectively.

The most eminent biosensors currently are the electrochemical glucose sensors. The quantification of glucose content by electrochemical analysis of the glucose oxidase biochemical reaction is the primary mechanism behind these sensors [36]. Shan et al. [37] and Kang et al. [38] reported the fabrication of graphene-based electrodes for electrochemical detection of glucose based on this strategy. Graphene-based nanocomposites perform extremely well in electrochemical sensing regimes owing to their enhanced electronic properties and diverse combinatorial compositions. For instance, several metal nanoparticles like gold and silver [39, 40] can be inserted into graphene matrix to make nanocomposites that facilitate the charge transport during electrochemical cycling. Auxiliary biomolecules such as chitosan can be coupled with graphene to catalytically activate the glucose oxidase biochemical reaction [41]. Horseradish peroxide (HRP), another important bio-enzyme, can be easily coupled with graphene sheets to catalyze the enzymatic glucose cascade reaction for electrochemical sensing [42].

The most important advantages that graphene offers in these nanocomposites are the high electrical conductivity and the extensive post-modification capability. Graphene-based hybrids can be easily deposited, drop-casted, or chemically coupled to the electrode surfaces. Graphene-based nanocomposites can also be spin-coated on the surface of glassy carbon electrodes (GCE) [36]. Moreover, there are several ways in which the electrochemical sensing can be conducted, such as

Table 2 List of the first, second, and third generation of graphene-based electrochemical glucose biosensors

Generation	Sensor Materials	Range (LOD)	Ref.
First	PVP-protected graphene-GOD-PFIL modified GC electrodes	2–14 mM (#)	[43]
	Reduced graphene sheets-AuPd alloy	0.01–3.5 mM (0.07 mM)	[44]
Second	NAD–GDH-FePhenTPy-SPCE	1.6–33 mM (0.6 mM)	[45]
	Graphene-PANI(COOH)-PEI-Fc/Cu-MCNB/GC electrodes	0.15–15 mM (0.16 mM)	[46]
Third	AgNP/F-SiO_2/GO	0–8 mM (#)	[47]
	AuNP_graphene_GOD	0.05–7.35 mM (0.02 mM)	[48]

PVP polyvinylpyrrolidone, *GOD* glucose oxidase, *PFIL* polyethylenimine-functionalized ionic liquid, *GC* glassy carbon, *NAD-GDH* nicotinamide adenine dinucleotide-dependent-glucose dehydrogenase, *FePhenTPy* 5-[2,5-di (thiophen-2-yl)-1H-pyrrol-1-yl]-1,10-phenanthroline iron (III) chloride, *SPCE* screen printed carbon electrode, *G-PANI(COOH)* poly(aniline-co-anthranilic acid)-grafted graphene, *PEI* polyethylene imine, *MCBN* redox polymer multicomponent nanobead, *Cu* copper, *Fc* ferrocene, *AgNP* silver nanoparticles, *F-SiO₂* functionalized SiO_2, *AuNP* gold nanoparticles
Not mentioned in the report

chronoamperometry, voltammetry, and impedance-based detection. Table 2 shows few examples of different types of electrochemical sensors fabricated by using the aforementioned strategies.

As illustrated in Table 2, several modifications to the initial strategy of electrochemical sensing have led to the development of numerous kinds of biosensors [43–48]. They are classified as the first, second, and third generation biosensors. In the first generation biosensors, the reaction between oxygen and glucose oxidase produces peroxide, which could be electrochemically detected on electrodes. The second generation sensors use auxiliary exogenous mediating agents to facilitate this process, whereas the third generation sensors involve a direct electrochemical reaction without the use of mediators [49]. Figure 6 clearly illustrates the differences between different generations of sensors.

Due to the success of graphene-based glucose biosensors, the electrochemical sensing was soon extended to other analytes. Easy synthesis of graphene hybrids and extensive capability of post-modification rendered the GBMs highly suitable for electrochemical sensing of analytes such as alcohol, ascorbic acid [50], and nucleic acid [51]. Figure 7 highlights some of the typical examples of electrochemical sensors based on graphene. The most common strategies for electrochemical sensing are cyclic voltammetry (Fig. 7a) and chronoamperometric method (Fig. 7b), as they directly correlate to the current generated over a range of applied voltage and at a specific voltage, respectively. Electrochemical sensing using resistance as a quantification tool is an indirect method, but it has been a powerful tool to develop sensors for pH and relative humidity [52, 53]. One such example of

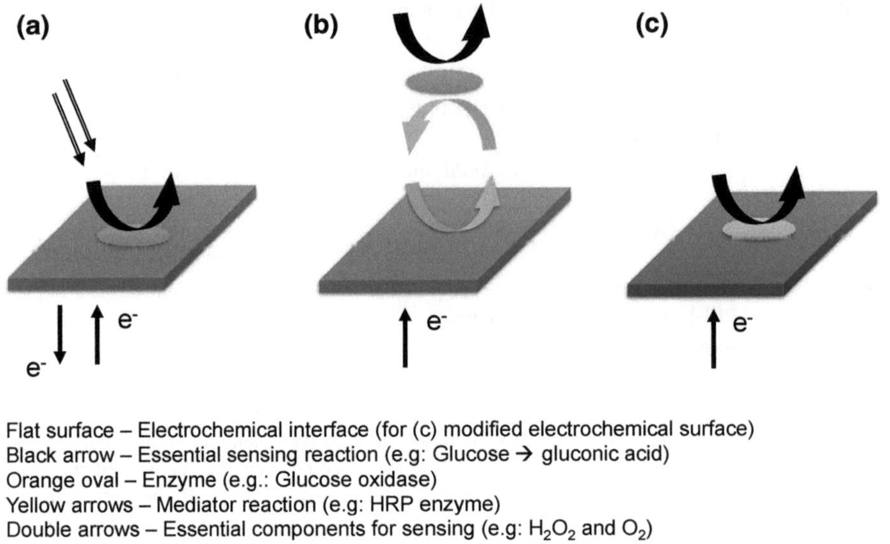

Fig. 6 Schematic representation of the electrochemical sensing mechanism involved in **a** the first, **b** second, and **c** third generation sensors. Only the effective reactions taking place at the electrode surface are illustrated in figure

electrochemical sensor for pH is shown in Fig. 7c [54]. The most important aspect about these sensors is the function of graphene. As stated earlier, due to the huge electron cloud of graphene, subtle changes in the electrochemical circuit can be detected very effectively, which allow for the development of sensitive sensors. Any perturbation on the graphene surface due to analyte attachment or because of the changes in refractive index of electrolyte can cause changes in the electrical parameters. Therefore, GBMs can be extensively used for the fabrication of sensitive, efficient, and highly selective sensors.

FRET-based detection is another important technique to develop biosensors with graphene. Graphene has a highly conjugated electronic system, which allows it to be a good acceptor of electrons. If coupled with an appropriate donor moiety at a specific distance (Förster distance), the hybrid may exhibit FRET, which can be used for biosensing. Graphene/GO offers a chemically susceptible 2D planar sheet for incorporating donors like quantum dots, nanoparticles, biomolecules, and organic probes [12, 55]. This strategy has been adopted to fabricate a wide range of FRET-based biosensors for the detection of analytes such as nucleic acid (DNA/RNA) [56], concanavalin A [57], bisphenol A [58], thrombin [59], protease [60], and kanamycin [61]. The schematic illustration for the basic mechanism of FRET-based biosensing with graphene is shown in Fig. 8. Myung et al. [62] reported a hybrid with Fe_3O_4 nanoparticles encapsulated by graphene for cancer detection. The encapsulation of nanoparticles increased the surface to volume ratio, which significantly improved its FRET-based detection limit for HER2 (human

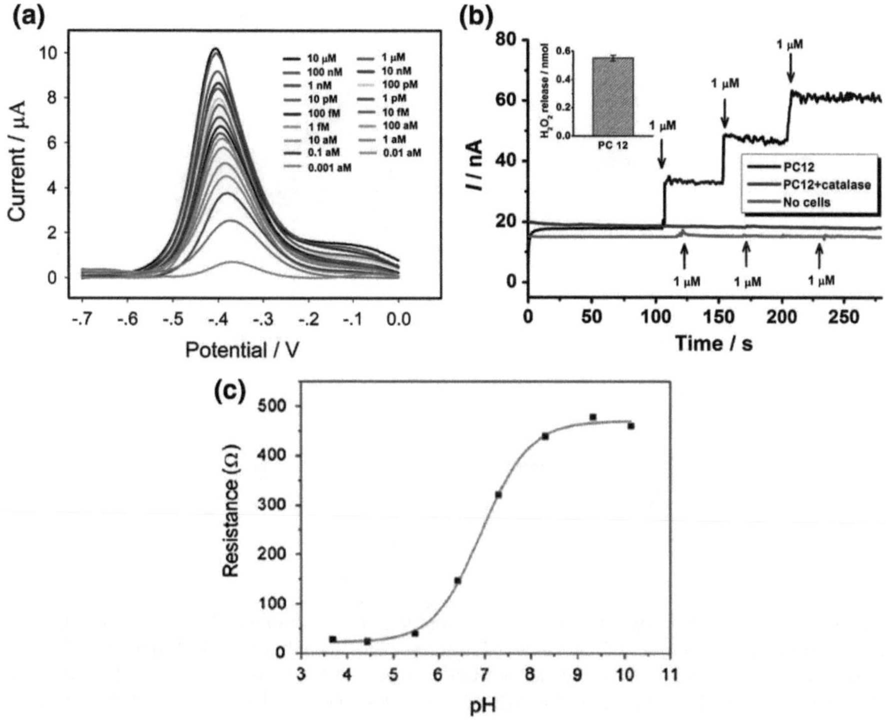

Fig. 7 **a** Square wave voltammograms of mercuric sensor versus [Hg^{2+}] ions in 20 mL of Tris containing 10 mM KCl (10 mM, pH 7.4) between −0.7 and 0 V under pulse amplitude of 25 mV and frequency of 10 Hz with a step potential of 4 mV. Reprinted with permission from [53]. © 2016 American Chemical Society. **b** Time course of the H_2O_2 release from PC 12 cells upon the successive addition of 1 μM ascorbic acid. *Inset* The release of H_2O_2 from 6.0×10^6 cells in 2 mL of deoxygenated PBS upon injection of 3 μM ascorbic acid. Reprinted with permission from [54]. © 2015 American Chemical Society. **c** Electron transfer resistance (R_{ct}) versus pH for polymer exfoliated graphene in solution. The frequency range to generate the R_{ct} values was 0.02 Hz to 100 kHz. Reprinted with permission from [55]. © 2016 American Chemical Society

epidermal growth factor receptor 2) and EGFR (Epidermal growth factor receptor) cells. Another typical example of GO/quantum dot-based FRET sensing was demonstrated by Dong et al. [63], where the CdTe quantum dots were modified with cyclin molecular beacon (MB) to selectively target DNA. The sensing mechanism was based on the FRET between GO and modified quantum dots. The schematic diagram showing the sensing activity of the hybrid is given in Fig. 8c, d. Graphene served as a very good mediator for the electron transfer from quantum dots via the FRET mechanism.

As seen from Fig. 8c, graphene is capable of quenching fluorescence from quantum dots in all three cases, which shows that the post-modification of donor motifs hardly interferes the FRET process between graphene and the quantum dots.

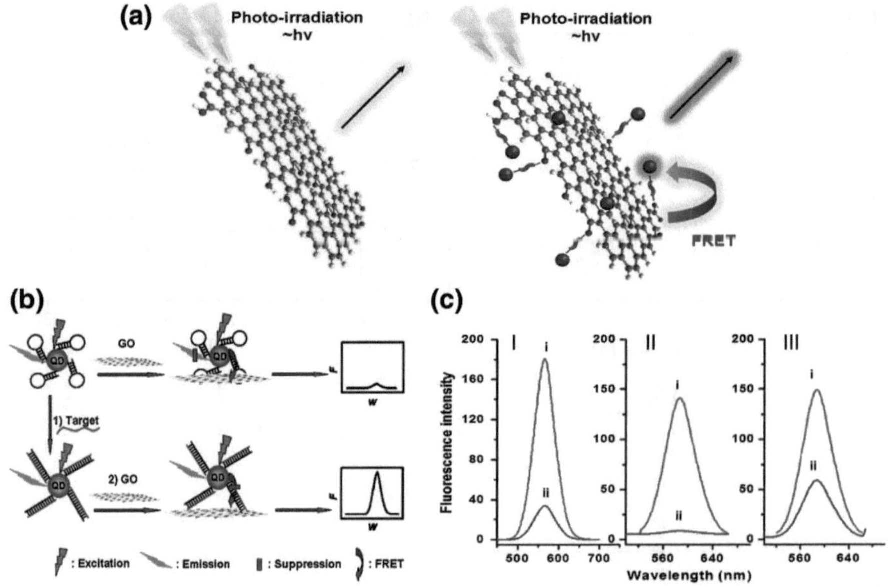

Fig. 8 **a** Scheme showing the mechanism of FRET in graphene-based nanocomposites. **b** Schematic representation of GO-induced fluorescence quenching of MB-quantum dots and biosensing mechanism. **c** Fluorescence emission spectra of (*I*) quantum dots (120 nM), (*II*) MB-quantum dots (120 nM), and (*III*) MB-quantum dots incubated with target (800 nM) in the buffer before (*i*) and after (*ii*) adding GO (0.1 μg/mL) for 5 min. Reprinted with permission from [63]. © 2010 American Chemical Society

Interestingly, graphene can also be used as an effective encapsulation agent that encloses virus for FRET-based detection of antibody as demonstrated by Bhatnagar et al. [64]. Due to the presence of a high electron cloud, graphene can also be used in electrical and SPR-based sensing of analytes. Huang et al. [65] fabricated a biosensor for detection of *E. coli* bacteria based on the changes in the conductance. Graphene typically offers a large surface area for detection and also provides homogeneous functionalization that helps in improving the performance of such biosensors. The enhanced electrochemical and electrical property of graphene facilitates the fabrication of graphene-based field effect transistors (FET) for analyte sensing. As graphene provides structural and morphological flexibility, it can be casted into several forms. For instance, Park et al. [66] reported the fabrication of FET-based bioelectronic nose that was highly selective toward an odorant amyl butyrate (AB) with a limit of detection (LOD) of 1.47×10^{-14} M. Graphene is typically used as a transducer and a bioconjugation site in FET sensors, and the most common analytes detected with these sensors are proteins, glucose, DNA, and pH.

Graphene-based hybrids encoded with gold nanostructures show remarkable SPR properties. The encapsulation of graphene over gold intensifies the surface electron density and enhances the performance of SPR-based sensors. Graphene is also useful in controlling the oxidation of metal nanoparticles, which contributes to increasing the lifetime of the biosensors. SPR sensing can be performed in typically two ways: (a) the guest molecule attaches on the host to perturb the electronic distribution, and (b) the guest molecule eliminates the host from the hybrid to cause changes in the electronic distribution (Fig. 9). Both the strategies can be effectively quantified in terms of analyte addition as shown in Fig. 9c, d. Another typical

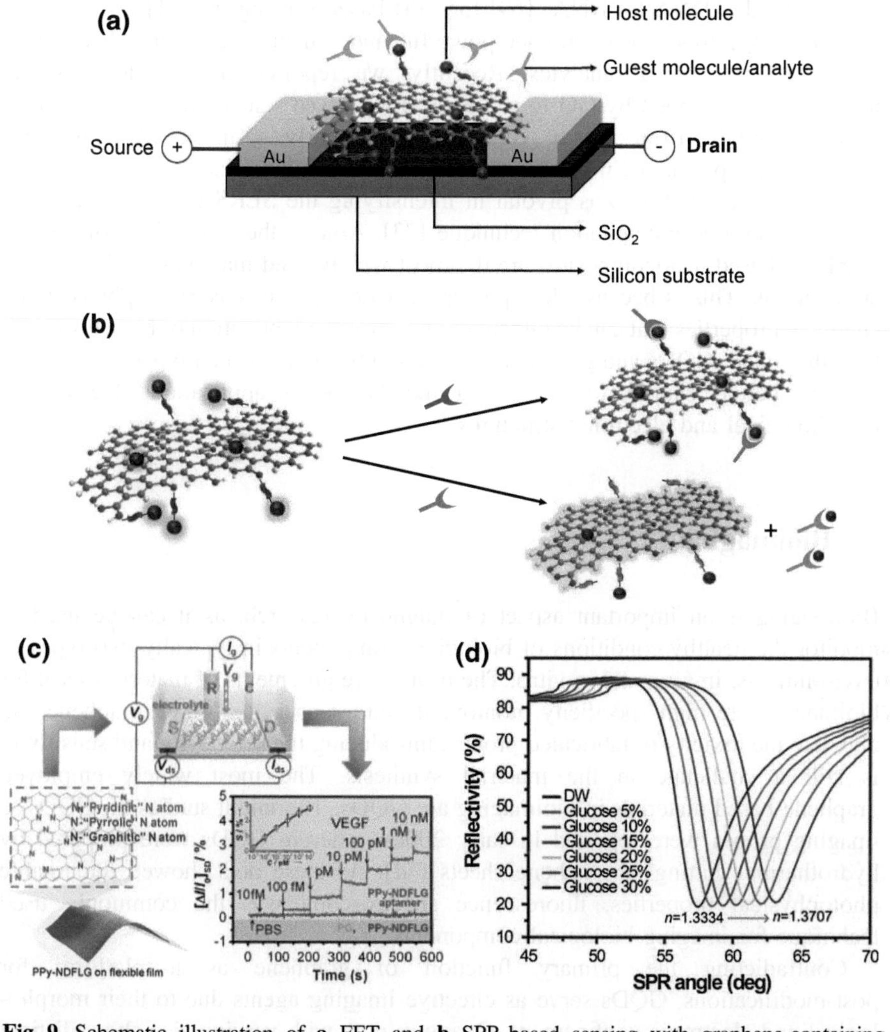

Fig. 9 Schematic illustration of **a** FET and **b** SPR-based sensing with graphene-containing nanocomposites. **c** Schematic diagram of a liquid-ion gated FET using aptamer-conjugated, polypyrrole-converted, nitrogen-doped few-layer graphene (PPy-NDFLG). *R* Ag/AgCl reference electrode; *C* platinum counter electrode; *S*, *D* source and drain electrodes). The real time response and calibration curves for the device were also shown. Reprinted with permission from [70]. © 2015 American Chemical Society. **d** Refractive index versus SPR angle plot for the [GO(+)/GO (−)]3-Au sample. Reprinted with permission from [71]. © 2015 American Chemical Society

example of SPR-based sensing is a biosensor developed by Fu et al. [67] for the detection of heparin by using the colorimetric quenching property of graphene. In this report, protamine was used as a mediator to quantitatively detect the changes in SPR and absorbance signals. Moreover, nanocomposites of graphene with silver and gold nanomaterials also showed very high selectivity and sensitivity to analytes like immunoglobulin class G (IgG) [68], immunoglobulin class E(IgE), and single-stranded DNA (ss-DNA) [69] for SPR-based sensing [69–71].

Raman spectroscopy is another powerful and sensitive technique that can be used for detection of analytes. Recently, we reported the synthesis of Au nanoparticles wrapped by GO to serve as SERS-based sensors in the detection of Raman signals from HeLa cells [72]. Similarly, Xu et al. developed graphene-encapsulated Cu nanoparticles for the detection of adenosine. The rigid graphene encapsulation was pivotal in intensifying the SERS signals, which was achieved via a co-precipitation technique [73]. Among the three kinds of GBMs, graphene-based nanocomposites are the most widely used materials for biosensing applications. This is because they possess a wide range of electrical, physical and chemical properties that can be effectively used for the fabrication of biosensors. On the other hand, GQDs and graphene-wrapped hybrids are special types of materials showing better usability in bioimaging and therapeutic applications due to their morphological and electronic structures.

4 Bioimaging

Bioimaging is an important aspect of diagnostic research, as it can be used to monitor the healthy conditions of biological components in typically two types of environments, in vivo and in vitro. The primary requirements of materials used for bioimaging are high specificity, nontoxicity and sensitivity. While graphene can alleviate the toxicity of fabricated probes, introducing the selectivity and sensitivity is still a challenge in the material synthesis. The most widely employed graphene-based materials in bioimaging are GQDs. The initial studies on GQDs as imaging probes were reported in early 2000s, wherein GQDs were prepared by hydrothermal cutting of graphene sheets [74]. As these dots showed remarkable photophysical properties, fluorescence spectroscopy was the commonly used technique for imaging biological components.

Contradicting the primary function of graphene as a platform for post-modifications, GQDs serve as effective imaging agents due to their morphological and electronic confinement. Quantum dots with varying size show distinct photophysical properties as shown in Fig. 4. GQDs are generally coupled with polymeric supports such as polypeptides [75], polyethyleneimine [76], polystyrene [77] and carbon nanotube-polymers [78] to effectively administer in biological environments. Though the optical properties can be tuned by controlling the size of GQDs, very recently Qu et al. [79] showed that GQDs can be co-doped with sulfur

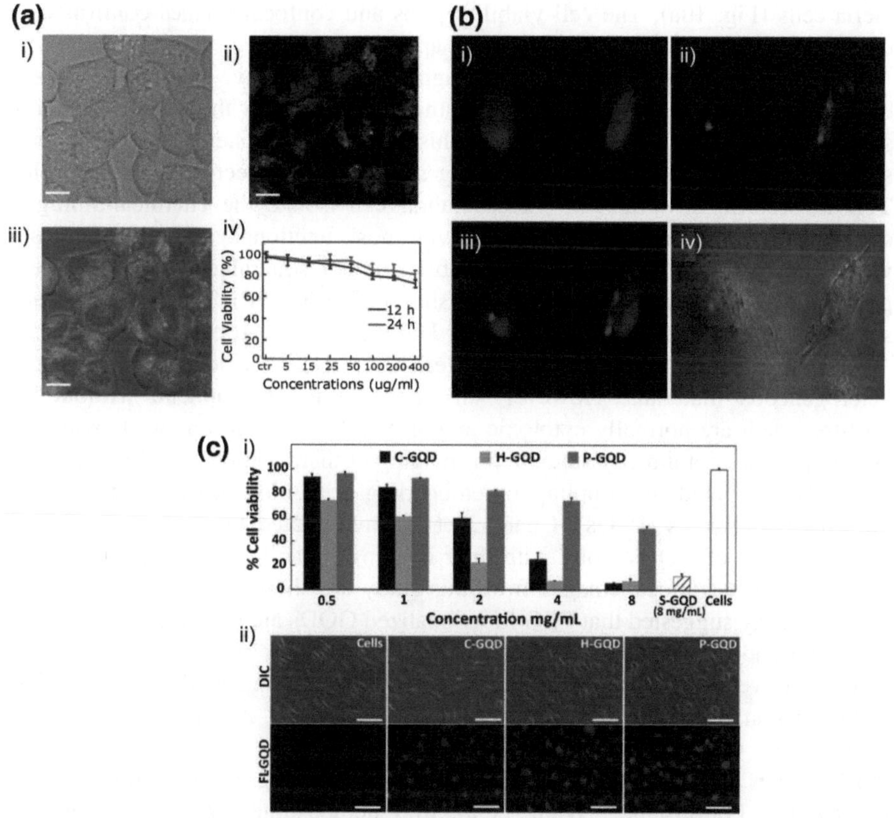

Fig. 10 **a** Two-photon cell imaging under (*i*) bright field and (*ii*) 800 nm excitation. (*iii*) The merged image of (*i*) and (*ii*). (*iv*) Cell viability after the incubation with N-GQD for 12 and 24 h. Reprinted with permission from [80]. © 2013 American Chemical Society. **b** Epifluorescence microscopy images of HeLa cancer cells internalized with GO–MSNP. The blue fluorescence is from 4',6-diamidino-2-phenylindole (DAPI) nuclear counterstain and the red fluorescence is from GO–MSNP. (*i*) Image of nuclei. (*ii*) Fluorescence of GO–MSNP (dark-field). (*iii*) Overlay of (*i*) and (*ii*). (*iv*) Overlay of (*iii*) with the phase contrast image. Reprinted with permission from [28]. © 2012 American Chemical Society. **c** (*i*) Cell viability for crude-GQDs (C-GQD), hydrothermally treated-GQDs (H-GQD), and GQDs embedded in a PEG matrix (P-GQDs) as assessed by 3-(4,5-dimethylthiazol-2-yl)-2,5-diphenyltetrazolium bromide (MTT) assay. Controls: GQDs with a PEG shell (S-GQD, 8 mg/mL); only cells. (*ii*) Cellular labeling of HeLa cells with GQDs. Blue: GQDs. Reprinted with permission from [81]. © 2014 American Chemical Society

(S) and nitrogen (N) to obtain distinctly emissive dots with different excitations. Moreover, the introduction of sophisticated techniques like positron emitting (PET), nonlinear optical spectroscopy, and photoacoustic imaging provides a versatile platform for the fabrication of new imaging probes. Recently, Liu et al. demonstrated the use of nitrogen-doped GQDs for two-photon imaging (excitation at 800 nm) of

HeLa cells (Fig. 10a). The cell viability plots and confocal images confirmed the efficacy of these materials for bioimaging applications [80].

The demerit of using GQDs for bioimaging is that extensive post-functionalizing capability is reduced. Hence, ease of synthesis associated with the introduction of selectivity is compromised. In light of this limitation, graphene-wrapped hybrids serve as viable substitutes for developing sensitive and selective imaging probes. This is because the graphene encapsulation can isolate the chemical/biological species in several environments and provide post-functionalization capabilities as well. Very recently, we reported the fabrication of squaraine dye-loaded mesoporous silica wrapped with ultrathin GO sheets (GO–MSNP). This hybrid material served as an excellent imaging probe for HeLa cells (Fig. 10b) [28].

The toxicity of graphene-based materials is relatively lower than most of heterogeneous materials. However, all three forms of graphene without any post-treatment are normally cytotoxic and/or genotoxic in vitro as well as in vivo. Hence, they are not directly suitable for imaging or therapeutics. This limitation can be easily alleviated by coupling or embedding with biocompatible species like polyethylene glycols (PEGs). Chandra et al. investigated the cytotoxic effect of different kinds of GQDs along with their imaging capability. The typical cell viability plots and confocal images from this study are shown in Fig. 10c [81]. The images clearly suggested that PEG functionalized GQDs are relatively nontoxic and provide a good image contrast.

In a conventional perspective, optical bioimaging is generally conducted by materials that exhibit stokes-shifted photoluminescence. An interesting class of compounds that display such a photoluminescence is the near-infrared (NIR) probes. These NIR dyes allow for in vivo imaging, with minimal medium interference. For instance, Wang et al. [82] demonstrated the application of a turn-on NIR probe for the detection of cancer cells composed of transferrin gold functionalized GO. GO was used as an acceptor to deactivate the probe by FRET. The function of graphene was similar to FRET-based biosensors. However, in this case its activity was not quantitative. Conventional photoluminescence has proven to be an effective tool for imaging. On the other hand, anti-stokes photoluminescence, a nonlinear optical phenomenon, is an emerging branch in bioimaging, which was proposed by Nicolaas Bloembergen as a theoretical possibility in 1954 [83]. The most widely used anti-stokes photoluminescence materials are the upconversion nanoparticles (commonly abbreviated as UCNPs) involving lanthanides. Graphene is generally employed a 2D support or host for these materials to be used in bioimaging [84]. As demonstrated by Zhu et al. [85], GQDs can also exhibit upconversion photoluminescence if they are modified adequately and excited at wavelengths in the NIR region.

Bimodal imaging is a new arena in which the materials are active to two kinds of imaging techniques to increase the sensitivity and efficiency of the probes. A typical example of this strategy is shown in Fig. 11a, where the synthesized probe is a two-photon active probe and additionally capable of generating photoacoustic signals. The schematic diagram of this material and its corresponding two-photon and photoacoustic activity is shown in Fig. 11b, c [86].

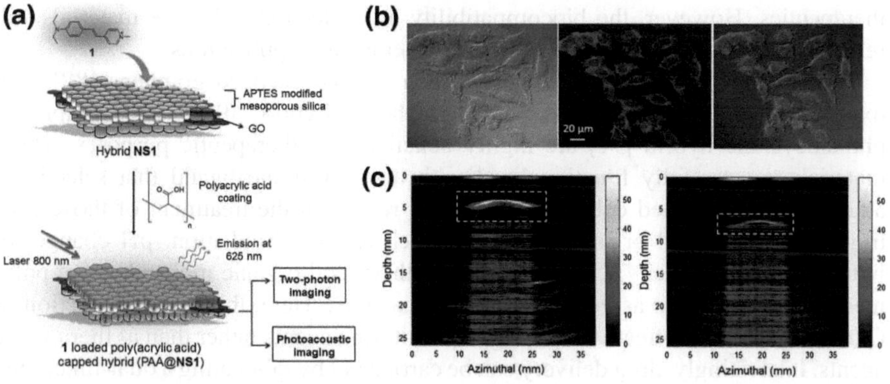

Fig. 11 **a** Schematic illustration for the preparation of (4-(4-diethylaminostyryl)-1-methyl pyridinium iodide) loaded GO-mesoporous silica followed by coating with polyacrylic acid (PAA@NS1). **b** Two-photon microscopic images of HeLa cancer cells incubated with PAA@NS1 (1 μg mL^{-1}): (*i*) bright field image, (*ii*) fluorescence image (λ_{ex} = 800 nm), and (*iii*) overlay image of (*i*) and (*ii*) showing the localization of the hybrid inside the cell cytoplasm. **c** Photoacoustic images of PAA@NS1-embedded gelatin inside a model chicken tissue. Reprinted with permission from [86]. © 2014 John Wiley & Sons, Inc

Another approach in the field of bioimaging is the graphene wrapping of biological components. As demonstrated by Mohanty et al. [30], the graphene shells used for the encapsulation were perfectly electron-transparent and did not impose any contrast artifacts, at least at the moderate resolution needed for the imaging of bacteria. This property could improve the visibility of bacteria during electron microscopy measurements. The encapsulated bacteria were exposed to different doses of electron irradiation in the microscope and compared with untreated bacteria under similar conditions. While the unwrapped bacteria shrank considerably and showed clear signs of damage due to the evaporation of water, the wrapped ones remained undamaged and stable even at a high irradiation dose. Such direct encapsulations of biological components by graphene are extremely useful in conducting mechanistic investigations and bioimaging studies.

5 Therapeutics

Therapeutics is the central area of biomedical research and has given rise to essential fields like bioimaging and biosensing. Therapeutics primarily deals with the treatment of diseased cells by selective administration of drugs in order to cure the affected cells or perform targeted cellular apoptosis to eliminate those cells [87]. Drug/gene delivery, photothermal therapy and chemotherapy are some of the ways by which these goals can be achieved. As GBMs show remarkable applicability in bioimaging and biosensing, these materials were frequently employed to the field of therapeutics. However, the biocompatibility and selectivity of these materials still remain the most important aspect of their therapeutic applications.

Nanocomposites of graphene sheets such as gold encoded graphene [88], iron oxide coated graphene [89], SiO_2-graphene hybrid [90], and graphene-phthalocyanine hybrid [91] are highly beneficial for therapeutic purposes. These materials are carefully functionalized with an appropriate ligand that selectively attaches to the affected cells and eventually results in the treatment of those cells triggered by external or internal stimuli such as light irradiation, pH change, or induced cellular toxicity. GQDs coupled with several organic molecules and polymers could also serve as good therapeutic materials. Here, the primary function of GQDs is to serve as effective imaging or monitoring agents rather than as therapeutic agents. Interestingly, drug delivery can be carried out by fabricating a dual-functional system having a drug delivery agent and an imaging probe. Nanocomposites of GQDs, graphene, GO and rGO can be well used for this purpose. For instance, nanographene sheets (NGS) post-functionalized by PEGs and Cy7 (NGS-PEG) can

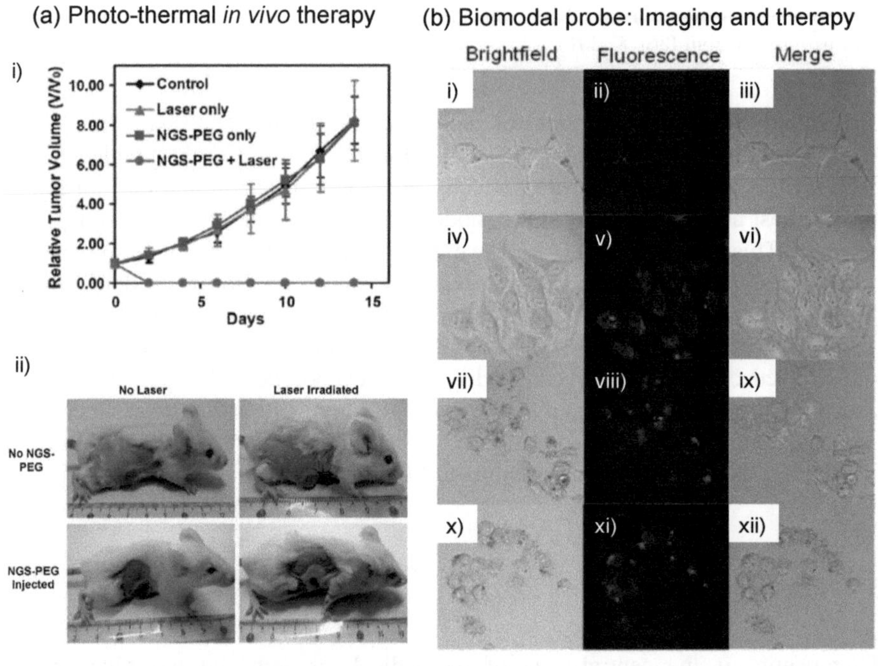

Fig. 12 **a** In vivo photothermal therapy using intravenously injected NGS-PEG. (*i*) Tumor growth curves of different groups after he treatment. The tumor volumes were normalized to their initial sizes. (*ii*) Representative photos of tumors on mice after various treatments indicated. The laser irradiated tumor on NGS-PEG injected mouse was completely destructed. Reprinted with permission from [92]. © 2010 American Chemical Society. **b** Targeting effect indicated by fluorescence microscopy images of (*i–iii*) HEK 293 normal cells and (*iv–vi*) HeLa cancer cells incubated with GSF@AuNPs for 4 h. Fluorescence microscopy images of HeLa cells incubated with (*vii–ix*) GSF@AuNPs with H_2O_2 (30 μM) and (*x–xii*) GSF@AuNPs with ascorbic acid (2 mM). Reprinted with permission from [93]. © 2010 American Chemical Society

be suitably employed as in vivo photothermal agents and fluorescence-based imaging probes [92]. The activity of these materials is shown in Fig. 12a.

In addition to high applicability in bioimaging, graphene-wrapped hybrids could also serve as good therapeutic materials. For instance, we reported the fabrication of gold nanoparticle-loaded mesoporous silica-coated graphene (GSF@AuNPs), and on the basis of its peroxidase activity, the hybrid was utilized as a selective, quantitative, and fast colorimetric probe for cancer cells. Concurrently, the hybrid was also used as ascorbic acid-mediated therapeutic agents (Fig. 12b) [93].

Under a similar strategy, Kim et al. [94] reported zinc phthalocyanine loaded graphene-wrapped gold nanoparticles that show excellent multifunctional properties for combinational treatment of cancer cells via photothermal and photodynamic therapy in addition to the Raman bioimaging. Table 3 summarizes some examples of GBMs capable of simultaneous therapy and bioimaging [75, 76, 92, 94–98]. The examples in the table elucidate extensive post-functionalization capability of graphene as well as its wide range of applicability in terms of bioimaging and therapeutics.

As evident from the table, a spectrum of GBMs that have varying sensitivity and selectivity in biological environments can lead to different combinations of imaging techniques, drug delivery, and treatment strategies. For instance, Moon et al. fabricated a dual-functional material by coating graphene on gold nanorods to serve as a photoacoustic imaging agent and a photothermal agent for therapeutics.

It can be evidently noted that photothermal therapy (PTT) and photodynamic therapy (PDT) are the most leading mechanisms for therapeutics. Essentially, PTT and PDT are two unique mechanisms that utilize light energy to destroy cell. PTT converts the incident photon energy into thermal energy that is utilized for cellular death by necrosis or apoptosis (Fig. 13) [99], whereas PDT is a relatively convoluted process in terms of the photophysical mechanism governing its therapeutic

Table 3 GBMs for bioimaging and therapeutics

GBMs	Bioimaging	Therapeutics	Ref.
QD_polypeptides_rGO	Fluorescence	Photothermal therapy	[75]
GO-BPEI	Photoluminescence	Gene delivery	[76]
NGS-PEGs	Fluorescence	Photothermal therapy (in vivo)	[92]
Zn–Au@GON	Raman	Photothermal and photodynamic therapy	[94]
rGO-AnNRs	Photoacoustic	NIR photothermal therapy	[95]
SiO_2@GN-Serum	Fluorescence	Chemo-thermal therapy (FA-DOX)	[96]
FAG-FMSNs-TA-Dox-PEG	FRET	ATP-assisted drug delivery	[97]
Nano-rGO-PEG	Fluorescence	NIR photothermal therapy	[98]

QD quantum dots, *BPEI* branched polyethylene imine, *NGS* nanographene sheets, *PEG* polyethylene glycols, *Zn* zinc, *Au* gold, *GON* graphene oxide nanocolloid, *AuNR* gold nanorods, *GN* graphene, *FAG* functionalized aptamer graphene quantum dots, *FMSN* fluorescent mesoporous silica nanoparticles, *TA* AS1411 aptamer, *Dox* Doxorubicin, *Nano-rGO* nanosized reduced graphene oxide

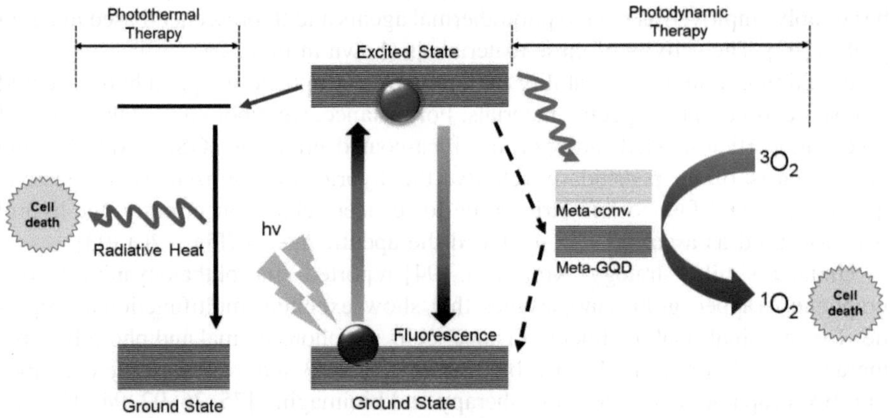

Fig. 13 Schematic illustration of photothermal therapy and photodynamic therapy mechanisms. *Left* photothermal therapy—the exited electron comes back to the ground state by the generation of radiative thermal energy for cellular death. *Right* photodynamic therapy—the electron in excited state relaxes to a metastable state where it can interact with triplet oxygen to induce cellular death. In GQDs, the interaction can occur during both processes, i.e., excited state to metastable state and metastable state to ground state

action. Herein, the incident photon excites the electron from the HOMO to the LUMO and then to a metastable long-lived triplet state. The electron in this state can interact with reactive moieties in its close proximity, such as triplet oxygen ($^3O_2$), and lead to cellular death by inducing reactive oxidative stress (ROS). The schematic representation of the mechanism is shown in Fig. 13. The conventional PDT agents generate singlet oxygen during the energy transfer from metastable state to ground state. However, GQDs can exhibit a slightly different mechanism of PDT. As shown by Ge et al. [100], the GQDs can generate singlet oxygen at two stages, first during electron transfer from LUMO to metastable state, and second during metastable state to ground state. This process helps in generating more amount of singlet oxygen. As stated in Sect. 4, NIR emissive dyes are effective tools for imaging. Similarly, NIR irradiation can serve as effective photothermal optical source for therapeutics, too [101, 102]. Another interesting use of photoirradiation, apart from the generation of heat generation or oxidative stress, could be photo-induced drug delivery by using photoacid generators (PAG). This unique property of a graphene-wrapped hybrid was demonstrated by He et al. [103], wherein the photoirradiation created a pH change leading to the breakage of boroester bond in the hybrid ultimately for the treatment of cancer cells.

GBMs can also be used as antimicrobial agents. There are several mechanisms that explain the antimicrobial activity of graphene (Fig. 14) [104]. The most widely accepted theory is the induced oxidative stress mechanism. Graphene interferes with the bacterial metabolism and disrupts the essential cellular functions, which ultimately lead to bacterial death. However, several reports suggest that the

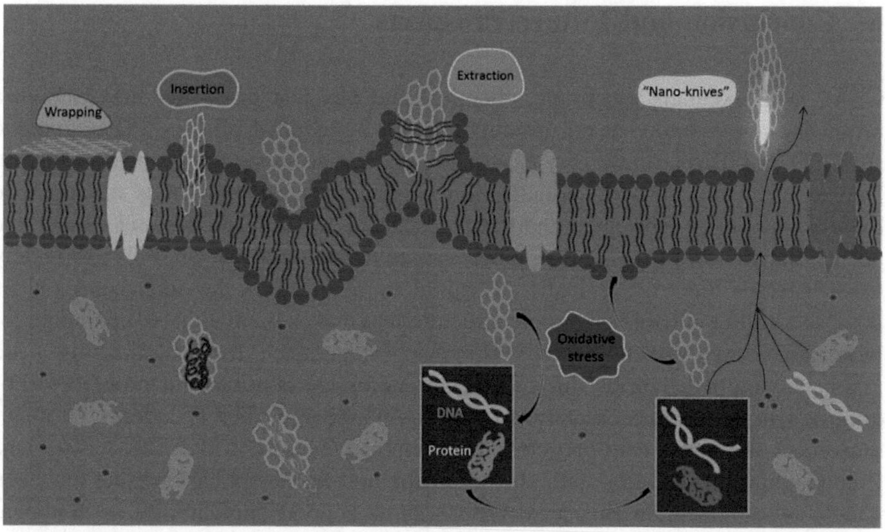

Fig. 14 Schematic illustration of the mechanisms governing the antimicrobial activity of graphene. Reprinted with permission from [104]. © 2016 American Chemical Society

antimicrobial activity of graphene is due to different types of perturbations to the phospholipid bilayer of bacteria. The first type of mechanism is the insertion, wherein graphene cuts through the bilayer due to the sharp molecular edges and eventually causes cellular death. The second type is a nano-knife action of graphene, a special case of insertion, where graphene inserts into the bilayer and causes bacterial death on account of the induced oxidative stress. In the case of extraction mechanism, the phospholipids are vigorously extracted from the lipid bilayer by graphene because of its strong hydrophobic interactions to expose the intracellular environment. Finally, for the wrapping mechanism, graphene covers the outer surface of the bilayer, restricts the access of essential nutrients, and ultimately leads to bacterial death. Each mechanism behind the antimicrobial activity of graphene was clearly illustrated in Fig. 14.

It is evident from the numerous reports and mechanistic studies on GBMs that the unique physical and chemical structure of graphene facilitates the development of efficient hybrid materials for targeted therapeutics including antimicrobial activities. The versatile post-functionalization capability allows GBMs to be highly selective and nontoxic, which can be capitalized for diverse therapeutic applications.

6 Conclusion and Future Prospects

GBMs with excellent electronic, chemical and physical properties have shown a promising applicability in all the directions of biomedical research. The synthetic strategies of GBMs can be very easy, allowing a wide scope for the post-treatment. Due to extensive post-functionalization capabilities, different kinds of graphene hybrids have been fabricated and used as sensors, imaging probes and therapeutic agents. In terms of biosensing, graphene-based nanocomposites are the most widely used materials due to chemically susceptible structures. On the other hand, GQDs possess a unique tunable electronic structure that makes them a viable candidate for bioimaging. However, more suitable materials for bioimaging are graphene-wrapped hybrids due to the morphological confinement properties. For drug delivery or therapeutics, heterogeneous nanocomposites encoded with quantum dots, polymers, nanoparticles, and biomolecules are widely used materials. Conventional therapeutics with chemotherapy and linear optical imaging has been widely investigated. In this area, GBMs show a good applicability in nonlinear imaging, photoacoustics and bimodal imaging. Therefore, GBMs could be extended to fabricate synergistic materials capable of demonstrating single imaging, bimodal imaging capabilities and therapeutics. On the other hand, the sensitivity, selectivity and efficacy of these materials should be further tuned to make them clinically viable. While GBMs provide a platform for the fabrication of hybrid materials with controlled morphology and tunable chemical properties for targeted applications in biomedical research, their toxicity, biocompatibility, and biodegradation need to be further investigated before translating toward future clinical uses.

Acknowledgments This work is supported by the NTU-A*Star Silicon Technologies Centre of Excellence under the program Grant No. 11235100003 and the NTU-Northwestern Institute for Nanomedicine.

References

1. Slowing II, Trewyn BG, Giri S, Lin VSY (2007) Mesoporous silica nanoparticles for drug delivery and biosensing applications. Adv Func Mater 17(8):1225–1236
2. Hazra S, Joshi H, Ghosh BK, Ahmed A, Gibson T, Millner P, Ghosh NN (2015) Development of a novel and efficient H_2O_2 sensor by simple modification of a screen printed Au electrode with Ru nanoparticle loaded functionalized mesoporous SBA15. RSC Adv 5 (43):34390–34397
3. Lehmana SE, Larsen SC (2014) Zeolite and mesoporous silica nanomaterials: greener syntheses, environmental applications and biological toxicity. Environ Sci Nano 1:200–213
4. Shen H, Zhang L, Liu M, Zhang Z (2012) Biomedical applications of graphene. Theranostics 2(3):283–294
5. Kelarakis A (2015) Graphene quantum dots: in the crossroad of graphene, quantum dots and carbogenic nanoparticles. Curr Opin Colloid Interface Sci 20:354–361
6. Luo S, Zhang E, Su Y, Cheng T, Shi C (2011) A review of NIR dyes in cancer targeting and imaging. Biomaterials 32(29):7127–7138

7. Galliford CV, Scheidt KA (2007) Pyrrolidinyl-spirooxindole natural products as inspirations for the development of potential therapeutic agents. Angew Chem Int Ed 46(46):8748–8758
8. Kannan RY, Salacinski HJ, Vara DS, Odlyha M, Seifalian AM (2006) Review paper: principles and applications of surface analytical techniques at the vascular interface. J Biomater Appl 21(1):5–32
9. Maxwell DJ, Taylor JR, Nie S (2002) Self-assembled nanoparticle probes for recognition and detection of biomolecules. J Am Chem Soc 124(32):9606–9612
10. Huang X, QiX Boey F, Zhang H (2012) Graphene-based composites. Chem Soc Rev 41:666–686
11. Liu J, Cui L, Losic D (2013) Graphene and graphene oxide as new nanocarriers for drug delivery applications. Acta Biomater 9:9243–9257
12. Mao HY, Laurent S, Chen W, Akhavan O, Imani M, Ashkarran AA, Mahmoudi M (2013) Graphene: promises, facts, opportunities, and challenges in nanomedicine. Chem Rev 113:3407–3424
13. Hummers WS, Offeman RE (1958) Preparation of graphitic oxide. J Am Chem Soc 80 (6):1339
14. Chen J, Yao B, Li C, Shi G (2013) An improved hummers method for eco-friendly synthesis of graphene oxide. Carbon 64:225–229
15. Marcano DC, Kosynkin DV, Berlin JM, Sinitskii A, Sun Z, Slesarev A, Alemany LB, Lu W, Tour JM (2010) Improved synthesis of graphene oxide. ACS Nano 4(8):4806–4814
16. Novoselov KS, Geim AK, Morozov SV, Jiang D, Zhang Y, Dubonos SV, Grigorieva Firsov AA (2004) Electric field effect in atomically thin carbon films. Science 306 (5696):666–669
17. Ham H, Park NH, Kang I, Kim HW, Shim KB (2012) Catalyst-free fabrication of graphene nanosheets without substrates using multiwalled carbon nanotubes and a spark plasma sintering process. Chem Commun 48:6672–6674
18. Schaffel F, Wilson M, Warner JH (2012) Motion of light adatoms and molecules on the surface of few-layer graphene. ACS Nano 5(12):9428–9441
19. Moon IK, Lee J, Ruoff RS, Lee H (2010) Reduced graphene oxide by chemical graphitization. Nat Commun 1:73
20. Mevold AH, Hsu WW, Hardiansyah A, Huang LY, Yang MC, Liu TY, Chan TY, Wang KS, Su YA, Jeng RJ, Wang JK, Wang YL (2015) Fabrication of gold nanoparticles/ graphene-PDDA nanohybrids for bio-detection by SERS nanotechnology. Nanoscale Res Lett 10:397
21. Tai H, Zhen Y, Liu C, Ye Z, Xie G, Du X, Jiang Y (2016) Facile development of high performance QCM Humidity sensor based on protonated polyethyleneimine-graphene oxide nanocomposite thin film. Sens Actuat B Chem 230:501–509
22. Darabdhara G, Boruah PK, Borthakur P, Hussain N, Das MR, Ahamad T, Alshehri SM, Malgras V, Wu KCW, Yamauchi Y (2016) Reduced graphene oxide nanosheets decorated with Au–Pd bimetallic alloy nanoparticles towards efficient photocatalytic degradation of phenolic compounds in water. Nanoscale 8(15):8276–8287
23. Panigrahy B, Sarmaa DD (2015) Enhanced photocatalytic efficiency of AuPd nanoalloy decorated ZnO-reduced graphene oxide nanocomposites. RSC Adv 5:8918–8928
24. Li L, Wu G, Yang G, Peng J, Zhao J, Zhu JJ (2013) Focusing on luminescent graphene quantum dots: current status and future perspectives. Nanoscale 5:4015–4039
25. Peng J, Gao W, Gupta BK, Liu Z, Romero-Aburto R, Ge L, Song L, Alemany LB, Zhan X, Gao G, Vithayathil SA, Kaipparettu BA, Marti AA, Hayashi T, Zhu JJ, Ajayan PM (2012) Graphene quantum dots derived from carbon fibers. Nano Lett 12:844–849
26. Zhang F, Liu F, Wang C, Xin X, Liu J, Guo S, Zhang J (2016) Effect of lateral size of graphene quantum dots on their properties and application. ACS Appl Mater Interfaces 8 (3):2104–2110
27. Ye R, Xiang C, Lin J, Peng Z, Huang K, Yan Z, Cook NP, Samuel ELG, Hwang CC, Ruan G, Ceriotti G, Raji ARO, Martí AA, Tour JM (2013) Coal as an abundant source of graphene quantum dots. Nat Commun 4:2943

28. Sreejith S, Ma X, Zhao Y (2012) Graphene oxide wrapping on squaraine-loaded mesoporous silica nanoparticles for bioimaging. J Am Chem Soc 134:17346–17349
29. Akhavan O, Ghaderi E, Esfandiar A (2011) Wrapping bacteria by graphene nanosheets for isolation from environment, reactivation by sonication, and inactivation by near-infrared irradiation. J Phys Chem B 115(19):6279–6288
30. Mohanty N, Fahrenholtz M, Nagaraja A, Boyle D, Berry V (2011) Impermeable graphenic encasement of bacteria. Nano Lett 11:1270–1275
31. Lanni F, Bailey B, Farkas DL, Taylor DL (1993) Excitation field synthesis as a means for obtaining enhanced axial resolution in fluorescence microscopes. Bioimaging 3(1):187–196
32. Kojima H, Nakatsubo N, Kikuchi K, Kawahara S, Kirino Y, Nagoshi H, Hirata Y, Nagano T (1998) Detection and imaging of nitric oxide with novel fluorescent indicators: diaminofluoresceins. Anal Chem 70(13):2446–2453
33. Koike C, Watanabe M, Oku N, Tsukada H, Irimura T, Okada S (1997) Tumor cells with organ-specific metastatic ability show distinctive trafficking in vivo: analyses by positron emission tomography and bioimaging. Cancer Res 15(57):3612–3619
34. Shang NG, Papakonstantinou P, McMullan M, Chu M, Stamboulis A, Potenza A, Dhesi SS, Marchetto H (2008) Catalyst-free efficient growth, orientation and biosensing properties of multilayer graphene nanoflake films with sharp edge planes. Adv Funct Mater 18(21):3506–3514
35. Mohanty N, Berry V (2008) Graphene-based single-bacterium resolution biodevice and DNA transistor: interfacing graphene derivatives with nanoscale and microscale biocomponents. Nano Lett 8(12):4469–4476
36. Wang J (2008) Electrochemical glucose biosensors. Chem Rev 108:814–825
37. Shan C, Yang H, Han D, Zhang Q, Ivaska A, Niu L (2010) Graphene/AuNPs/chitosan nanocomposites film for glucose biosensing. Biosens Bioelectron 25(5):1070–1074
38. Kanga X, Wang J, Wu H, Aksay IA, Liu J, Lin Y (2009) Glucose oxidase–graphene–chitosan modified electrode for direct electrochemistry and glucose sensing. Biosens Bioelectron 25(12):901–905
39. Liu B, Tang D, Tang J, Su B, Lia Q, Chen G (2011) A graphene-based Au(111) platform for electrochemical biosensing based catalytic recycling of products on gold nanoflowers. Analyst 136:2218–2220
40. Lina D, Wua J, Jua H, Yan F (2014) Nanogold/mesoporous carbon foam-mediated silver enhancement for graphene-enhanced electrochemical immunosensing of carcinoembryonic antigen. Biosens Bioelectron 52:153–158
41. Nguyena KT, Zhao Y (2014) Integrated graphene/nanoparticle hybrids for biological and electronic applications. Nanoscale 6:6245–6266
42. Tian J, Huang T, Wang P, Lu J (2015) GOD/HRP bienzyme synergistic catalysis in a 2-D graphene framework for glucose biosensing. J Electrochem Soc 162(12):B319–B325
43. Shan C, Yang H, Song J, Han D, Ivaska A, Niu L (2009) Direct electrochemistry of glucose oxidase and biosensing for glucose based on graphene. Anal Chem 81:2378–2382
44. Yang J, Deng S, Lei J, Ju H, Gunasekaran S (2011) Electrochemical synthesis of reduced graphene sheet–AuPd alloy nanoparticle composites for enzymatic biosensing. Biosens Bioelectron 29(1):159–166
45. Kim DM, Kim MY, Reddy SS, Cho J, Cho CH, Jung S, Shim YB (2013) Electron-transfer mediator for a NAD-glucose dehydrogenase-based glucose sensor. Anal Chem 85 (23):11643–11649
46. Gopalan AI, Muthuchamy N, Komathi S, Lee KP (2015) A novel multicomponent redox polymer nanobead based high performance non-enzymatic glucose sensor. Biosens Bioelectron 15:30540–30546
47. Jang HD, Kim SK, Chang H, Roh KM, Choi JW (2012) A glucose biosensor based on TiO_2-graphene composite. Biosens Bioelectron 38:184–188
48. Chen Y, Li Y, Sun D, Tian D, Zhang J, Zhu JJ (2011) Fabrication of gold nanoparticles on bilayer graphene for glucose electrochemical biosensing. J Mater Chem 21:7604–7611

49. Montornes JM, Vreeke MS, Katakis I (2008) Glucose biosensors. In: Bartlett PN (ed) Bioelectrochemistry: fundamentals, experimental techniques and applications. Wiley, Chichester. doi:10.1002/9780470753842.ch5
50. Zhou M, Zhai Y, Dong S (2009) Electrochemical sensing and biosensing platform based on chemically reduced graphene oxide. Anal Chem 81(14):5603–5613
51. Hua Y, Li F, Han D, Wu T, Zhang Q, Niu L, Bao Y (2012) Simple and label-free electrochemical assay for signal-On DNA hybridization directly at undecorated graphene oxide. Anal Chim Acta 753:82–89
52. Ali S, Hassan A, Hassan G, Bae J, Lee CH (2016) All-printed humidity sensor based on gmethyl-red/methyl-red composite with high sensitivity. Carbon 106:23–32
53. Zhang Y, Zeng GM, Tang L, Chen J, Zhu Y, He XX, He Y (2015) Electrochemical sensor based on electrodeposited graphene-Au modified electrode and nanoau carrier amplified signal strategy for attomolar mercury detection. Anal Chem 87:989–996
54. Zhang Y, Bai X, Wang X, Shiu KK, Zhu Y, Jiang H (2014) Highly sensitive graphene–Pt nanocomposites amperometric biosensor and its application in living cell H_2O_2 detection. Anal Chem 86:9459–9465
55. Wang H, Xia B, Yan Y, Li N, Wang JY, Wang X (2013) Water-soluble polymer exfoliated graphene: as catalyst support and sensor. J Phys Chem B 117:5606–5613
56. He S, Song B, Li D, Zhu C, Qi W, Wen Y, Wang L, Song S, Fang H, Fan CA (2010) Graphene nanoprobe for rapid, sensitive, and multicolor fluorescent DNA analysis. Adv Funct Mater 20:453–459
57. Chen Q, Wei W, Lin JM (2011) Homogeneous detection of concanavalin a using pyrene-conjugated maltose assembled graphene based on fluorescence resonance energy transfer. Biosens Bioelectron 26(11):4497–4502
58. Zhu Y, Cai Y, Xu L, Zheng L, Wang L, Qi B, Xu C (2015) Building An Aptamer/graphene oxide FRET biosensor for one-step detection of bisphenol A. ACS Appl Mater Interfaces 7 (14):7492–7496
59. Chang H, Tang L, Wang Y, Li JJJ (2010) Graphene fluorescence resonance energy transfer aptasensor for the thrombin detection. Anal Chem 82(6):2341–2346
60. Zhang M, Yin BC, Wang XF, Ye BC (2011) Interaction of peptides with graphene oxide and its application for real-time monitoring of protease activity. Chem Commun 47:2399–2401
61. Li H, Sun DE, Liu Y, Liu Z (2014) An ultrasensitive homogeneous aptasensor for kanamycin based on upconversion fluorescence resonance energy transfer. Biosens Bioelectron 55:149–156
62. Myung S, Solanki A, Kim C, Park J, Kim KS, Lee KB (2011) Graphene-encapsulated nanoparticle-based biosensor for the selective detection of cancer biomarkers. Adv Mater 23:2221–2225
63. Dong H, Gao W, Yan F, Ji H, Ju H (2010) Fluorescence resonance energy transfer between quantum dots and graphene oxide for sensing biomolecules. Anal Chem 82:5511–5517
64. Bhatnagar D, Kumar V, Kumar A, Kaur I (2016) Graphene quantum dots FRET based sensor for early detection of heart attack in human. Biosens Bioelectron 79:495–499
65. Huang Y, Dong X, Liu Y, Lic LL, Chen P (2011) Graphene-based biosensors for detection of bacteria and their metabolic activities. J Mater Chem 21:12358–12362
66. Park SJ, Kwon OS, Lee SH, Song HS, Park TH, Jang J (2012) Ultrasensitive flexible graphene based field-effect transistor (FET)-type bioelectronic nose. Nano Lett 12(10):5082–5090
67. Fu X, Chen L, Li J, Lin M, You H, Wang W (2012) Label-free colorimetric sensor for ultrasensitive detection of heparin based on color quenching of gold nanorods by graphene oxide. Biosen Bioelectron 34(1):227–231
68. Huang KJ, Liu YJ, Cao JT, Wang HB (2014) An aptamer electrochemical assay for sensitive detection of immunoglobulin e based on tungsten disulfide-graphene composites and gold nanoparticles. RSC Adv 4:36742–36748
69. Zagorodko O, Spadavecchia J, Serrano AY, Larroulet I, Pesquera A, Zurutuza A, Boukherroub R, Szunerits S (2014) Highly sensitive detection of DNA hybridization on

commercialized graphene-coated surface plasmon resonance interfaces. Anal Chem 86 (22):11211–11217
70. Kwon OS, Park SJ, Hong JY, Han AR, Lee JS, Lee JS, Oh JH, Jang J (2012) Flexible FET-Type VEGF aptasensor based on nitrogen-doped graphene converted from conducting polymer. ACS Nano 6(2):1486–1493
71. Chung K, Rani A, Lee JE, Kim JE, Kim Y, Yang H, Kim SO, Kim D, Kim DH (2015) Systematic study on the sensitivity enhancement in graphene plasmonic sensors based on layer-by-layer self-assembled graphene oxide multilayers and their reduced analogues. ACS Appl Mater Interfaces 7:144–151
72. Ma X, Qu Q, Zhao Y, Luo Z, Zhao Y, Ng KW, Zhao Y (2013) Graphene oxide wrapped gold nanoparticles for intracellular raman imaging and drug delivery. J Mater Chem B 1:6495–6500
73. Xu S, Man B, Jiang S, Wang J, Wei J, Xu S, Liu H, Gao S, Liu H, Li Z, Li H, Qiu H (2015) Graphene/Cu nanoparticle hybrids fabricated by chemical vapor deposition As surface-enhanced raman scattering substrate for label-free detection of adenosine. ACS Appl Mater Interfaces 7:10977–10987
74. Shen J, Zhu Y, Yang X, Li C (2012) Graphene quantum dots: emergent nanolights for bioimaging, sensors, catalysis and photovoltaic devices. Chem Commun 48:3686–3699
75. Hu SH, Chen YW, Hung YT, Chen IW, Chen SY (2012) Quantum-dot-tagged reduced graphene oxide nanocomposites for bright fluorescence bioimaging and photothermal therapy monitored in situ. Adv Mater 24:1748–1754
76. Kim H, Namgung R, Singha K, Oh IK, Kim WJ (2011) Graphene oxide−polyethylenimine nanoconstruct as a gene delivery vector and bioimaging tool. Bioconjugate Chem 22:2558–2567
77. Oh SD, Kim J, Lee DH, Kim JH, Jang CW, Kim S, Choi SH (2016) Structural and optical characteristics of graphene quantum dots size-controlled and well-aligned on a large scale by polystyrene-nanosphere lithography. J Phys D Appl Phys 49:025308
78. Bartelmess J, Quinn SJ, Giordani S (2015) Carbon nanomaterials: multi-functional agents for biomedical fluorescence and raman imaging. Chem Soc Rev 44:4672–4698
79. Qu D, Sun Z, Zheng M, Li J, Zhang Y, Zhang G, Zhao H, Liu X, Xi Z (2015) Three colors emission from S, N Co-doped graphene quantum dots for visible light H_2 production and bioimaging. Adv Optical Mater 3:360–367
80. Liu Q, Guo B, Rao Z, Zhang B, Gong JR (2013) Strong two-photon-induced fluorescence from photostable, biocompatible nitrogen-doped graphene quantum dots for cellular and deep-tissue imaging. Nano Lett 13:2436–2441
81. Chandra A, Deshpande S, Shinde DB, Pillai VK, Singh N (2014) Mitigating the cytotoxicity of graphene quantum dots and enhancing their applications in bioimaging and drug delivery. ACS Macro Lett 3:1064–1068
82. Wang Y, Chen JT, Yan XP (2013) Fabrication of transferrin functionalized gold nanoclusters/graphene oxide nanocomposite for turn-on near-infrared fluorescent bioimaging of cancer cells and small animals. Anal Chem 85:2529–2535
83. Bloembergen N (1959) Solid state infrared quantum counters. Phys Rev Lett 2:84–85
84. Yin PT, Shah S, Chhowalla M, Lee KB (2015) Design, synthesis, and characterization of graphene − nanoparticle hybrid materials for bioapplications. Chem Rev 115(7):2483–2531
85. Zhou B, Shi B, Jin D, Liu X (2015) Controlling upconversion nanocrystals for emerging applications. Nat Nanotechnol 10:924–936
86. Nguyen KT, Sreejith S, Joseph J, He T, Borah P, Guan EY, Lye SW, Sun H, Zhao Y (2014) Poly(acrylic acid)-capped and dye-loaded graphene oxide-mesoporous silica: a nano-sandwich for two-photon and photoacoustic dual-mode imaging. Part Part Syst Charact 31:1060–1066
87. Shi X, Gong H, Li Y, Wang C, Cheng L, Liu Z (2013) Graphene-based magnetic plasmonic nanocomposite for dual bioimaging and photothermal therapy. Biomaterials 34(20):4786–4793

88. Turcheniuk K, Boukherroub R, Szunerits S (2015) Gold-graphene nanocomposites for sensing and biomedical applications. J Mater Chem B 3:4301–4324
89. Mendes RG, Bachmatiuk A, El-Gendy AA, Melkhanova S, Klingeler R, Büchner B, Rümmeli MH (2012) A facile route to coat iron oxide nanoparticles with few-layer graphene. J Phys Chem C 116(44):23749–23756
90. Kumar KS, Kumar VB, Paik P (2013) Recent advancement in functional core-shell nanoparticles of polymers: synthesis, physical properties, and applications in medical biotechnology. J Nanopart 2013:672059
91. Jiang BP, Hu LF, Wang DJ, Ji SC, Shen XC, Liang H (2012) Graphene loading water-soluble phthalocyanine for dual-modality photothermal/photodynamic therapy via a one-step method. J Mater Chem B 2:7141–7148
92. Yang K, Zhang S, Zhang G, Sun X, Lee ST, Liu S (2010) Graphene in mice: ultrahigh in vivo tumor uptake and efficient photothermal therapy. Nano Lett 10:3318–3323
93. Maji SK, Mandal AK, Nguyen KT, Borah P, Zhao Y (2015) Cancer cell detection and therapeutics using peroxidase-active nanohybrid of gold nanoparticle-loaded mesoporous silica-coated graphene. ACS Appl Mater Interfaces 7:9807–9816
94. Kim YK, Na HK, Kim S, Jang H, Chang SJ, Min DH (2015) One-pot synthesis of multifunctional Au@graphene oxide nanocolloid Core@Shell nanoparticles for raman bioimaging, photothermal, and photodynamic therapy. Small 11(21):2527–2535
95. Moon H, Kumar D, Kim H, Sim C, Chang JH, Kim JM, Kim H, Lim DK (2015) Amplified photoacoustic performance and enhanced photothermal stability of reduced graphene oxide coated gold nanorods for sensitive photoacoustic imaging. ACS Nano 9(3):2711–2719
96. Liu Y, Bai J, Jia X, Jiang X, Guo Z (2015) Fabrication of multifunctional SiO_2@GN-serum composites for chemo-photothermal synergistic therapy. ACS Appl Mater Interfaces 7(1):112–121
97. Zheng FF, Zhang PH, Xi Y, Chen JJ, Li LL, Zhu JJ (2015) Aptamer/graphene quantum dots nanocomposite capped fluorescent mesoporous silica nanoparticles for intracellular drug delivery and real-time monitoring of drug release. Anal Chem 87(23):11739–11745
98. Robinson JT, Tabakman SM, Liang Y, Wang H, Casalongue HS, Vinh D, Dai H (2011) Ultrasmall reduced graphene oxide with high near-infrared absorbance for photothermal therapy. J Am Chem Soc 133(17):6825–6831
99. Melamed JR, Edelstein RS, Day ES (2015) Elucidating the fundamental mechanisms of cell death triggered by photothermal therapy. ACS Nano 9(1):6–11
100. Ge J, Lan M, Zhou B, Liu W, Guo L, Wang H, Jia Q, Niu G, Huang X, Zhou H, Meng X, Wang P, Lee C-S, Zhang W, Han X (2014) A graphene quantum dot photodynamic therapy agent with high singlet oxygen generation. Nat Commun 5:4596
101. Li M, Yang X, Ren J, Qu K, Qu X (2012) Using graphene oxide high near-infrared absorbance for photothermal treatment of alzheimer's disease. Adv Mat 24(13):1722–1728
102. Bian X, Song ZL, Qian Y, Gao W, Cheng ZQ, Chen L, Liang H, Ding D, Nie XK, Chen Z, Tan W (2014) Fabrication of graphene-isolated-Au-nanocrystal nanostructures for multimodal cell imaging and photothermal-enhanced chemotherapy. Sci Rep 4:6093
103. He D, He X, Wang K, Zou Z, Yang X, Li X (2014) Remote-controlled drug release from graphene oxide-capped mesoporous silica to cancer cells by photoinduced pH-jump activation. Langmuir 30:7182–7189
104. Zou X, Zhang L, Wang Z, Luo Y (2016) Mechanisms of the antimicrobial activities of graphene materials. J Am Chem Soc 138(7):2064–2077

Part II
Medicine: Biosensors

Hybrid Graphene Metallic Nanoparticles for Biodetection

Manos Gkikas

Abstract Sensitive and accurate techniques for detection in water, environment and issues related to health, have been a great challenge among scientists. To date, many analytical tools based on different physical, chemical, and biological phenomena have been developed for detection of biomolecules, biomedical imaging, and biosensing, including fluorescence spectroscopy, surface-enhanced Raman scattering (SERS), electrochemistry, and techniques that are based on a specific biological recognition. However, due the demands of the new era and the advances in modern biomedicine, materials that combine relatively low detection limits, easiness of application, as well as low cost, are of great challenge. Herein, the progress towards graphene/metallic nanoparticle hybrids is discussed, emphasizing on the advances of different synthetic methods, the decoration with well-dispersed metallic particles (Au, Ag, Pt, Pd, Cu, and QDs) on graphene, as well as different bioapplications.

Keywords Graphene · Metal nanoparticles · Biosensors · Bioimaging · FRET · Electrocatalysis · Biodetection

1 Introduction

Graphene oxide (GO) has excellent mechanical properties, large specific area with both sides been accessible, abundance of functional groups and can be easily converted to reduced graphene oxide (rGO) following reduction, improving the electrical properties. Metallic nanoparticles assembled onto graphene-based materials not only enhance the carbon support with the properties of the metals, but also prevent restacking of the graphene sheets as inorganic spacers. The shape and size of the inorganic particles can be synthetically tuned for utilization in specific

M. Gkikas (✉)
Department of Chemistry, University of Massachusetts Lowell,
Lowell, MA 01854, USA
e-mail: manos_gkikas@uml.edu

G. Gonçalves et al. (eds.), *Graphene-based Materials in Health and Environment*,
Carbon Nanostructures, DOI 10.1007/978-3-319-45639-3_3

applications, either industrial or biomedical, allowing for the development of the next generation materials as biosensors, biocatalytic templates, and contrast agents in biomedicine. Additionally, the recent significant advances in optics, laser technology, surface plasmon resonance (SPR) spectroscopy, and nanofabrication, have arisen as a versatile tool that offers excellent sensitivity, selectivity, and structural information in biological media.

Inorganic nanoparticles in carbon templates have been utilized as "contrast agents" and "exothermic reactors". Lately, an increasing number of publications highlight the symbiosis of carbon-metal materials [1–4], and a plethora of bioapplications using either graphene [5–11] or metallic nanoparticles [12–16]. While microorganisms have been observed to secrete secondary metabolites to assist in their survival against metals, such as gold [17], naked Au-nanoparticles (Au-NPs) are reported to have low cytotoxicity in human cells [18, 19]. Moreover, they serve as contrast agents in biomedicine [20], producing significant photothermal ablation in cancer cells using near-IR light as an external stimulus, or as high payload doxorubicin carriers for near-IR light-triggered drug release [21]. Therefore, the combination of graphene with metallic nanoparticles seems to be extremely interesting for biomedical applications and biodetection.

In most of the cases, the synthesis of metal particles occurs through an initial reduction step of metal salts with an organic compound or a polymer, and sequential stabilization with a different (or the same) organic compound or polymer. Materials that have a dual role, serving both as a reductant and a stabilizer seem to be more challenging in terms of cost for large-scale applications. Tuning the size, the shape, and the dispersion of the inorganic particles onto graphene defines the sensing properties and applications to a great extent. Towards that direction, the initial concentration of graphene, the metal salt feed, the reaction temperature, as well as the utilization of different reductants and stabilizers (Table 1), have been found to affect the size and shape of the metal particles onto graphene and their stability. Hybrid structures of graphene with gold, silver, copper, platinum, and palladium nanoparticles as well with quantum dots are herein discussed, emphasizing on the role as detection probes, biosensors, and bioimaging tools, depending on the detection method; either surface-enhanced Raman spectroscopy (SERS), electrochemistry, fluorescence, or techniques that are based on a specific biological recognition.

2 Detection of Graphene/Metal Hybrids Using Surface-Enhanced Raman Spectroscopy (SERS)

The SERS effect results from the combination of two main mechanisms designated by electromagnetic and chemical mechanisms. The electromagnetic enhancement involves the excitation of surface plasmons on the metal structures, whereas the chemical enhancement involves the formation of charge-transfer complexes between the metal and the analyte [22]. SERS can boost the Raman signal of neat

Table 1 Reaction conditions for the synthesis of graphene/metal nanohybrids

Hybrid structure	Graphene (mg/mL)[a]	Salt (moles)	Reductant	Stabilizer	T (°C)	D (nm)	Ref.
GO/Au-spheres	1.50 (2.5 mL)	12×10^{-6} (50 mL)[b]	Sodium citrate	–	80 °C (30 min)	21	[23]
rGO/Au-spheres	3.86 (1 mL)	5×10^{-6} (20 mL)[b]	Sodium citrate	2-Mercapto pyridine	100 °C (15 min)	20–40	[25]
rGO/Ag-spheres	0.20 (5 mL)	9.5×10^{-6}(0.1 mL)[c]	$NaBH_4$	–	25 (-)	5–60	[28]
GO/Au-spheres	2.00 (1 mL)	10×10^{-6} (0.1 mL)[b]	PLP	PLP	50 °C (20 h)	77	[30]
GO/Au-spheres	3.86 (1 mL)	5×10^{-6} (20 mL)[b]	Sodium citrate	Dimercapto succinic acid	100 °C (15 min)	20	[32]
GO/CdSe-spheres	1.59 (6.3 mL)	85 & 117×10^{-6} (–)[d]	–	$CH_3(CH_2)_{14}COO^-$	240 °C (20 min)	5–6	[58]
rGO/Au-spheres	0.25 (4 mL)	3.2×10^{-6} (–)[b]	IL-NH_2[e]	IL-NH_2[e]	25 °C (20 h)	3	[62]

[a]The graphene suspension volumes reported in the table refer to the overall volume prior to mixing with the metal salt
[b]Refers to $HAuCl_4.3H_2O$
[c]Refers to $Cd(O_2C_{16}H_{31})_2$ and SeO_2
[d]Refers to $AgNO_3$
[e]IL-NH_2: 1-(2-aminoethyl)-3-imidazolium bromide

graphene or neat metal particles by 10^8 times or more, which has exhibited potential for ultrasensitive analytical applications. However, the high sensitivity of SERS suffers from its poor reproducibility. Metallic-NPs assembled on graphene nanosheets (and supported on Si wafers) have been used to detect various aromatic molecules. The detection of different biomolecules onto graphene–metal hybrids using SERS is analytically described in Table 2.

Aqueous graphene/gold suspensions have been successfully produced by the group of Marques [23], using an initial "aging" step, where the oxygen functional groups at the surface of GO coordinate with Au(III) ions, bringing the two materials in close in proximity, followed by sodium citrate reduction for 1 h at 80 °C. Different surface treatments of graphene were examined (GO, thermally reduced graphene, hydrazine-reduced graphene), showing the important role of the oxygen groups at the surface of graphene on the nucleation and growth of particles. The "aging" treatment of GO with gold ions led to a homogeneous distribution of small-sized nanospheres (~21 nm diameter) anchored on graphene (Fig. 1), while the thermally reduced graphene led to agglomeration of particles, yielding Au-NPs with ~300 nm diameter. No evidence of the formation of Au-NPs was shown in the case of hydrazine-reduced graphene. UV-Vis characterization for the GO/Au hybrids revealed a peak at 540 nm, while that of the thermally reduced graphene/Au hybrids was broader and shifted to ~590 nm. The GO/Au-NPs were tested as a substrate for SERS, for detection of the fluorescent dye rhodamine 6G (Rh6G), after deposition of the graphene/gold dispersion onto silica substrates, and sequentially the SERS probe. Enhanced Raman spectrum was obtained for Rh6G adsorbed at the surface of the GO/Au-NPs nanocomposite, in comparison with Rh6G on neat graphene, where a broad band was obtained in the region of 1600–1300 cm^{-1},

Table 2 Different biomolecules detection onto graphene/metal hybrids using SERS

Hybrid structure	SPR band (nm)	Substrate	Ref.
GO/Au-NPs	~540	Rhodamine 6G	[23]
rGO/Au-NPs	~539	Rhodamine 6G	[24]
rGO/Ag-NPs	~405	Rhodamine 6G	[24]
rGO/2-MPy/Au-NPs	~590	*p*-Aminothiophenol	[25]
rGO/Au-NPs	~539	*p*-Aminothiophenol	[24]
GO/PVP/Ag-NPs	~450	*p*-Aminothiophenol	[26]
rGO/Ag-NPs	~405	*p*-Aminothiophenol	[24]
GO/PVP/PDDA/Ag-NPs	–	Folic acid	[27]
GO/Au-NPs	~524	Rhodamine B	[28]
GO/Ag-NPs	~390	TMPyP[a]	[29]
GO/Ag-NPs	~390	Crystal violet	[29]
rGO/Au-NPs	~539	Nile blue	[24]
rGO/Ag-NPs	~405	Nile blue	[24]
GO/Au-NPs	~540	Cellular uptake[b]	[32]

[a] 5,10,15,20-tetrakis(1-methyl-4-pyridinio)porphyrin tetra(*p*-toluenesulfonate).
[b] Cellular uptake and internalization of the hybrids on Ca Ski cells

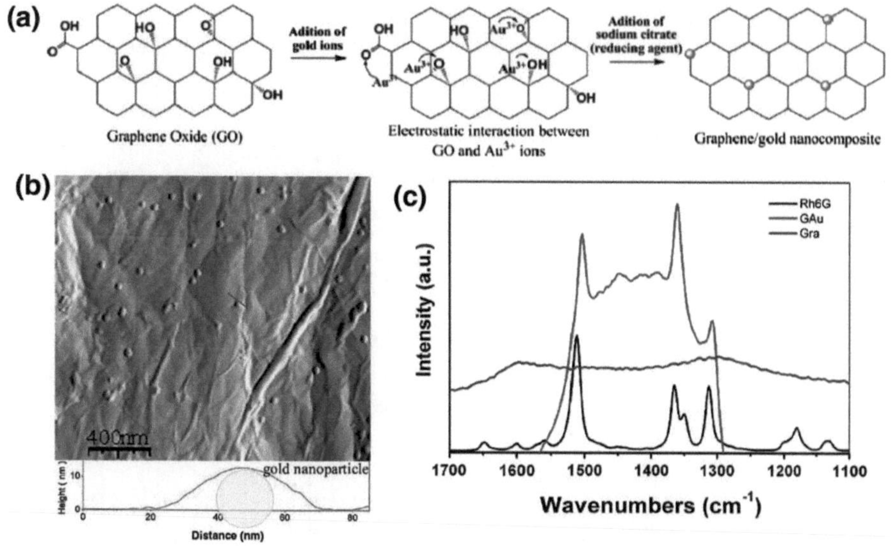

Fig. 1 **a** Schematic of the "aging" treatment and reduction of Au(III) ions onto graphene. **b** AFM image of the GO/Au-NPs of approximately 8.4 % gold nanoparticles per area of graphene. **c** SERS spectrum of Rhodamine 6G adsorbed at a single graphene sheet (*blue line*) and at GO/Au-NP (*red line*) as well as the Raman spectra of solid Rh6G (*black line*) (reproduced from Goncalves et al. [23])

characteristic of the s G and D peaks of graphene (Fig. 1). This work was pioneering for the time, showing that Au(III) cannot be reduced by GO, unless a reductant is used, while different surface treatments of graphene can affect the nucleation of gold nanoparticles and therefore their final size.

Graphene/metal hybrid nanostructures prepared with Au-NPs or Ag-NPs were fabricated by the group of Wu [24] as SERS probes for detection of molecules such as rhodamine 6G, nile blue, and *p*-aminothiophenol (Fig. 2). Rhodamine 6G is a cationic dye possessing a maximum absorption band at ~530 nm (non-resonant excitation), nile blue is also a cationic dye with a larger macrocyclic conjugated structure (able to form p–p stacking) and a maximum absorption band at ~625 nm (resonant excitation), while *p*-aminothiophenol has the weakest conjugated structure among the probe molecules and binds Au-NPs and Ag-NPs via Au–S and Ag–S bonds. Following the sodium citrate reduction method for 24 h at 100 °C, the authors claimed complete reduction of GO to rGO, as well as metallic particles anchored on graphene with a diameter of 10–30 nm (rGO/Au-NPs) and 10–35 nm (rGO/Ag-NPs). After the decoration of graphene with the metallic particles, an enhancement factor of 3.6 was found for graphene/Au and a 13.1 for graphene/Ag, according to the G peak, which mainly arises from the electromagnetic effect based on the charge transfer.

For all the three probe molecules, a SERS effect of Au-NPs or Ag-NPs on the probe molecules was found when the probe concentration was in the range of 1×10^{-4} M while at concentrations below 1×10^{-5} M, only the SERS spectra of graphene was

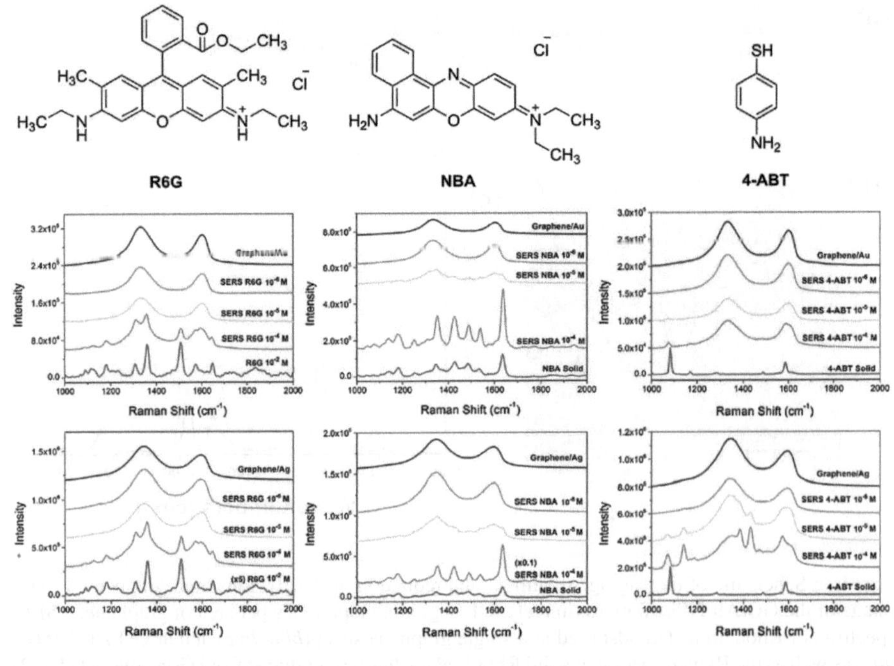

Fig. 2 Competitive surface-enhanced Raman scattering effects in rGO/Au-NPs and rGO/Au-NPs hybrids using different SERS probes: rhodamine 6G, nile blue, and *p*-aminothiophenol (reproduced from Sun et al. [24])

observed. Though the SERS effect of the Au-NPs or Ag-NPs on graphene was present in all the cases competitive SERS effects were obtained depending on the probe. The chemical mechanism attributed to p–p stacking was found to dominate in the case of the macrocyclic nile blue (Raman signal of probe affected by graphene), while for *p*-aminothiophenol and rhodamine 6G the electromagnetic-based SERS effect dominated (Raman signal of probe affected by the anchored metallic particles). Overall, it was shown that nile blue had the strongest suppressing effect on the SERS of graphene (high charge transfer due to large amounts of probe absorbed on graphene), wherein the D and G bands of graphene could not even be detected at 1×10^{-4} M probe concentration. Rhodamine 6G had a moderate effect, while *p*-aminothiophenol had the weakest suppressing effect on the SERS of graphene, since the probe molecules cannot be absorbed on the graphene sheets, but only bind to the metallic-NPs. Therefore, the Raman spectrum could be considered a superimposition of the spectrum of 4-aminothiophenol and that of graphene, enhanced by the metallic-NPs individually (Fig. 2).

Nanocomposites of GO and rGO with Au-NPs thermally reduced by sodium citrate and capped with 2-mercaptopyridine were reported to demonstrate excellent SERS [25]. The authors claimed the absence of gold nanoparticles anchored on graphene in the absence of the stabilizer. It was supported that the aromatic pyridine stabilizers led to physisorption on the graphene sheets and could further facilitate electron transfer between the metal and graphene. The deposition of gold particles on the graphene matrix led to broadening of the surface plasmon resonance and to a red shift from 520 nm up to ~580 nm, indicating the formation of aggregation (Fig. 3). The GO/Au-NPs showed stronger SERS intensity than neat Au-NPs, using *p*-aminothiophenol as a SERS probe, possibly due to the aggregation of Au-NPs onto the graphene sheets. That could lead to a coupled electromagnetic effect, and explain the enhanced Raman scattering, in combination with the strong electronic interactions of Au-NPs and GO.

GO/Poly(*N*-vinyl-2-pyrrolidone)/Ag-NPs (GO/PVP/Ag-NPs) hybrids prepared by simultaneous reduction and stabilization of Ag-NPs by PVP, showed superior SERS over *p*-aminothiophenol [26]. New bands in the Raman spectrum of the hybrids denoted charge transfer from Ag-NPs to *p*-aminothiophenol. Using higher weight ratios between $AgNO_3$ and GO (from 24 to 384), it was managed to obtain higher SERS with higher weight ratios, in accordance with a red shift in the UV-Vis spectrum of Ag-NPs (~450 nm) and band broadening that denotes increase in size.

PVP-capped GO functionalized with poly(diallyldimethyl ammonium chloride) (PDDA) through electrostatics led to GO/PVP/PDDA/Ag-NPs assembled structures after mixing the cationic polymer-functionalized graphene with as-prepared Ag-NPs (negatively charged, ~35 nm) [27]. The obtained hybrid nanostructures exhibited strong SERS activity in the addition of folic acid, resulting from the anchored Ag-NPs and the enrichment of folic acid molecules onto GO due to electrostatic interactions (overall positive zeta potential of the hybrid), finally leading to an ultrasensitive, label-free detection of folic acid by SERS (Fig. 4). Since folic acid receptors are found on the surfaces of various human tumor cells and folic acid has been considered a possible targeting agent of cancer cells, enhancing its signal is of paramount importance in biomedicine and biotechnology. The SERS spectra of 10^{-4} M folic acid showed a strong signal on GO/PVP/PDDA/Ag-NPs in comparison with neat Ag-NPs, confirming also the main vibrations of folic acid. Focusing on the folic acid vibration peak at 1595 cm^{-1} (signature peak), the authors found an increase in the intensity of

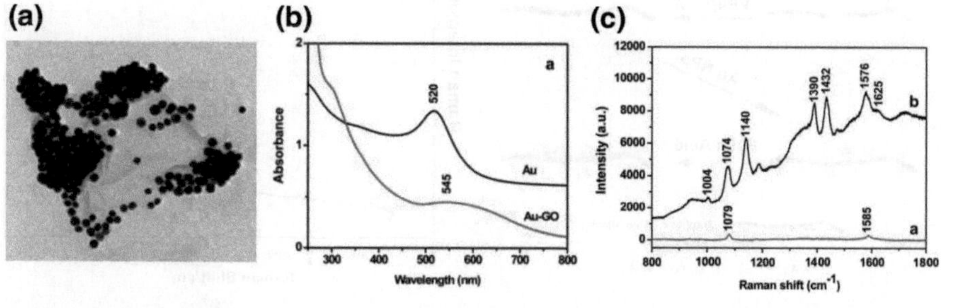

Fig. 3 **a** TEM images of Au-NPs deposited onto GO sheets. **b** UV-Vis spectra of 40 nm pre-synthesized Au-NPs before (*black line*) and after (*red line*) attachment onto GO sheets. **c** SERS spectra of *p*-aminothiophenol using Au-NPs (*red line*) and GO/Au-NPs (*black line*) as SERS substrates (reproduced from Huang et al. [25])

the SERS by increasing the folic acid concentration, with a detection limit of 9 nM and linearity between 9 and 180 nM in water. The sensing probe was also effective in detecting known concentrations of folic acid in dilute human serum, extending the sensing applicability to potential biological samples.

rGO/Ag-NPs composites were prepared by the group of Kamat [28] and were used to boost the SERS sensitivity of the porphyrin derivative 5,10,15,20-tetrakis (1-methyl-4-pyridinio)porphyrin tetra(*p*-toluenesulfonate) (TMPyP). The hybrids were synthesized by simultaneous reduction of $AgNO_3$ and GO by $NaBH_4$, yielding a surface plasmon band at ~390 nm. The enhanced performance of the rGO/Ag-NPs hybrid was attributed to the complexation between the porphyrin and the composite (both electrostatic and π–π stacking interactions between TMPyP and rGO), supported by a red shift in the porphyrin absorption band, which allowed for target molecule detection at low concentration levels. A nearly linear dependence of the SERS signal strength on TMPyP concentration was found using the rGO/Ag-NPs composites, rendering it as a potential probe for determination of the TMPyP concentration (Fig. 5). By monitoring the integrated intensity of the TMPyP Raman band at 1560 cm^{-1} as a function of [Ag^0], it was found that mixing ~0.26 mM Ag^0 with 0.03 mg/mL RGO led to the highest SERS signal.

GO/Au-NPs hybrids prepared by UV-irradiation of a GO suspension containing $HAuCl_4$ and a 25 % NH_3 solution (with or without the addition of citric acid) served as a SERS substrate for Rhodamine B (RhB) detection [29]. Casting an aqueous solution of the fluorescent dye molecule onto SiO_2/Si substrates covered with a film containing the graphene–gold hybrids led to enhancement in the Raman spectrum, higher than that of neat Au-NPs onto the same substrate.

Grafting the water-soluble, left-handed helical polypeptide, poly(proline) (PLP) to GO, Gkikas et al. [30] managed to functionalize the 2D carbon support with a biocompatible, non-charged polypeptide, and decorate it with Au-NPs,

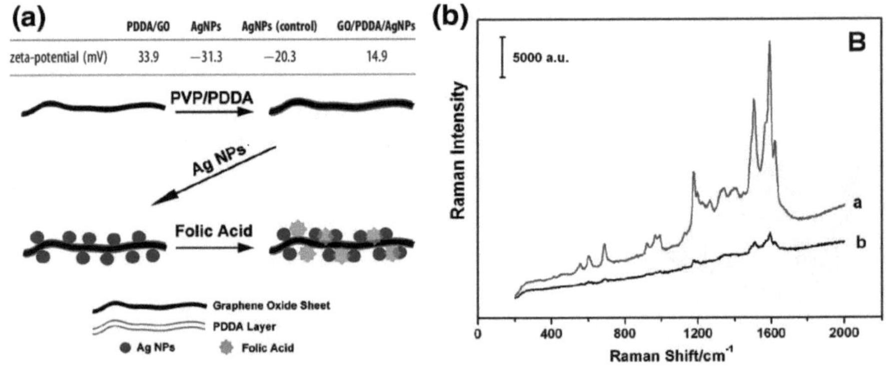

Fig. 4 **a** Fabrication of GO/PVP/PDDA/Ag-NPs for SERS detection of folic acid. **b** SERS spectra of 10^{-4} M folic acid onto the hybrid (*red line*) and onto neat Ag-NPs colloids (*black line*) (reproduced from Ren et al. [27])

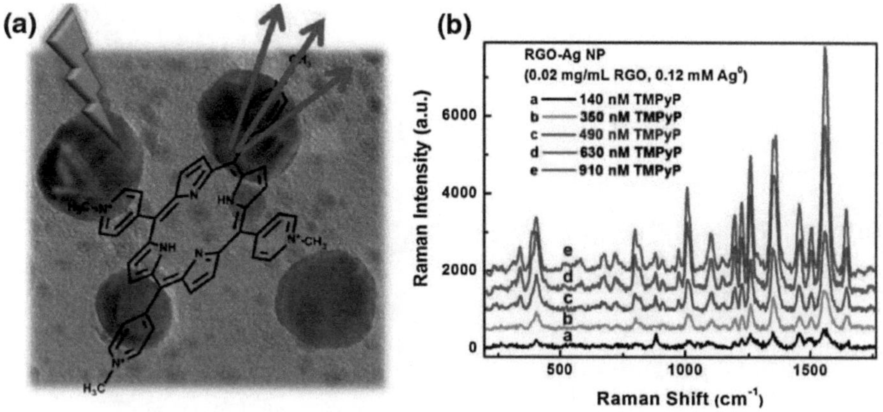

Fig. 5 **a** Schematic representation of rGO/Ag-NPs composites for SERS detection of the porphyrin TMPyP. Both electrostatic interactions and π–π stacking of TMPyP with RGO lead to a flattening of the porphyrin on the rGO surface. **b** SERS spectra of the hybrids using different TMPyP concentrations. The background of the rGO Raman scattering has been subtracted (reproduced from Murphy et al. [28])

showing capability for SERS effect, due to the high aggregation of Au-NPs onto graphene The helical hybrid material served as simultaneous reductant and stabilizer for Au-NPs, due to the dual role of the synthetic PLP homopolypeptide [31]. PLP's rigid backbone, containing consecutive pyrrolidine rings, led to efficient restacking of graphene sheets, and enabled in situ reduction of Au(III) ions by the polymer and simultaneous stabilization of the as-formed Au-NPs in water in one-step, without the formation of toxic by-products. Unlike PLP/Au-NPs though where the band was centered at 530 nm [31], the surface plasmon band of GO/PLP/Au-NPs [30] was centered at ~600 nm (Fig. 6), confirming aggregation of metallic particles when attached onto graphene (size is proportional to a red shift at higher wavelengths). Kinetic measurements of the GO/PLP/Au-NPs hybrid by UV-Vis showed a surface plasmon resonance peak emerging at 580 nm during the first h, which was shifted to higher wavelengths during the second and the third h. The band centered at 600 nm however narrowed after the fourth h and increased its intensity by the sixth h (Fig. 6). The size and shape of Au-NPs still remained spherical, while their average size onto the wrinkled graphene sheets was ~77 nm, as revealed by HRTEM, making it a suitable substrate for SERS in vitro and in vivo.

Graphene/Au(0) hybrids were recently introduced in biomedicine. Ren and Zhang [32] examined the cellular uptake and intracellular pathway of GO/Au-NPs inside cells via SERS. Gold particles were used as the SERS-active substrate and GO as the delivery scaffold. Au-NPs (~20 nm in size) prepared by sodium citrate reduction and capped with dimercaptosuccinic acid were chemically attached onto GO/PEG-NH_2

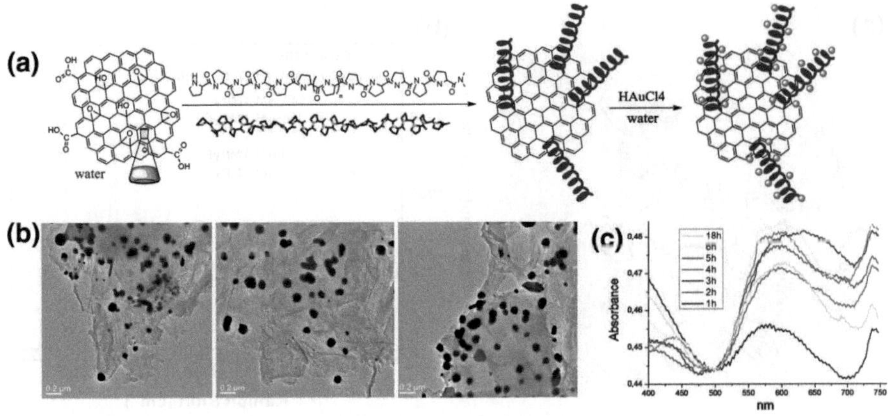

Fig. 6 **a** Schematic of gold nanoparticles anchored on GO/PLP through grafting PLP onto graphene sheets, in situ reduction and stabilization of Au-NPs by PLP in water in one-step. **b** HRTEM images of GO/PLP/Au-NPs showing decorated wrinkled graphene structures. **c** UV-Vis kinetics of formed Au-NPs onto GO/PLP with the surface plasmon resonance band shifted to higher wavelengths due to the attachment of the polymer onto the carbon matrix (reproduced from Gkikas et al. [30])

hybrids through EDC coupling. The enhanced by the metal nanoparticles SERS signal of graphene provided a "molecular fingerprint" for identifying molecules from different spots on Ca Ski cells (cytoplasm or nucleus). Incubation of cells with GO/PEG/Au-NPs revealed a strong Raman signal of D and G bands of graphene, indicating internalization of the gold hybrids in Ca Ski cells (Fig. 7). On the other hand, negligible SERS signal was obtained using neat GO.

In order to understand the internalization process of GO/Au-NPs in Ca Ski cells, different incubation times were tested. Incubation for 4 h showed a strong signal that started to decrease after 8 h and minimized after 12 h, possibly due to the release of Au-NPs from GO. Even for that duration of time though (4 h), gold particles were detected by TEM in the cytoplasm, while some particles were shown to be entrapped by cell organelles (endosomes or lysosomes). Metal-NPs generally entering cells via pinocytosis, through an energy-dependent process. A significant decrease in the SERS signal of GO/PEG/Au-NPs hybrids was observed by NaN_3 incubation, a factor that decreases ATP. Similar results were obtained by GO itself. The authors used several types of endocytic inhibitors which selectively block specific uptake pathways and found out that the cell entry of GO-PEG/Au-NPs was mainly promoted by clathrin-mediated endocytosis. Their work opened a new window in using GO/Au-NPs for biomedicine by studying the changes in the Raman spectrum for in vitro and possibly in vivo studies.

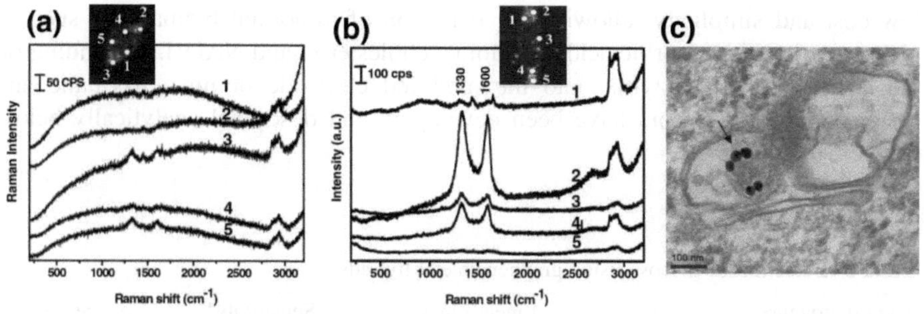

Fig. 7 Raman spectra and dark-field microscopic images (*inset*) at different spots of Ca Ski cells after incubation for 4 h with **a** GO and **b** GO-PEG/Au-NPs. The two prominent peaks at 1330 and 1600 cm^{-1} are assigned to the D and G bands of graphene. **c** TEM image of a Ca Ski cell incubated for 4 h with GO/Au-NPs showing gold particles in the cytoplasm (reproduced from Huang et al. [32])

3 Graphene/Metal Hybrids as Novel Biosensors (Enzymatic and Non-enzymatic)

Biosensors are becoming increasingly important due to their applications in biological and chemical analyses, clinical detection, and environmental monitoring. Graphene exhibits significantly lower charge-transfer resistance compared to graphite and carbon glass electrodes (GCEs), while graphene-based electrodes have been shown superior electrocatalysis over carbon nanotubes, comparable to graphite and GCE. Such EC behavior attests fast electron transfer of graphene owing to its unique electronic structure and high surface area, ranging between 1500 and 2500 m^2/g. The transfer though of electrons from/to graphene to/from molecules is related to the target analyte as well as the amount of defects, functional groups, and impurities on graphene. The oxygen functional groups on graphene are oxidized/reduced at mild EC potentials, while at more extreme pH and potentials (below −2 V) the carboxyl groups can also be reduced. GO has an oxygen fraction of ~50 % and is non-conductive, while reduced graphene oxide (rGO) has an oxygen fraction of <10 % (edge functionalization) and restores the conductivity, along with the electrical, thermal, and mechanical properties, that are similar to GO. Most common reducing agents, including hydrazine, do not reduce the carboxyl groups, which are left as remaining oxygen functionalities in rGO.

Addition of metallic nanoparticles onto either GO or rGO improves the electrical conductivity and enhances the electron transfer between graphene and the analyte. In electrochemistry, the metallic nanoparticles, M(0), are oxidized to M(+x) while the analytes act as reducing agents and auto-oxidized. For sensing biomolecule analytes, voltammetry and amperometry are widely used, due to their sensitivity, low cost and simplicity, allowing for detection of important biomarkers, such as glucose, uric acid, ascorbic acid, dopamine, cholesterol, and NADH. Depending on the addition of an enzyme onto the modified electrode or not, enzymatic and nonenzymatic biosensors have been developed, and described analytically below (Table 3).

Table 3 Detection of glucose using graphene/metal hybrids

Hybrid structure	Linear range	Sensitivity[a]	Voltage	Ref.
Non-enzymatic				
CGE/Graphene/Cu-NPs	0.5 μM–4.5 mM (0.5 μM)	–	+0.50	[33]
CGE/GO-thionine/Au-NPs	200 μM–13.4 mM (50 μM)	–	–	[34]
CGE/GO/Pd-NPs/Nafion	200 μM–10.0 mM (200 μM)	–	+0.40	[35]
CGE/GO/Pt-NPs/Nafion	2.0 μM–10.3 mM	1.26 μA mM^{-1} cm^{-2}	+0.47	[36]
	10.3–20.3 mM	0.64 μA mM^{-1} cm^{-2}		
ITO-GE/GO/PdCu/Nafion	1.0–18.0 mM (20 μM)	48.0 μA mM^{-1} cm^{-2}	−0.40	[37]
Enzymatic				
CGE/Graphene/Pt-NPs/GOx/Nafion	1.0 μM–20.0 mM (1.0 μM)	61.5 μA mM^{-1} cm^{-2}	+0.70	[38]
CGE/Graphene/Pd-NPs/GOx/Nafion	4.0 μM–10.0 mM (4.0 μM)	61.5 μA mM^{-1} cm^{-2}	+0.70	[38]
CGE/GO/Au-NPs/GOx/Nafion	1.0 μM–30.0 mM (1.0 μM)	–	+0.80	[39]
CGE/GO/AuPt/GOx/Nafion	1.0 μM–25.0 mM (1.0 μM)	–	+0.80	[39]
CGE/Graphene/Au-NPs/GOx/Nafion	15.0 μM–5.8 mM (5.0 μM)	–	+0.40	[40]
CGE/rGO/PAMAM/Ag-NPs/GOx	32 μM–1.9 mM (4.5 μM)	75.7 μA mM^{-1} cm^{-2}	−0.25	[41]
CGE/rGO/Ag-NPs/GOx	0.5–12.5 mM (160 μM)	3.84 μA mM^{-1} cm^{-2}	−0.49	[42]
Au electrode/GO/PDOPA/GOx	0.1 μM–4.7 mM (0.1 μM)	28.4 μA mM^{-1} cm^{-2}	+0.70	[43]
CGE/Graphene/Cu-NPs/GOx/Nafion	50 μM–12.0 mM (5 μM)	34.0 μA mM^{-1} cm^{-2}	–	[44]

[a] Values in parenthesis represent the lower limit of detection (LOD)

3.1 Nonenzymatic Glucose Sensors

One of the major challenges in nonenzymatic glucose detection is the interfering electrochemical signals caused by some other coexisting organic substances, such as uric acid (UA) and ascorbic acid (AA). The normal physiological level of glucose is 3–8 mM, which is much higher than those of interfering species of AA (~0.1 mM) and UA (~0.02 mM). In order to mitigate interfering effects, nafion is commonly used in modified electrodes, as an additional material to provide a repelling effect towards negative charged interfering species, resulting thus in high selectivity to glucose.

A stable nonenzymatic glucose sensor was developed by Luo and Liu [33], by electrochemical deposition of metallic Cu-NPs onto graphene sheets. The modified graphene/Cu(0) electrode displayed a synergistic effect of graphene sheets and Cu-NPs towards the oxidation of glucose in alkaline solution, showing higher oxidation current and a negative shift in peak potential. A linear detection range up to 4.5 mM glucose was managed, with a detection limit of 0.5 μM, and a fast amperometric response to glucose (<2 s). Additionally, the graphene/Cu(0) electrode exhibited high selectivity to glucose, with low levels of interference by other chemicals such as ascorbic acid, dopamine, uric acid, carbohydrate, and chloride ions.

High loading of Au-NPs onto graphene sheets was obtained by the group of Xu [34], using thionine-functionalized GO and sodium citrate reduction of the metal precursor. Glassy carbon electrodes (CGE) modified with the obtained nanocomposites showed remarkably electrocatalytic activity towards the oxidation of glucose, leading to a nonenzymatic sensor with a linear range between 200 μM−13.4 mM and a detection limit of 50 μM [34]. Well-dispersed Pd-NPs on GO have been also used as a nonenzymatic glucose sensor. Using a simple ultrasonication method, Cui and Zheng [35] managed to decorate graphene with Pd(0) particles. The formatted biosensor electrode (surface-covered by nafion) showed an effective amperometric response at 0.4 V, with a linear correlation to glucose between 0.2 −10 mM in alkaline solution, high selectivity, and high response even after 2 weeks of storage (Fig. 8).

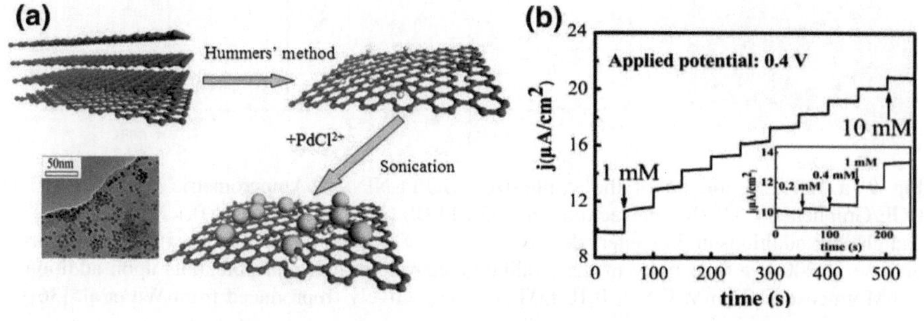

Fig. 8 **a** Preparation method of GO/Pd-NPs using ultrasonication. **b** Current density over time response of CGE/GO/Pd-NPs/Nafion electrode towards successive addition of 1.0 mM glucose in 0.1 NaOH at 0.4 V. *Inset* amplified response curve at low glucose concentrations (reproduced from Wang et al. [35])

Using ethanol as a reductant, the group of Chen [36] managed to assemble Pt-NPs with a flower-like structure onto graphene sheets, showing enhanced electrocatalysis of glucose. The modified CGE/Graphene/Pt-NPs/Nafion electrode exhibited a fast response at 0.47 V within 5 s, with two linear regions, one between 2 μM and 10.3 mM, and one between 10.3 and 20.3 mM, and a detection limit (LOD) of 2 μM, which was lower than many nonenzymatic electrochemical glucose sensors. In addition, sufficient sensitivity in the presence of high concentrations of chloride ions was managed, as well as low interference effects of AA (12.0 %) and UA (3.9 %) at 0.47 V (Fig. 9).

Following a hydrothermal method, the group of Liu [37] managed to assemble bimetallic PdCu nanoparticles onto GO, fabricating PdCu/GE hybrid materials with 3D micropores as nonenzymatic glucose biosensors. It was supported that the addition of glutamate assisted in the coordination of Pd^{2+} and Cu^{2+} ions, while ethylene glycol served as the reductant, leading to assembled bimetallic nanostructures onto graphene. The fabricated ITO glass electrode/GO/PdCu-NPs/Nafion electrode showed significant electrocatalytic activity toward glucose oxidation in alkaline solution, presenting a substantial increase in the oxidation current and a decrease in the onset potential of oxidation compared to the mono-metallic modified GE hybrids [37]. The modified electrode presented a quick respond to glucose oxidation at −0.4 V applied potential, with a linear range up to 18 mM (covering blood glucose levels in diabetic patients) and a detection limit about of ~20 μM. Additionally, a high selectivity to glucose was accomplished with significant resistance against poisoning by commonly interfering species such as dopamine, ascorbic acid, uric acid, acetamidophenol, and fructose at the same concentration (1 mM; current responses of ~10–16 % of that of glucose). The current response of the graphene/PdCu electrode remained stable throughout a 60 days examined period, showing a long-term stability.

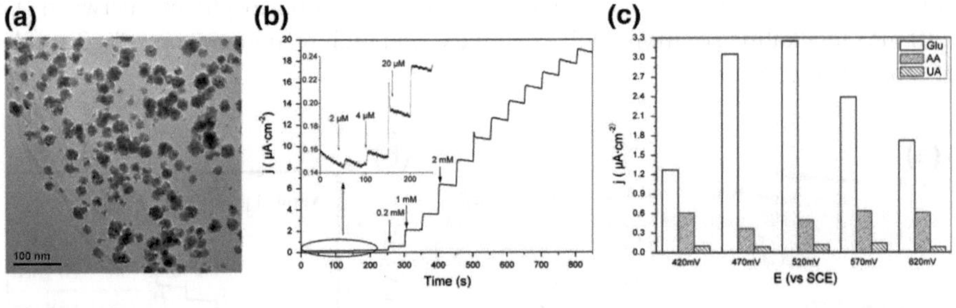

Fig. 9 **a** HRTEM images of the synthesized GO/Pt-NPs. **b** Amperometric responses of the CGE/Graphene/Pt-NPs/Nafionelectrode in 0.05 M PBS (pH 7.4) containing 0.1 M NaCl at 0.47 V and glucose additions at 50 s intervals. *Inset* shows the amperometric responses of four sensors to glucose in 200 s. **c** Selectivity of the modified electrode at different potentials upon addition of 1 mM glucose, 0.02 mM UA or 0.10 mM AA, respectively (reproduced from Wu et al. [36])

3.2 *Enzymatic Glucose Sensors*

The immobilization of enzymes on graphene/metal modified electrodes has attracted significant attention, since enzymes are selective and highly responsive to specific substrates. Graphene has excellent electrical conductivity, and thus metal-dispersed graphene can improve the electron transfer between the enzyme and the electrode, allowing for direct electron communication between the electrode and the active center of the enzyme.

Exfoliated graphite nanoplatelets (xGnPs) decorated with uniform, small-sized Pd-NPs and Pt-NPs were reported by the group of Lee as a cheap alternative instead of carbon nanotubes [38]. The xGnPs were synthesized by a microwave process and sonication, and had an average size of ~1 μm and a thickness of 10 nm (about 30 graphene sheets, considering an interlayer distance of 0.335 nm). xGnPs/Pd(0) and xGnPs/Pt(0) hybrids were prepared by microwave-assisted polyol process, where H_2PtCl_6 or $Pd(NO_3)_2$ precursors were added to xGnPs dispersed in ethylene glycol, followed by microwave-heating and short irradiation to reduce the metal particles. Utilization of the ionic liquid, 1-butyl-3-methylimidazolium acetate, in ethylene glycol at RT was reported to lead to even smaller sized metal-NPs. TEM images of xGnPs/Pt-NPs showed particles of 1–4 nm with a high surface coverage, while the Pd-NPs had a higher dispersity and appeared as worm-shaped aggregates. SEM images of the xGnPs/Pt-NPs/Nafion composite (Nafion was used to solubilize and drop-cast the metal/graphite suspension onto the CGE) showed that the graphite layers were densely packed and exhibited high surface roughness (Fig. 10).

Modified CGE electrodes of xGnPs/Pd-NPs and xGnPs/Pt-NPs hybrids were tested as glucose biosensors. Glucose oxidase (GOx) was added to an 85 % iso-propanol solution of xGnPs/metal-NPs/Nafion or xGnPs/Nafion, and the suspension was added to a CGE electrode after sonication, finally yielding a GOx-modified electrodes. For both metal-coated glucose biosensors, no more than 2 s were needed to achieve 95 % steady current, while the glucose detection limits were 4 μM

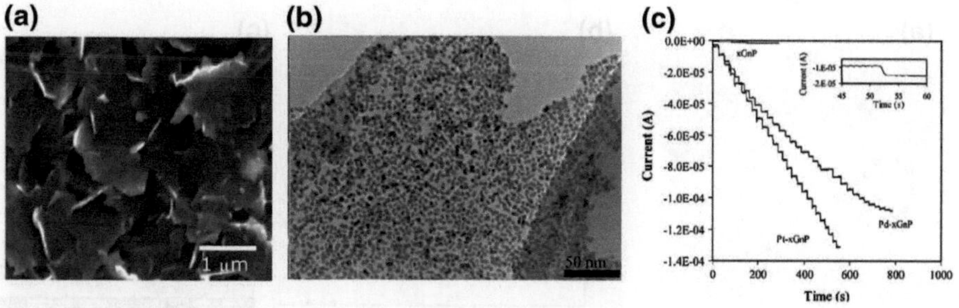

Fig. 10 a SEM image of the xGnPs/Pt-NPs/Nafion nanocomposite showing densely packed graphene sheets with high surface roughness. **b** TEM image of 20 wt% xGnPs/Pt-NPs. **c** Amperometric responses of the biosensors upon addition of 0.5 mM (10 times) and subsequently 1.0 mM glucose solution in 50 mM phosphate buffer at +0.7 V. *Inset* showing the fast response of the xGnPs/Pt-NPs glucose biosensor upon 0.5 mM glucose (reproduced from Lu et al. [38])

(Pd-NPs) and 1 μM (Pt-NPs) respectively at an applied potential of +0.7 V. The linear detection ranges of the glucose biosensors were up to about 10 mM for xGnP/Pd and 20 mM for xGnP/Pt respectively. The enhanced performance of platinum was attributed to the smaller size of the Pt-NPs. The xGnPs/Pt-NPs/Nafion glucose biosensor had a higher sensitivity than MWNTs/Pt-NPs sensors, and a much quicker response, being among the best glucose biosensors reported so far [38]. The maximum sensitivity was observed with a 20 wt% Pt-loading, while higher loadings were reported to reduce the surface coverage and the effective contact between Pt-NPs and the enzyme. Interfering effects using 0.1 mM AA and 0.2 mM UA in the presence of 4 mM glucose (average physiological concentration of blood glucose), showed only 8 and 18 % signal increase respectively, in comparison with 57 % (AA) and 125 % (UA) obtained for a Pt-free electrode.

The immobilization of Au-NPs onto graphene was reported to assist directly to the electron transfer between glucose oxidase (GOx) and the modified hybrid glassy carbon electrode, by increasing the electroactive area of the electrode, and accelerating the response to glucose [39]. The enzyme immobilization on the CGE/graphene/Au electrode was proposed to occur via adsorption through the carboxylic groups of graphene, enabling improvement of the relative activity of GOx in the presence of metallic particles (10 wt% metal loading). Taking advantage of the graphene's large specific area, a high amount of GOx could be immobilized within the graphene nanosheets. Using nafion layers between each step in the modification of the electrode, high sensitivity response was obtained, with linearity up to 30 mM for glucose at 10 wt% Au and 25 mM glucose using 10 wt% Pt–Au, as well as detection limit of 1 μM (Fig. 11). Interference by other electroactive substances such as ascorbic acid (0.2 mM) and uric acid (0.2 mM) were supported to be eliminated in the presence of an extra layer of 0.5 % Nafion. Similar results were obtained by the group of Li [40], where GOx was immobilized in GO/Au-NPs/Nafion composites to fabricate a glucose biosensor electrode, with a steady state current at 5 s at 0.4 V, detection limit of 5 μM, and linearity in the range of 15 μm–5.8 mM.

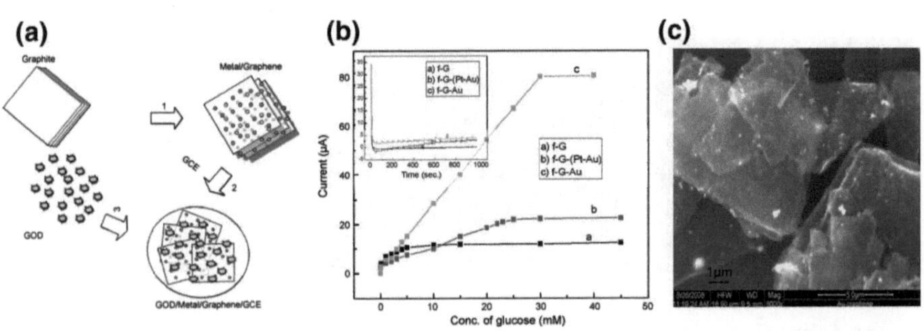

Fig. 11 **a** Schematic of the fabricated graphene/Au-NPs/GOx/Nafion electrode for glucose sensing. **b** Amperometric responses of modified electrodes at different glucose concentrations at 0.8 V. *Inset* chronoamperometric response of 1.0 mM glucose. **c** SEM image of graphene/Au-NPs (reproduced from Baby et al. [39])

Poly(amidoamine)-reduced and stabilized Ag-NPs were assembled onto graphene oxide, under microwave irradiation for 60 min at 80 °C, and the hydrazine-reduced nanocomposite was used as a novel immobilization matrix for glucose oxidase [41]. The hybrid material exhibited excellent direct electron transfer properties for GOx with a rate constant of 8.59 s^{-1}, while the fabricated glucose biosensor displayed satisfactory analytical performance, with high sensitivity (75.72 μA mM^{-1} cm^{-2}), low detection limit (4.5 μM), and linear range from 32 μM to 1.89 mM. Alternately, electrochemically reduced Ag^{+} ions onto a metal-anchored, GO-modified CGE electrode, followed by drop-casting of glucose oxidase, led to a fast direct electron transfer of GOx at the nanocomposite, with a rate constant of 5.27 s^{-1} [42]. The GOx immobilized rGO/Ag nanocomposite electrode exhibited good electrocatalytic activity toward glucose, with a linear range between 0.5−12.5 mM detection limit of 160 μM, and an acceptable sensitivity and selectivity for glucose.

A glucose biosensor based on graphene/poly(dopamine)/GOx onto Au electrode was fabricated by Shi and Sun [43]. Addition of dopamine and glucose oxidase onto a graphene oxide suspension, followed by an electrochemical oxidation polymerization of dopamine onto the electrode, lead to a modified GO/poly(-dopamine)/GOx/Au electrode with short response periods (< 4 s), detection sensitivity of 28.4 μA mM^{-1} cm^{-2}, and a low Michaelis–Menten constant (6.77 mM). Additionally, a linear glucose detection range between 0.1 μM and 4.7 mM was obtained, with a low limit of detection (0.1 μM), probably due to the biocompatibility of poly(dopamine), which enhanced the enzyme absorption and promoted direct electron transfer between the redox enzymes and the surface of electrodes. A graphene/Cu-NPs/GOx platform for sensing glucose was fabricated by the group of Huang [44], combining the high conductivity of graphene and the large surface area of Cu-NPs. The biosensor showed high stability and sensitivity, with a linear detection range between 50 μM and 12 mM, and a detection limit of 5 μM.

3.3 Uric Acid, Ascorbic Acid, and Dopamine Sensors

The sensing of other important biomolecules such as uric acid, ascorbic acid, dopamine, NADH, serotonin, and cholesterol is herein described. Uric acid is the primary end product of purine metabolism. Abnormal levels of UA are symptoms of several diseases, including gout, hyperuricemia, and Lesch Nyan disease. Ascorbic acid is a vital vitamin and a very popular antioxidant, preventing common cold, infertility, cancer, and AIDS. Dopamine is a neurotransmitter and one of the most significant catecholamines. It plays a very important role in the functioning of the central nervous, cardiovascular, renal and hormonal systems as well as in drug addiction and Parkinson's disease. Cholesterol is a lipid that composes about 30 % of all animal cell membranes, and is required to build and maintain membranes (structural integrity), in addition to modulating the membrane fluidity over the range of physiological temperatures. Serotonin is a neurotransmitter that derives from tryptophan, and is primarily found in the central nervous system, in the gastrointestinal tract, and in blood platelets. Its role has been connected to the feeling of happiness. Finally, nicotinamide adenine dinucleotide (NAD) is a coenzyme found in all living cells, with an essential role in metabolism, involving redox reactions and carrying electrons from one reaction to another. Linear detection range and limits, as well as advantages of the different hybrid sensors on the detection of important analytes are described analytically in Table 4.

Graphene/Au-NPs biosensors have been also prepared for sensing uric acid. Using sodium borohydride as a reductant and 4-dimethylaminopyridine as a stabilizer, the group of Shi [45] managed to assemble small-sized (2–6 nm) Au-NPs onto 1-pyrenebutyric acid-functionalized graphene (PFG) at different weight ratios

Table 4 Detection of other biomolecules using graphene/metal hybrids

Hybrid structure	Linear range	Voltage	Analyte	Ref.
CGE/rGO/Au-NPs	2.0–62 μM (0.2 μM)	+0.80	Uric acid	[45]
CGE/rGO/Ag-NPs/Nafion	10.0–800 μM (8.2 μM)	+0.62	Uric acid	[46]
CGE/Graphene ink/Pt-NPs/Nafion	0.05–11.9 μM (0.05 μM)	–	Uric acid	[47]
CGE/rGO/Pt-NPs	10.0–130 μM (0.45 μM)	+0.40	Uric acid	[50]
CGE/GO/Pd-NPs	20.0–2280 μM (–)	+0.10	Ascorbic acid	[48]
CGE/rGO/Ag-NPs/Nafion	10.0–800 μM (9.6 μM)	+0.23	Ascorbic acid	[46]
CGE/Graphene ink/Pt-NPs/Nafion	0.15–34.4 μM (0.15 μM)	+0.20	Ascorbic acid	[47]
CGE/rGO/poly(pyrrole)/Au-NPs	0.0001–5.0 μM (0.0018 μM)	+0.25	Dopamine	[49]
CGE/rGO/Pt-NPs	10.0–170 μM (0.25 μM)	+0.18	Dopamine	[50]
CGE/rGO/Ag-NPs/Nafion	10.0–800 μM (5.4 μM)	+0.49	Dopamine	[46]
CGE/Graphene ink/Pt-NPs/Nafion	0.03–8.1 μM (0.03 μM)	+0.30	Dopamine	[47]
CGE/Graphene/Au-nanorods	20.0–60.0 μM (6.0 μM) 160–480 μM	–	NADH	[51]
CGE/Graphene/Au-nanorods	5.0–377 μM (1.5 μM)	–	Ethanol	[51]
CGE/Graphene/Chitosan/PtPd-NPs/ChOx	2.2–520 μM (0.75 μM)	–	Cholesterol	[52]
CGE/Graphene/PLA/Pd-NPs	0.1–100 μM (0.08 μM)	+0.35	Serotonin	[53]

(10–400:1). The PFG sheets showed a great loading capacity of Au-NPs, due to the high specific surface area of graphene and the additional negative charges introduced by the pyrenebutyrate moieties, which in combination with the negative charges of GO led to strong electrostatic interactions. Low Au(0)/graphene weight ratios (10:1) result in stable carbon solutions, while higher weight ratios (more than 20:1) led to precipitation of all the components, with discoloration of the supernatant (Fig. 12). Though the anchored Au-NPs on graphene were much denser at 300:1 weight ratio (TEM images), the hybrid materials formed at 10:1 showed better electrocatalytic activity. The CGE/PFG/Au-NPs electrodes showed

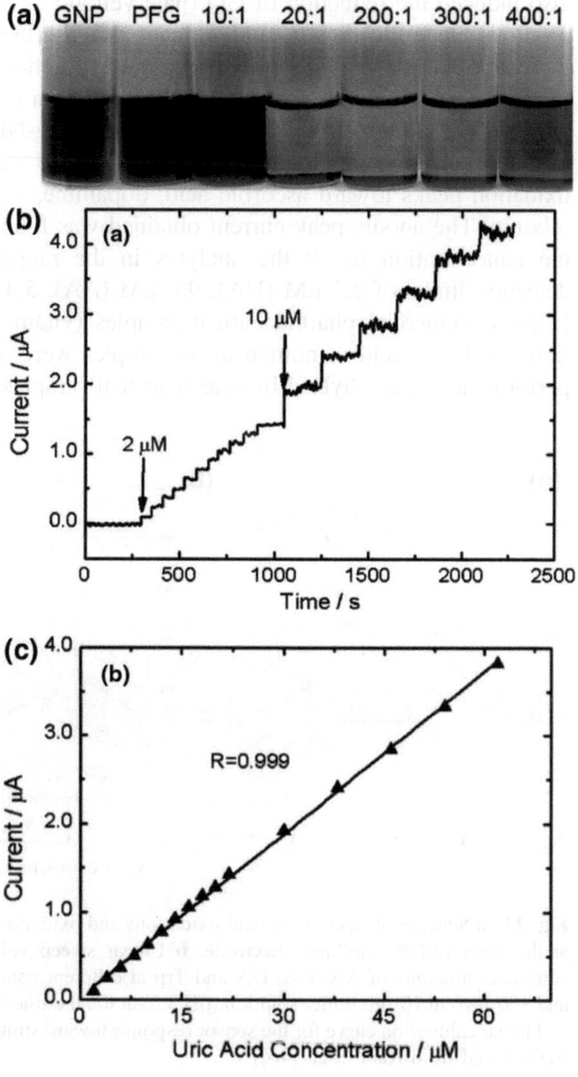

Fig. 12 **a** Photos of Au-NPs with 1-pyrenebutyric acid-modified graphene at different weight ratios. **b** Amperometric responses of the CGE/PFG/Au-NPs electrode upon subsequent addition of uric acid in 50 mM PBS solution at 0.8 V, along with **c** the calibration curve (reproduced from Hong et al. [45])

amperometric responses within 2 s after injection of UA, and a linear relationship between oxidation current between 2 and 62 μM UA at 0.8 V potential. The detection limit was 0.2 μM, comparable to other reported UA electrochemical sensors. The stability of the modified CGE/PFG/Au-NPs electrode was also much higher than the bare CGE/Au-NPs electrode.

The group of Srivastava [46] fabricated a modified electrode composed of reduced graphene oxide and Ag-NPs, achieving simultaneous detection of different biomolecules (Fig. 13). A one-step synthesis was used, where GO, $AgNO_3$ and NaOH were heated at 80 °C for 15 min to obtain rGO/Ag-NPs hybrids as a precipitate from aqueous solution. From XRD and UV characterization, it was shown that NaOH not only facilitated the reduction of Ag^+ to form Ag-NPs, but also induced the reduction of GO (pale yellow) to rGO (black). The rGO/Ag-NPs modified electrode showed a significant increase in oxidation peak current, which indicated that the Ag-NPs provided a catalytic effect on the individual electrocatalytic oxidation of different analytes (Fig. 13), in comparison with the bare GCE and rGO-modified electrode. The CGE/rGO/Ag-NPs/Nafion electrode exhibited excellent electrocatalytic activity, stability, and selectivity with well-separated oxidation peaks toward ascorbic acid, dopamine, uric acid, and tryptophan in the mixture. The anodic peak current obtained was found to be linearly dependent on the concentration of all the analytes in the range of 10–800 μM, with lower detection limits of 8.2 μM (UA), 9.6 μM (AA), 5.4 μM (dopamine), and 7.5 μM (Trp). Commercial pharmaceutical samples (vitamin C tablets), dopamine injections, and uric acid in human urine samples were also tested to show successful performance of the hybrid biosensor in real samples.

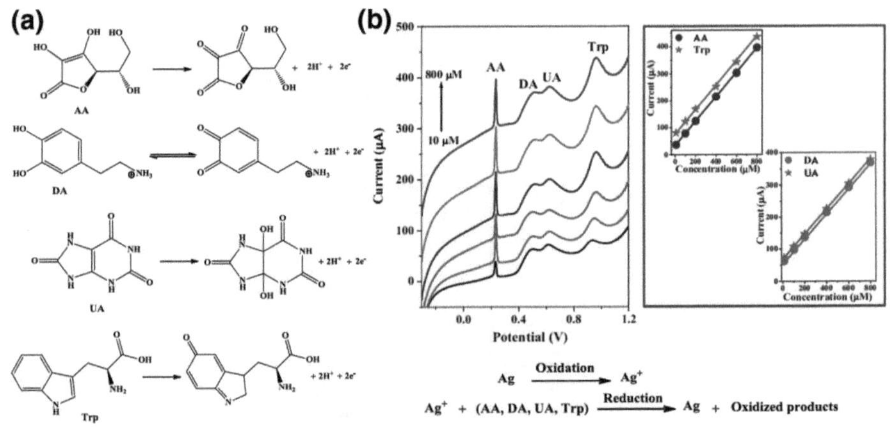

Fig. 13 **a** Scheme of electrochemical oxidations and oxidation products of AA, DA, UA, and Trp at the rGO/Ag-NPs modified electrode. **b** Linear sweep voltammetry of a mixture containing equimolar amounts of AA, UA, DA and Trp at different concentrations (10, 100, 200, 400, 600 and 800 μM) in 10 mL buffer solution (pH 3.5) at the modified electrode at a scan rate of 50 mV/s, and linear calibration curve for the sensor response toward simultaneous electrocatalytic oxidations (reproduced from Kaur et al. [46])

A graphene/Pt(0)/Nafion modified electrode prepared by a graphene ink (paste of exfoliated oxidized graphene, ethanol, water and nafion), ultrasonically treated with a colloidal solution of Pt-NPs in 2 M sulfuric acid/ethylene glycol 1:1, led to simultaneous detection of ascorbic acid, dopamine, and uric acid [47]. The prepared graphene/Pt ink with graphene-supported Pt-NPs of ~1.7 nm demonstrated electrochemical peak potential separations of 185 mV (AA to DA), 144 mV (DA to UA), and 329 mV (AA to UA) respectively. The authors obtained linearity between 0.05−11.85 μM for UA (LOD: 0.05 μM), 0.15–34.4 μM for AA (LOD: 0.03 μM), and 0.03–8.13 μM for dopamine (LOD: 0.03 μM), in 0.10 M PBS solution that also contained 0.1 M KCl. The high performance of the graphene/Pt-NPs/Nafion electrode, in comparison with bare GCE and graphene electrodes, was attributed to the low oxidation potential for AA and the high oxidation current for UA and dopamine, due to the synergistic effect of graphene and the well-dispersed Pt nanoparticles [47]. On the other hand, Pd-NPs (~2.6 nm) supported on GO were shown to be effective ascorbic acid sensors. Following an auto-redox method, where thermally reduced GO was mixed with K_2PdCl_4 for 30 min at 30 °C, the group of Chen [48] managed to obtain a CGE/rGO/Pd-NPs modified electrode with a rapid response to AA within 5 s, and a good linear correlation to AA concentration in the range of 20–2280 μM, as well as good sensing selectivity towards dopamine and uric acid.

Reduced graphene oxide/poly(pyrrole) (rGO/PPy) hybrids were decorated with Au-NPs and used as dopamine sensors [49]. Following in situ chemical oxidative polymerization of pyrrole on the surface of GO sheets by Au(III) initiation and sequential reduction by hydrazine, the group of Shen obtained composite wrinkled graphene structures, with flower-like Au-NPs, sandwiched by the rGO/PPy hybrid sheets. This novel sensor exhibited high sensitivity, with a linear range of 0.1–5000 μM and detection limit of 18.29 pM at pH = 6.5 (PBS), along with good repeatability, with only 3.1 % deviation after eight repeated dopamine analyses. Additionally, the modified electrode showed selectivity for dopamine (5 μM) even after using much higher amounts of AA (125 μM) and UA (330 μM) in the mixture.

Pt(IV) ions assembled on GO and sequentially reduced by $NaBH_4$ to obtain rGO/Pt-NPs hybrids [50]. The authors managed to obtain three times higher peak currents for dopamine and uric acid than those obtained on the bare GCE due to the high surface area of the rGO/Pt-NPs composites. The modified CGE/rGO/Pt-NPs electrode showed excellent electrocatalytic activity towards simultaneous oxidation of dopamine and uric acid, in the presence of 1 mM ascorbic acid, with linearity between 10.0−170 μM for dopamine (0.25 μM LOD) and 10.0–130 μM for uric acid (0.45 μM LOD), providing a promising alternative for clinical and routine sensing applications (Fig. 14).

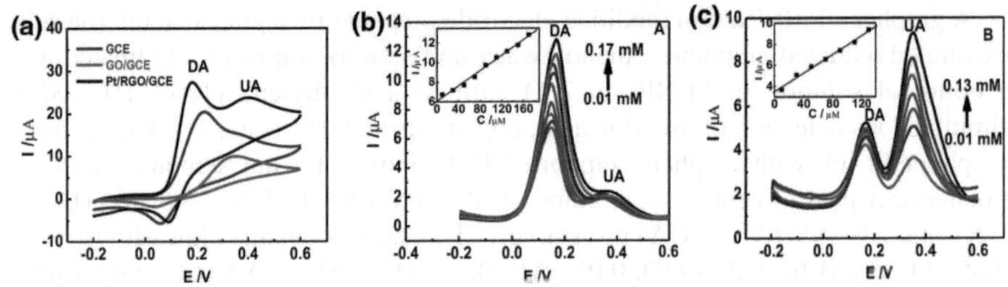

Fig. 14 **a** Cyclic voltammograms of bare CGE, CGE/GO and CCE/GO/Pt-NPs in 0.1 M phosphate solutions (pH 7.0) containing 1.0 mM AA, 0.5 mM DA, and 0.5 mM UA at a scan rate of 50 mV s^{-1}. Differential pulse voltammetry of CCE/GO/Pt-NPs in 0.1 M phosphate solutions containing **b** 1.0 mM AA and 0.1 mM UA at different concentrations of DA: 0.01–0.17 mM, and **c** 1.0 mM AA and 0.1 mM DA at different concentrations of UA: 0.01–0.13 mM (reproduced from Xu et al. [50])

3.4 Biosensing Other Analytes

Graphene/Au-nanorods hybrid nanostructures were shown to exhibit excellent performance toward dihydronicotinamide adenine dinucleotide (NADH) oxidation, with a low detection limit of 6 μM and linearity between 20–60 and 160–480 μM [51]. Addition of the enzyme alcohol dehydrogenase (ADH) onto the modified electrode enabled fast electron transfer between the electrode and the enzyme, due to the enhanced electroactive area of the modified electrode, and was utilized as an additional platform for sensing ethanol. The response displayed a good linear range from 5−377 μM with detection limit 1.5 μM. Furthermore, the interference effects of redox active substances, such as uric acid, ascorbic acid, and glucose for the proposed biosensor were negligible.

An biosensor with enhanced sensitivity for detection of cholesterol was developed by the group of Yuan [52], using cholesterol oxidase (ChOx) immobilized onto a CGE/graphene/chitosan/PtPs-NPs electrode. The nanocomposites were prepared by an electrodeposition method, yielding PtPd-NPs-doped chitosan–graphene hybrids. The fabricated biosensor exhibited a linear response to cholesterol in the range of 2.2–520 μM, with response in less than 7 s, and detection limit of 0.75 μM, while the Michaelis–Menten constant was found to be 0.11 mM. Apart from the excellent reproducibility and stability, the biosensor exhibited high specificity to cholesterol with complete elimination of interference molecules such as glucose, UA, and AA.

Graphene-poly(lactic acid)/Pd-NPs (Graphene/PLA/Pd-NPs) hybrids were synthesized by the group of Jeon [53], and the modified electrode was utilized as a serotonin biosensor. The anchoring –OH groups of PLA and the –COOH and –OH groups on the surface of GO assisted in the coordination of the metal salts, while thermal reduction with $NABH_4$ led to the formation of anchored metallic nanoparticles and partial reduction of GO. The electrical conductivity of the CGE/Graphene/PLA/Pd-NPs electrode largely enhanced compared with those of

GO and GO/PLA. The modified electrode showed good sensitivity and selectivity for serotonin, with a linear range of 0.1–100 μM, fast response time, and a detection limit of 0.08 μM.

4 Photoluminescent/Fluorescent Graphene/Metal Hybrids for Bioimaging

Graphene/quantum dots (QDs) have been also fabricated as probes for biosensing applications, bioimaging, and as emerging diagnostic tools. QDs have high photoluminescence quantum yields with narrow emission spectra (and broad absorption spectra), stable fluorescence, extinction coefficients several times higher than conventional organic fluorophores, and low photobleaching [54, 55]. These characteristics have made them excellent donors of fluorescence resonance energy transfer (FRET). Anchoring QDs and other metallic-NPs onto graphene combines the properties of the materials, and has been utilized as a powerful imaging tool.

A sensitive and selective platform for sensing biomolecules through FRET (from QDs to graphene oxide) was designed by Yan and Ju [56]. QDs were first modified (capped) with a molecular beacon (MB) that recognizes the target analyte. The hybrid material contained a fluorophore (QDs) and a quencher (GO), while as a MB, a single-stranded oligonucleotide hybridization probe with a stem-loop structure (hairpin loop) was used. The loop contained a probe sequence that was complementary to a target oligonucleotide sequence. Binding of the probe to a complementary target, led to disruption of the stem and restoration of the fluorescence of the QDs (Fig. 15), due to the increasing QDs–GO distance and the weakened DNA–GO interaction. The signal increase was found to be depended on the concentration and sequence of the target analyte. The GO-quenching approach could be used for detection of DNA sequences, with high quenching efficiency and sensitivity, and good specificity. The binding-induced fluorescence restoration produced a novel method for detection of target molecules, and opened new avenues for biorecognition using FRET analysis.

An aptamer-functionalized graphene oxide biosensor was developed by the group of Wu [57] for detection of Pb^{2+} ions through fluorescence restoration of QDs (Fig. 16). Covalently modified with single-stranded DNA (ssDNA), GO sheets were mixed with cDNA-functionalized CdSe/ZnS quantum dots (prepared through carbodiimide coupling of mercaptopropionic acid-functionalized QDs), allowing for quenching of the QDs fluorescence, due to energy transfer from QDs to the GO sheets. The GO/aptamer/QDs ensemble acted as a “turn-on” fluorescent sensor for Pb^{2+} detection. Upon interaction of Pb^{2+} ions with the aptamer, a conformational change was induced, that led to G-quadruplex/Pb^{2+} complexes, detachment from the GO sheets, and fluorescence recovery of the QDs in a “turn-on” mechanism. The fabricated sensor had a detection limit of 90 pM, excellent selectivity toward Pb^{2+}, and was examined over a wide range of metal

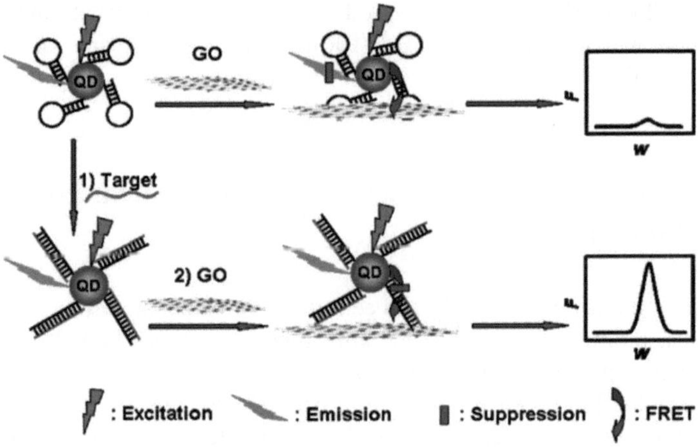

Fig. 15 Upon recognition of the MB to the oligonucleotide target, FRET is hindered due to the increasing QDs-GO distance, increasing in this way the fluorescence of QDs (reproduced from Dong et al. [56])

ions. Additionally, it was shown to work effectively even in the utilization of river water, exhibiting sensitivity irrespectively of the source of the media.

Exfoliated GO reduced by *p*-phenylene diamine (graphene zeta potential of +39.37 mV) and dispersed in 1-octadecene, was decorated with 5–6 nm CdSe QDs, after thermal reduction of the corresponding $Cd(O_2C_{16}H_{31})_2$ and SeO_2 salts [58]. The long, negatively charged alkane surfactant molecules of $CH_3(CH_2)_{14}COO^-$, were supported to be released from the cadmium salt, leading to electrostatic interactions and adsorption onto the positively charged graphene nanosheets, leading to stabilization of the assembled CdSe QDs by the surfactants. The group of Yu [58] managed to obtain graphene/QDs nanocomposites of increased stability in organic solvents and fluorescent quantum yields of 50 % of that of monodispersed QDs. An additional increase in the size of the particles from 3−4 nm to 7−8 nm was found by increasing the reaction times from 6 to 60 min at 240 °C, as well as fluorescence emission at different colored light (from blue to red) upon UV-irradiation (Fig. 17).

A strongly fluorescent, nontoxic, QD-tagged rGO nanocomposite was developed by the group of Chen [59], combining the capability of cell/tumor bioimaging with photothermal therapy. By introducing long alkyl tail spacers between QDs, they managed to assemble QD arrays onto rGO, reducing in this way the fluorescence quenching between QDs and graphene. This was accomplished by mixing 11-mercaptoundecanoic acid (MUA)-coated CdSe/ZnS core-shell QDs with reduced graphene oxide, containing surface-adsorbed poly(L-lysine) (PLL), in TBE buffer (Tris/Borate/EDTA) in the presence of 0.1 M NaCl (Fig. 18). It was found that the MUA surfactant created an effective gap between QDs onto graphene, reducing in this way the fluorescence quenching, while utilization of smaller surfactants (C_3 vs. C_{11}) led to higher quenching. The fluorescence signal of QDs supported on graphene

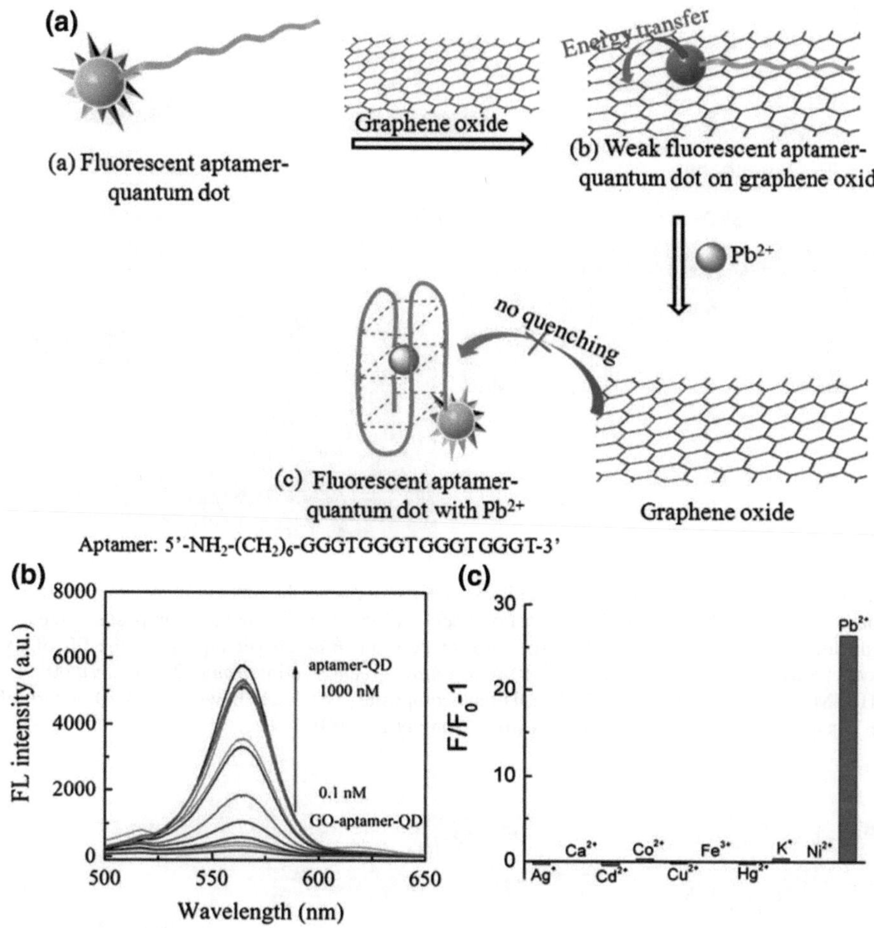

Fig. 16 **a** Schematic of the "turn-on" fluorescent graphene/DNA/QDs biosensor for Pb^{2+} detection. **b** Fluorescence spectra of the GO/aptamer/QDs ensemble assay upon addition of Pb^{2+} from 0 to 1000 nM (using 300 nM aptamer–QDs and 200 mg/mL GO at pH = 7.0). **c** Selectivity of GO/aptamer/QDs toward 50 nM Pb^{2+} over other metal ions at 500 nM. (Reproduced from Li et al. [57])

was also found to depend strongly on the size of reduced graphene, allowing for lower quenching by using smaller sized reduced graphene nanomaterials ($\sim$38 nm vs. $\sim$260 nm).

The reduced graphene/metallic nanoparticle hybrids were further tailored to enter "target cells" via modification with folic acid, through complexation with poly (L-lysine). Internalization via the folate receptor-mediated endocytosis mechanism showed much higher uptake for the smaller sized graphene ($\sim$38 nm), providing 20 times higher fluorescence intensity. By examining two different cell lines (MCF-7

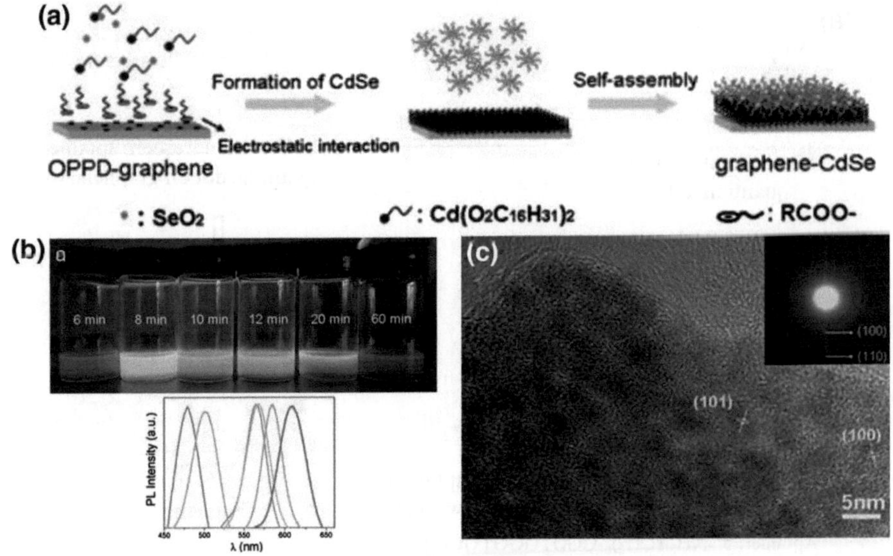

Fig. 17 **a** Schematic illustration of the self-assembly process for the fabrication of graphene/CdSe nanocomposites. **b** Photos and PL spectra of a series of graphene/CdSe nanocomposites prepared with different reaction times under irradiation of 365 nm. *Blue curves* represents the PL of the product after 6 min; *cyan* 8 min; *green* 10 min; *yellow-green* 12 min; *orange* 20 min; *red* 60 min. **c** HTEM image of the graphene/CdSe QD nanocomposite. The *inset* shows the SAED pattern of the hexagonal CdSe phase (reproduced from Wang et al. [58])

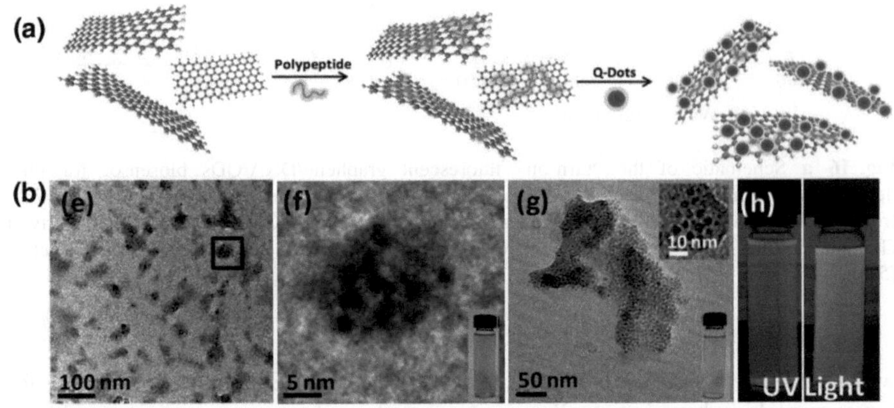

Fig. 18 **a** Schematic representation of poly(L-lysine) adsorption onto reduced graphene oxide, allowing subsequent QDs adsorption. **b** TEM image of ~38 nm graphene/QDs at lower and higher magnification (5 nm scale bar), and ~260 nm graphene/QDs (50 nm scale bar). The *insets* at the lower right of the figures show the sample suspensions under visible light. The last image illustrates suspensions of two graphene/QDs suspensions under UV light (reproduced from Hu et al. [59])

and HeLa cells), a high portion of the ~38 nm graphene/QDs were found in the cytoplasm (allowing for photothermal ablation studies), while most of the ~260 nm graphene/QDs were found at the cell surface, (Fig. 16). Cell viability decreased to less than 5 % using the ~38 nm nanomaterial, after 9 min of irradiation. This was accomplished through the generated heat that was provided by the NIR-irradiated QDs on graphene, causing targeted cell killing (Fig. 19) and simultaneous fluorescence quenching of QDs (additional monitoring probe) [59].

Graphene oxide/transferrin/Au-NPs (GO/Tf/Au-NPs) hybrid nanostructures were fabricated by the group of Yan as a NIR fluorescent probe for imaging cancer cells and small animals [60]. The iron-binding, blood plasma glycoprotein, transferrin was used a reductant and stabilizer of Au-NPs, as well as a functional ligand for targeting specifically the transferrin receptor (TfR). XPS analysis revealed a covalent Au–S interaction for the Tf/Au-NPs, probably due to binding to the sulfhydryl groups of transferrin. By simply mixing GO (~195 nm by DLS) with Tf/Au-NP in water, it was managed to obtain the nanocomposite hybrid structures, through π–π stacking and hydrogen bonding interactions between the protein and the oxygenated lattice of GO. The GO/Tf/Au-NPs hybrids were used as a "turn-off/turn-on" NIR fluorescent probe (710 nm emission) upon specific recognition by TfR, due to donor/acceptor fluorescence energy transfer (FRET) between Tf/Au-NPs and the highly planar GO (Fig. 20). Dynamic light scattering results showed an induced aggregation of Tf/Au-NPs upon recognition by TfR, which probably denoted detachment of graphene and allows for fluorescence recovery of the probe.

The selectivity of the probe was tested among several proteins (bovine and human serum albumin, trypsin, and immunoglobulin G), showing significant fluorescence recovery only for TfR. Fluorescence quenching was increased by increasing the GO concentration from 10 to 60 μg/mL (~100 % quenching) or by increasing the TfR concentration. In addition, the NIR fluorescent probe showed no cytotoxicity in normal and primary cancer cells (4 days incubation), and was utilized for imaging TfR in cancer cells. Human cancer cells with high expression levels of TfR (such as Hela) showed significant fluorescence recovery after 4 h of incubation with the probe as well as high uptake, while no fluorescence was observed for normal cell lines. Real-time imaging was also performed in mice, using the graphene hybrid probe. Intratumorally injected Hela tumor-bearing mice showed enhanced fluorescence inside the tumor after 45 min (detectable signal for up to 12 h), in comparison with signal-free, normal nude mice, showing potential for effective in vivo imaging of cancer cells using the probe.

A multifunctional nanocomposite for cancer theranostics was developed by the group of Liu [61], enabling for both magnetic resonance (MR) and X-ray dual-modal imaging. GO decorated with both iron oxide and gold nanoparticles, showed strong superparamagnetism and enhanced photothermal ablation in vivo. The cationic polymer poly(ethyleneimine) (PEI) adsorbed on negatively charged GO/Fe_2O_3–NPs hybrid by electrostatic interactions (the zeta potential was increased from −35 to +35 mV), served as the platform for the decoration with Au-NPs (Fig. 21). Addition of the graphene/iron oxide composite to a gold seed solution

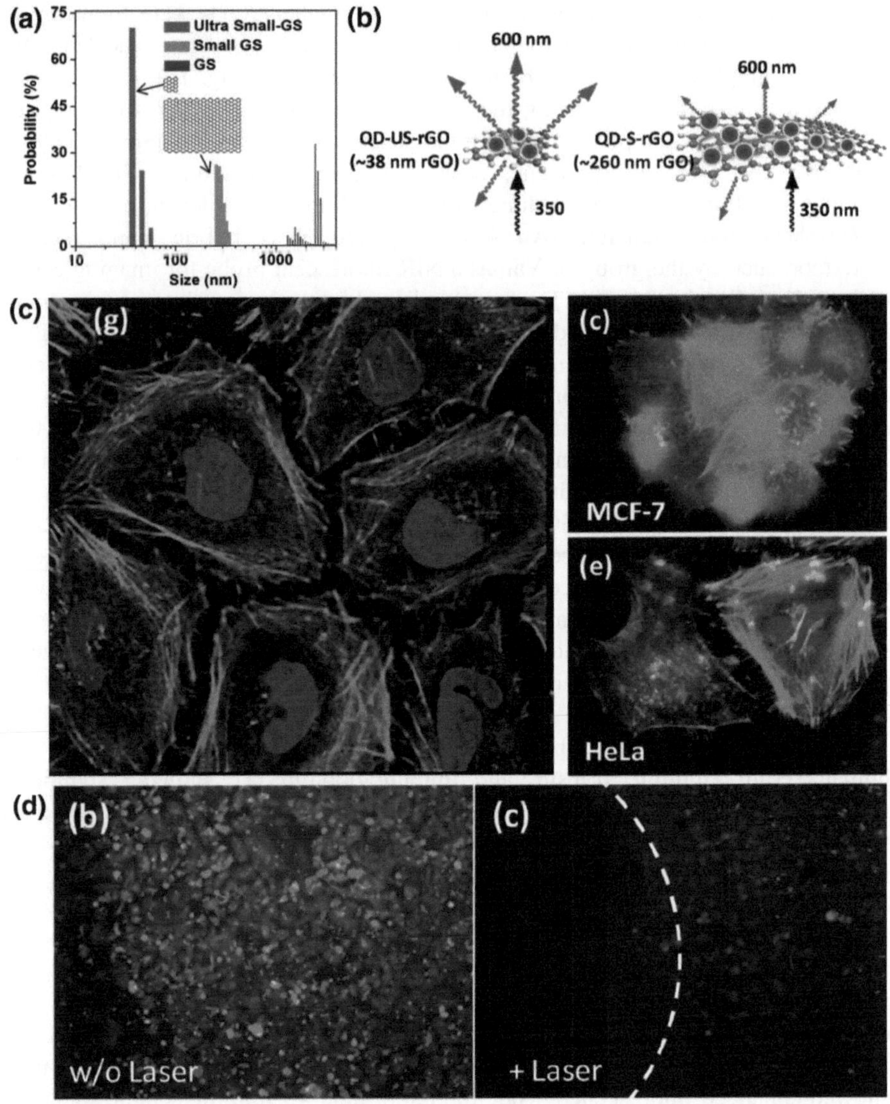

Fig. 19 **a** Size distributions of ultra-small and small graphene nanosheets, and as-prepared graphene nanosheets, measured by DLS. **b** Schematic representation of the shielding effect of the two graphene/QDs hybrids. The larger graphene hybrid stimulates less QD fluorescence (600 nm). **c** Confocal images showing the cellular uptake of MCF-7 cells and HeLa cells incubated with ~38 nm graphene/QDs (folic acid-conjugated). QD fluorescence (*red-orange*) of the particles was exhibited in many regions, including the cytoplasm (*blue*). **d** Thermal ablation using the ~38 nm graphene/QDs. Before irradiation, the cells were viable (*green*) and the internalized QDs were fluorescent (*red*). After irradiation (808 nm for 4 min), essentially all of the cells were killed and the fluorescence was absent (reproduced from Hu et al. [59])

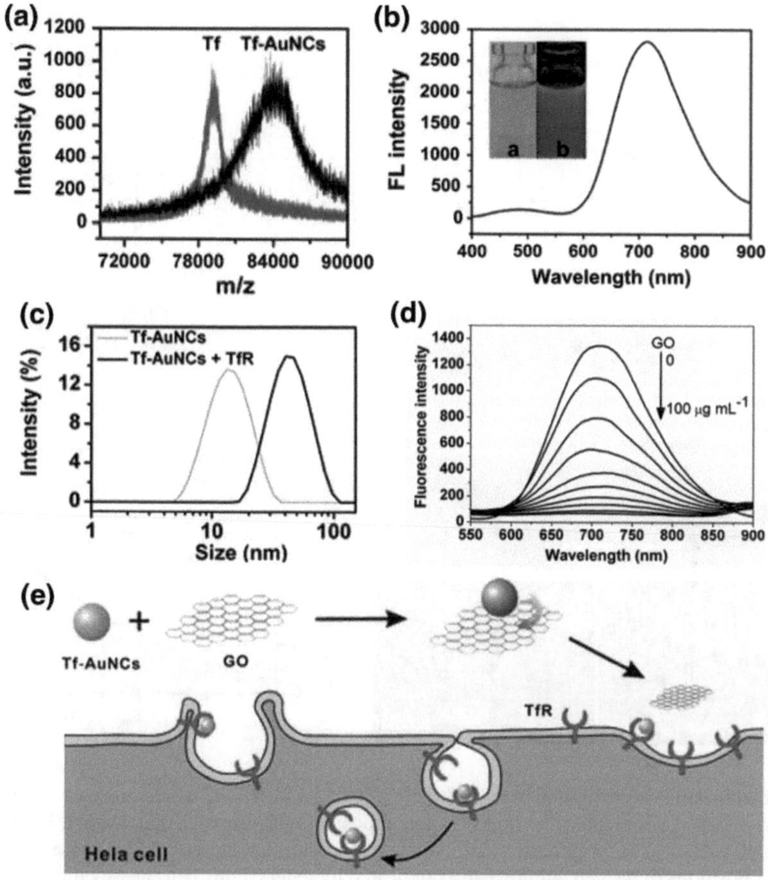

Fig. 20 **a** MALDI-TOF-MS of Tf (*gray*) and Tf-Au NCs (*black*). **b** Emission spectra of the as-prepared Tf/Au-NPs. *Inset*: Tf/Au-NPs under ambient light and under 365 nm irradiation. **c** DLS spectra of Tf/Au-NPs (*gray*) and Tf/Au-NPs in the presence of 5 μg/mL TfR (*black*). **d** Dependence of the fluorescence quenching of Tf/Au-NPs in the presence of various GO concentrations (0, 10, 20, 30, 40, 50, 60, 70, 80, 90, 100 μg/mL). **e** Schematic illustration of the GO/Tf/Au-NPs hybrid composites as a "turn-on" NIR fluorescent probe for bioimaging TfR over-expressed cancer cells (reproduced from Wang et al. 2013) [60]

and sequentially mixing with a gold growth solution, followed by formaldehyde reduction, led to GO/Fe_xO_y-NPs/Au-NPs nanohybrids that can simultaneous provide the material with magnetic and plasmonic properties. To improve water solubility and stability, the hybrid structure was further functionalized with PEG, end-capped with lipoic acid (cyclic disulfide structure) via Au-S chemical attachment. The material with a GO/Fe_xO_y-NPs/Au-NPs weight ratio of 1:2.11:1.57 also showed NIR absorbance and was further used for animal studies.

In vitro tests with 4T1 cancer cells after 2 h incubation showed enhanced cell killing using a NIR laser with GO/Fe_xO_y-NPs/Au-NPs-PEG hybrids, in comparison

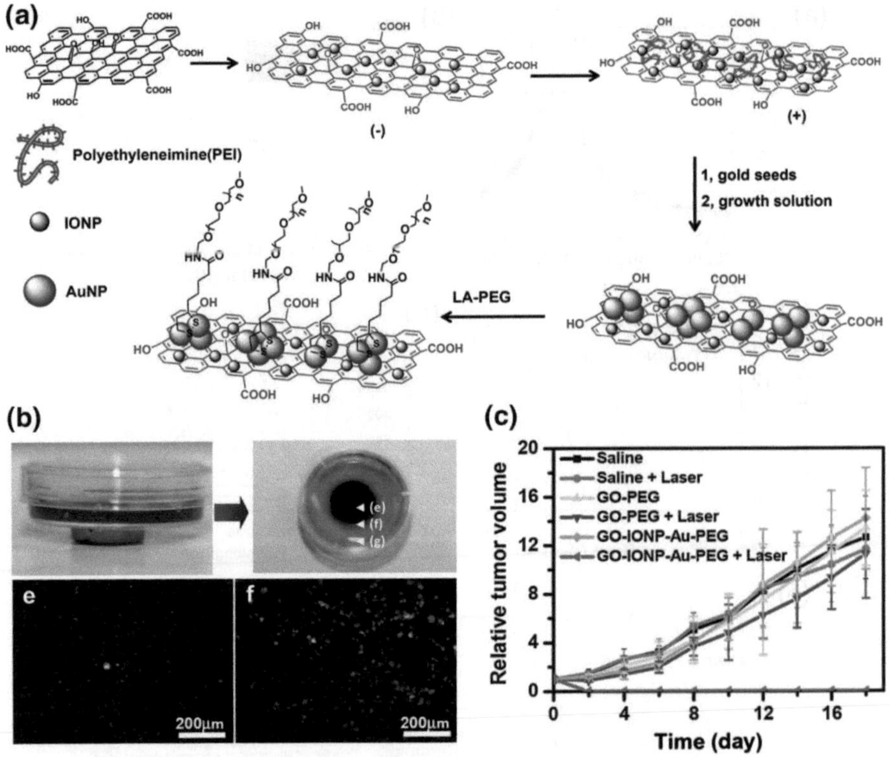

Fig. 21 **a** Schematic illustration of the GO/Fe_xO_y-NPs/Au-NPs-PEG nanocomposite synthesis and **b** Photos of 4T1 cells incubated with the nanocomposite in the presence of a magnet. Fluorescence images of calcein AM/propidium iodide co-stained cells after laser exposure were collected at different positions of the cell culture dish as indicated by letters. Only cells nearby the magnet were destructed after NIR laser irradiation. **c** Tumor growth curves of BALB/c mice bearing 4T1 tumors using various treatments. The tumor volumes were normalized to their initial sizes. Error bars were based on standard deviations of 7 tumors per group (reproduced from Shi et al. [61])

with GO/PEG or GO/Fe_xO_y-NPs/PEG at the same, low GO concentration (10 μg/mL). Additionally, by accumulating the 4T1 cells in the center of a cell culture dish using a magnet, after incubation with the hybrid, selective cell killing was achieved by using NIR irradiation (Fig. 21). Furthermore, intratumorally injected with GO/Fe_xO_y-NPs/Au-NPs-PEG, BALB/c mice bearing 4T1 tumors, displayed a darkening effect under an MR scanner, and a significant tumor-contrast using X-ray imaging. Real-time temperature-monitoring using an infrared thermal camera demonstrated a rapid increase on the surface temperature of the tumor to ~55 °C within 5 min of NIR laser irradiation, in contrast with tumors treated with GO/PEG (45 °C) or PBS (38 °C) respectively. The tumor growth was significantly ablated only using the graphene/Fe/Au material, while the GO/PEG hybrids showed slightly delayed growth upon laser irradiation.

5 Other Biodetection Applications of Graphene/Metal Hybrids

A graphene/Au-NPs platform was fabricated for DNA detection/hybridization via gold nanoparticles [62]. Modified with 3,4,9,10-perylene tetracarbroxylic acid (PTCA) graphene was reacted with 1-(2-aminoethyl)-3-metylimidazolium bromide at RT, yielding an ionic liquid (IL)-derivative anchored on graphene, which was finally used as a simultaneous reductant and stabilizer for Au-NPs. Immobilization of pDNA on the CGE/graphene-PTCA/Au-NPs electrode was achieved through coordination of the guanidine bases of DNA with Au(0), while the formed recognition layer, enabled hybridization with complementary (target) DNA of appropriate concentration (Fig. 22a). The amount of pDNA on the electrode was evaluated using methylene blue (MB), due to the specific affinity of MB for the guanine bases of pDNA. The reduction process of MB (measured by cyclic voltammetry) was proportional to the moles of the guanine bases, allowing for estimation of the surface density of pDNA. Lower detection limits for DNA hybridization were obtained for the CGE/graphene-PTCA/Au-NPs electrode, with a wider detection range than other Au-based, non-graphene, DNA biosensors, as well as high biosensing reproducibility [62]. Graphene/metal-NPs with enzymes on the surface of electrodes can enhance biosensing. Utilization of a fusion enzyme of organophosphorus hydrolase (OPH) with a gold-binding polypeptide (GBP-OPH) [63], enabled the immobilization of the enzyme onto gold-anchored graphene sheets through affinity interactions, achieving a high biocatalytic degradation of paraoxon, with a linear response between 2–20 μM (Fig. 22b).

Metallic-NPs including Au, Pt, Ag, and Pd as well as latex (polystyrene spheres) were successfully incorporated into GO and rGO sheets through the mediation of

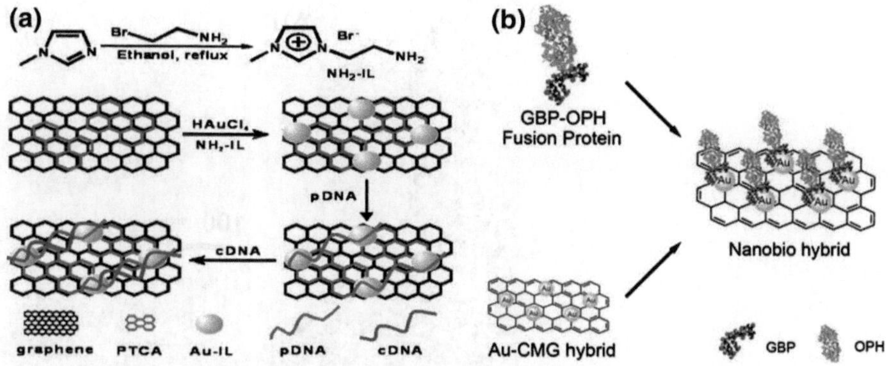

Fig. 22 **a** Schematic of a graphene/Au-NPs platform fabricated for DNA hybridization detection. The anchored gold particles are used for immobilization of pDNA while the graphene/Au/pDNA hybrid serves as the hybridization platform for complementary DNA (reproduced from Hu et al. 2010) [62]. **b** OPH-functionalized graphene/Au-NPs serving as hybrid biocatalytic platform for the degradation of organophosphorus compounds (reproduced from Yang et al. [63])

bovine serum protein (BSA), which served as the metal particle template [64]. BSA has hydrophilic and hydrophobic patches on its protein surface, that are able to adsorb onto the graphene sheets, possibly through π–π interactions (for rGO/BSA), or hydrogen bonding and π–π stacking in the case of GO/BSA conjugates (Fig. 23). Tyr-residues of BSA were reported to serve as a reductant for GO, either partially (at 25 °C and pH = 8) or fully (rGO: 55–90 °C and pH = 12). Pre-synthesized metallic particles (with sodium citrate and $NaBH_4$) were mixed with the GO/BSA or rGO/BSA templates, and adsorbed onto the hybrid template possibly through histidines or amines or thiol groups of BSA. Pd-NPs assemblies on the GO/BSA template had a unique worm-like shape. The hybrid structures were purified with gel electrophoresis. Metal particle-decorated rGO/BSA sheets were trapped in the

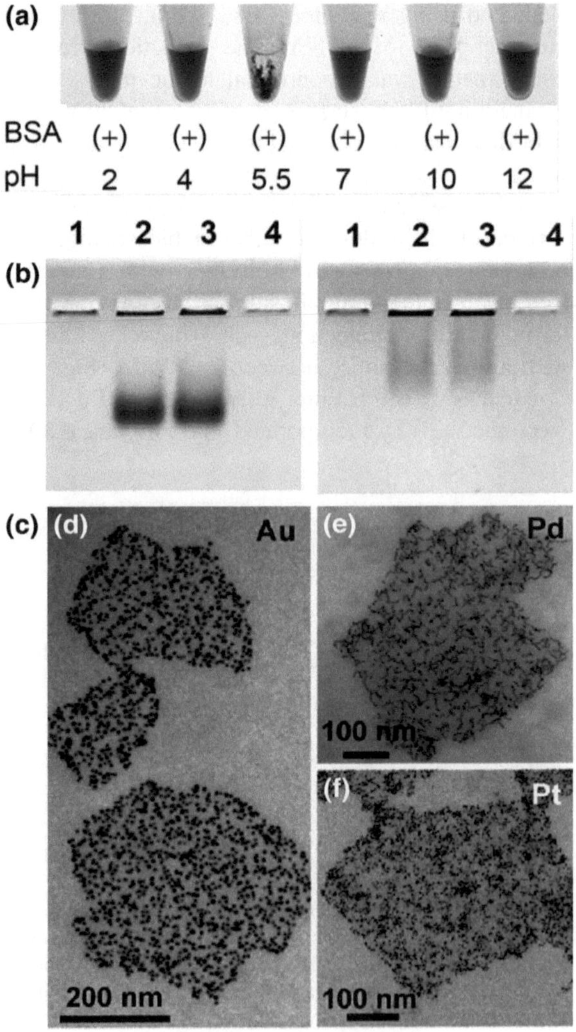

Fig. 23 **a** pH-dependent solubility of rGO/BSA reflecting the isoelectric pH of BSA (~5). **b** Purification of rGO/BSA/Au-NPs and rGO/BSA/Pd-NPs on a 2 % agarose gel. *Lanes 1–4* correspond to (1) RGO/BSA, (2) rGO/BSA + NPs, (3) GO/BSA + NPs, and (4) GO/BSA. **c** Zoomed TEM images of isolated rGO/BSA sheets decorated with Au-NPs, Pt-NPs and Pd-NPs (reproduced from Liu et al. [64])

gel due to their bulky size, purifying the excess materials, and leading to highly pure graphene/metal hybrids with a uniform particle distribution. Addition of salt, such as 0.1 M NaCl, resulted in neutralization of the surface charge of Au-NPs (increase in z-potential), leading to much higher loadings of gold particles onto the GO/BSA templates, due to reduced electrostatic repulsions. Co-assembled metallic particles with latex particles were also achieved in this way [64].

6 Conclusion and Outlook

Taking advantage of the mechanical properties and the large specific area of GO, with both sides been accessible, metallic particles can be easily adsorbed on graphene, after reduction and stabilization with organic compounds or polymers, increasing the distance between graphene sheets (metal spacers) and providing the "carbon material" with "inorganic" properties. The preparation and dispersion of metal nanoparticles (Au, Ag, Pt, Cu, Pd, QDs) on graphene is analytically described in this Book Chapter, as well as their utilization in applications relative to biodetection, biomedicine, electrocatalysis, fluorescence resonance energy transfer, and bioimaging. Immobilization of chemically active compounds or enzymes on graphene/metal nanohybrids leads to signal enhancement (SERS) or to a selective response to specific substrates. Since graphene has excellent electrical conductivity, metal-dispersed graphene can lead to charge-transfer complexes between the metal and the analyte, improving the electron transfer between the enzyme and the substrate onto electrodes, and enhancing the sensing capability. Additionally, fluorescence restoration of graphene/metals hybrids upon complexation with affinity biomolecules, leading to "turn-on" mechanisms from the inactive state (due to FRET), renders them excellent probes for bioimaging. Furthermore, DNA hybridization applications through graphene/metallic particles, enhanced antibacterial properties, and enzyme performance after attachment onto gold-anchored graphene are herein discussed, in order to understand the current materials and design the next generation hybrid materials.

References

1. Grzelczak M, Vermant J, Furst EM, Liz-Marzan LM (2010) Directed self-assembly of nanoparticles. ACS Nano 7:3591–3605
2. Lee SH, Lee DH, Lee WJ, Kim SO (2011) Tailored assembly of carbon nanotubes and graphene. Adv Funct Mater 21:1338–1354
3. Turcheniuk K, Boukherroub R, Szunerits S (2015) Gold-graphene nanocomposites for sensing and biomedical applications. J Mater Chem B 3:4301–4324
4. Gkikas M (2015) Metallic particle assemblies on graphene. Curr Org Chem 19:1773–1790
5. Shen H, Zhang L, Liu M, Zhang Z (2012) Biomedical applications of graphene. Theranostics 2:283–294

6. Zhang Y, Nayak TR, Hong H, Cai W (2012) Graphene: a versatile nanoplatform for biomedical applications. Nanoscale 4:3833–3842
7. Chung C, Kim YK, Shin D, Ryoo SR, Hong BH, Min DH (2013) Biomedical applications of graphene and graphene oxide. Acc Chem Res 46:2211–2224
8. Yang Y, Asiri AM, Tang Z, Du D, Lin Y (2013) Graphene based materials for biomedical applications. Mater Today 16:365–373
9. Bitounis D, Ali-Boucetta H, Hong BH, Min DH, Kostarelos K (2013) Prospects and challenges of graphene in biomedical applications. Adv Mater 25:2258–2268
10. Yang K, Feng L, Hong H, Cai W, Liu Z (2013) Preparation and functionalization of graphene nanocomposites for biomedical applications. Nat Protoc 8:2392–2403
11. Sydlik SA, Jhunjhunwala S, Webber MJ, Anderson DG, Langer R (2015) In vivo compatibility of graphene oxide with differing oxidation states. ACS Nano 9:3866–3874
12. Jain PK, El-Sayed IH, El-Sayed M (2007) Au Nanoparticles target cancer. Nano Today 2:18–29
13. De M, Ghosh PS, Rotello VM (2008) Applications of nanoparticles in biology. Adv Mater 20:4225–4241
14. Bhattacharya R, Mukherjee P (2008) Biological properties of "naked" metal nanoparticles. Adv Drug Deliv Rev 60:1289–1306
15. Luo Z, Zheng K, Xie J (2014) Engineering ultrasmall water-soluble gold and silver nanoclusters for biomedical applications. Chem Commun 50:5143–5155
16. Giner-Casares JJ, Liz-Marzan LM (2014) Plasmonic nanoparticles in 2D for biological applications: Toward active multipurpose platforms. Nano Today 9:365–377
17. Johnston CW, Wyatt MA, Li X, Ibrahim A, Shuster J, Southam G, Magarvey NA (2013) Gold biomineralization by a metallophore from a gold-associated microbe. Nat Chem Biol 9:241–245
18. Connor EE, Mwamuka J, Gole A, Murphy CJ, Wyatt MD (2005) Gold nanoparticles are taken up by human cells but do not cause acute cytotoxicity. Small 1:325–327
19. Alkilany AM, Murphy CJ (2010) Toxicity and cellular uptake of gold nanoparticles: what we have learned so far? J Nanopart Res 12:2313–2333
20. Yang X, Yang M, Pang B, Vara M, Xia Y (2015) Gold nanomaterials at work in biomedicine. Chem Rev 115:10410–10488
21. You J, Zhang G, Li C (2010) Exceptionally high payload of doxorubicin in hollow gold nanospheres for near-infrared light-triggered drug release. ACS Nano 2:1033–1041
22. Alvarez-Puebla RA, Liz-Marzán LM (2010) SERS-based diagnosis and biodetection. Small 6:604–610
23. Goncalves G, Marques PAAP, Granadeiro CM, Nogueira HIS, Singh MK. Gracio J (2009) Surface modification of graphene nanosheets with gold nanoparticles: the role of oxygen moieties at graphene surface on gold nucleation and growth. Chem Mater 21:4796–4802
24. Sun S, Wu P (2011) Competitive surface-enhanced Raman scattering effects in noble metal nanoparticle-decorated graphene sheets. Phys Chem Chem Phys 13:21116–21120
25. Huang J, Zhang L, Chen B, Ji N, Chen F, Zhang Y, Zhang Z (2010) Nanocomposites of size-controlled gold particles and graphene oxide: Formation and applications in SERS and catalysis. Nanoscale 2:2733–2738
26. Zhang Z, Xu F, Yang W, Guo M, Wang X, Zhang B, Tang J (2011) A facile one-pot method to high-quality Ag-graphene composite nanosheets for efficient surface-enhanced Raman scattering. Chem Commun 47:6440–6442
27. Ren W, Fang Y, Wang E (2011) A binary functional substrate for enrichment and ultrasensitive SERS spectroscopic detection of folic acid using graphene oxide/Ag nanoparticle hybrids. ACS Nano 5:6425–6433
28. Murphy S, Huang L, Kamat PV (2012) Reduced graphene oxide–silver nanoparticle composite as an active SERS material. J Phys Chem C 117:4740–4747
29. Guardia L, Villar-Rodil S, Paredes JI, Rozada R, Martinez-Alonso A, Tascon JMD (2012) UV light exposure of aqueous graphene oxide suspensions to promote their direct reduction, formation of graphene–metal nanoparticle hybrids and dye degradation. Carbon 50:1014–1018

30. Gkikas M, Theodosopoulos G, Das BP, Tsianou M, Iatrou H, Sakellariou G (2014) Gold-decorating graphene nanosheets composed of a biocompatible non-charged water-soluble polypeptide. Eur Polym J 60:106–113
31. Gkikas M, Timonen J, Ruokolainen J, Alexandridis P, Iatrou H (2013) Facile aqueous synthesis and stabilization of nearly monodispersed gold nanospheres by poly(L-proline). J Polym Sci Part A 51:1448–1456
32. Huang J, Zong C, Shen H, Liu M, Chen B, Ren B, Zhang Z (2012) Mechanism of cellular uptake of graphene oxide studied by surface-enhanced Raman spectroscopy. Small 16:2577–2584
33. Luo J, Jiang S, Zhang H, Jiang J, Liu X (2012) A novel non-enzymatic glucose sensor based on Cu nanoparticle modified graphene sheets electrode. Anal Chim Acta 709:47–53
34. Kong FY, Li XR, Zhao WW, Xu JJ, Chen HY (2012) Graphene oxide–thionine–Au nanostructure composites: preparation and applications in non-enzymatic glucose sensing. Electrochem Commun 14:59–62
35. Wang Q, Cui X, Chen J, Zheng X, Liu C, Xue T, Wang H, Jin Z, Qiao L, Zheng W (2012) Well-dispersed palladium nanoparticles on graphene oxide as a non-enzymatic glucose sensor. RCS Adv 2:6245–6249
36. Wu GH, Song XH, Wu YF, Chen XM, Luo F, Chen X (2013) Non-enzymatic electrochemical glucose sensor based on platinum nanoflowers supported on graphene oxide. Talanta 105:379–385
37. Yuan M, Liu A, Zhao M, Dong W, Zhao T, Wang J (2014) Bimetallic PdCu nanoparticle decorated 3D graphene hydrogel for non-enzymatic amperometric glucose sensor. Sens Actuators B Chem 190:707–714
38. Lu J, Do I, Drzal LT, Worden RM, Lee I (2008) Nanometal-decorated exfoliated graphite nanoplatelet based glucose biosensors with high sensitivity and fast response. ACS Nano 9:1825–1832
39. Baby TT, Jyorthimayee Aravind SS, Arockiadoss RB, Rakhi RB, Ramaprabhu S (2010) Metal decorated graphene nanosheets as immobilization matrix for amperometric glucose biosensor. Sens Actuators B Chem 145:71–77
40. Zhou K, Zhu Y, Yang X, Li C (2010) Electrocatalytic oxidation of glucose by the glucose oxidase immobilized in graphene-Au-nafion biocomposite. Electrocatalysis 22:259–264
41. Luo Z, Yuwen L, Han Y, Tian J, Zhu X, Weng L, Wang L (2012) Reduced graphene oxide/PAMAM–silver nanoparticles nanocomposite modified electrode for direct electrochemistry of glucose oxidase and glucose sensing. Biosens Bioelectron 36:179–185
42. Palanisamy S, Karuppiah C, Chen SM (2014) Direct electrochemistry and electrocatalysis of glucose oxidase immobilized on reduced graphene oxide and silver nanoparticles nanocomposite modified electrode. Colloids Surf B Biointerfaces 114:164–169
43. Ruan C, Shi W, Jiang H, Sun Y, Liu X, Zhang X, Sun Z, Dai L, Ge D (2013) One-pot preparation of glucose biosensor based on polydopamine–graphene composite film modified enzyme electrode. Sens Actuators B Chem 173:826–832
44. Huang K-J, Wang L, Li J, Gan T, Liu W-M (2013) Glassy carbon electrode modified with glucose oxidase–graphene–nano-copper composite film for glucose sensing. Measurement 1:378–383
45. Hong W, Bai H, Xu Y, Yao Z, Gu Z, Shi G (2010) Preparation of gold nanoparticle/graphene composites with controlled weight contents and their applications in biosensors. J Phys Chem 114:1822–1826
46. Kaur B, Pandiyan T, Satpati B, Srivastava R (2013) Simultaneous and sensitive determination of ascorbic acid, dopamine, uric acid, and tryptophan with silver nanoparticles-decorated reduced graphene oxide modified electrode. Colloid Surface B 111:97–106
47. Sun CL, Lee HH, Yang JM, Wu CC (2011) The simultaneous electrochemical detection of ascorbic acid, dopamine, and uric acid using graphene/size-selected Pt nanocomposites. Biosens Bioelectron 26:3435–3450
48. Wu GH, Wu YF, Liu XW, Rong MC, Chen XM, Chen X (2012) An electrochemical ascorbic acid sensor based on palladium nanoparticles supported on graphene oxide. Anal Chim Acta 745:33–37

49. Qian T, Yu C, Zhou X, Wu S, Shen J (2014) Au nanoparticles decorated polypyrrole/reduced graphene oxide hybrid sheets for ultrasensitive dopamine detection. Sensor Actuator B Chem 193:759–763
50. Xu T-Q, Zhang Q-L, Zheng J-N, Lv Z-Y, Wei J, Wang A-J, Feng J-J (2014) Simultaneous determination of dopamine and uric acid in the presence of ascorbic acid using Pt nanoparticles supported on reduced graphene oxide. Electrochim Acta 115:109–115
51. Li L, Lu H, Deng L (2013) A sensitive NADH and ethanol biosensor based on graphene–Au nanorods nanocomposites. Talanta 113:1–6
52. Cao S, Zhang L, Chai Y, Yuan R (2013) Electrochemistry of cholesterol biosensor based on a novel Pt–Pd bimetallic nanoparticle decorated graphene catalyst. Talanta 109:167–172
53. Han HS, You J-M, Jeong H, Jeon S (2013) Synthesis of graphene oxide grafted poly(lactic acid) with palladium nanoparticles and its application to serotonin sensing. Appl Surf Sci 284:438–445
54. Bruchez M Jr, Moronne M, Gin P, Weiss S, Alivisatos AP (1998) Semiconductor nanocrystals as fluorescent biological labels. Science 281:2013–2016
55. Michalet X, Pinaud FF, Bentolila LA, Tsay JM, Doose S, Li JJ, Sundaresan G, Wu AM, Gambhir SS, Weiss S (2005) Quantum dots for live cells, in vivo imaging, and diagnostics. Science 307:538–544
56. Dong H, Gao W, Yan F, Ji H, Ju H (2010) Fluorescence resonance energy transfer between quantum dots and graphene oxide for sensing biomolecules. Anal Chem 82:5511–5517
57. Li M, Zhou X, Guo S, Wu N (2013) Fluorescent aptamer-functionalized graphene oxide biosensor for label-free detection of mercury(II). Biosens Bioelectron 43:69–74
58. Wang Y, Yao H-B, Wang X-H, Yu S-H (2011) One-pot facile decoration of CdSe quantum dots on graphene nanosheets: novel graphene-CdSe nanocomposites with tunable fluorescent properties. J Mater Chem 21:562–566
59. Hu SH, Chen YW, Hung WT, Chen IW, Chen SY (2012) Quantum-dot-tagged reduced graphene oxide nanocomposites for bright fluorescence bioimaging and photothermal therapy monitored in situ. Adv Mater 24:1748–1754
60. Wang Y, Chen JT, Yan XP (2013) Fabrication of transferrin functionalized gold nanoclusters/graphene oxide nanocomposite for turn-on near-infrared fluorescent bioimaging of cancer cells and small animals. Anal Chem 85:2529–2535
61. Shi X, Gong H, Li Y, Wang C, Cheng L, Liu Z (2013) Graphene-based magnetic plasmonic nanocomposite for dual bioimaging and photothermal therapy. Biomaterials 34:4786–4793
62. Hu Y, Hua S, Li F, Jiang Y, Bai X, Li D, Niu L (2011) Green-synthesized gold nanoparticles decorated graphene sheets for label-free electrochemical impedance DNA hybridization biosensing. Biosens Bioelectron 26:4355–4361
63. Yang M, Choi BG, Park TJ, Heo NS, Hong WH, Lee SY (2011) Site-specific immobilization of gold binding polypeptide on gold nanoparticle coated graphene sheet for biosensor application. Nanoscale 3:2950–2960
64. Liu J, Fu S, Yuan B, Li Y, Deng Z (2010) Toward a universal "adhesive nanosheet" for the assembly of multiple nanoparticles based on a protein-induced reduction/decoration of graphene oxide. J Am Chem Soc 132:7279–7281

Part III
Medicine: Tissue Engineering

Stem Cells Commitment on Graphene-Based Scaffolds

Maurizio Buggio, Marco Tatullo, Stefano Sivolella, Chiara Gardin, Letizia Ferroni, Eitan Mijiritsky, Adriano Piattelli and Barbara Zavan

Abstract In the last years, a rapid development in production, and functionalization of graphene give rise to several products that have shown great potentials in many fields, such as nanoelectronics, energy technology, sensors, and catalysis. In this context we should not forget the biomedical application of graphene that became a new area with outstanding potential. The first study on graphene for biomedical applications has been performed by Dai in 2008 that reported the use of graphene oxide as an efficient nanocarrier for drug delivery. This pioneristic study opened the doors for the use of graphene in widespread biomedical applications such as drug/gene delivery, biological sensing and imaging, antibacterial materials, but also as biocompatible scaffold for cell culture and tissue engineering. The application of graphene-based scaffolds for tissue engineering applications is confirmed by the many exciting and intriguing literature reports over the last few years, that clearly confirm that graphene and its related substrates are excellent platforms for adhesion, proliferation, and differentiation of various cells such as human

M. Buggio
Nanomedicine Lab, Faculty of Medical and Human Sciences, Institute of Inflammation and Repair, University of Manchester, Manchester M13 9PT, UK

M. Tatullo
Research and Development in Biomedicine, Stem Cells Unit, Tecnologica Research Institute, Crotone, Italy

S. Sivolella
Department of Neurosciences, University of Padova, Padua, Italy

C. Gardin · L. Ferroni · B. Zavan (✉)
Department of Biomedical Sciences, University of Padova, Via Ugo Bassi 58/B, 35131 Padua, Italy
e-mail: barbara.zavan@unipd.it

E. Mijiritsky
Department of Oral Rehabilitation, the Maurice and Gabriela Goldschleger School of Dental Medicine, Tel-Aviv University, Tel-Aviv, Israel

A. Piattelli
Department of Medical, Oral and Biotechnological Sciences, University "G. d'Annunzio", Via dei Vestini 31, 66100 Chieti, Italy

G. Gonçalves et al. (eds.), *Graphene-based Materials in Health and Environment*, Carbon Nanostructures, DOI 10.1007/978-3-319-45639-3_4

Mesenchymal stem cells, human neuronal stem cells, and induced pluripotent stem cells. Since most of the papers on this fields are related to in vitro studies, several future in vivo investigations need to be conducted in order to lead to its utilization as implantable tissue engineering material.

1 Tissue Engineering

Tissue engineering is a field of biotechnology which combined biomaterials properties with cells, and biologically active molecules in order to produce to functional tissues. Aim of tissue engineering is to reconstruct functional constructs that are able to restore, maintain, or improve damaged tissues or whole organs. Example of these applications is referred to artificial skin and cartilage that are engineered tissues approved by the FDA; and are currently use in human patients.

Regenerative medicine uses tissue engineering strategies in order to promote self-healing—in order to stimulate the body to use its own system, in order to regenerate cells and reconstruct tissues and organs. These terms "tissue engineering" and "regenerative medicine" have become largely interchangeable, since both hope to improve the treatments for complex, often chronic, diseases. In any case, tissue engineering and regenerative medicine work with cells that are the building blocks of tissue, and tissues are the basic units of function in the body. Cells are able to produce their own 3D support structures, that is called extracellular matrix (ECM) [1]. This scaffold does more than just support the cells; because it also acts as a relay station for several biological molecules. Indeed, cells receive messages from many sources that are available to the local environment. Each signal induces a chain of responses that regulate what happens to the cell [2]. The interaction with their environment is fundamentals for the cells to organize a tissue, to respond to an insult and to induce tissue regeneration. The study of these mechanisms includes: how individual cells respond to signals, produce novel information to control, and manipulate these processes or even create new ones [3].

The in vitro reconstruction of tissue for clinical application begins with building a scaffold using several types of biomaterial, and cells of the patients that are expanded in plastics cultures plates. Once scaffold are created, cells alone or in presence of several growth factors are incorporated. If this microenvironment is appropriated, a novel tissue will be generated. In several cases, the cells scaffold. Growth factors are all mixed together at once, allowing the tissue to "self-assemble" [4].

Another strategy for direct to generate new tissue is based on the use of an existing scaffold. The organ or a tissue of a donor are treated in order to eliminate all the cells and the remaining ECM formed by collagen is used to grow new tissue [5]. This last process has been largely used to bioengineer heart, liver, lung, and kidney tissue, holding great promise to make customized organs that would not be rejected by the immune system.

According to its legal definition (Clemson Advisory Board for Biomaterials "Definition of the word 'Biomaterials'," the Sixth Annual International Biomaterial Symposium, April 20–24, 1974), "a biomaterial is a systemically, pharmacologically inert substance designed for implantation within or incorporation with living systems."

One of the main topics in tissue engineering is to produce and characterize biocompatible materials suitable for the growth and the proliferation of specific tissue types. The study of biocompatible scaffolds includes the characterization of the mechanical properties, chemical and biological compatibility and toxicity and the degradation in an appropriate time window. The final goal is a better understanding of the complexity of the native extracellular matrix (ECM) through the study of in vitro models, in order to create scaffold properly designed to modulate cell differentiation (Fig. 1). To this end, in the last decades numerous types of biomaterials were deeply investigated and in this group the carbon-based materials (e.g., carbon nanotubes or graphene) have received a particular attention [6].

In recent years, we have seen a significant increase of novel approaches related to synthesis and functionalization of graphene and its related derivatives increasing their potentials in many fields, such as sensors, composite materials, nanoelectronics, catalysis, and energy technology [7].

In addition to all the applications that we have just mentioned, its biomedical application represents a relative new area with high potential. The first study on

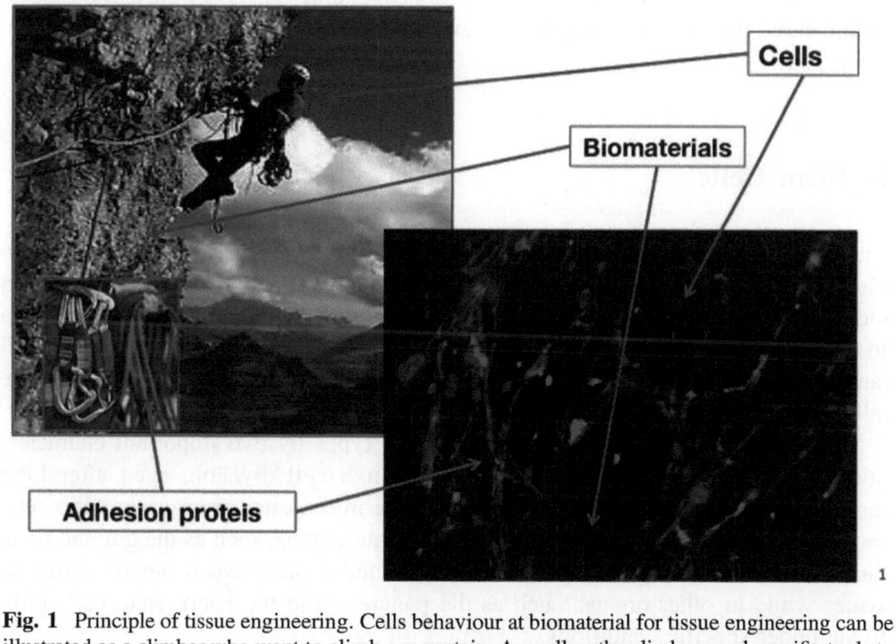

Fig. 1 Principle of tissue engineering. Cells behaviour at biomaterial for tissue engineering can be illustrated as a climber who want to climb a mountain. As well as the climber need specific tools to climb, in the same matter the cells need of specific tools that are the adhesion proteins

graphene for biomedical applications has been reported by Liu et al. [8], that described the use of graphene oxide (GO) as an efficient nanocarrier for drug delivery. Subsequently a lot of interesting papers have been developed in order to explore the use of graphene for wide range of biomedical applications: biological sensing and imaging, drug or gene delivery, biocompatible scaffold for cell culture and as antibacterial materials [9]. This interest on the graphene and its derivatives bioapplications is mainly due to its many intriguing properties, such as thermal conductivity ($\sim$5000 W/m/K), exceptional electronic conductivity (mobility of charge carriers, 200,000 cm^2 V^{-1} s^{-1}), high specific surface area (2630 m^2/g mechanical strength (Young's modulus, $\sim$1100 Gpa), biocompatibility, low cost and scalable production, and facile biological/chemical functionalization of GO [10].

The tremendous recent interest in the use of graphene-based nanomaterials for tissue engineering applications has culminated in many exciting and intriguing literature reports over the last few years, clearly indicating that graphene and its related substrates are excellent platforms for promoting the adhesion, proliferation, and differentiation of several cells such as human mesenchymal stem cells (hMSCs), human neuronal stem cells (hNSCs), and induced pluripotent stem cells (iPSCs) [11].

All these characteristics make this material potentially suitable for a range of applications for tissue engineering, regenerative medicine, drug delivery, differentiation of human neural stem cells, and osteogenic differentiation of human stem cells. However, the shape and physical chemical characteristics of the carbon-based nanomaterial play an extremely important role on the interaction with cells, tissues, and organs.

2 Stem Cells

Since the 1960s, a remarkable potential of a small group of cells to develop into many different cell types was described, these cells were identified as stem cells. In addition, in many tissues they are like a sort of internal repair system, activating a no limit division in order to replenish other cells. During this division, each new cell can either remain in a stemness state or differentiate into a more specialized mature cell.

Stem cells are distinguished from other cell types by two important characteristics: they are capable of auto-renewing through cell division, even after long periods of quiescence; and they can differentiate into tissue- or organ-specific cells under certain physiologic conditions [12]. In some organs, such as the gut and bone marrow, stem cells regularly divide to repair and replace worn out or damaged tissues while in other organs, such as the pancreas and the heart, stem cells only divide under specific conditions.

Based on these unique regenerative abilities, stem cells are described as a new potential tool for treating diseases such as diabetes or heart disease [13].

Mainly there are two different types of stem cells: embryonic stem cells and adult stem cells. As the name suggests, the embryonic stem cells are derived from embryos. Most embryonic stem cells are derived from an in vitro fertilization and donated for research purposes with informed consent of the donor. As long as the embryonic stem cells are grown in vitro under appropriate conditions, they can remain undifferentiated, but when the cells are allowed to clump together to form embryoid bodies start the consequential beginning of the spontaneous differentiation. On the other side, an adult stem cell is defined as an undifferentiated cell, found among differentiated cells in a tissue or organ. The adult stem cell can renew itself and can differentiate into some or all of the major specialized cell types of the tissue or organ. The primary roles of these cells in a living organism are to maintain and repair the tissue in which they are found.

The history of research on adult stem cells began in the 1950s when it was described that the bone marrow contains at least two kinds of stem cells [14]. The first, called hematopoietic stem cells, can differentiate into all the types of blood cells. The second one, called mesenchymal stem cells was discovered a few years later [15]. The nonhematopoietic stem cells belong to a small proportion of the stromal cell population in the bone marrow and can generate bone, cartilage, and fat cells that support the formation of blood and fibrous connective tissue. Since then, the mesenchymal stem cells (MSCs) have been grown from other tissues and it has been described that they could even have immunomodulatory properties.

2.1 *Mesenchymal Stromal Cells (MSCs)*

Mesenchymal stromal cells (MSCs) are multipotent adult stem cells, they are nonhematopoietic, and they have mesodermal and neuroectodermal origin. MSC can be found in several organs and tissues such as adipose tissue, dental pulp, and especially in the bone marrow (BM). MSCs are able to differentiate into cells of mesodermal origin like adipocytes, chondrocytes or osteocytes, but they can also give rise to several other phenotypes. For instance, they are able to differentiate in vitro into non-mesodermal cell types such as neurons and astrocytes. More over it is well known that MSCs posses an extended degree of plasticity compared to other adult stem cell populations (Fig. 2).

MSCs can easily be isolated and amplificated in vitro thanks to their properties, i.e., ability to growth on plastic monolayer. They could be identified by the presence of positive and negative surface markers such as the presence of CD73, CD90, CD105 and lack of characteristic hematopoietic markers such as CD14, CD19, CD34, CD45, and HLA-DR, in addition to endothelial markers like CD318-10 [16]. Indeed, MSCs were first described by Friendenstein as hematopoietic supportive cells of bone marrow [17]. MSCs not only differentiate to bone in vitro and into a subset of the cells, but they also show a high proliferative potential when plated at low density in tissue culture. Besides, the presence in vivo of this stromal stem cell population is able to maintain the marrow microenvironment [18]. The definition of

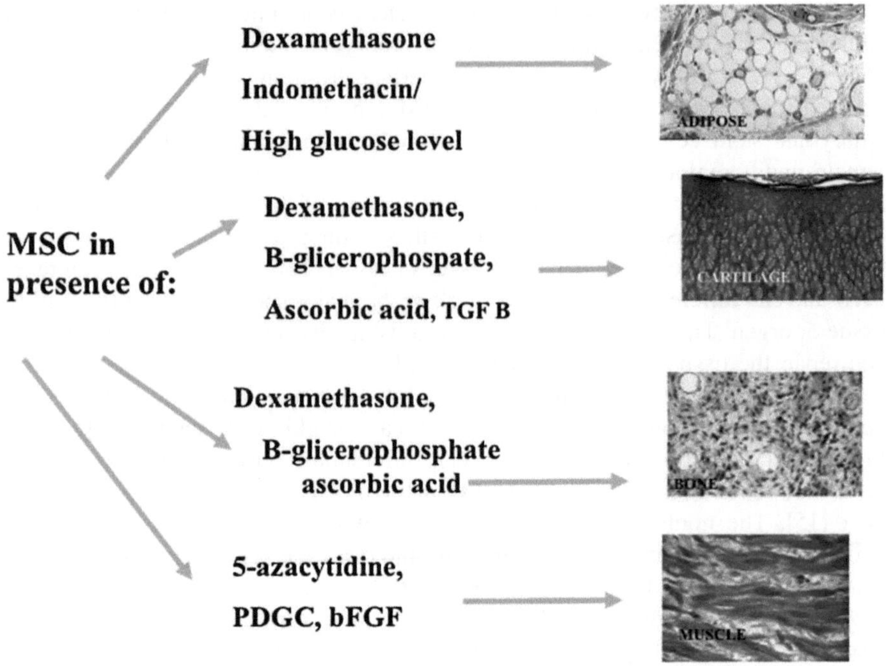

Fig. 2 Ability of MSC to commit into different phenotype

mesenchymal stem cell has been popularized by Arnold Caplan that established that MSCs gave rise to bone, cartilage, tendon, ligament, marrow stroma, adipocytes, dermis, muscle, and connective tissue. However, incontestable data have been published to support the "this stemness ability" of these cells [19].

2.2 *Induced Pluripotent Stem Cells (iPSCs)*

The remarkable event in the history of stem cell development is represented by the discovery of the ability of somatic cells to be reprogrammed into embryonic state-like cells [20]. These novel cells obtained by the reprogramming procedure are now referred as induced pluripotent stem cells (iPSCs) showing great similar properties to the embryonic stem cells (ESCs) [21]. The principal characteristic of the ESCs that justifies their wide differentiation potential is their highest level of the stemness hierarchy [22]. This unique pluripotent property has placed iPSCs as the most suitable candidate for clinical applications and laboratory settings. Moreover, they also offer an interesting beneficial model system to study mammalian embryogenesis and disease processes [23]. Despite this remarkable advantages offered by ESCs in addressing the advances in human medicine, unfortunately, there are several concerns and controversies that have hindered their use in clinical settings. The major ethical controversy in ESC research strongly related to the current methods to produce embryonic stem cell lines that require the destruction of living human embryos. In this context today, the only ESC lines currently available for studies are derived from blastocyst generated for in vitro fertilization (IVF). The main ethical consideration is raised upon the morality of destroying embryos based on the argument that human life begins when an egg is fertilized.

For these reasons, several researchers are focusing their attention in developing alternative mechanisms for generating ESC lines or in founding non-embryonic sources of pluripotent stem cells (PSC). Some authors developed novel methods based on transferring somatic nuclei into enucleated oocytes (SCNT) or embryonic stem cell cytoblasts inducing genetic instability with low probability [24]. This accomplishment confers the exploitation of the intracellular environment thanks to the modification of cellular epigenetics and reprogramming in order to establish PSC [25]. In this view Yamanaka developed novel finding that has received a lot of attentions, direct to the production of cells with the same pluripotency of ESCs from adult somatic cells thanks to the induction of four genes that are normally expressed in ESCs [26]. These novel cells are the induced pluripotent stem, iPSCs. In few words, iPSCs are adult cells that have been genetically reprogrammed to an embryonic stem cell-like state through the ectopic introduction of transcriptional factor genes critical for maintaining the properties of ESCs. This return match process of adult cells to a state like to ESCs offers unheard-of capability to produce patient and disease specific tissues for targeted research, as well as drug screening methods for the development of new therapies.

The iPSCs are resembling to ESC in terms of cell behavior, gene expression, morphology, epigenetic status, and differentiation potential both in culture and in vivo [27]. Moreover, although iPCS have all the criteria to be identified as PSCs, it is yet unclear whether iPSCs and ESCs differ in clinically significant ways and indeed, this is an area of investigation actively being pursued. In this view, we here underline the historical background of the generation of iPSCs and highlight their properties and characteristics as well as their potential therapeutic applications.

The first time on which adult mouse fibroblasts have been successful reprogramming of into iPSCs has been announcement by Shinya Yamanaka and his team at Kyoto University in 2006 [28]. This revolutionary technique was obtained through a combination of reprogramming factors that actually worked, even with modest efficiency. This approach required a retrovirus to transduce mouse fibroblasts with the selected genes and cells that are subsequently isolated by antibiotic selection of Fbx15 positive cells. Unfortunately, in this preliminary work, compared to original patterns in ESC lines, this iPSC line showed DNA methylation errors and failed to produce viable chimeras when injected into developing embryos. Subsequently in 2007, Yamanaka's group together with two other independent research groups from Harvard, MIT, and UCLA showed iPSCs reprogramming by mouse fibroblasts able to produce viable chimeras [29–31]. As in the earlier work, these cell lines were also derived from mouse fibroblasts by retroviral mediated reactivation of the same four endogenous pluripotent factors (Oct4, Sox2, Klf4 and c-Myc), but a different marker for detection was utilized. Moreover, Nanog, an important gene in ESCs and which has been shown to be a major determinant of cellular pluripotency was used instead of Fbx [32, 33].

Far away form these pioneristic works, various methods have been successfully applied to derive iPSCs from different cellular source, and in all cases the properties and characteristics of the established iPSCs are very much like the naturally isolated pluripotent stem cells, such as ESCs. Indeed, all of them form tightly packed and flat colonies with high nucleo-cytoplasmic ratios, defined borders and prominent nucleoli. Morphology of iPSCs, are also indistinguishable from ESC at ultra-structural level when viewed under electron microscopy, showing the similarities between these two populations of pluripotent cells [34]. The main important characteristic of IPSC is represented by their ability to differentiate into the cells/tissues representative of the three primary germ layers, the ectoderm, endoderm, and mesoderm [35] iPSCs have been successfully differentiated into a number of different differentiated cell types including haematopoietic, keratinocytes, chondrocytes, neurons, endothelial cells, osteocytes, pancreatic-insulin producing cells, hepatocyte-like cells, and retinal cells [36]. In addition, iPSCs also have been shown to differentiate into functional cells able to organize themselves into tissues [37].

Genomic stability is critically important for any therapeutic application, this is because genomic modifications may result in the development of certain disease even if the established human and mouse iPS cell lines demonstrate a stable genomic integrity, which is a crucial aspect in generating a high-quality iPSCs. For this reason, in vitro differentiation and teratoma formation have been widely used to assess the developmental potential of iPSCs [38]. Important results shown that iPSCs obtained from dermal fibroblasts or bone marrow-derived mesenchymal cells from patients with specific disease such as Duchenne muscular dystrophy, Down syndrome, type 1 diabetes mellitus, Parkinson disease and also demonstrated similar cell morphology and physiology [32, 33].

2.3 *Neural Stem Cells (NSCs)*

Since the first characterization, adult stem cells have been discovered in many organs and tissues and it has been hypothesized that they reside in a specific area of each tissue called niche [39]. Adult stem cells may remain in a nondividing state for long period of time until they get the input to divide and differentiate by signals coming from the belonging tissue [40]. In addition, a population of adult stem cells has been characterized in the central nervous system, the neural stem cells (NSCs). These cells are defined multipotent cells thanks to their ability to be committed into neurons, astrocytes, and oligodendrocytes. The properties to generate functionally mature neurons capable of joining existing populations of cells makes NSCs interesting candidates to potentially restore function to damaged brain tissue by rewiring broken connections.

3 Graphene and Osteogenic Differentiation

Injury, cancer, inflammation, etc., are able to produce large bone defects which generally have great difficulty in healing. Indeed, this event mainly depends on the size of a bone defect, influencing the repairing rate of a bone defect. By definition indeed, a critical-sized bone defect is the prototype of discontinuity defects, with a failed spontaneous repair. When the size of the defect is greater compare to the healing capacity of bone tissues, the fibrous connective tissues, migrating faster than osteoblasts, dominantly occupy the bone defects [41]. The self-repairing ability of the bone can be moreover negatively influenced when a bacterial activity occurs in bone defects [42]. Infected bone defects can be resulted from acute high-energy injuries and chronic infectious diseases [43]. These adverse events have important consequences if we are talking about maxillofacial region. In this context newly regenerated bone tissue with an adequate volume is indispensable to restore the maxillofacial esthetics and musculoskeletal functions. According to the World Health Organization, injuries are a global public health problem. There are approximately 3–9 million injuries recorded annually in developed countries [44] and with the development of economy and transportation, road traffic-related high-energy injuries account for nearly 1.2 million deaths [45]. This number unfortunately has doubled in the last 30 years in European countries [46]. Within the trauma, injuries to the extremities can result in open comminuted fractures that are often associated with severe soft tissue damages, bone defects, infections, and ultimately nonunion [47]. In this view searching for novel strategies direct to improve large bone defects has obtained grat attention.

Graphene-based materials have been described to support the proliferation of mesenchymal stem cells (MSCs) and promote the osteogenic differentiation in vitro [48]. Graphene and its derivatives are interesting materials for biomedical applications since carbon is the basis of organic chemistry. However, the shape and physical and chemical characteristics of carbonaceous nanomaterials play an extremely important role in how they interact with cells, tissues, and organs [49]. Anchorage-dependent cells need to adhere to substrates in order to spread, proliferate, and perform their functions. Although substrates coated with graphene-based materials are not cytotoxic [50], the use of the material in solutions might pose hazards to cells and tissues. As graphene-based materials can be functionalized, there is an increased interest in using them for biomedical applications. In fact, the surface functionalization may be an important step for pacifying its strong hydrophobicity that may be associated with toxic effects. Nonetheless, the potential long-term adverse effects of functionalized graphene cannot be neglected [51]. Further studies regarding the safety, biodistribution, and adverse effects are needed before the material can be used at large in biological systems. In bone reconstruction stem cell-based therapy might be a promising solution but it requires the constant development of biocompatible platforms that can promote and enhance cell viability, attachment, migration, and differentiation. Several materials such as poly-L-lactic acid (PLLA), chitosan, polycaprolactone (PCL), and composites based on these materials are constantly developed and improved to match some properties of native bone [52]. However, fine-tuning the mechanical properties and chemical and physical characteristics to match native bone properties is rather challenging [53]. In some polymers, such as PLLA and PCL, the lack of sites for cell adhesion may require chemical modification to provide such cues to allow stem cell adhesion. Furthermore, their by-products upon degradation can trigger immune responses [54]. Bioactive inorganic materials are also widely used in bone research. However, due to their brittle nature, they often fail to match the fracture toughness of bone and may not be suitable for load bearing applications [55]. Graphene-based materials allow stem cell attachment and growth and spontaneous osteogenic differentiation, supporting its introduction as an alternative material for bone regeneration research (Fig. 3) [56].

Cell adhesion, viability, and proliferation rate are directly related to the biocompatibility of the substrate [57]. In fact, cell attachment and differentiation are greatly affected by the surface characteristics of materials and by forces generated at

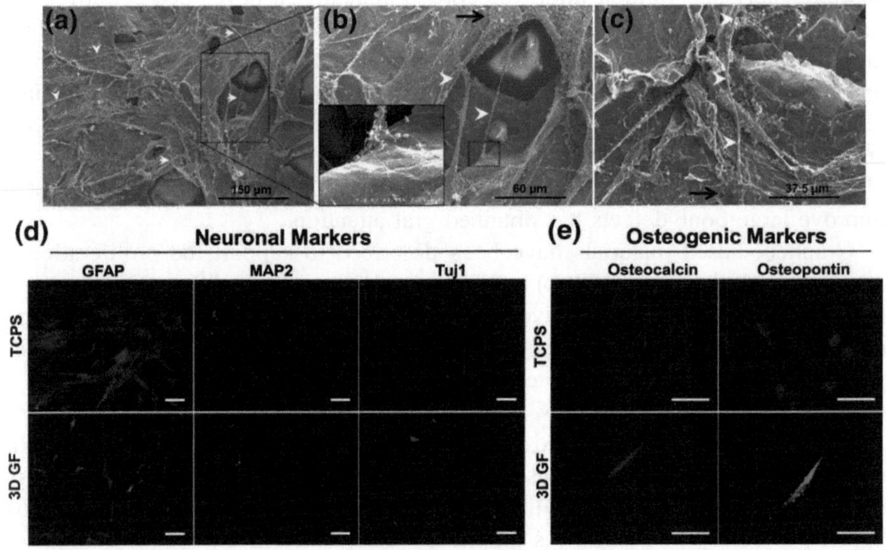

Fig. 3 Phenotypic and morphologic changes on human MSCs cultured on graphene foams (GFs) show a spontaneous osteogenic differentiation. **a–c** After for four days of cultures on GFs, hMSCs formed protrusions up to 100 mm in length (*yellow arrowheads*) that extended from small cell bodies (*black arrows*). To great interest several of these protrusions spanned large pores in the GF and interacted directly with the material surface. After seven days of cultured hMSCs on GFs or in tissue culture polystyrene (TCPS) (control) cells immunostaining has been performed for neuronal phenotype (**d**) and for osteogenic (**e**) markers. Results showed that cultures of hMSCs on GFs stimulated de novo expression of the osteogenic marker osteocalcin and upregulated the expression of osteopontin and did not affect the expression of neuronal markers, but these data confirm that 3D GFs promote the spontaneous osteogenic differentiation of hMSCs. **d** *Scale bars* 100 μm and **e** *scale bars* = 50 μm. Reproduced from Qi et al. [56] with permissions, © 2014, Royal Society of Chemistry

the cell/material interfaces. Graphene-based coatings are noncytotoxic and allow the attachment and proliferation of fibroblasts, osteoblasts, and mesenchymal stem cells (MSC) and have been shown to enhance stem cell differentiation [54]. Other factors for the increased differentiation may be attributed to the presence of wrinkles and ripples on graphene [58]. These are created during the production of graphene. The CVD graphene is usually synthesized at high temperatures (~ 1000 °C) and it experiences negative thermal expansion while cooling. Thus, graphene expands laterally while the metal used as the sacrificial substrate shrinks, resulting in the formation of those wrinkles and ripples [11]. It is known that the transport phenomena of cytokines, chemokines, and growth factors are drastically different between two- and three-dimensional (3D) microenvironments interfering in signaling transduction, cell–cell communications, and tissue development [59]. Graphene 3D construct (3DGp) can be synthesized via CVD using a nickel foam as template and are capable of inducing spontaneous neuronal and osteogenic differentiation of MSC [60]. Cells in 3DGp presented a spindle shaped and elongated morphology with thin and aligned nuclei, typical of osteoprogenitor cells and expressed osteogenic markers OCN and osteopontin (OPN) even without the use of osteogenic medium [61]. Recently, Vinicius's group has succeeded in culturing periodontal ligament stem cells (PDLSC) in 3DGp [62]. After 5 days, the surface of 3DGp was covered by cells having an elongated shape, showing that 3DGp is a suitable substrate for PDLSC attachment and proliferation. Although graphene holds the potential to induce spontaneous osteogenic differentiation of stem cells, this property is significantly enhanced by the use of chemical inductors for osteogenic differentiation. These higher levels of differentiation are possible due to the capability of graphene-based materials to adsorb typical osteogenic inducers such as dexamethasone and β-glycerophosphate Dexamethasone can be adsorbed due to—stacking between the aromatic rings in the molecules and the graphene basal plane. GO is prone to bind to ascorbic acid due to the degree of hydrogen bonding that is formed between the OH moieties of the acid and GO. Hence, graphene and its derivatives allow the loading and release of drugs and proteins that can enhance the osteogenic differentiation of stem cells. A novel strategy is combining Graphene and various materials to enhance Osteogenic Differentiation. Although graphene has great benefits for osteogenic differentiations due its excellent physical properties, it can also be chemically modified [63, 64] or combined with other materials like polymers, ceramics, and metals to further improve the differentiative potential. GO is a widely used form of graphene due to the presence of carboxylic, epoxy, and hydroxide groups, which allow wide range of reactions and functionalization opportunities ceramic/functionalized graphene composites can improve biological outcomes of ceramic-based materials [65]. Hydroxyapatite (HA), for example, is a calcium phosphate ceramic commonly used for bone repair or regeneration due to its chemical similarity to that of natural apatite in bones. The addition of graphene nanoplatelets (GNP) to 45S5 Bioglass results in a composite with high electrical conductivity and increased concentration of GNP [65]. The electrically conductive biomaterials can be used in bone tissue engineering to facilitate cell growth and tissue regeneration with physioelectrical signal transfer [66]. The addition of GO to HA coatings can increase the coating adhesion strength on titanium sheets [67]. The GO/HA composite coating also exhibits higher corrosion resistance than pure HA coatings (Fig. 3). Furthermore, the GO-modified coating presents higher cell viability in comparison with titanium substrate regardless of the coating of HA [68]. Polymers have also been modified with graphene to provide better environments for cell survival and differentiation. Due to the large surface area and delocalized electrons, GO and rGO have the potential to bind and solubilize molecules acting as drug delivery vehicles [69]. Poly (l-lysine-graft-ethylene glycol)-(PLL-g-PEG-) coated PEDOT electrodes can be used as electroactive device for spatial-temporal controlled drug release. Such devices can be used for long-term cell culturing and controlled differentiation of MSC through electrical stimulation [70]. These findings corroborate graphene as a promising material that can increment bioactivity and differentiative potential of candidate materials for bone tissue regeneration. In their study, for example, Elkhenany et al. [71] described for the first time that GO films promote cell proliferation and enhance osteogenesis of MSCs isolated form goat bone marrow (BMMSCs), showing a strong potential of graphene in bone regeneration. In another study, Duan et al. [72] studied the incorporation of carbon nanomaterials (CNMs) with poly(L-lactide) (PLLA) and proposing this composition as potential scaffold for stem cell application in bone tissue engineering. They showed that the scaffolds prepared with PLLA/CNM could support the cell adhesion, proliferation, and differentiation both in vitro and in vivo. As in the previous study, the authors highlighted that CNMs could improve the osteocompatibility of PLLA scaffolds by promoting the osteogenic differentiation of BMMSCs showing that this effect was closely related to the concentration of CNM.

Further studies as the one by Kumar et al. [73] showed a correlation between the chemical functionalization of graphene and the increased stem cell osteogenesis. In this study they produced polycaprolactone (PCL) composites incorporated rGO, GO and amino-functionalized GO (AGO) of different filler (1, 3 and 5 %). They showed that the presence of the amine group on AGO surface was promoting more proliferation and osteogenesis of human MSC compared to the other two functionalized materials. In addition, the authors described bactericidal properties of the graphene and also in this aspect, AGO was the best at inhibiting formation of biofilm.

More recently, Nair et al. [74] incorporated GO nanoflakes (0.5 and 1 wt%) into a gelatine-hydroxyapaptite (GHA) matrix and they studied the effect to enhance mechanical strength and osteogenic differentiation. They showed no significant difference in mechanical strength between $GOGHA_{0.5}$ and $GOGHA_1$ scaffolds and in a system that was mimicking physiological condition for 60 days, around the half of GO loaded was released in sustained and linear manner. Moreover, $GOGHA_{0.5}$ scaffolds were used for in vitro studies, which have shown to induce osteogenic differentiation of human adipose-derived mesenchymal stem cells (hADMSCs) without any osteogenic supplements in the cell culture media like dexamethasone,

L-ascorbic acid and β-glycerolphosphate and these levels of differentiation were comparable to the cells on GHA scaffolds provided with osteogenic supplements [74]. Lyu et al. [75] described a similar study where they examined osteoinductivity of a self-supporting graphene hydrogel (SGH) film as an experimental platform for human adipose-derived stem cells (hADSCs). In this platform hADSCs showed better adaptation, proliferation, and differentiation than on graphene and carbon fiber films.

The osteogenic differentiation potential of MSCs on graphene substrates was confirmed also by the study of Luo et al. [76], where they showed that GO incorporated poly(lactic-*co*-glycolic acid) (PLGA) nanofibrious mats afford extracellular matrix biomimetic microenvironment for hMSCs adhesion and proliferation. They demonstrated that these scaffolds induce expression of osteogenic markers like alkaline phosphatase (ALP), collagen type I (Col I), and osteocalcin (Ocn). Similarly, Lee et al. [77], investigating on the expression of proteins including osteopontin (OPN), demonstrated that the osteogenic differentiation of human MSCs was increased by rGO-coated hydroxyapatite (HAp) when incubated in basal media without any osteoinductive agents and the osteogenic activity is further enhanced when these agents are added to the culture media. In a second study, Lee et al. [78] examined the capability of nanocomposites of rGO and HAp to increase the osteogenesis of MC3T3-E1 preosteoblasts and promote new bone formation. Moreover, the osteogenic responses mediated by these nanocomposites were more stimulated when the cells where incubated in the presence of ostegenic agents. In addition, all the potential in bone regeneration described in vitro were also confirmed in vivo showing that the rGO/HAp grafts significantly enhance the bone formation in full-thickness calvarial defect. Owing the incredible properties of graphene, Raucci et al. [79] described the possibility to produce biomineralized HAp nanocrystals-GO. They used the sol-gel method in order to obtain HAp nanoparticles (diameter 5 nm, length 70 nm) intercalated uniformly and strongly with GO sheets. The interaction of the HAp with GO improved the bioactivity of the materials with the formation of HAp layer on the material surface after biomimetic treatment. In addition, they showed a high viability of hMSCs on HAp-GO hybrids supports with enhanced osteogenic differentiation in a basal medium without the addition of any bone-specific factors.

Not only the osteogenic differentiation capability of graphene was studied, for example Olivares-Navarrete et al. [80] studied the effect of surface chemistry on osteoblast maturation using Ti and carbon-coated surfaces. They described that in integrin β1-silenced MG63 osteoblast-like cells there was less differentiation on rough surfaces independent on the chemical composition of the surface. They also showed that silencing of integrin α1 and α2 affected the osteoblast maturation on Ti surfaces, but not on the carbon-coated surfaces. On the other hand, the integrin αv was the only subunit that affected the osteoblast maturation on carbon-coated substrates. This study suggests a major role of the integrin β1in roughness recognition, and that the different alpha subunits are involved in the surface chemistry recognition.

4 Cartilage Regeneration

Under an histological point of view articular cartilage is the unique tissue with lacks vascularization and that contains only a sparse population of a single cell type called chondrocyte. These cells reside within a prestressed special extracellular matrix structure formed by collagen-proteoglycan giving to the tissue its compressive strength and enables frictionless motion during habitual loading [81]. These features severely influence the ability of cartilage to be regenerate after injury. Indeed if the lesion is relatively small due to trauma or disease, and even, it can progress rapidly and lead to the destruction of cartilage structure and to the lost of its mechanical function. The absence of self-repair has promoted the development of various interventions in order to facilitate regeneration of cells and cartilaginous matrix [82]. In this view usually the process direct to improve cartilage repair is pursued by the application of two treatment methods, both of which have some drawbacks. In case of the presence of severely damaged and that it means that the majority of the articulating surface is disabled, a whole joint surgery will be performed replacing the living biological tissue with a prosthetic device. These surgeries are quite successful and provide many years of joint function, since the synthetic material used for the device is not fully able to show all the properties of the native extracellular matrix [83]. When cartilage injury is small and localized, an autograft or allograft devices-based surgery will be trimmed to size and fit into the defect. However these grafting solutions often provide a limited-term benefit, due to the cartilage properties and because of the scarcity of cells that can facilitate graft integration with the host tissues [84]. In light of these considerations a definite need exists for the development of more effective methods to stimulate cartilage regeneration and integration, and to provide a durable, long-lasting replacement for the original cartilage tissue. To response to this problem, novel tissue engineering approaches direct to induce and enhance cartilage regeneration have been developed. These approaches have the aim to recapitulate the developmental blueprints when placed into the context of natural cartilage achieving different landmarks in the process of cartilage formation, and then improving the regeneration process. It is well defined that cartilage formation begins with MSC condensation that induce their chondrogenic differentiation [85]. Cells produce a dense matrix that serves as cartilage template for the subsequent generation of both the articular cartilage and the subchondral bone [86]. In this view the development of biomaterials to promote MSC commitment into mature chindrocytes is of a growing interest.

Lee et al. [87] demonstrated the ability of graphene, GO and porous GO (PGO) to support pellet formation and chondrogenic differentiation of hMSCs. These cells were assembled with graphene flakes in the solution to form graphene-cell biocomposites. They showed while loading GO and its derivates in the composites there was initially a fast chondrogenic differentiation, but excessive amounts are affected the viability of the cells (Fig. 4). However, the use of a porous graphene for the biocomposites formation allowed a higher loading of GO and further enhanced the differentiation rate of the cells.

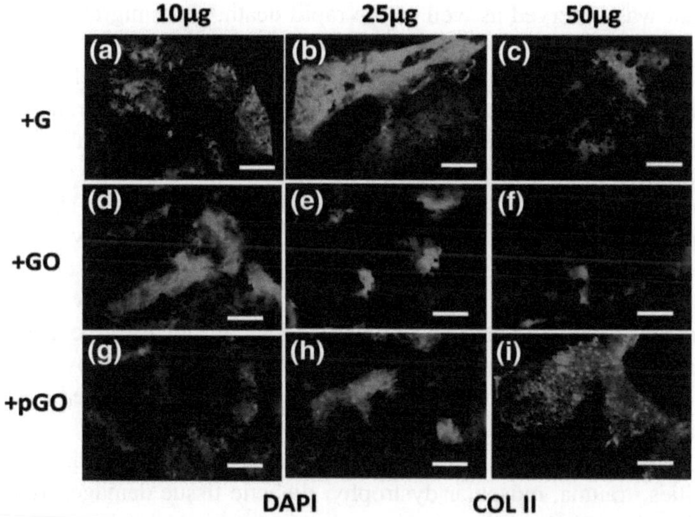

Fig. 4 Immunostaining type II collagen produced by stem cells incubated with G (**a–c**), GO (**d–f**) and pGO (**g–i**) after 4 weeks of culture. *Blue* DAPI, *Green* type II collagen (*Scale bars* 200 μm). When MSC is incubated with 25 μg of G produced the largest amount of type II collagen that is a specific marker of mature cartilage (**b**) while increasing concentration of GO induces a decreasing on the amount of type II collagen (**d–f**). In contrast, the extent of type II collagen produced increased with increasing concentration of pGO (**g–i**). Reproduced from Lee et al. [87] with permissions, © 2015, Wiley-VCH Verlag

5 Graphene and Myoblast Differentiation

The human body mass is formed mostly by skeletal muscle that comprises a large percentage of (40–50 %) playing essential role in postural support, locomotion, and breathing. In presence of to minor injuries, i.e., as contusions, lacerations, or, exercise-induced tears, skeletal muscle show a remarkable capacity for regeneration [88]. These types of injuries account for up to 55 % of all sports-related injuries and do not result in significant loss of muscle mass and are able to heal without therapeutic intervention [89]. By contrast when severe injuries occurs, such as the loss of muscle mass greater than 20 %, extensive and irreversible fibrosis, scarring, and loss of muscle function could occur [90]. Common clinical situations of traumatic injuries occurring after motor vehicle accidents, aggressive tumor ablation, and by prolonged denervation are lead to the volumetric muscle loss. Surgical reconstruction is not able to fully regenerate the lost muscle tissue and often leads to donor site morbidity [91]. As a result, also for this type of treatment the development of novel therapeutic strategies to treat these severe skeletal muscle injuries is an area of active investigation. Great results are reported in early clinical trials based on cell-based treatment, showing the safety of intramuscular cell injections and their preponderant role in muscle regeneration [92]. In all cases, donor cell engraftment was observed as well as the rapid death, poor migration, and immune rejection of the injected cells.

In this context tissue engineering strategies for skeletal muscle reconstruction presents a novel promising therapeutic avenue. Several biomaterial scaffolds has been developed as synthetic extracellular matrices to provide localized delivery of different cell populations and growth factors to the injured skeletal muscle site [93]. Further optimization of these biomaterial platforms may lead to significant functional improvement of severely damaged skeletal muscle. In this way, well-defined biomaterial meshes are designed in order to incorporate biological and biochemical factors that mimic the in vivo satellite cell niche which may generate great improvements in cell regenerative potential both during ex vivo expansion and in vivo regeneration. Moreover, biomaterial scaffolds could provide spatial and temporal appropriate microenvironment cues that could provide the best support for the normal regenerative mechanism.

As above reports several are the causes of skeletal muscle such as developmental abnormalities, trauma, muscular dystrophy, diabetic tissue damage, irradiative and physical injuries. Skeletal muscle tissue is characterized by limited regeneration ability and in presence of these conditions often end up in permanent damages and loss of physical mobility [94]. In light of such consideration, cell-based therapy researchers for skeletal muscle engineering have been increased in the last decades and they are mainly focused on hMSCs applications thanks to their capabilities of self-renewal for prolonged time and multi-lineage differentiation under specific stimuli [95]. It has been established that the myogenic differentiation of hMSCs occurs when culturing these cells in a media supplemented with specific inducers like dexamethasone, hydrocortisone and 5-azacytidine. Unfortunately, these chemical drugs can cause unexpected effects on the differentiation of hMSCs [96].

Another important aspect in the skeletal muscle tissue engineering is the need to find a close imitation of the extracellular matrix (ECM) that provide a proper environment to support the adhesion, migration, proliferation, and differentiation of stem cells. In the group of nanomaterial, carbon nanotubes (CNTs) emerged as versatile candidates to resemble the ECM. Zhao et al. reported the fabrication and characterization of polyethylene glycol-linked multiwalled carbon nanotube (PEG-CNT) films their properties to direct the spontaneous skeletal myogenic differentiation of hMSCs. The study of gene expression of non-induced hMSCs plated on PEG-CNT films showed that these cells presented a significant upregulation of general myogenic markers including early stage markers of myoblast differentiation protein-1 (MyoD) and desmin, as well as a late phase marker of myosin heavy chain-2 (MHC). Moreover, they reported a significant increase of skeletal muscle-specific markers like fast skeletal troponin-C (TnC) and ryanodine receptor-1 (Ryr) and the completed absence of adipogenic, chondrogenic, and osteogenic markers in the non-induced hMSCs on PEG-CNT. Based on these results, the authors can conclude that the PEG-CNT films help hMSCs specifically differentiate into myoblasts [97]. Similar results were described by Chaudhuri et al. [98] where they studied the differentiation of human cord blood derived MSCs into skeletal muscle cells on spin coated thin GO sheets and electrospun fibrous meshes of GO-PCL composite. Myoblast differentiation capability of GO sheets was attributed to its surface change and nanostructured surface morphology.

6 Neuronal Differentiation

Annually more than half a million peripheral nerve injury cases are reported worldwide [99]. Nerve autografting is the current "gold standard" technique to repair a completely transected nerve with gap size larger than 30 mm [100]. However, anesthesia, denervation and numbness of the donor site, painful neuroma formation and time consuming operations are main limitations of this technique [101]. In the last few years, numerous microsurgical techniques [102] have been developed to reconstruct a long segment of a damaged peripheral nerve. Advancements in the field of tissue engineering and biomaterial science have led researchers to develop synthetic nerve conduits as an alternative to nerve autografting. However, to date, the clinical use of the clinically approved nerve conduits is limited to small diameter nerves with short gaps (<30 mm) as their basic hollow tube designs fail to mimic extracellular matrix (ECM) nanostructure. So, they become incompetent to support axonal regeneration in defects with longer gaps and larger diameter [103].

Extracellular matrix (ECM) is one of the important components that influences neural repair and regeneration. ECM molecules regulate Schwann cells (SCs) morphology, migration, and myelination by providing support and anchorage site for these cells. It is well documented that ECM regulates axonal growth via providing binding sites and guides the growing axons during development and regeneration [104]. Furthermore, it has been discovered that the interaction between SCs and ECM molecules is essential for the release of diffusible nerve growth factors from SCs which are crucial for neurite outgrowth [105].

Damage to the central nervous system (CNS) from degenerative disease or traumatic injuries is particularly devastating and at the moment the regeneration of a damaged nervous system is on of the hot topics in the field of the neuroscience. The stem cell-based tissue engineering is one of the most promising approaches, showing promising achievements in the functional recovery front thanks to the self-renewal and differentiation capabilities of stem cells [106]. However, the first barrier that appeared in the application of stem cells in the neuronal tissue engineering is the low survival rate of the cells after transplantation [107]. Based on this consideration, different types of natural and synthetic biomaterial scaffolds that should be analogous to the ECM with appropriate mechanical properties, biocompatibility, controllable degradability with nontoxic degradation products and it has to permit the cell adhesion and proliferation.

Shah et al. have demonstrated the capability of a graphene-nanofiber hybrid scaffold that direct neuronal stem cells through a selective differentiation into mature olygodendrocytes, in a culture medium differentiation inducers free. Moreover, the authors described this hybrid scaffold as a combination of the morphological features of the nanofibers (ideal for fabricating nerve guidance conduits) and the unique properties of the graphene-based nanomaterial (permissive surface for protein and cell adhesion, as well as high conductivity to mediate electrical stimulation) in a single culture platform [108].

Other studies described the potential neurogenesis of graphene, Weaver et al. explored the potential of a nanocomposite film composed of conducting polymer poly(3,4-ethylenedioxythiophene)and GO nanosheet (GO/PEDOT) as a scaffolding material to influence the behavior of NSCs in vitro. In this study they showed, first, that GO/PEDOT material in nontoxic improves NSC differentiation toward the neural differentiation; secondly, that using the carboxylic acid groups donated by GO, biomolecules like interferon-ϒ (IFNϒ) and platelet-derived growth factor (PDGF), were covalently immobilized at the film surface in order to preferentially promote either neuronal or oligodendrocyte lineage differentiation (Fig. 5) [109]. In a similar study, Guo et al. used a bioactive 3D porcine acellular dermal matrix (PADM) as a basic material with a layer of rGO nanosheets on the surface of the PADM channels to obtain a porous scaffold (PADM-rGO) suitable for neuro-tissue engineering applications. Their findings revealed that the PADM-rGO scaffold not only could support MSC attachment but also maintain the cells in a more active proliferative and neural differentiation state, characteristics confirmed by the upregulation at the gene as well as protein level of Nestin, Neuronal class III β-Tubulin (Tuj1), and microtubule-associated protein 2 (MAP2) [110].

All of the mentioned studies have shown the potentiality of graphene as substrate for the growth of NSC, neurons and glia in in vitro models, but reports testing the effect directly with adult neural cells in vivo are rare. Defterali et al. studied the biocompatibility of thermal reduced graphene (TRG) with neurons and glia, as well as with the generation of new neurons in the adult brain. TRG injected in the brain combined with a retroviral vector expressing a reporter gene (in this case GFP) to label dividing progenitor cells in the core of the adult olfactory bulb (OB) did not alter de novo neurogenesis. They found an increased number of microglia, however, since this was also observed in the no-TRG injected group (negative control) they concluded that this reaction was due to the injection itself. At the end of the study the authors concluded that TRG directly injected into the brain does not have a deleterious effect on adult OB cells and it may be a biocompatible material with neuronal and glial cells in vivo supporting its use in studies of brain repairs and function [110].

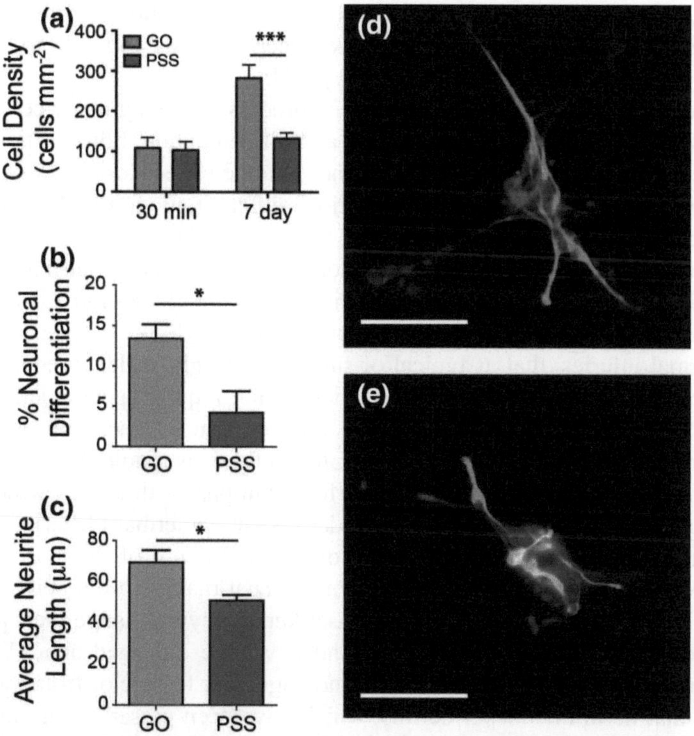

Fig. 5 Biological results of neural stem cell physiology after seeding and differentiation on poly (3,4-ethylenedioxythiophene) (PEDOT) scaffolds. **a** Total cell density present on PEDOT films growth in graphene oxide (GO) or poly(sodium-4-styrenesulfonate) (PSS) after 30 min and 7 day in culture in differentiation media. As reported no differences are detectable between the groups during the initial attachment. The presence of the GO/PEDOT nanocomposite at 7 day (***$p < 0.001$; $n = 9$) indicates a significantly increasing in the density of cells. **b** Percentage of neuronal differentiation and **c** average length of neurite outgrowth from differentiated neurons on PEDOT films after 7 day. The GO/PEDOT film meshes are able to induce higher value of differentiated neurons and a longer average neurite length (*$p < 0.05$; $n = 3$). Representative fluorescent images of differentiated NSCs growing on the **d** GO/PEDOT nanocomposite and the **e** PEDOT/PSS substrate. Cells were stained by immunolabeling for neuron-specific β-III-tubulin (*green*), astrocyte-specific glial fibrillary acidic protein (*red*), and nuclei (*blue*). *Scale bar* measures 50 μm in **d**, **e**. Reproduced from Weaver et al., [109] with permissions, © 2015, Wiley-VCH Verlag

7 Skin Regeneration

Skin is the tissue whose principal function is to protect the human body from surrounding environment. Skin is the biggest organ of the body. Injury of skin can occur for many factors, such as genetic disorders (bullous conditions), acute trauma, chronic wounds or even surgical interventions. Among these the most common event for major skin loss is thermal trauma, because substantial areas of skin are damaged, often without the possibility of skin regeneration. Burns and scorch sometimes could induce rapid, extensive, deep wounds difficult to treat with common strategies, and can lead to death [111].

Skin wound healing process is a complex process involving many cell types and processes, such as epidermal proliferation and differentiation, fibroblastic migration and activation to produce novel ECM and endothelial cell organization for vascularization. Process such as cell migration, differentiation, ECM synthesis and wound contraction are the main actors of this process [112].

Wounds defects can be divided into epidermal, superficial partial-thickness, deep partial-thickness and full-thickness in relationship depth of the injury. The treatment approaches therefore require different approaching [113].

Epidermal injuries, that as typical of sunburns of light scalds, are characterized by erythema and minor pain. These injuries do not require specific surgical treatment because only the upper epidermis is affected. The regeneration of this tissue rapidly occurs without scarring, and no ECM synthesis is required to contribute the scar tissue [114]. By contrary when superficial but partial-thickness wounds affect the epidermis and superficial parts of the dermis, an epidermal blistering is present and severe pain accompanying this type of injury. In case of thermal trauma this process is evident [115]. Such type of wounds heal thanks to the epithelialization from the margins of the wound, where basal keratinocytes change their phenotype into a proliferating migratory cell type and cover the damaged area [116]. Cells migrate from different sites, from the wound edge, hair follicle or from sweat gland remnants that lie in the deeper dermis, which have been preserved in this depth of injury [45, 117]. In lined with epithelial cells, each hair follicle and sweat gland are capable to contribute in epithelial regeneration across the wounded surface. Moreover, the hair follicles of human skin contain a reserve of stem cells, that located in the bulge region of the follicle, capable of self-renewal [118]. In presence of deep partial-thickness injuries involving greater dermal damage could affect the skin appendages remaining and therefore they take longer to heal. In this kind of injury scarring is more pronounced and defects fibroplasia is more intensive when compared with the superficial partial-thickness wounds.

Full-thickness is an injury characterized by the complete destruction of the related elements of the epithelial regeneration process. This type of damage heals thanks to contraction, followed by the epithelialization starting from the edge of the wound, leading to cosmetic and functional defects. All full-thickness skin wounds with more than 1 cm in diameter require the use of skin grafting because they are not able to epithelialize on their own. This process moreover could lead to extensive scarring, and in the end resulting in limitations in joint mobility and unfortunately with severe cosmetic deformities [119].

Currently, full-thickness injuries clinical "gold standard" treatments are represented by split-thickness autologous skin grafting. The procedure requires a dermatome harvesting a small biopsy of epidermis formed with a superficial part of the dermis and applied to the full-thickness wound [120]. Once applied to the wound, capillaries of the split skin graft (SSG) form anastomoses with the existing capillary network to provide nutrients for graft survival; this process is known as graft "take"

[121]. If presence of extensive injury, and if the donor sites are extremely limited, the clinical strategies require the application of meshing techniques. This strategy could be used by means of the perforation of grafted and subsequent stretching to cover greater areas of the wound [122]. Although the advantage of this method allows to cover a great area and to reduce mortality rates, the cosmetic and functional outcomes of such a treatment are inferior when compared with the standard application. This is due to the lack of dermis in the interstices of the stretched meshed skin graft, and the process related to epithelialization from graft margins across interstices is slow. This induces delayed healing, a greater graft contraction, scar tissue formation and pronounced "crocodile skin" appearance of the scar [123]. Alternative approaches life-saving in the treatment of extensive full-thickness wounds, in presence of difficulties such as donor sites for skin harvesting are not available, include the use of cultured autologous keratinocytes and/or bioengineered skin substitutes. In these last years significant progress has been obtained in the development and the further clinical application of these products [46]. In light of the great demand for skin-substitutes products, there is a long history of material development, and many research groups worldwide have focused on creating biomaterials for skin substitution [124]. Skin substitutes are described as skin equivalents or as tissue-engineered skin, tissue-engineered skin constructs, living skin replacements and, more recently, as bioengineered alternative tissue. Since these terms differ slightly from each other, they are considered to be equal by the majority of the researchers. All these substitute need to comply with three major requirements: (i) be safe for the patient, (ii) be clinically effective; (iii) and be cost-effective in application. The biomaterial for skin reconstruction must be biodegradable, repairable, and able to support the reconstruction of normal tissue. They must have the same physical and mechanical properties to the skin it replaces. It should provide prevent fluid, pain relief, and heat loss from the wound surface. Moreover they must protect the wound from infection. It is also of great advantage if the skin substitute bioconstruct is, readily available, user-friendly, and possesses a long shelf life.

Until now, no currently commercially available tissue-engineered skin equivalent possesses all the above-mentioned properties, since most of them are able to replace the functional and anatomical properties of the native skin. There are, however, a number of bioengineered skin-replacement products which are currently available to clinicians and are used for wound healing purposes [125].

To overcome these issues, Li et al. proposed the combination of a 3D graphene foam (GF) scaffold loaded with MSCs to improve skin wound healing. The 3D-GF scaffold showed a good biocompatibility end promoted cell growth and proliferation. Moreover, this system reduced scarring in animal model, potentially due to the biomechanical and biochemical signal from 3D-GFs. The combination of the 3D-GF scaffold with MSCs provided an upregulation of vascular endothelial growth factor (VEGF) and basic fibroblast growth factor (bFGF), which may lead neo-vascularization. Therefore, the authors concluded that 3D-GF loaded with MSCs provided the necessary cues to improve skin wound healing with a reduced scar formation (Fig. 6) [126].

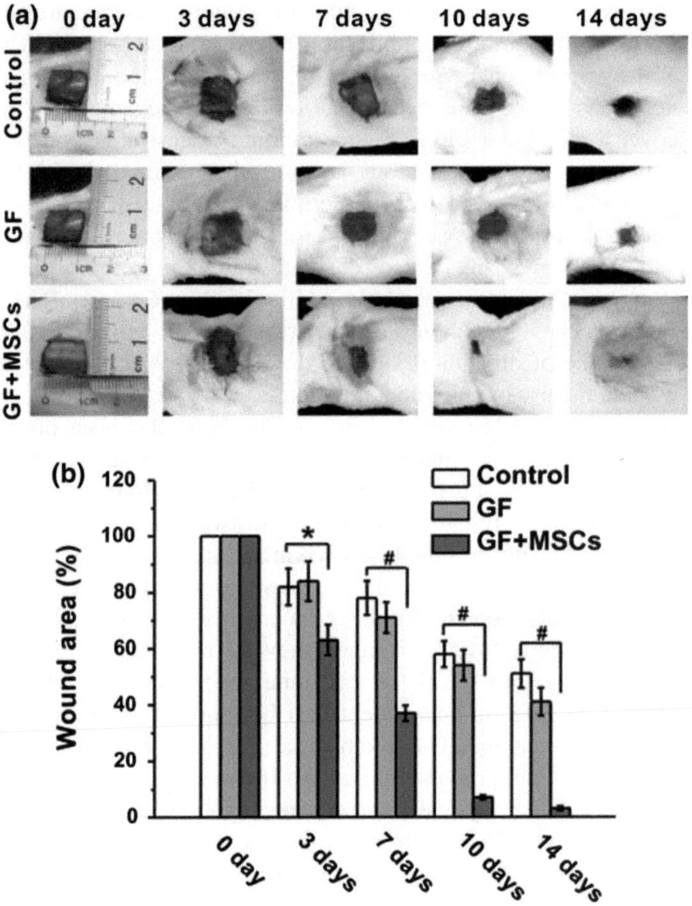

Fig. 6 Graph related to the effect of MSC on GF and GF + MSCs during wound closure. **a** Images related to of the experimental group of wounds post-transplantation of 3, 7, 10 and 14 days. **b** % of wound area in the experimental groups along the implantation time. Data were presented by means ± SEM. $*p < 0.05$ and $^{\#}p < 0.01$. Reproduced from Li et al. [126] with permissions, © 2015, Elsevier

8 Cardiomyogenic Differentiation

The heart is a complex of chemo-mechano-electrical system that acts as actuator and sensor in our biological system. This organ can transduce electrical inputs in order to generate electrical wave propagation needed for the mechanical contraction. Heart arises from multipart events that interact with one another to self-regulate. More then chemical signaling, that are mediated by soluble growth factors, it provides the foundation for stem cell differentiation (SCD) process [127], there is growing evidence that environmental cues are also of prime importance in guiding differentiation events. Indeed, it is clear that the whole stem cell niche is important in determining cell fate. Many factors are also involved on stem such extracellular matrix (ECM) proteins, mechanical loading oxygenation, and innervation [128]. Thanks to these factors, cardiac cells form anisotropic layers to contract in response to electrical signals in the heart [129]. All mechanical properties are moreover thought to contribute to differentiation and further maturation during embryogenesis [130]. In light of such consideration, we focus the attention on the impact of graphene-based scaffolds and their mechanical and topographical properties on stem cells commitment into cardiomyocytes. Correct differentiation of stem cells into cardiomyocytes has important medical applications. Indeed it could offer the possibility to recreate cardiac-like tissue from patient- specific in vitro drug toxicity assay. Moreover it can be designed cell-based therapies for the treatment of cardiac diseases [131].

The first strategy stem cell followed by researchers is represented by the possibility of provide cell–cell communication [132]. The embryonic stem cells of mouse model have been the first to trigger spontaneous differentiation at early time, such as the second passage. They are able to form adherent dishes to exhibit positively charged functional groups. Although these cells can form adhesions, they are not able to recapitulate the signals given by different ECM components during embryogenesis. More over, the high variability in terms of surface charge and nanotopography could show severe effects on cell behavior [133]. Most of the ECM proteins have electrostatic properties and then they can be adsorbed on the surface of tissue culture. In the large panorama of stem cells for tissue repair process, MSCs have greater potential to repair heart diseases. However, it has been described that the use of MSCs as cardiac therapy are limited due to their low ability to differentiate into cardiomyocytes in vivo [134]. Whether, embryonic stem cells (ESCs) are able to differentiate spontaneously into cardiomyocytes [135]. It has been known that ESCs, transplanted into a damaged heart, are able to be integrated into the recipient tissue and to improve its functions. Lee et al. demonstrated that the culture of hESCs on graphene promotes the progressive differentiation of these cells into mesodermal and endodermal cells and their subsequent cardiomyogenic differentiation. Additionally, the analysis of the cadiomyogenic differentiation of hESCs cultured on graphene or Matrigel (substrate used in conventional cadiomyogenic differentiation) revealed that graphene was superior to Matrigel. The authors hypothesized that the increased cardiomyogenic differentiation of graphene is induced by the nanoroughness of the material, which enhanced hESCs adhesion and the upregulation of ERK signaling. However, they did not observe functional cells (beating) in the graphene system and this indicated that the method does not induce full differentiation into contractile cardiomyocytes [136].

9 Discussion

In conclusion, it can resume that the presence of graphene for regenerative medicine:

1. does not alternate the morphology and the proliferation of the stem cell in normal stem cell media,
2. is biocompatible;
3. Its presence in medical device or injured tissues would do not affect the physiological environment.
4. Is able to induce the commitment of stem cells.

Bibliography on the interactions of graphene and stem cells is growing rapidly, and includes studies primarily motivated by biomedical applications, and environmental health and safety. Moreover the correct mechanism with GO and its derivatives composites in SCs commitment is still undefined. For this reason the issue of its potential toxicity must occur.

In this view, we have here reported a review of the international literature in order to produce a state-of-the-art of graphene applied to stem cells in regenerative medicine. Several researchers confirmed that graphene-based scaffolds are able to act on the final destiny human stem cells clearly appearing to be a step forward in the field of tissue engineering.

In these last year a novel existing new application has emerged thanks to the combination of graphene with others nanomaterials. These novel strategies have great potential for several future technologies ranging from biotechnological and biomedical applications such as drug delivery, and genetical targeting. The novel setting in graphene production provides different levels of dispersibility and impurities within the nanomaterials. Among the differen type of graphene, GO is considered more biocompatible than graphene thanks to its greater solubility/dispersibility. This, results in less damage and toxicity in several cell type such as skin fibroblasts and red blood cells, and bacteria. However, they are less effective in liver, kidney, and spleen. In all this in vitro system, cells exhibit time and dosage dependency, and this is depending on protein adsorption and reactions. However, before reach the clinic, the toxicology profile and safety efficacies of such materials are essential steps in their development. Nevertheless, its great potential on regenerative medicine will offer a variety of new and existing tools based for use in the areas of advanced disease diagnosis, imaging, and targeted therapies for the treatment of a wide of severely debilitating diseases.

References

1. Khalili AA, Ahmad MR (2015) A review of cell adhesion studies for biomedical and biological applications. Int J Mol Sci 16(8):18149–18184
2. Chandra P, Lee SJ (2015) Synthetic extracellular microenvironment for modulating stem cell behaviors. Biomark Insights 10(Suppl 1):105–116
3. Martino MM, Brkic S, Bovo E, Burger M, Schaefer DJ, Wolff T, Gürke L, Briquez PS, Larsson HM, Gianni-Barrera R, Hubbell JA, Banfi A (2015) Extracellular matrix and growth factor engineering for controlled angiogenesis in regenerative medicine. Front Bioeng Biotechnol 3:45
4. Knothe Tate ML, Detamore M, Capadona JR, Woolley A, Knothe U (2016) Engineering and commercialization of human-device interfaces, from bone to brain. Biomaterials 95:35–46
5. Hendow EK, Guhmann P, Wright B, Sofokleous P, Parmar N, Day RM (2016) Biomaterials for hollow organ tissue engineering. Fibrogenesis Tissue Repair 9:3
6. Bressan E, Ferroni L, Gardin C, Sbricoli L, Gobbato L, Ludovichetti FS, Tocco I, Carraro A, Piattelli A, Zavan B (2014) Graphene based scaffolds effects on stem cells commitment. J Transl Med 12(1):296
7. Nayak TR, Andersen H, Makam VS, Khaw C, Bae S, Xu X, Ee PL, Ahn JH, Hong BH, Pastorin G, Özyilmaz B (2011) Graphene for controlled and accelerated osteogenic differentiation of human mesenchymal stem cells. ACS Nano 5:4670–4678
8. Liu Z, Robinson JT, Sun XM et al (2008) PEGylated nanographene oxide for delivery of water-insoluble cancer drugs. J Am Chem Soc 130(33):10876–10877
9. Shen H, Zhang L, Liu M, Zhang Z (2012) Biomedical applications of graphene. Theranostics 2(3):283–294
10. Xuan Y, Wu YQ, Shen T et al (2008) Atomic-layer-deposited nanostructures for graphene-based nanoelectronics. Appl Phys Lett 92(1):013101–013103
11. Hu S, Zeng Y, Yang S, Qin H, Cai H, Wang J (2015) Application of graphene based nanotechnology in stem cells research. J Nanosci Nanotechnol 15(9):6327–6341 (Review)
12. Zavan B, Vindigni V, Gardin C, D'Avella D, Della Puppa A, Abatangelo G, Cortivo R (2010) Neural potential of adipose stem cells. Discov Med 10(50):37–43 (Review)
13. Bressan E, Carraro A, Ferroni L, Gardin C, Sbricoli L, Guazzo R, Stellini E, Roman M, Pinton P, Sivolella S, Zavan B (2013) Nanotechnology to drive stem cell commitment. Nanomedicine 8(3):469–486
14. Hao J, Zhang Y, Jing D, Shen Y, Tang G, Huang S, Zhao Z (2015) Mechanobiology of mesenchymal stem cells: perspective into mechanical induction of MSC fate. Acta Biomater 20:1–9
15. Kfoury Y, Scadden DT (2015) Mesenchymal cell contributions to the stem cell niche. Cell Stem Cell 16(3):239–253
16. D'souza N, Rossignoli F, Golinelli G, Grisendi G, Spano C, Candini O, Osturu S, Catani F, Paolucci P, Horwitz EM, Dominici M (2015) Mesenchymal stem/stromal cells as a delivery platform in cell and gene therapies. BMC Med 13:186
17. Schipani E, Kronenberg HM (2008) Adult mesenchymal stem cells. StemBook [Internet]. Harvard Stem Cell Institute, Cambridge
18. Owen M (1988) Marrow stromal stem cells. J Cell Sci Suppl 10:63–76 (Review)
19. Caplan AI (2016) MSCs: the sentinel and safe-guards of injury. J Cell Physiol 231 (7):1413–1416
20. González F, Huangfu D (2016) Mechanisms underlying the formation of induced pluripotent stem cells. Wiley Interdiscip Rev Dev Biol 5(1):39–65
21. Takahashi K, Yamanaka S (2015) A developmental framework for induced pluripotency. Development 142(19):3274–3285
22. Raab S, Klingenstein M, Liebau S, Linta L (2014) A Comparative view on human somatic cell sources for iPSC generation. Stem Cells Int 2014:768391

23. Inoue H, Nagata N, Kurokawa H, Yamanaka S (2014) PS cells: a game changer for future medicine. EMBO J 33(5):409–417
24. Fulka J Jr, Fulka H (2007) Somatic cell nuclear transfer (SCNT) in mammals: the cytoplast and its reprogramming activities. Adv Exp Med Biol 591:93–102 (Review)
25. Tweedell KS (2008) New paths to pluripotent stem cells. Curr Stem Cell Res Ther 3(3):151–162
26. Wu M, Chen G, Hu B (2013) Induced pluripotency for translational research. Genom Proteom Bioinform 11(5):288–293
27. Okano H, Nakamura M, Yoshida K, Okada Y, Tsuji O, Nori S, Ikeda E, Yamanaka S, Miura K (2013) Steps toward safe cell therapy using induced pluripotent stem cells. Circ Res 112(3):523–533
28. Chari S, Mao S (2016) Timeline: iPSCs—the first decade. Cell 164(3):580
29. Wan W, Cao L, Kalionis B, Xia S, Tai X (2015) Applications of induced pluripotent stem cells in studying the neurodegenerative diseases. Stem Cells Int 2015:382530
30. Olariu V, Lövkvist C, Sneppen K (2016) Nanog, Oct4 and Tet1 interplay in establishing pluripotency. Sci Rep 6:25438
31. Takahashi K, Yamanaka S (2016) A decade of transcription factor-mediated reprogramming to pluripotency. Nat Rev Mol Cell Biol 17(3):183–193
32. Liu Z, Skamagki M, Kim K, Zhao R (2015) Canonical MICRORNA activity facilitates but may be dispensable for transcription factor-mediated reprogramming. Stem Cell Rep 5 (6):1119–1127
33. Porciuncula A, Kumar A, Rodriguez S, Atari M, Araña M, Martin F, Soria B, Prosper F, Verfaillie C, Barajas M (2016) Pancreatic differentiation of Pdx1-GFP reporter mouse induced pluripotent stem cells. Differentiation S0301–4681(16):30015–30019. doi:10.1016/j.diff.2016.04.005
34. Chanana AM, Rhee JW, Wu JC (2016) Human-induced pluripotent stem cell approaches to model inborn and acquired metabolic heart diseases. Curr Opin Cardiol 31(3):266–274
35. El Khatib MM, Ohmine S, Jacobus EJ, Tonne JM, Morsy SG, Holditch SJ, Schreiber CA, Uetsuka K, Fusaki N, Wigle DA, Terzic A, Kudva YC, Ikeda Y (2016) Tumor-free transplantation of patient-derived induced pluripotent stem cell progeny for customized islet regeneration. Stem Cells Transl Med 5(5):694–702
36. Di Foggia V, Makwana P, Ali RR, Sowden JC (2016) Induced pluripotent stem cell therapies for degenerative disease of the outer retina: disease modeling and cell replacement. J Ocul Pharmacol Ther 32(5):240–252. doi:10.1089/jop.2015.0143. Epub 2016 Mar 30
37. Riera M, Fontrodona L, Albert S, Ramirez DM, Seriola A, Salas A, Muñoz Y, Ramos D, Villegas-Perez MP, Zapata MA, Raya A, Ruberte J, Veiga A, Garcia-Arumi J (2016) Comparative study of human embryonic stem cells (hESC) and human induced pluripotent stem cells (hiPSC) as a treatment for retinal dystrophies. Mol Ther Methods Clin Dev 3:16010
38. Gabr MM, Zakaria MM, Refaie AF, Khater SM, Ashamallah SA, Ismail AM, El-Badri N, Ghoneim MA (2014) Generation of insulin-producing cells from human bone marrow-derived mesenchymal stem cells: comparison of three differentiation protocols. Biomed Res Int 2014:832736
39. Reinhard J, Brösicke N, Theocharidis U, Faissner A (2016) The extracellular matrix niche microenvironment of neural and cancer stem cells in the brain. Int J Biochem Cell Biol S1357–2725(16):30107–30108. doi: 10.1016/j.biocel.2016.05.002. [Epub ahead of print]
40. Belenguer G, Domingo-Muelas A, Ferrón SR, Morante-Redolat JM, Fariñas I (2016) Isolation, culture and analysis of adult subependymal neural stem cells. Differentiation 91 (4-5):28–41. doi:10.1016/j.diff.2016.01.005. Epub 2016 Mar 23
41. Zhao H, Chai Y (2015) Stem cells in teeth and craniofacial bones. J Dent Res 94 (11):1495–1501
42. Aly LA (2015) Stem cells: sources, and regenerative therapies in dental research and practice. World J Stem Cells 7(7):1047–1053

43. Padial-Molina M, O'Valle F, Lanis A, Mesa F, Dohan Ehrenfest DM, Wang HL, Galindo-Moreno P (2015) Clinical application of mesenchymal stem cells and novel supportive therapies for oral bone regeneration. Biomed Res Int 2015:341327
44. Tatullo M, Marrelli M, Paduano F (2015) The regenerative medicine in oral and maxillofacial surgery: the most important innovations in the clinical application of mesenchymal stem cells. Int J Med Sci 12(1):72–77
45. Liu Y, Hu J, Wang S (2014) Mesenchymal stem cell-mediated treatment of oral diseases. Histol Histopathol 29(8):1007–1015
46. Egusa H, Sonoyama W, Nishimura M, Atsuta I, Akiyama K (2012) Stem cells in dentistry–part II: clinical applications. J Prosthodont Res 56(4):229–248
47. Dawsonand I, Oreffo ROC (2008) Bridging the regenerationgap:stem cells, biomaterials and clinical translation in bone tissue engineering. Arch Biochem Biophys 473(2):124–131
48. Morikawa S, Ouchi T, Shibata S, Fujimura T, Kawana H, Okano H, Nakagawa T (2016) Applications of mesenchymal stem cells and neural crest cells in craniofacial skeletal research. Stem Cells Int 2016:2849879
49. Fisher JN, Peretti GM, Scotti C (2016) Stem cells for bone regeneration: from cell-based therapies to decellularised engineered extracellular matrices. Stem Cells Int 2016:9352598
50. Yousefi AM, James PF, Akbarzadeh R, Subramanian A, Flavin C, Oudadesse H (2016) Prospect of stem cells in bone tissue engineering: a review. Stem Cells Int 2016:6180487
51. Liu X, Ma PX (2004) Polymeric scaffolds for bone tissue engineering. Ann Biomed Eng 32 (3):477–486
52. Rezwan K, Chen QZ, Blaker JJ, Boccaccini AR (2006) Biodegradable and bioactive porous polymer/inorganic composite scaffolds for bone tissue engineering. Biomaterials 27 (18):3413–3431
53. Dubey N, Bentini R, Islam I, Cao T, Castro Neto AH, Rosa V (2015) Graphene: a versatile carbon-based material for bone tissue engineering. Stem Cells Int 2015:1–12
54. Crowder SW, Prasai D, Rath R et al (2013) Three-dimensional graphene foams promote osteogenic differentiation of human mesenchymal stem cells. Nanoscale 5(10):4171–4176
55. Rosa V, Della Bona A, Cavalcanti BN, Nör JE (2012) Tissue engineering: from research to dental clinics. Dent Mater 28(4):341–348
56. Qi WY, Yuan W, Yan J, Wang H (2014) Growth and accelerated differentiation of mesenchymal stem cells on graphene oxide/poly-L-lysine composite films. J Mater Chem B 2:5461–5467
57. Nayak TR, Andersen H, Makam VS et al (2011) Graphene for controlled and accelerated osteogenic differentiation of human mesenchymal stem cells. ACS Nano 5(6):4670–4678
58. Tang LAL, Lee WC, Shi H et al (2012) Highly wrinkled crosslinked graphene oxide membranes for biological and chargestorage applications. Small 8(3):423–431
59. Jin G, Li K (2014) The electrically conductive scaffold as the skeleton of stem cell niche in regenerative medicine. Mater Sci Eng C Mater Biol Appl 45:671–681
60. Feng L, Wu L, Qu X (2013) New horizons for diagnostics and therapeutic applications of graphene and graphene oxide. Adv Mater 25(2):168–186
61. Li M, Liu Q, Jia Z et al (2014) Graphene oxide/hydroxyapatite composite coatings fabricated by electrophoretic nanotechnology for biological applications. Carbon 67:185–197
62. Cheng C, Li D (2013) Solvated graphenes: an emerging class of functional softmaterials. Adv Mater 25(1):13–30
63. Bernhard JC, Vunjak-Novakovic G (2016) Should we use cells, biomaterials, or tissue engineering for cartilage regeneration? Stem Cell Res Ther 7:56
64. Hoppe A, Güldal NS, Boccaccini AR (2011) A review of the biological response to ionic dissolution products from bioactive glasses and glass-ceramics. Biomaterials 32 (11):2757–2774
65. Baino F, Vitale-Brovarone C (2011) Three-dimensional glass-derived scaffolds for bone tissue engineering: current trends and forecasts for the future. J Biomed Mater Res A 97(4):514–535

66. Baino F, Novajra G, Vitale-Brovarone C (2015) Bioceramics and scaffolds: a winning combination for tissue engineering. Front Bioeng Biotechnol 3:202
67. Jung HS, Lee T, Kwon IK, Kim HS, Hahn SK, Lee CS (2015) Surface modification of multipass caliber-rolled Ti alloy with dexamethasone-loaded graphene for dental applications. ACS Appl Mater Interfaces 7(18):9598–9607
68. Shi YY, Li M, Liu Q, Jia ZJ, Xu XC, Cheng Y, Zheng YF (2016) Electrophoretic deposition of graphene oxide reinforced chitosan-hydroxyapatite nanocomposite coatings on Ti substrate. J Mater Sci Mater Med 27(3):48
69. Schroeder KL, Goreham RV, Nann T (2016) Graphene quantum dots for theranostics and bioimaging. Pharm Res 33(10):2337–2357. doi:10.1007/s11095-016-1937-x. Epub 2016 May 20
70. Hsiao YS, Kuo CW, Chen P (2013) Multifunctional graphene-PEDOT microelectrodes for on chip manipulation of human mesenchymal stem cells. Adv Funct Mater 23(37):4649–4656
71. Elkhenany H, Amelse L, Lafont A, Bourdo S, Caldwell M, Neilsen N, Dervishi E, Derek O, Biris AS, Anderson D, Dhar M (2015) Graphene supports in vitro proliferation and osteogenic differentiation of goat adult mesenchymal stem cells: potential for bone tissue engineering. J Appl Toxicol 35(4):367–374
72. Duan S, Yang X, Mei F, Tang Y, Li X, Shi Y, Mao J, Zhang H, Cai Q (2014) Enhanced osteogenic differentiation of mesenchymal stem cells on poly(L-lactide) nanofibrous scaffolds containing carbon nanomaterials. J Biomed Mater Res, Part A 103(4):1424–1435
73. Kumar S, Raj S, Kolanthai E, Sood AK, Sampath S, Chatterjee K (2015) Chemical functionalization of graphene to augment stem cell osteogenesis and inhibit biofilm formation on polymer composites for orthopedic applications. ACS Appl Mater Interfaces 7 (5):3237–3252
74. Nair M, Nancy D, Krishnan AG, Anjusree GS, Vadukumpully S, Nair SV (2015) Graphene oxide nanoflakes incorporated gelatin-hydroxyapatite scaffolds enhance osteogenic differentiation of human mesenchymal stem cells. Nanotechnology 26(16):161001
75. Lyu CQ, Lu JY, Cao CH, Luo D, Fu YX, He YS, Zou DR (2015) Induction of osteogenic differentiation of human adipose-derived stem cells by a novel self-supporting graphene hydrogel film and the possible underlying mechanism. ACS Appl Mater Interfaces 7 (36):20245–20254
76. Luo Y, Shen H, Fang Y, Cao Y, Huang J, Zhang M, Dai J, Shi X, Zhang Z (2015) Enhanced proliferation and osteogenic differentiation of mesenchymal stem cells on graphene oxide-incorporated electrospun poly(lactic-co-glycolic acid) nanofibrous mats. ACS Appl Mater Interfaces 7(11):6331–6339
77. Lee JH, Shin YC, Jin OS, Kang SH, Hwang Y-S, Park J-C, Hong SW, Han DW (2015) Reduced graphene oxide-coated hydroxyapatite composites stimulate spontaneous osteogenic differentiation of human mesenchymal stem cells. Nanoscale 7(27):11642–11651
78. Lee JH, Shin YC, Lee S-M, Jin OS, Kang SH, Hong SW, Huh JB, Han D-W (2015) Enhanced osteogenesis by reduced graphene oxide/hydroxyapatite nanocomposites. Sci Rep 5(November):18833
79. Raucci MR, Giugliano D, Longo A, Zeppetelli S, Carotenuto G, Ambrosio L (2016) Comparative facile methods for preparing graphene oxide-hydroxyapatite for bone tissue engineering. J Tissue Eng Regen Med. doi:10.1002/term.2119
80. Olivares-Navarrete R, Rodil SE, Hyzy SL, Dunn GR, Almaguer-Flores A, Schwartz Z, Boyan BD (2015) Role of integrin subunits in mesenchymal stem cell differentiation and osteoblast maturation on graphitic carbon-coated microstructured surfaces. Biomaterials 51:69–79
81. Brun P, Dickinson SC, Zavan B, Cortivo R, Hollander AP, Abatangelo G (2008) Characteristics of repair tissue in second-look and third-look biopsies from patients treated with engineered cartilage: relationship to symptomatology and time after implantation. Arthritis Res Ther 10(6):R132

82. Huang BJ, Hu JC, Athanasiou KA (2016) Cell-based tissue engineering strategies used in the clinical repair of articular cartilage. Biomaterials 98:1–22
83. Chen C, Bang S, Cho Y, Lee S, Lee I, Zhang S, Noh I (2016) Research trends in biomimetic medical materials for tissue engineering: 3D bioprinting, surface modification, nano/micro-technology and clinical aspects in tissue engineering of cartilage and bone Curr Opin Neurol 25(5):597–603. doi:10.1097/WCO.0b013e328357f288
84. Should Bernhard JC, Vunjak-Novakovic G (2016) We use cells, biomaterials, or tissue engineering for cartilage regeneration? Stem Cell Res Ther 7(1):56. doi:10.1186/s13287-016-0314-3
85. Nazempour A, Van Wie BJ (2016) Chondrocytes, mesenchymal stem cells, and their combination in articular cartilage regenerative medicine. Ann Biomed Eng 44(5):1325–1354
86. Panadero JA, Lanceros-Mendez S, Ribelles JL (2016) Differentiation of mesenchymal stem cells for cartilage tissue engineering: Individual and synergetic effects of three-dimensional environment and mechanical loading. Acta Biomater 33:1–12
87. Lee WC, Lim CH, Su KC, Loh KP, Lim CT (2015) Cell-assembled graphene biocomposite for enhanced chondrogenic differentiation. Small 11(8):963–969
88. Qazi TH, Mooney DJ, Pumberger M, Geissler S, Duda GN (2015) Biomaterials based strategies for skeletal muscle tissue engineering: existing technologies and future trends. Biomaterials 53:502–521
89. Tedesco FS, Cossu G (2012) Stem cell therapies for muscle disorders. Curr Opin Neurol 25:597–603
90. Smith BD, Grande DA (2015) The current state of scaffolds for musculoskeletal regenerative applications. Nat Rev Rheumatol 11(4):213–222
91. Gentile NE, Stearns KM, Brown EH, Rubin JP, Boninger ML, Dearth CL, Ambrosio F, Badylak SF (2014) Targeted rehabilitation after extracellular matrix scaffold transplantation for the treatment of volumetric muscle loss. Am J Phys Med Rehabil 93(11 Suppl 3):S79–S87
92. Cezara CA, Mooneya DJ (2015) Biomaterial-based delivery for skeletal muscle repair. Adv Drug Deliv Rev 84:188–197
93. Mertens JP, Sugg KB, Lee JD, Larkin LM (2014) Engineering muscle constructs for the creation of functional engineered musculoskeletal tissue. Regen Med 9(1):89–100
94. Duffy RM, Feinberg AW (2014) Engineered skeletal muscle tissue for soft robotics: fabrication strategies, current applications, and future challenges. Wiley Interdiscip Rev Nanomed Nanobiotechnol 6(2):178–195
95. Sicari BM, Dearth CL, Badylak SF (2014) Tissue engineering and regenerative medicine approaches to enhance the functional response to skeletal muscle injury. Anat Rec 297(1):51–64
96. Davies BM, Morrey ME, Mouthuy PA, Baboldashti NZ, Hakimi O, Snelling S, Price A, Carr A (2013) Repairing damaged tendon and muscle: are mesenchymal stem cells and scaffolds the answer? Regen Med 8(5):613–630
97. Zhao C, Andersen H, Ozyilmaz B, Ramaprabhu S, Pastorin G, Ho HK (2015) Spontaneous and specific myogenic differentiation of human mesenchymal stem cells on polyethylene glycol-linked multi-walled carbon nanotube films for skeletal muscle engineering. Nanoscale 7(43):18239–18249
98. Chaudhuri B, Bhadra D, Moroni L, Pramanik K (2015) Myoblast differentiation of human mesenchymal stem cells on graphene oxide and electrospun graphene oxide–polymer composite fibrous meshes: importance of graphene oxide conductivity and dielectric constant on their biocompatibility. Biofabrication 7(1):015009
99. Sedaghati T, Seifalian AM (2015) Nanotechnology and bio-functionalisation for peripheral nerve regeneration. Neural Regen Res 10(8):1191–1194. doi:10.4103/1673-5374.162678
100. Matsumoto K, Ohnishi K, Sekine T, Ueda H, Yamamoto Y, Kiyotani T, Nakamura T, Endo K, Shimizu Y (2000) Use of a newly developed artificial nerve conduit to assist peripheral nerve regeneration across a long gap in dogs. ASAIO J 46:415–420

101. Takahashi K, Yamanaka S (2006) Induction of pluripotent stem cells from mouse embryonic and adult fibroblast cultures by defined factors. Cell 126(4):663–676
102. Hood B, Levene HB, Levi AD (2009) Transplantation of autologous Schwann cells for the repair of segmental peripheral nerve defects. Neurosurg Focus 26:E4
103. Tsintou M, Dalamagkas K, Seifalian AM (2015) Advances in regenerative therapies for spinal cord injury: a biomaterials approach. Neural Regen Res 10:726–742
104. Ferroni L, Gardin C, Tocco I, Epis R, Casadei A, Vindigni V, Mucci G, Zavan B (2013) Potential for neural differentiation of mesenchymal stem cells. Adv Biochem Eng Biotechnol 129:89–115
105. Zavan B, Michelotto L, Lancerotto L, Della Puppa A, D'Avella D, Abatangelo G, Vindigni V, Cortivo R (2010) Neural potential of a stem cell population in the adipose and cutaneous tissues. Neurol Res 32(1):47–54
106. Gardin C, Piattelli A, Zavan B (2016) Graphene in regenerative medicine: focus on stem cells and neuronal differentiation. Trends Biotechnol 34(6):435–437
107. Gardin C, Vindigni V, Bressan E, Ferroni L, Nalesso E, Puppa AD, D'Avella D, Lops D, Pinton P, Zavan B (2011) Hyaluronan and fibrin biomaterial as scaffolds for neuronal differentiation of adult stem cells derived from adipose tissue and skin. Int J Mol Sci 12 (10):6749–6764
108. Shah S, Yin PT, Uehara TM, Chueng STD, Yang L, Lee KB (2014) Guiding stem cell differentiation into oligodendrocytes using graphene-nanofiber hybrid scaffolds. Adv Mater 26(22):3673–3680
109. Weaver CL, Cui XT (2015) Directed neural stem cell differentiation with a functionalized graphene oxide nanocomposite. Adv Healthc Mater 4:1408–1416
110. Guo W, Wang S, Yu X, Qiu J, Li J, Tang W, Li Z, Mou X, Liu H, Wang Z (2016) Construction of a 3D rGO–collagen hybrid scaffold for enhancement of the neural differentiation of mesenchymal stem cells. Nanoscale 8(4):1897–1904
111. Shevchenko RV, James SL, James SE (2010) A review of tissue-engineered skin bioconstructs available for skin reconstruction. J R Soc Interface 7:229–258
112. Tonello C, Zavan B, Cortivo R, Brun P, Panfilo S, Abatangelo G (2003) In vitro reconstruction of human dermal equivalent enriched with endothelial cells. Biomaterials 24(7):1205–1211
113. Papini R (2004) Management of burn injuries of various depths. Br Med J 329:158–160
114. Pham C, Greenwood J, Cleland H, Woodruff P, Maddern G (2007) Bioengineered skin substitutes for the management of burns: a systematic review. Burns 33:946–957
115. Mandrycky C, Wang Z, Kim K, Kim DH (2016) 3D bioprinting for engineering complex tissues. Biotechnol Adv 34(4):422–434
116. Santema TB, Poyck PP, Ubbink DT (2016) Skin grafting and tissue replacement for treating foot ulcers in people with diabetes. Cochrane Database Syst Rev 2:CD011255
117. Mitsukawa N, Higaki K, Ito N, Muramatsu H, Karube D, Akita S, Kubota Y, Satoh K (2016) Combination treatment of artificial dermis and basic fibroblast growth factor for skin defects: a histopathological examination. Wounds 28(5):158–166
118. Tonello C, Vindigni V, Zavan B, Abatangelo S, Abatangelo G, Brun P, Cortivo R (2005) In vitro reconstruction of an endothelialized skin substitute provided with a microcapillary network using biopolymer scaffolds. FASEB J 19(11):1546–1548 Epub 2005 Jun 21
119. Garwood CS, Steinberg JS, Kim PJ (2015) Bioengineered alternative tissues in diabetic wound healing. Clin Podiatr Med Surg 32(1):121–133
120. Figallo E, Flaibani M, Zavan B, Abatangelo G, Elvassore N (2007) Micropatterned biopolymer 3D scaffold for static and dynamic culture of human fibroblasts. Biotechnol Prog 23(1):210–216
121. Nyame TT, Chiang HA, Leavitt T, Ozambela M, Orgill DP (2015) Tissue-engineered skin substitutes. Plast Reconstr Surg 136(6):1379–1388
122. Dąbrowska AK, Rotaru GM, Derler S, Spano F, Camenzind M, Annaheim S, Stämpfli R, Schmid M, Rossi RM (2016) Materials used to simulate physical properties of human skin. Skin Res Technol 22(1):3–14

123. Rowan MP, Cancio LC, Elster EA, Burmeister DM, Rose LF, Natesan S, Chan RK, Christy RJ, Chung KK (2015) Burn wound healing and treatment: review and advancements. Crit Care 12(19):243
124. Wang HY, Zhang YQ (2015) Processing silk hydrogel and its applications in biomedical materials. Biotechnol Prog 31(3):630–640
125. Sun BK, Siprashvili Z, Khavari PA (2014) Advances in skin grafting and treatment of cutaneous wounds. Science 346(6212):941–945
126. Li Z, Wang H, Yang B, Sun Y, Huo R (2015) Three-dimensional graphene foams loaded with bone marrow derived mesenchymal stem cells promote skin wound healing with reduced scarring. Mater Sci Eng, C 57:181–188
127. Farouz Y, Chen Y, Terzic A, Menasché P (2015) Concise review: growing hearts in the right place: on the design of biomimetic materials for cardiac stem cell differentiation. Stem Cells 33(4):1021–1035
128. Sun X, Altalhi W, Nunes SS (2016) Vascularization strategies of engineered tissues and their application in cardiac regeneration. Adv Drug Deliv Rev 96:183–194
129. Alrefai MT, Murali D, Paul A, Ridwan KM, Connell JM, Shum-Tim D (2015) Cardiac tissue engineering and regeneration using cell-based therapy. Stem Cells Cloning 8:81–101
130. Huyer LD, Montgomery M, Zhao Y, Xiao Y, Conant G, Korolj A, Radisic M (2015) Biomaterial based cardiac tissue engineering and its applications. Biomed Mater 10 (3):034004
131. Lee T-J, Park S, Bhang SH, Yoon J-K, Jo I, Jeong G-J, Hong BH, Kim B-S (2014) Graphene enhances the cardiomyogenic differentiation of human embryonic stem cells. Biochem Biophys Res Commun 452(1):174–180
132. Ahmed M, Yildirimer L, Khademhosseini A, Seifalian AM (2012) Nanostructured materials for cardiovascular tissue engineering. J Nanosci Nanotechnol 12(6):4775–4785
133. Martinelli V, Cellot G, Fabbro A, Bosi S, Mestroni L, Ballerini L (2013) Improving cardiac myocytes performance by carbon nanotubes platforms. Front Physiol 4:239
134. Sreejit P, Verma RS (2013) Natural ECM as biomaterial for scaffold based cardiac regeneration using adult bone marrow derived stem cells. Stem Cell Rev 9(2):158–171
135. Zwi-Dantsis L, Gepstein L (2012) Induced pluripotent stem cells for cardiac repair. Cell Mol Life Sci 69(19):3285–3299
136. Kim T, Kahng YH, Lee T, Lee K, do Kim H (2013) Graphene films show stable cell attachment and biocompatibility with electrogenic primary cardiac cells. Mol Cells 36 (6):577–582

Graphene: An Emerging Carbon Nanomaterial for Bone Tissue Engineering

Nileshkumar Dubey, Fanny Esther Denise Decroix and Vinicius Rosa

Abstract The development of materials and strategies that can promote faster bone healing and improved regeneration of bony defects is of high interest. Graphene and its derivatives (graphene oxide and reduced graphene oxide) have remarkable mechanical properties, can be chemically modified and allow the attachment of molecules and proteins. Due to these characteristics, these carbon-based materials have received increasing attention for several biomedical applications. As graphenes can improve mechanical properties of several biomaterials, induce, and increase cell differentiation toward osteoblasts, they have emerged as interesting alternatives for to promote bone regeneration. Herein, the key achievements made with graphenes for bone tissue engineering are presented with particular emphasis on their combination with biomaterials for bone regeneration and as coatings for biomedical implants.

1 Introduction

Bone is a calcified connective tissue providing structural support and protection to the body. In addition, bone offers an environment for hemopoiesis and acts as a storage for minerals such as calcium [1]. Throughout the life course, bone undergoes remodeling by a process of resorption and formation in order to maintain its health and promote the repair of injuries. However, conditions like tumor resections, trauma, fractures, and congenital skeletal defects can result in inadequate healing depending on the extension of the bone loss or defect. Furthermore, the bone healing may be compromised by conditions such as infection, aging, diabetes,

N. Dubey · V. Rosa (✉)
Faculty of Dentistry, National University of Singapore, Singapore, Singapore
e-mail: denvr@nus.edu.sg

F.E.D. Decroix · V. Rosa
Centre for Advanced 2D Materials and Graphene Research Centre, National University of Singapore, Singapore, Singapore

G. Gonçalves et al. (eds.), *Graphene-based Materials in Health and Environment*,
Carbon Nanostructures, DOI 10.1007/978-3-319-45639-3_5

and impaired blood supply [2]. In such circumstances, some form of grafting or bone substitute material is required to fill the defects created and promote bone formation [3, 4].

Bone grafting is a surgical procedure aiming at filing a void created by a bone loss or defect. The two most common types of bone grafts are autologous and allogeneic grafts [5]. Autologous bone grafts are considered the 'gold standard' for bone regenerative treatment. They involve the harvesting of 'donor' bone, typically from the anterior and posterior iliac crest of the pelvis, followed by transplantation into the defect site [6]. Nevertheless, the use of autologous graft has disadvantages such as the need for additional surgical procedures, high cost, and limited availability of graft. In order to overcome some of these limitations, allogeneic bone grafts obtained from bone banks may alternatively be used [2]. Allogeneic grafts, like demineralized bone matrix or cortico-cancellous grafts, are devitalized via irradiation or freeze-drying. However, these processes often result in reduced osteoinductive properties because of the absence of viable cells [5]. Other limitations of allogenic graft include batch-to-batch variability, risk of immunogenicity, and infection transmission [7]. To overcome these problems, bone substitute materials are being constantly developed with new components and structural properties to allow cell infiltration, bone growth, and vascularization.

Numerous bone substitute materials can be used for bone tissue regeneration and repair [8, 9]. Chitosan, for example, is a natural bone substitute material which has shown to facilitate the osteogenic differentiation of mesenchymal stem cells (MSCs) in vitro [10]. Nonetheless, chitosan itself is not osteoconductive and, due to its poor mechanical properties, it is unable to support load bearing bone implants [11]. In addition to polymers, bioceramics allow attachment and differentiation of stem cells or osteoprogenitor cells into osteoblasts stimulating bone formation [9]. Though ceramics possess excellent bioactivity and biocompatibility, they are exhibit brittle, difficult to shape may have slow resorption rates [12]. Synthetic polymers can overcome some of these disadvantages [13]. Nonetheless, polymers fall short of achieving satisfactory results in bone regeneration treatment because of low mechanical stability, lack of remodeling, and potential to induce inflammatory reactions [5]. Alternatively, ceramic and polymer composite materials are interesting candidates for bone tissue engineering since their properties can be optimized to suit the mechanical and physiologic demands of the host tissue [14]. However, the mechanical properties, structural stability and bioactive potential largely depend on the interactions between the matrix and filler in the composite materials. For instance, the lack of adhesion between nanohydroxiapatite and polymer matrix result in early failure at the interface compromising the structural ability, mechanical property, and bioactivity of the composite material [15]. Hence, the constant development and improvements of biomaterials for bone tissue engineering is of great interest.

Graphene is a carbon-based material with promising applications in a number of fields including material sciences, chemistry, and electronics [16–18]. Moreover, it has demonstrated a positive impact on attachment, growth, and osteogenic differentiation of MSCs [19, 20]. Graphene can be manufactured in relatively pure form and combined with other materials to improve their mechanical and biological properties [16, 18, 21]. The present chapter highlights key achievements obtained with graphene for bone tissue engineering [19, 22, 23].

2 Graphene and Its Derivatives for Osteogenesis

Graphene is a single atomic sheet of conjugated sp^2 carbon arranged in a honeycomb pattern [24]. The presence of free π electron, reactive sites for surface reactions and strong C–C bond make graphene a material with unique properties. It is electrically superconductive, chemically stable, and has exceptionally high mechanical strength and modulus of elasticity [25–27]. Moreover, graphene offers a large surface area that can be chemically modified to enhance its capabilities [18].

Essentially, there are three types of graphene. The first is pristine graphene, a thin layer of pure carbon obtained mainly by chemical vapor deposition (CVD), by exfoliation of graphite and other methods. The second is graphene oxide (GO) which is a hydrophilic oxidized form of graphene laced with oxygen-containing groups. This amphiphilic compound allows surface functionalization and can be dispersed in aqueous solution, making it an attractive candidate for gene or drug delivery and substrate modification [16, 28, 29]. The third type is reduced graphene oxide (rGO) which can be obtained from the reduction of GO by removal of the oxygen-containing groups [16, 28].

Graphene and its derivatives, when used as coatings allow stem cell attachment, growth, and enhance osteogenic differentiation which makes them promising alternative for bone tissue engineering [20].

Graphene, both in two-and three-dimensional forms, accelerates the differentiation of MSCs into osteoblasts [20]. MSCs cultured on graphene-coated silicon dioxide (SiO_2) formed a confluent monolayer with spindle-shaped cells, whereas those cultured on the pure SiO_2 substrate formed separate islands of polygonal cells [22]. Considering that the cell shape is correlated with the differentiation of MSCs to osteogenic lineage [30], the spindle-shaped morphology of MSCs on graphene reflects an enhanced potential for osteoblastic differentiation than pure SiO_2 [22]. MSCs cultured on polydimethyl siloxane (PDMS), polyethylene terephthalate (PET), glass, and SiO_2 coated with CVD-grown graphene showed no significant difference in cell morphology or viability compared to uncoated surfaces suggesting good cytocompatibility of graphene. Nonetheless, the graphene-coated substrates enhanced stem cell differentiation toward osteoblastic lineage. The rate of differentiation induced by the graphene-coated substrates was comparable to the differentiation observed on the controls under the influence of bone morphogenic protein-2 (BMP-2), a classical inducer for osteogenic differentiation [19]. Furthermore, in another study, bone marrow mesenchymal stem cells (BMMSCs) cultured on PDMS coated with graphene for 12 days with osteogenic medium showed a ~7-fold increase in mineral deposit formation compared to those on PDMS only [31]. Similarly, periodontal ligament stem cells (PDLSCs) cultured on glass coated with graphene presented a ~2-fold increase in the absorbance reading for alizarin red compared to glass in the presence of osteogenic medium. Remarkably, PDLSCs cultured on graphene with basal medium demonstrated higher mineralization compared to those cultured on glass with osteogenic medium [23].

Graphene three-dimensional constructs (3DGp) are also capable of inducing spontaneous osteogenic differentiation of MSCs [23, 31, 32]. The MSCs seeded on 3DGp constructs showed spindle-shaped and elongated morphology with thin, aligned nuclei, typical of osteoprogenitor cells [32]. The scaffolds also induced high genomic and protein expressions of osteogenic markers, such as runt-related transcription factor 2 (RUNX2) and osteocalcin [23, 32]. The first is a transcription factor important for osteogenesis and skeletal morphogenesis while the second is expressed in mature osteoblasts. In fact, PDLSCs culture in graphene scaffolds for 28 days with osteogenic medium presented a ~4-fold increase in osteocalcin gene expression compared to those cultured in a polystyrene scaffold designed for three-dimensional cell culture [23].

Graphene derivatives like GO and rGO have been of particular interest to promote osteogenic differentiation of stem cells [31, 33–35]. For instance, GO-coated tissue culture plates promoted the growth and maintained metabolic activity and viability of BMMSCs. In addition, cells grown on GO-coated tissue culture plates in complete medium for 21 days demonstrated a five-fold higher mineralization compared to cells grown on GO-coated tissue culture plates in differentiation medium [34]. BMMSCs cultured on GO-coated PDMS substrates showed a ~2.5-fold increase in mineralized deposition compared to cells cultured on uncoated PDMS substrates in osteogenic medium [31]. Likewise, MSCs cultured on rGO-coated PDMS substrates demonstrated a 7.4-fold increase in alizarin red content compared to cells grown on PDMS substrates in osteogenic medium [33]. Graphene nanogrids, namely graphene oxide nanoribbon (GONR) and reduced graphene oxide nanoribbon (rGONR), were used as two-dimensional patterns to investigate the osteogenic differentiation of MSCs. MSCs cultured on GONR and rGONR-based grids in osteogenic medium exhibited 6.4-and 16.3-fold higher alizarin red contents than obtained for PDMS substrates with osteogenic medium after 7 days [33]. Interestingly, GONR and rGONR showed patterned osteogenic differentiation significantly higher than GO and rGO sheets alone in absence of osteogenic inducers [33].

Although several studies show that graphene and its derivatives can promote and enhance osteogenic differentiation, the mechanisms involved remain largely unknown. Nonetheless, it is very likely that both chemical and physical characteristics of graphenes act as driving forces to enhance osteoblastic differentiation.

It is widely known that the physical characteristics of substrates (e.g., roughness, elastic modulus, and superficial patterns) affect stem cells differentiation toward different lineages. The unique elastic properties of graphene could provide cues to mechanosensitive pathways that trigger the osteogenic differentiation events [27, 36]. In fact, PDLSCs displayed a high expression of two myosin heavy chain genes when cultured on pristine graphene with basal or osteogenic medium [23]. In addition to its elastic properties, the presence of morphological features on surface of graphene (e.g., wrinkles and ripples) can also influence the osteogenic differentiation [19, 20]. For CVD-grown graphene, these features are created during the cooling stage of the deposition when the material undergoes a negative thermal expansion that is restrained by the shrinkage of the metal substrate. The projection of graphene structures out of the bidimensional plane can assist cell anchorage and increase the cytoskeletal tension, which may contribute to the higher osteoblastic differentiation observed [19, 23].

Besides the physical features, the chemical characteristics of graphene are certainly also responsible for the enhanced osteogenic differentiation. Due to the presence of reactive groups on the surface of the graphenes, the materials can adsorb proteins and biomolecules that may promote differentiation [16]. Graphene was found to be an effective platform for adsorptions of dexamethasone, β-glycerophosphate and ascorbic acid that are typical osteogenic inducers [31, 33, 37]. Dexamethasone modulates the phosphorylation of RUNX2, a transcription factor vital for osteogenesis, while ß-glycerophosphate serves as a source of phosphate to promote bone matrix mineralization [38, 39]. Graphene's capability to adsorb osteogenic inducers through π–π stacking interactions increases the local concentration of osteogenic inducers from culture medium, enhancing the differentiation toward osteogenic lineage and mineralization [31, 33]. Due to differences in its chemical structure, graphene adsorbs higher amounts of dexamethasone, and ß-glycerophosphate than GO [31]. In contrast, it adsorbs less ascorbic acid than GO because of a lack of –OH moieties which form hydrogen bonds with the acid. However, this did not abate osteogenic differentiation on graphene as ascorbic acid affects mainly mature osteoblasts [40].

Collectively, these data demonstrate that the surface properties of graphenes [19, 33] and their ability to adsorb proteins [31, 33] can govern the behaviour of stem cells, indicating the potential of graphene platforms for bone tissue engineering and regeneration.

3 Graphene and Derivatives: Composites for Bone Tissue Engineering

Composite materials are combination of at least two materials that differ in composition or morphology. The result is a material with improved chemical, physical, and/or mechanical properties in comparison to their single constituents [41].

Graphenes have emerged as promising materials for bone tissue engineering due to their unique physical, chemical, and mechanical properties [20]. Pristine graphene has been repeatedly shown to allow cell attachment, proliferation, and enhance osteogenic differentiation [19, 23, 33]. Nonetheless, the majority of research to improve physicochemical properties and bioactivity of biomaterials focus on GO and rGO. The first is a highly hydrophilic entity that presents several carboxyl, epoxy, and hydroxyl groups present on its surface [42]. These functional groups promote interfacial interactions between GO and other moieties [28] offering a large-scale and cost-effective way to prepare graphene-based composites [43, 44]. As such, it has been used to sustain the release of proteins, to improve mechanical properties and enhance bioactivity of several biomaterials. Alternatively, rGO can be produced by the reduction of GO. Although this process results in a material that resembles pristine graphene, oxygen groups, and defects in different proportions are found on its surface [28]. Considering the structural differences of graphenes, they offer interesting opportunities to be combined with biomaterials for bone tissue engineering.

(a) Graphene–Polymer Composites

Synthetic and natural polymers are attractive alternatives for bone tissue engineering due to the possibility to process them into several sizes or shapes [45, 46]. Nonetheless, in order to achieve the aforementioned benefit, the mechanical behavior and degradation rate of the polymers are often compromised [47].

The modifications of polymers with graphene-based materials often result in composites with improved characteristics. For instance, both GO and rGO have been shown to improve physicomechanical properties of polymers [48, 49]. Moreover, the adsorption of proteins promoted by the surface polarity of GO creates conditions that enhance the bioactivity of polymer-based materials [50, 51]. Also, the increased surface roughness of polymeric scaffolds promoted by the addition of rGO improves cells spreading and osteogenic differentiation [52].

In fact, graphenes present characteristics that can improve several aspects of polymeric materials. Chitosan, for example, is a natural polymer extracted from microorganisms, crustaceans and the cell membrane of fungi. It is biocompatible, biodegradable and it is a potential candidate to repair osseous and chondral defects. In spite of these qualities, its poor mechanical properties and low bioactivity limit its use for bone tissue engineering [48, 49].

Several improvements have been reported with the use of graphene to tune biomaterials. The addition of 3 wt% GO to chitosan scaffolds was able to increase the modulus of elasticity from 2.65 to 6.74 GPa and hardness from 0.31 to 1.12 GPa, respectively [49]. The GO-modified chitosan scaffold can also present a higher adsorption of serum albumin protein compared to chitosan alone. This is a result of GO's negative charge and polarity, allowing electrostatic and Van der Waals force interactions with the protein's functional groups [50]. These characteristics also improved the attachment and proliferation of preosteoblasts MC3T3-E1 cells in GO-modified chitosan [49, 50]. Carboxymethyl-chitosan, a derivative of chitosan with higher water solubility, also experienced enhancements in modulus of elasticity and hardness when combined to 0.2w/v% of GO. Apart from these enhancements, the modification with GO led to remarkable improvements in the osteogenic potential of a chitosan-based substrate. After 7 days from seeding BMMSC, there was upregulation of osteogenesis-related genes, such as osteopontin, osteocalcin, and

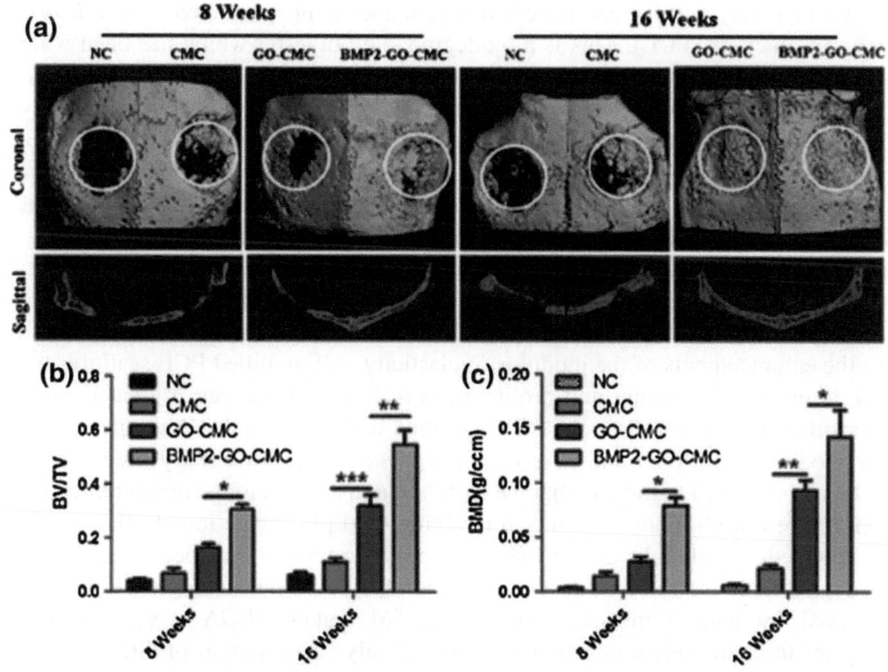

Fig. 1 Micro-CT imaging analysis of calvarial defect treated with chitosan (CMC), chitosan-GO composite (GO-CMC), or with the composite loaded with BMP-2 (BMP-2-GO-CMC). **a** Representative coronal and sagittal images of calvarial bone defects 8 and 16 weeks postimplantation; **b** after 16 weeks, both GO-modified scaffolds presented higher bone volume to tissue volume (**b**) and bone mineral density (**c**) as compared to CMC alone (*$p < 0.05$; **$p < 0.01$; ***$p < 0.001$). Adapted from Ruan J et al. © 2015 John Wiley and Sons

alkaline phosphatase (ALP). The osteoinduction/conduction of GO/chitosan composite was further confirmed in vivo using calvarial defects in rats. After 16 weeks from implantation of the scaffolds, the new bone volume to tissue volume ratio (BV/TV) was 28.6 % for the GO/chitosan composite and 11 % for the group treated with chitosan only (Fig. 1) [48].

The combination of rGO with chitosan influenced positively stem cells functions through nanotopographic cues provided by rGO. Interestingly, more MSCs attached on 5 % rGO/chitosan substratum compared to the unmodified chitosan. Nonetheless, the proliferation observed after 5 days in the chitosan modified with 5 % rGO was significantly lower than those with smaller amount of incorporated rGO (0.05 and 0.5 %) and unmodified chitosan which can be related to possible cytotoxicity induced by rGO or due to the cell differentiation process. The second reason finds support in the fact that the chitosan with 5 % rGO presented the highest calcium deposition and high osteocalcin protein expression [52].

Another polymer that can benefit from enhancements promoted by graphenes is polycaprolactone (PCL). This is a biodegradable polyester which has been widely studied for various biomedical applications such as sutures, subdermal contraceptive devices and wound dressings. Its use for orthopedic applications is however limited by its poor mechanical performances [53]. It has been combined with GO, rGO, and amine-functionalized GO (AGO) in order to assess the impact of each composite-specific chemical interactions, bioactivity, and mechanical properties. The addition of graphenes to PCL resulted in composites with increased modulus of elasticity. The original material experienced increases from 344 to 497 MPa with the addition of 5 wt% AGO and to 626 MPa with 5 % GO. Indeed, the GO particles tend to increase the crystallinity of the PCL which may be one of the key reasons for the enhancements of the modulus of elasticity. All modified PCL scaffolds were able to ensure attachment and proliferation of MSCs. However, a greater mineral deposition was observed on the PCL modified with GO and AGO compared to the one modified with rGO and pure PCL [54]. The higher bioactivity observed with GO may be related to the fact that the carboxyl and amine groups present in GO and AGO have a high-binding affinity for calcium and phosphate ions [55].

Polylactic-*co*-glycolic acid (PLGA) is a polymer with good biocompatibility and biodegradability. The fact that the products obtained from its degradation can be removed by natural metabolic pathways [56] makes PLGA a very attractive polymer for bone tissue engineering. Interestingly, the addition of GO to electrospun PLGA nanofibrous mats resulted in decreased mechanical properties, such as tensile strength, ultimate strain and Young's modulus. Nonetheless, GO allowed higher adsorption of dexamethasone compared to the unmodified version of the nanofibrous substrate. The higher adsorption may have contributed to the higher gene expression of ALP, osteocalcin and collagen I by MSCs in the presence of dexamethasone. Remarkably, after 28 days, the GO/PLGA scaffolds displayed significantly higher amount of osteocalcin (ng/ng of DNA) as compared to the control, regardless of the presence of dexamethasone in the culture medium [57].

In a similar perspective, poly-dopamine (PDA) is a polymer that contains catechol and amine functional groups. GO can be simultaneously reduced and modified by the dopamine present in the polymer, allowing a better dispersion and interfacial bonding of the flakes [58]. It has been shown that rGO/PDA compounds display a high propensity to induce nucleation of hydroxyapatite in simulated body fluid compared to rGO alone. This effect is mainly due to the high amount of catecholamine moieties promoting GO-PDA bonds and acting as a CA^{2+} binder [59]. rGO/PDA substrates promote higher adhesion and proliferation of osteoblastic MC3T3-E1 cells compared to glass and GO substrates. Cells seeded on rGO/PDA spread to a projected area of approximately 30×10^3 μm^2 while cells on GO and glass reached 14.5 and 8.6×10^3 μm^2, respectively. Additionally, the rGO/PDA substrate promoted higher ALP activity indicating an enhanced osteogenic differentiation compared to GO [60].

Collectively, GO and rGO offer functionalities (e.g., oxygen-containing groups) that permit to combine them with a wide range of polymeric materials. The composites obtained usually display enhanced physicomechanical properties and an improved osteogenic potential widening the possible applications of graphenes with polymers in bone tissue engineering.

(b) Graphene–Inorganic Materials Composites

Although the use of polymers-based scaffolds in bone tissue engineering has shown some interesting successes in terms of tissue ingrowth and differentiation, the majority of polymeric scaffolds lack sufficient mechanical strength to be used in load bearing sites [61]. Moreover the degradation of some biopolymers induces an autocatalytic ester breakdown which lowers the pH in the microenvironment posing difficulties to cell survival and differentiation [62].

Alternatively to polymers, inorganic materials (or bioceramics) demonstrate a great potential in bone tissue engineering. Bioceramics (e.g., hydroxyapatite, calcium phosphate, and others) usually present high tissue compatibility and osteoconductivity [63, 64]. Nonetheless, bioceramics often have a low fracture toughness and can be very brittle. The composites arising from the combination of bioceramics with graphenes have the potential to overcome some of these challenges. In fact, some of the composites obtained present considerable enhancements in the mechanical properties and improved the adhesion, proliferation and osteodifferentiation of cells in vitro and in vivo.

Hydroxyapatite (HAp) is a bioceramic found in natural bone within the collagenous matrix. HAp particles have a great osteoinduction potential and induce uncommitted cells to differentiate into the osteolineage [63, 65]. Nonetheless, HAp has low fracture toughness, is very brittle and difficult to shape. The addition of carbon nanotubes or alumina oxide can reinforce the material but can potentially impair its biological properties, induce adverse events in surrounding tissues or lead to the decomposition of HAp [66, 67].

HAp has been combined with graphenes for scaffolds and coatings with enhanced capabilities. For instance, the addition of GO to HAp coating can increase the coating adhesion strength to titanium in a concentration-dependent manner. While the pure HAp presented adhesion strength of 1.5 MPa, the addition of 2 wt% GO increased the adhesion to 2.7 while 5 wt% GO resulted in 3.3 MPa. In addition, the GO/HAp composite coatings exhibited higher corrosion resistance in comparison with pure HAp [68]. Although these are interesting findings, the most remarkable effect arising from this combination is the improvement of the mechanical properties. A nanocomposite of HAp containing 1 wt% rGO showed an elastic modulus 47 % higher than the unmodified HAp. Moreover, the nanocomposite had a fracture toughness 203 % greater than the control [69].

Besides the improvements in the mechanical properties, the combination of HAp and rGO can lead to an enhanced bioactivity of the resultant composites. When colloidal dispersion of rGO-coated HAp was used to culture MSCs for 21 days, there were increases in the ALP activity and amount of calcium deposits compared to solutions of HAp and rGO alone after 21 days. In addition, the rGO-coated HAp composites increased significantly the expression of proteins such as osteopontin and osteocalcin. This confirms that the synergistic effect provided by rGO/HAp is able to promote osteogenic differentiation in the absence of osteogenic inducers (Fig. 2) [70]. The enhanced bioactivity of HAp/rGO- composites was further confirmed in vivo by treating critical-sized calvarial defects in rabbits. After 4 weeks, the new bone density was substantially greater for the rGO/HAp graft (52.9 %) compared to HAp alone (26.8 %) and untreated control (17.7 %) (Fig. 3) [71].

Similar trends were observed with the combination of HAp and 1 wt% of graphene nanosheets (GNS) which increased the hardness of HAp from 5.5 to 7.2 GPa and fracture toughness from 0.58 to 1.06 MPa $m^{1/2}$ [72]. Furthermore, the osteoblastic cells formed a confluent layer on the surface of the GNS/HAp composites compared to the separate islands formed on the pure HAp surface. When soaked in simulated body fluid for 7 days, the 1 wt% GNS/HAp composite displayed an apatite layer more uniform and thicker than the one observed on HAp alone. Not only the amount of apatite changed with the presence of GNS, but also the localization of the crystals grown. While on pure HAp the mineralization took place near and/or inside the pores, it permeated the whole surface of the GNS/HAp composites [72].

Some hydrophilic components used for tissue engineering research contain NH_2 or OH groups (e.g., polyethylene glycol, chitosan, hyaluronic acid, and gelatin)

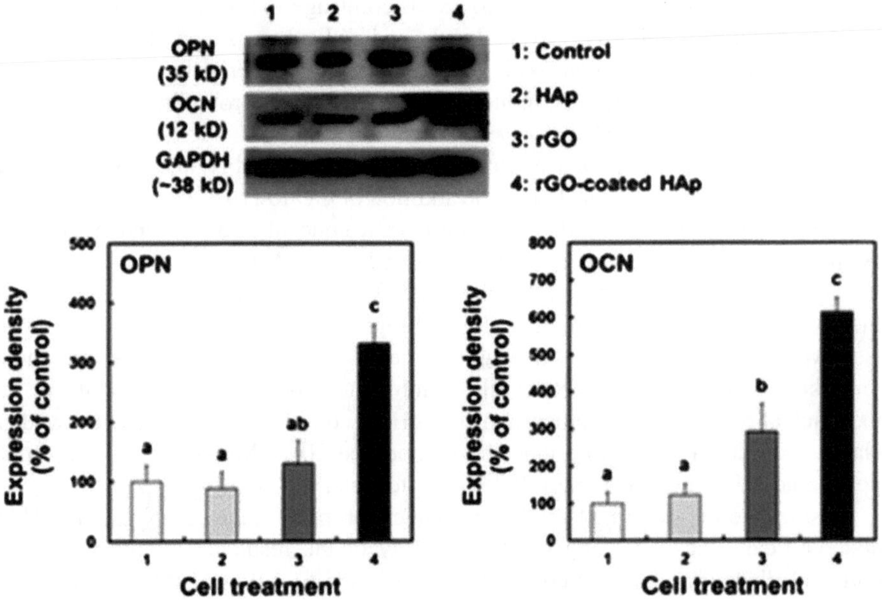

Fig. 2 Protein expressions of osteopontin and osteocalcin in MSCs incubated with a colloidal dispersion of HAp microparticles, rGO, or rGO-coated HAp composites (*top*). The high expression density observed for the rGO-coated HAp composites after 21 days (*bottom*) suggest that GO and HAp act synergistically to increase osteoblastic differentiation. Adapted from Lee JH et al. © 2015, Royal Society of Chemistry

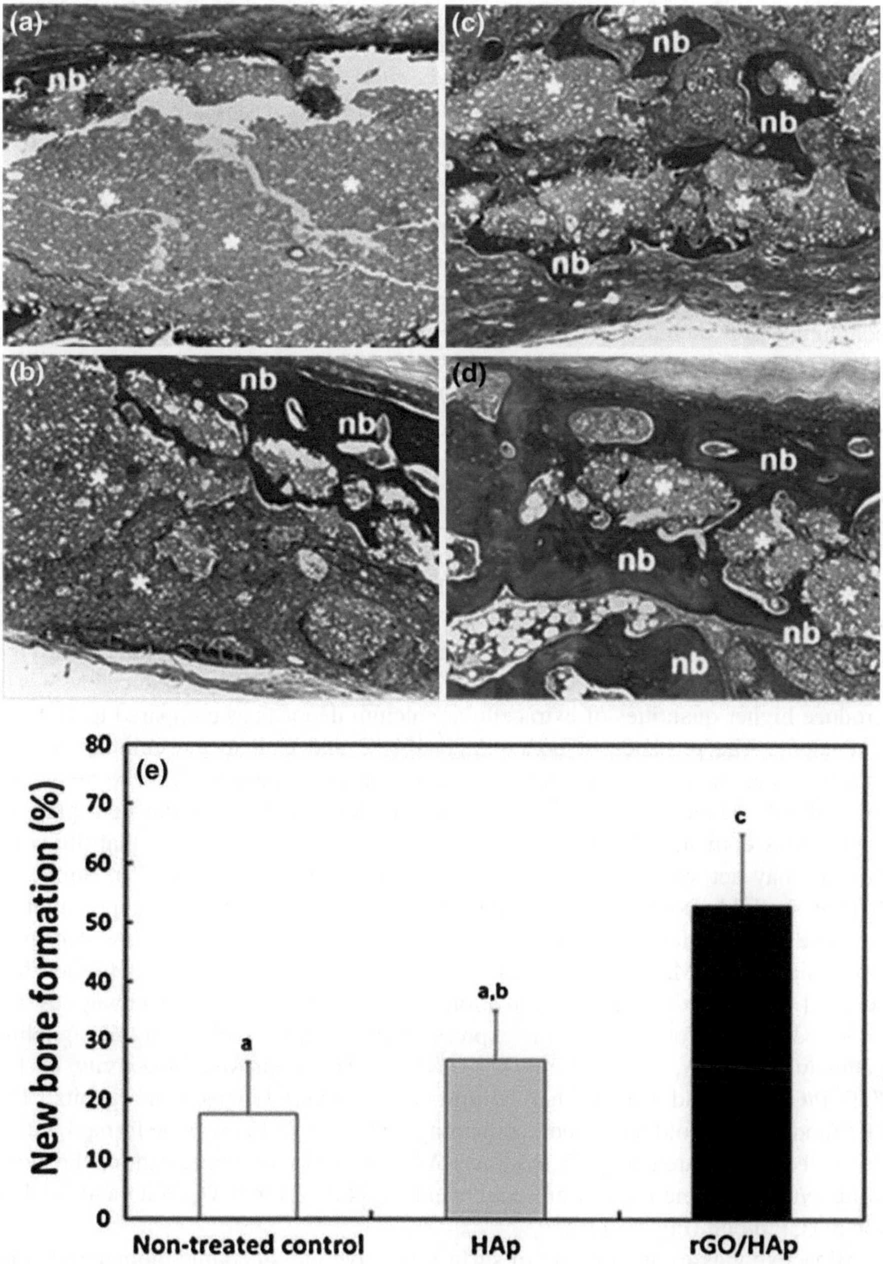

Fig. 3 Defects filled with HAp grafts (**a**, **b**) presented lower quantity of newly formed bone as compared to rGO/HAp grafts (**c**, **d**). The percentage of new bone formation (**e**) was also higher for rGO/HAp as compared to HAp alone and non-treated control. Lee JH et al. © 2015, Rights Managed by Nature Publishing Group

enabling good dispersions of GO. Hence, these materials can be combined with GO and HAp to further improve their bioactivity. Actually, chitosan/GO/HAp nanocomposites soaked in simulated body fluid were capable of releasing more Ca and P ions compared to pure HAp particles. The higher release of these ions by the nanocomposite may be related to the lower crystallinity of the materials provided by the presence of chitosan that increases the rate of dissolution of the nanocomposite [42]. Similar effects were observed for polyethylene glycol/GO/HAp nanocomposites which also released higher quantities of Ca and P ions compared to the pure HAp nanorods [73]. In addition to these materials, gelatin has also been combined with both GO and HAp (0.5 and 1 wt%). The composites containing GO increased significantly both the compressive and yield strengths compared to the gelatin combined with HAp only. Most importantly, after 24 h from seeding, MSCs on the gelatin/GO/HAp scaffold were able to adhere and spread, forming a cell sheet like appearance, while the cells seeded on gelation/HAp scaffolds presented round morphology. Lastly, the GO-modified scaffold induced higher ALP activity after 7 days compared to the gelatin/HAp. Although the presence of GO failed to induce higher ALP activity after 21 days, it induced a high-protein expression of osteopontin compared to the scaffolds without GO [74].

Calcium phosphate (CaP) and β-tri-calcium-phosphate (β-TCP) are bioceramics with good biocompatibility, osteoconductivity, and osteoinductivity potentials. Similarly to HAp, CaP can also be combined to graphene to improve its bioactivity. The combination of CaP and GO resulted in a nanocomposite that induced MSCs to produce higher quantities of extracellular calcium deposits as compared to GO and CaP alone. The presence of nonconjugated GO and CaP in the culture medium induced a calcification level similar to the one observed only for CaP. Nonetheless, the CaP/GO nanocomposite increased calcification, ALP and osteocalcin proteins expressions compared to GO and CaP alone. These results confirm that the compounds may act synergistically in order to enhance osteogenesis [75]. Similarly, β-TCP scaffolds combined with GO flakes displayed a remarkable potential to promote cell proliferation and increase ALP activity and osteogenic-related gene expression of BMMSCs. The enhanced differentiation obtained in the GO-modified scaffold may be due to the activation of the canonical Wnt pathway as the GO-modified scaffold induced the expression of proteins involved in this signaling pathway (WNT3A, LPP5, AXIN2, CTNNB). The improved bioactivity of the β-TCP/GO scaffold was further confirmed in calvarial defects in rabbits. The GO-modified scaffold increased the quantity and quality of new bone formed within the defects compared to β-TCP alone. After 8 weeks of implantation, the bone volume/total volume ratios were equivalent to 44.8 % for β-TCP/GO and 30.4 % for β-TCP alone (Fig. 4) [76].

Bioactive glasses are a group of surface reactive glass-ceramic biomaterials that can induce osteogenic differentiation and stimulate bone [77]. As previously observed in other bioceramics, the combination of graphenes to bioglasses can improve mechanical properties. The addition of 0.5 wt% of GO to 58S bioglass scaffold increased the compressive strength from 23.6 to 48.7 MPa and fracture

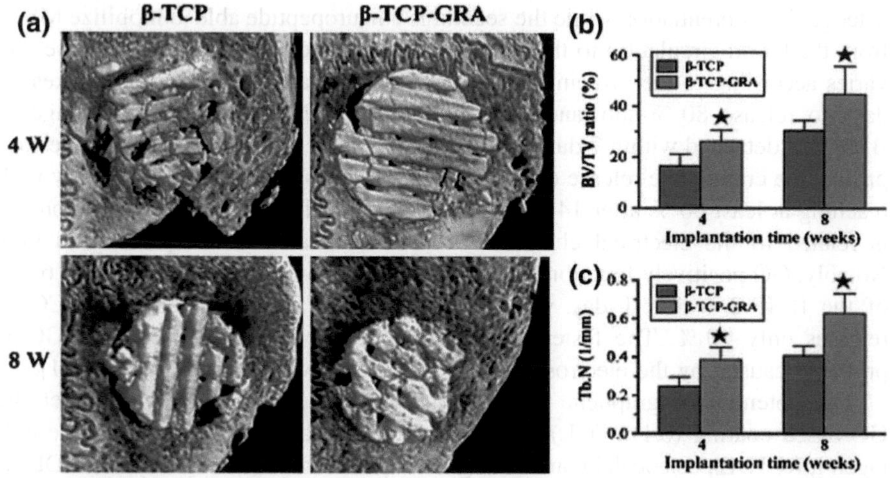

Fig. 4 Micro-CT analysis of in vivo bone formation ability for β-TCP and β-TCP-GRA scaffolds after being implanted in the cranial bone defects of rabbits for 4 and 8 weeks. (*$p < 0.05$). Adapted from Wu C et al. © 2015, Elsevier

toughness from 1.4 to 1.9 MPa $m^{1/2}$. The 58S/GO scaffold allowed the formation of a hydroxyl-carbonated apatite layer on the surface of 58S bioglass after 7 days of immersion in simulated body fluid. After one week from seeding, MG-63 cells were successfully attached to the scaffold and displayed round shape with mineralized nodules [78].

Numerous types of bioceramics are constantly developed and introduced in clinics to promote bone tissue regenerations. The addition of graphenes to bioceramics often results in composites with improved mechanical properties. In some cases, the presence of graphenes boosts the bioactivity of bioceramics even further, making these composites even more interesting for bone tissue regeneration.

4 Graphene as Implant Coating Material to Enhance Osteogenic Differentiation

The development of coatings and surface modifications of implant materials aims at increasing cell attachment, accelerating differentiation and promoting the osseointegration [79]. In this perspective, graphenes may be considered as good alternatives since they can be deposited onto surfaces with different geometries and have the ability to bind to biomolecules. This last characteristic makes graphene a potential candidate for the adsorption and release of drugs from implant surfaces at the surgical site. GO-based coatings have been shown to efficiently attach osteogenic inducers, such as BMP-2 and substance P. The first is a classical inducer of osteogenic differentiation while the second is a neuropeptide able to mobilize MSCs from the blood circulation to the implant site. It was observed that the release rate varies according to the protein attached to GO's surface. For instance, it takes 10 days to release 80 % substance P bound to GO while the same percentage of BMP-2 is detached within 7 days [80]. Although there are differences in the release profile, the cumulative release of both BMP-2 and substance P in vitro is very high reaching at least 80 % after 14 days [80, 81]. The differences in the release profile is related to the electrical character of the chemical moieties present on GO. Notably, GO positively functionalized with –NH3+ (GO+) releases more than 65 % of the BMP-2 within 1 day while GO negatively functionalized with –COO– releases only 40 %. The faster release of BMP-2 by positively charged GO is probably caused by the electrostatic due to the positive nature of BMP-2 [81].

The potential of graphene on implants can be exemplified by the fact the GO-based coating (0.1 mg/mL) on sodium titanate can significantly increase collagen I, RUNX2, osteocalcin and ALP gene expression (14 and 21 days) in PDLSC compared to Na–Ti alone [82]. GO sheets showed enhanced interactions with proteins due to hydrophobic π domains present in its core region and ionized groups around the edges which may enhance the differentiation of PDLSCs [35, 82]. Likewise, titanium substrates coated with negatively charged GO sheets ($GOCOO^-$) were capable of adsorbing BMP-2 and promote its sustained release in vitro. This can be attributed to the electrostatic and hydrophilic interactions between $GOCOO^-$ and BMP2. The Ti/GO-implants conjugated with BMP-2 demonstrated extensive bone formation in mouse calvarial defects after 8 weeks compared to Ti/BMP-2 substrates [81]. The joint delivery of BMP-2 and substance P by GO lead to a greater bone regeneration in vivo compared to BMP-2 release only. Indeed, titanium coated with BMP-2 and substance P only did not show improvements in bone formation (Fig. 5) [80]. Similar to GO, a rGO-based coating loaded with dexamethasone on commercially pure titanium has accelerated the bone tissue regeneration compared to rGO alone. Further in vivo analysis showed that after 8 weeks from implantation, the regenerated bone area was 93.38 % for rGO loaded with dexamethasone and 68.72 % for rGO alone [83].

The functionalization of GO may also offer additional opportunities to improve osteogenesis. GO/titanate nanowire scaffolds modified with discrete functional terminal groups demonstrated improved osteogenesis and mineralization of MG-63 cells compared to unmodified titanate scaffolds [84]. GO functional groups, such as –COOH and –OH serve as an anchor for binding bioactive groups and protein on graphene surface, leading to modulation of the cell proliferation, differentiation and mineralization [31]. In particular, –OH groups provide favorable sites for calcium phosphate nucleation and increased cellular functions [84]. This can be attributed to selective binding of $\alpha_5\beta_1$ integrin peptide to –OH groups which support higher recruitment levels of proteins for the cytoskeleton and signaling pathways [84, 85].

Hence, graphenes present promising physicochemical characteristics that make them exciting materials to be used as coatings of biomedical implants [86, 87].

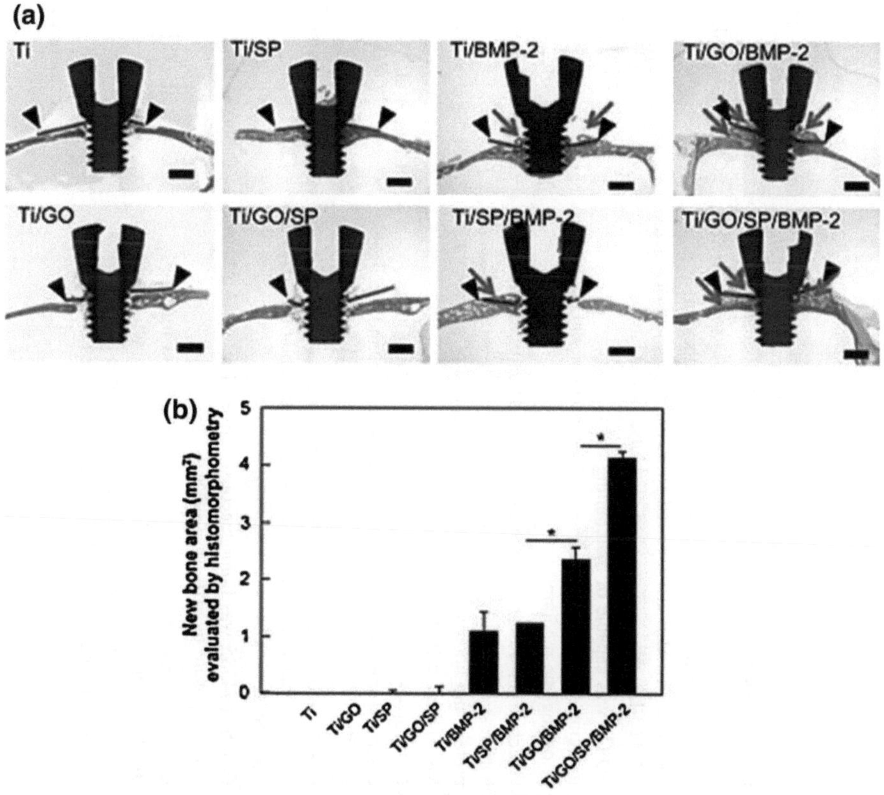

Fig. 5 **a** Histological analysis with Goldner's trichrome staining showing that titanium implants coated with GO and loaded with BMP-2 (Ti/GO/BMP-2) presented higher bone formation as compare with implants loaded with BMP-2 only (Ti/BMP-2). Similar effects were observed with the addition of substance P to the set (Ti/GO/SP/BMP-2 vs. TI/SP/BMP-2). **b** The higher bone are formed with the presence of GO was confirmed using histomorphometric analysis (*$p < 0.05$). Adapted from La WG et al. © 2014 La et al.

5 Conclusion

Graphene allows cell attachment, proliferation and can also induce and enhance differentiation toward osteoblasts. Moreover, its different forms can be further functionalized or combined with biomolecules and biomaterials. As a result, graphenes emerged as potential candidates to improve the physicomechanical properties and bioactivity of materials used for bone tissue engineering (Table 1).

Despite their promising capabilities, there are several challenges that need to be addressed before graphenes become a clinical reality. There is a need to determine crucial parameters related to biosafety, such as toxic levels, biodistribution, and excretion. Moreover, the cellular mechanisms underlying the enhanced osteogenesis promoted by graphenes remain largely unknown and will demand considerable

Table 1 Osteogenic potential of graphene-based materials

Graphene type	Graphene manufacturing technique	Main outcomes	References
GONR, rGONR	Unzipping of CNT	GONR and rGONR resulted in good adsorption of osteogenic inducers and differentiation of MSCs into osteogenic lineage	Akhavan et al. [33]
Reduced graphene oxide/polydopamine	Dopamine treatment	rGO/PDA induced a higher nucleation of hydroxyapatite in SBF compared to rGO alone. It promoted higher adhesion, proliferation, ALP activity, and spreading of osteoblastic MC3T3-E1 cells compared to glass and GO substrates	Cheng et al. [60]
3D Gp	CVD	3D Gp was capable of maintaining the viability of MSC and inducing spontaneous osteogenic differentiation without the use of osteogenic medium	Crowder et al. [32]
Graphene oxide/chitosan	–	GO/chitosan was able to increase the modulus of elasticity and hardness of the polymer. Higher adsorption of serum albumin protein and improved attachment/proliferation of preosteoblasts MC3T3-E1 cells were also observed	Depan et al. [49]
Graphene oxide	Hummer's method	GO/titanate scaffolds functionalized with –OH groups, exhibited improved cell viability, proliferation, and osteogenic activity	Dong et al. [84]
Graphene oxide	Hummer's technique	Graphene oxide is a biocompatible platform for BMMSCs inducing mineralized nodule formation at the end of 21 days for BMMSCs	Elkhenany et al. [34]
Graphene/45S5 bioglass	Graphene from from Cheap Tubes Inc.	Graphene/45S5 was homogeneously covered by a HAp layer upon exposure to SBF for 7 days	Fabbri et al. [88]
Graphene oxide/58S bioglass	GO from Nanjing JCNANO Tech Co.	GO/58S increased the compressive strength and fracture toughness of 58S. It allowed formation of an apatite layer after immersion in SBF and MG-63 cells were able to attach and display a round shape with mineralized nodules after 7 days	Gao et al. [78]

(continued)

Table 1 (continued)

Graphene type	Graphene manufacturing technique	Main outcomes	References
Graphene sheet	CVD	CVD grown graphene was shown to be cytocompatible with human osteoblastic cells and MSCs. The cells adhered and proliferated more on CVD-grown graphene than when grown on SiO_2 substrate	Kalbacova et al. [22]
Reduced graphene oxide/chitosan	Hummer's method—hydrazine treatment	rGO/chitosan enhanced the attachment of MSCs while their proliferation was lower. This would indicate a cell differentiation process which was supported by results showing a higher calcium deposition and OCN expression	Kim et al. [52]
Graphene oxide, reduced graphene oxide, amine graphene oxide/polycaprolactone	Hummer's method—hydrazine treatment	All modified PCL scaffolds increased the modulus of elasticity of the polymer and were able to ensure attachment and proliferation of MSCs. Greater mineral deposition was observed on the GO-PCL and AGO/PCL probably due to the high-binding affinity of carboxyl and amine groups for Ca and P ions	Kumar et al. [54]
Graphene oxide	Hummer's method	The delivery of BMP-2 and Substance-P using GO enhanced the bone formation on titanium implanted in mouse calvaria	La et al. [81]
Reduced graphene oxide/hydroxyapatite	Modified Hummers and Offeman method—Hydrazine treatment	In vivo data (rabbit model) showed that the new bone density was substantially greater for the rGO/HAp composite after 4 weeks	Lee et al. [71]
Reduced graphene oxide/hydroxyapatite	Modified Hummers and Offeman method—Hydrazine treatment	rGO/HAp induced an increased ALP activity and amount of calcium deposits in MSCs cultures. The expressions of OPN and OCN were also promoted	Lee et al. [70]
Graphene and graphene oxide	CVD and Hummer's technique	G and GO were shown to be effective pre-concentration platforms hence demonstrating accelerated growth and differentiation of stem cells along with facilitating stem cell differentiation toward other lineages	Lee et al. [31]

(continued)

Table 1 (continued)

Graphene type	Graphene manufacturing technique	Main outcomes	References
Graphene oxide/polylactic-*co*-glycolic acid	Modified Hummer's method	GO/PLGA resulted in decreased mechanical properties but allowed higher adsorption of osteogenic inducers compared to unmodified PLGA. MSCs showed an enhanced osteogenic genes expression with and without dexamethasone	Luo et al. [57]
Graphene oxide/hydroxyapatite/chitosan	Improved Hummer's method	Chitosan/GO/HAp was capable of releasing more Ca and P ions in SBF compared to HAp	Mohandes and Salavati-Niasari [42]
Graphene oxide/hydroxyapatite/polyethylene glycol	Modified Hummer's method	PEG/GO/HAp released higher quantities of Ca and P ions compared to pure HAp	Mohandes and Salavati-Niasari [73]
Graphene oxide/hydroxyapatite/gelatin	Modified Hummer's method	GO/HAp/gelatin increased the compressive and yield strengths significantly. MSCs seeded on the scaffold were able to adhere and spread while higher ALP activity was observed after 7 days as well as higher OPN expression after 21 days	Nair et al. [74]
Graphene sheet	CVD	Graphene coated on PDMS, glass, PET, and SiO_2 can be a biocompatible scaffold for MSCs and induce osteogenic lineage in MSCs	Nayak et al. [19]
Graphene oxide/carboxymethyl chitosan	Hummer's method	GO/carboxymethyl-chitosan enhanced the modulus of elasticity and hardness of the polymer. BMMSCs showed an upregulation of osteogenesis-related genes and improved bone regeneration was achieved in vivo	Ruan et al. [48]
Graphene oxide/calcium phosphate	GO from Nanocs Inc.	GO/CaP induced higher quantities of extracellular calcium deposits in MSCs compared to GO and CaP alone. ALP and OCN proteins expressions were also enhanced	Tatavarty et al. [75]
Graphene oxide	Hummer's method	GO is an efficient delivery carrier for BMP-2, a classical inducer for osteogenic differentiation both in vitro and in vivo	La et al. [80]

(continued)

Table 1 (continued)

Graphene type	Graphene manufacturing technique	Main outcomes	References
Graphene oxide/β-tri-calcium-phosphate	GO from SIMBATT Energy Technology Co.	GO/β-TCP promoted cell proliferation, increase of ALP activity, osteogenic gene expression, and activation of the Wnt pathway in BMMSCs. The GO-composite was also able to enhance the quantity and quality of new bone formed in a calvarial defect study.	Wu et al. [76]
2D and 3D Graphene	CVD	The high elastic modulus of graphene could be a potential factor for increased cellular differentiation and proliferation	Xie et al. [23]
Graphene/hydroxyapatite	GNS from ACS Material	GNS/HAp increased the hardness and fracture toughness of HAp. The composite enabled a uniform spreading of osteoblastic cells as well as a stimulated apatite deposition (quantity and quality)	Zhang et al. [72]
Graphene oxide	Hummer's method	GO-coated sodium-titanate substrate increased proteins and genes expression of osteogenesis-related markers compared to sodium titanate of periodontal ligament cells	Zhou et al. [82]

research efforts in the years to come. Finally, more in vivo testing comparing the outcomes promoted by graphenes with currently used biomaterials is necessary to benchmark the clinical performances of these carbonaceous materials.

Nonetheless, the promising findings obtained so far attest that graphenes may positively impact and move the field of bone tissue engineering forward.

Acknowledgments The authors were supported by National University Health System (NUHSRO/2014/017/B2B/02), National University of Singapore (R-221-000-091-112) and National Research Foundation CRP award "Novel 2D materials with tailored properties: beyond graphene" (R-144-000-295-281).

References

1. Henkel J, Woodruff MA, Epari DR, Steck R, Glatt V, Dickinson IC, Choong PF, Schuetz MA, Hutmacher DW (2013) Bone regeneration based on tissue engineering conceptions—a 21st century perspective. Bone Res 1(3):216
2. Oryan A, Alidadi S, Moshiri A (2013) Current concerns regarding healing of bone defects. Hard Tissue 2:13
3. Bao CLM, Teo EY, Chong M, Liu Y, Choolani M, Chan J (2013) Advances in bone tissue engineering, Regenerative Medicine and Tissue Engineering. In: Prof. Jose A. Andrades (ed) InTech, doi:10.5772/55916. Available at: http://www.intechopen.com/books/regenerative-medicine-and-tissue-engineering/advances-in-bone-tissue-engineering
4. Bose S, Roy M, Bandyopadhyay A (2012) Recent advances in bone tissue engineering scaffolds. Trends Biotechnol 30(10):546–554
5. Amini AR, Laurencin CT, Nukavarapu SP (2012) Bone tissue engineering: recent advances and challenges. Crit Rev Biomed Eng 40(5):363–408
6. Finkemeier CG (2002) Bone-grafting and bone-graft substitutes. J Bone Joint Surg Am 84 (3):454–464
7. Gazdag AR, Lane JM, Glaser D, Forster RA (1995) Alternatives to autogenous bone graft: efficacy and indications. J Am Acad Orthop Surg 3(1):1–8
8. Kolk A, Handschel J, Drescher W, Rothamel D, Kloss F, Blessmann M, Heiland M, Wolff K-D, Smeets R (2012) Current trends and future perspectives of bone substitute materials–from space holders to innovative biomaterials. J Cranio Maxillofac Surg 40(8):706–718
9. Stevens MM (2008) Biomaterials for bone tissue engineering. Mater Today 11(5):18–25
10. Lai G-J, Shalumon K, Chen S-H, Chen J-P (2014) Composite chitosan/silk fibroin nanofibers for modulation of osteogenic differentiation and proliferation of human mesenchymal stem cells. Carbohydr sPolym 111:288–297
11. Croisier F, Jérôme C (2013) Chitosan-based biomaterials for tissue engineering. Eur Polym J 49(4):780–792
12. Porter JR, Ruckh TT, Popat KC (2009) Bone tissue engineering: a review in bone biomimetics and drug delivery strategies. Biotechnol Prog 25(6):1539–1560
13. Liu X, Ma PX (2004) Polymeric scaffolds for bone tissue engineering. Ann Biomed Eng 32 (3):477–486
14. Yunos DM, Bretcanu O, Boccaccini AR (2008) Polymer-bioceramic composites for tissue engineering scaffolds. J Mater Sci 43(13):4433–4442
15. Sahoo NG, Pan YZ, Li L, He CB (2013) Nanocomposites for bone tissue regeneration. Nanomedicine 8(4):639–653
16. Goenka S, Sant V, Sant S (2014) Graphene-based nanomaterials for drug delivery and tissue engineering. J Controlled Release 173:75–88

17. Song Y, Wei W, Qu X (2011) Colorimetric biosensing using smart materials. Adv Mater 23 (37):4215–4236
18. Loh KP, Bao Q, Ang PK, Yang J (2010) The chemistry of graphene. J Mater Chem 20 (12):2277–2289
19. Nayak TR, Andersen H, Makam VS, Khaw C, Bae S, Xu X, Ee P-LR, Ahn J-H, Hong BH, Pastorin G (2011) Graphene for controlled and accelerated osteogenic differentiation of human mesenchymal stem cells. ACS Nano 5(6):4670–4678
20. Dubey N, Bentini R, Islam I, Cao T, Neto AHC, Rosa V (2015) Graphene: a versatile 531 carbon-based material for bone tissue engineering. Stem Cells Int 2015:804213
21. Tang Z, Wu H, Cort JR, Buchko GW, Zhang Y, Shao Y, Aksay IA, Liu J, Lin Y (2010) Constraint of DNA on functionalized graphene improves its biostability and specificity. Small 6(11):1205–1209
22. Kalbacova M, Broz A, Kong J, Kalbac M (2010) Graphene substrates promote adherence of human osteoblasts and mesenchymal stromal cells. Carbon 48(15):4323–4329
23. Xie H, Cao T, Gomes JV, Neto AHC, Rosa V (2015) Two and three-dimensional graphene substrates to magnify osteogenic differentiation of periodontal ligament stem cells. Carbon 93:266–275
24. Geim AK, Novoselov KS (2007) The rise of graphene. Nat Mater 6(3):183–191
25. Neto AC, Guinea F, Peres N, Novoselov KS, Geim AK (2009) The electronic properties of graphene. Rev Mod Phys 81(1):109
26. Bunch JS, Verbridge SS, Alden JS, Van Der Zande AM, Parpia JM, Craighead HG, McEuen PL (2008) Impermeable atomic membranes from graphene sheets. Nano Lett 8 (8):2458–2462
27. Lee C, Wei X, Kysar JW, Hone J (2008) Measurement of the elastic properties and intrinsic strength of monolayer graphene. Science 321(5887):385–388
28. Dreyer DR, Park S, Bielawski CW, Ruoff RS (2010) The chemistry of graphene oxide. Chem Soc Rev 39(1):228–240
29. Liu J, Cui L, Losic D (2013) Graphene and graphene oxide as new nanocarriers for drug delivery applications. Acta Biomater 9(12):9243–9257
30. Matsuoka F, Takeuchi I, Agata H, Kagami H, Shiono H, Kiyota Y, Honda H, Kato R (2013) Morphology-based prediction of osteogenic differentiation potential of human mesenchymal stem cells. PLoS ONE 8(2):e55082
31. Lee WC, Lim CHY, Shi H, Tang LA, Wang Y, Lim CT, Loh KP (2011) Origin of enhanced stem cell growth and differentiation on graphene and graphene oxide. ACS Nano 5(9):7334–7341
32. Crowder SW, Prasai D, Rath R, Balikov DA, Bae H, Bolotin KI, Sung H-J (2013) Three-dimensional graphene foams promote osteogenic differentiation of human mesenchymal stem cells. Nanoscale 5(10):4171–4176
33. Akhavana OG, Shahsavara M (2013) Graphene nanogrids for selective and fast osteogenic differentiation of human mesenchymal stem cells. Carbon 59:200–211. doi:10.1016/j.carbon.2013.03.010
34. Elkhenany H, Amelse L, Lafont A, Bourdo S, Caldwell M, Neilsen N, Dervishi E, Derek O, Biris AS, Anderson D (2015) Graphene supports in vitro proliferation and osteogenic differentiation of goat adult mesenchymal stem cells: potential for bone tissue engineering. J Appl Toxicol 35(4):367–374
35. Shi X, Chang H, Chen S, Lai C, Khademhosseini A, Wu H (2012) Regulating cellular behavior on few-layer reduced graphene oxide films with well-controlled reduction states. Adv Funct Mater 22(4):751–759
36. Engler AJ, Sen S, Sweeney HL, Discher DE (2006) Matrix elasticity directs stem cell lineage specification. Cell 126(4):677–689
37. Chen D, Feng H, Li J (2012) Graphene oxide: preparation, functionalization, and electrochemical applications. Chem Rev 112(11):6027–6053
38. Langenbach F, Handschel J (2013) Effects of dexamethasone, ascorbic acid and β-glycerophosphate on the osteogenic differentiation of stem cells in vitro. Stem Cell Res Ther 4(5):117

39. Phillips JE, Gersbach CA, Wojtowicz AM, García AJ (2006) Glucocorticoid-induced osteogenesis is negatively regulated by Runx2/Cbfa1 serine phosphorylation. J Cell Sci 119 (3):581–591. doi:10.1242/jcs.02758
40. Quarles LD, Yohay DA, Lever LW, Caton R, Wenstrup RJ (1992) Distinct proliferative and differentiated stages of murine MC3T3-E1 cells in culture: an in vitro model of osteoblast development. J Bone Miner Res 7(6):683–692
41. Nicolais L, Gloria A, Ambrosio L (2010) 17—The mechanics of biocomposites. In: Biomedical composites. Woodhead Publishing, pp 411–440. doi:10.1533/9781845697372.3.411
42. Mohandes F, Salavati-Niasari M (2014) Freeze-drying synthesis, characterization and in vitro bioactivity of chitosan/graphene oxide/hydroxyapatite nanocomposite. RSC Adv 4 (49):25993. doi:10.1039/c4ra03534h
43. Compton OC, Nguyen ST (2010) Graphene oxide, highly reduced graphene oxide, and graphene: versatile building blocks for carbon-based materials. Small 6(6):711–723
44. Zhu Y, Murali S, Cai W, Li X, Suk JW, Potts JR, Ruoff RS (2010) Graphene and graphene oxide: synthesis, properties, and applications. Adv Mater 22(35):3906–3924
45. Hayashi T (1994) Biodegradable polymers for biomedical uses. Prog Polym Sci 19(4):663–702
46. Rosa V, Zhang Z, Grande R, Nör J (2013) Dental pulp tissue engineering in full-length human root canals. J Dent Res. doi:10.1177/0022034513505772
47. Dhandayuthapani B, Yoshida Y, Maekawa T, Kumar DS (2011) Polymeric scaffolds in tissue engineering application: a review. Int J Polym Sci 290602
48. Ruan J, Wang X, Yu Z, Wang Z, Xie Q, Zhang D, Huang Y, Zhou H, Bi X, Xiao C, Gu P, Fan X (2015) Enhanced physiochemical and mechanical performance of chitosan-grafted graphene oxide for superior osteoinductivity. Adv Funct Mater. doi:10.1002/adfm.201504141
49. Depan D, Girase B, Shah JS, Misra RD (2011) Structure-process-property relationship of the polar graphene oxide-mediated cellular response and stimulated growth of osteoblasts on hybrid chitosan network structure nanocomposite scaffolds. Acta Biomater 7(9):3432–3445. doi:10.1016/j.actbio.2011.05.019
50. Depan D, Misra RD (2013) The interplay between nanostructured carbon-grafted chitosan scaffolds and protein adsorption on the cellular response of osteoblasts: structure-function property relationship. Acta Biomater 9(4):6084–6094. doi:10.1016/j.actbio.2012.12.019
51. Depan D, Pesacreta TC, Misra RDK (2014) The synergistic effect of a hybrid graphene oxide–chitosan system and biomimetic mineralization on osteoblast functions. Biomater Sci 2 (2):264–274
52. Kim J, Kim Y-R, Kim Y, Lim KT, Seonwoo H, Park S, Cho S-P, Hong BH, Choung P-H, Chung TD, Choung Y-H, Chung JH (2013) Graphene-incorporated chitosan substrata for adhesion and differentiation of human mesenchymal stem cells. Mater Chem B 1(7):933. doi:10.1039/c2tb00274d
53. Woodruff MA, Hutmacher DW (2010) The return of a forgotten polymer—polycaprolactone in the 21st century. Prog Polym Sci 35(10):1217–1256
54. Kumar S, Raj S, Kolanthai E, Sood AK, Sampath S, Chatterjee K (2015) Chemical functionalization of graphene to augment stem cell osteogenesis and inhibit biofilm formation on polymer composites for orthopedic applications. ACS Appl Mater Interfaces 7(5):3237–3252. doi:10.1021/am5079732
55. Landis WJ, Jacquet R (2013) Association of calcium and phosphate ions with collagen in the mineralization of vertebrate tissues. Calcif Tissue Int 93(4):329–337
56. Gloria A, De Santis R, Ambrosio L (2010) Polymer-based composite scaffolds for tissue engineering. J Appl Biomater Biomech 8(2):57–67
57. Luo Y, Shen H, Fang Y, Cao Y, Huang J, Zhang M, Dai J, Shi X, Zhang Z (2015) Enhanced proliferation and osteogenic differentiation of mesenchymal stem cells on graphene oxide-incorporated electrospun poly(lactic-*co*-glycolic acid) nanofibrous mats. ACS Appl Mater Interfaces 7(11):6331–6339. doi:10.1021/acsami.5b00862

58. Xu LQ, Yang WJ, Neoh KG, Kang ET, Fu GD (2010) Dopamine-induced reduction and functionalization of graphene oxide nanosheets. Macromolecules 43(20):8336–8339
59. Nassif N, Martineau F, Syzgantseva O, Gobeaux F, Willinger M, Coradin T, Cassaignon S, Azaïs T, Giraud-Guille MM (2010) In vivo inspired conditions to synthesize biomimetic hydroxyapatite. Chem Mater 22(12):3653–3663
60. Cheng J, Liu H, Zhao B, Shen R, Liu D, Hong J, Wei H, Xi P, Chen F, Bai D (2015) MC3T3-E1 preosteoblast cell-mediated mineralization of hydroxyapatite by poly-dopamine-functionalized graphene oxide. J Bioactive Compat Polym 30(3):289–301. doi:10.1177/0883911515569918
61. Oh S, Oh N, Appleford M, Ong JL (2006) Bioceramics for tissue engineering applications—a review. Am J Biochem Biotechnol 2(2):49–56
62. Agrawal C, Ray RB (2001) Biodegradable polymeric scaffolds for musculoskeletal tissue engineering. J Biomed Mater Res 55(2):141–150
63. Tonetto A, Lago PW, Borba M, Rosa V (2015) Effects of chrondro-osseous regenerative compound associated with local treatments in the regeneration of bone defects around implants: an in vivo study. Clin Oral Investig 20(2):267–274
64. Zyman Z, Ivanov I, Glushko V, Dedukh N, Malyshkina S (1998) Inorganic phase composition of remineralisation in porous CaP ceramics. Biomaterials 19(14):1269–1273
65. Lin L, Chow KL, Leng Y (2009) Study of hydroxyapatite osteoinductivity with an osteogenic differentiation of mesenchymal stem cells. J Biomed Mater Res, Part A 89(2):326–335
66. Choi JW, Kong YM, Kim HE, Lee IS (1998) Reinforcement of hydroxyapatite bioceramic by addition of Ni3Al and Al2O3. J Am Ceram Soc 81(7):1743–1748
67. Lahiri D, Ghosh S, Agarwal A (2012) Carbon nanotube reinforced hydroxyapatite composite for orthopedic application: a review. Mater Sci Eng, C 32(7):1727–1758
68. Li M, Liu Q, Jia Z, Xu X, Cheng Y, Zheng Y, Xi T, Wei S (2014) Graphene oxide/hydroxyapatite composite coatings fabricated by electrophoretic nanotechnology for biological applications. Carbon 67:185–197
69. Yi L, Jing H, Hua L (2013) Synthesis of hydroxyapatite–reduced graphite oxide nanocomposites for biomedical applications: oriented nucleation and epitaxial growth of hydroxyapatite. J Mater Chem B 1:1826–1834
70. Lee JH, Shin YC, Jin OS, Kang SH, Hwang YS, Park JC, Hong SW, Han DW (2015) Reduced graphene oxide-coated hydroxyapatite composites stimulate spontaneous osteogenic differentiation of human mesenchymal stem cells. Nanoscale 7(27):11642–11651. doi:10.1039/c5nr01580d
71. Lee JH, Shin YC, Lee SM, Jin OS, Kang SH, Hong SW, Jeong CM, Huh JB, Han DW (2015) Enhanced osteogenesis by reduced graphene oxide/hydroxyapatite nanocomposites. Sci Rep 5:18833. doi:10.1038/srep18833
72. Zhang L, Liu W, Yue C, Zhang T, Li P, Xing Z, Chen Y (2013) A tough graphene nanosheet/hydroxyapatite composite with improved in vitro biocompatibility. Carbon 61:105–115
73. Mohandes F, Salavati-Niasari M (2014) In vitro comparative study of pure hydroxyapatite nanorods and novel polyethylene glycol/graphene oxide/hydroxyapatite nanocomposite. J Nanopart Res 16(9):1–12
74. Nair M, Nancy D, Krishnan AG, Anjusree GS, Vadukumpully S, Nair SV (2015) Graphene oxide nanoflakes incorporated gelatin-hydroxyapatite scaffolds enhance osteogenic differentiation of human mesenchymal stem cells. Nanotechnology 26(16):161001. doi:10.1088/0957-4484/26/16/161001
75. Tatavarty R, Ding H, Lu G, Taylor RJ, Bi X (2014) Synergistic acceleration in the osteogenesis of human mesenchymal stem cells by graphene oxide-calcium phosphate nanocomposites. Chem Commun (Camb) 50(62):8484–8487. doi:10.1039/c4cc02442g
76. Wu C, Xia L, Han P, Xu M, Fang B, Wang J, Chang J, Xiao Y (2015) Graphene-oxide-modified β-tricalcium phosphate bioceramics stimulate in vitro and in vivo osteogenesis. Carbon 93:116–129. doi:10.1016/j.carbon.2015.04.048

77. Rahaman MN, Day DE, Bal BS, Fu Q, Jung SB, Bonewald LF, Tomsia AP (2011) Bioactive glass in tissue engineering. Acta Biomater 7(6):2355–2373
78. Gao C, Liu T, Shuai C, Peng S (2014) Enhancement mechanisms of graphene in nano-58S bioactive glass scaffold: mechanical and biological performance. Sci Rep 4:4712. doi:10.1038/srep04712
79. Liu X, Chu PK, Ding C (2004) Surface modification of titanium, titanium alloys, and related materials for biomedical applications. Mater Sci Eng R Rep 47(3):49–121
80. La WG, Jin M, Park S, Yoon HH, Jeong GJ, Bhang SH, Park H, Char K, Kim BS (2014) Delivery of bone morphogenetic protein-2 and substance P using graphene oxide for bone regeneration. Int J Nanomed 9(Suppl 1):107–116. doi:10.2147/IJN.S50742
81. La WG, Park S, Yoon HH, Jeong GJ, Lee TJ, Bhang SH, Han JY, Char K, Kim BS (2013) Delivery of a therapeutic protein for bone regeneration from a substrate coated with graphene oxide. Small 9(23):4051–4060
82. Zhou Q, Yang P, Li X, Liu H, Ge S (2016) Bioactivity of periodontal ligament stem cells on sodium titanate coated with graphene oxide. Scientific Reports 6:19343
83. Jung HS, Y-j Choi, Jeong J, Lee Y, Hwang B, Jang J, Shim J-H, Kim YS, Choi HS, Oh SH, Lee CS, Cho D-W, Hahn SK (2016) Nanoscale graphene coating on commercially pure titanium for accelerated bone regeneration. RSC Adv 6(32):26719–26724. doi:10.1039/C6RA03905G
84. Dong W, Hou L, Li T, Gong Z, Huang H, Wang G, Chen X, Li X (2015) A dual role of graphene oxide sheet deposition on titanate nanowire scaffolds for osteo-implantation: mechanical hardener and surface activity regulator. Sci Rep 5:18266
85. Keselowsky B, Wang L, Schwartz Z, Garcia A, Boyan B (2007) Integrin α5 controls osteoblastic proliferation and differentiation responses to titanium substrates presenting different roughness characteristics in a roughness independent manner. J Biomed Mater Res, Part A 80(3):700–710
86. Podila R, Moore T, Alexis F, Rao A (2013) Graphene coatings for biomedical implants. JoVE 73:e50276
87. Zhao C, Lu X, Zanden C, Liu J (2015) The promising application of graphene oxide as coating materials in orthopedic implants: preparation, characterization and cell behavior. Biomed Mater 10(1):015019
88. Fabbri P, Valentini L, Hum J, Detsch R, Boccaccini AR (2013) 45S5 bioglasss-derived scaffolds coated with organic–inorganic hybrids containing graphene. Mater Sci Eng C 33:3592–3600

Potentiality of Graphene-Based Materials for Neural Repair

María Teresa Portolés and María Concepción Serrano

Abstract The use and interest of graphene-based materials for neural repair is still in its infancy. In the last years, a more and more solid body of work is being published on the ability of these materials to create biocompatible and biofunctional substrates able to promote the in vitro growth of neural cells, often supporting enhanced neural differentiation of stem/progenitor cells. Although in vivo studies with these materials are rare, encouraging pioneer works in the brain and the spinal cord might impulse the research community to translate their potentiality from cell cultures to animal models, a closer scenario for their potential use in human healthcare in the future. In this chapter, we first describe some relevant generalities regarding the nervous tissue and approaches to accomplish neural repair. Then, we expose the literature published to date on the use of graphene-based materials for neural repair and neural-related applications and discuss their potentiality in the field.

Abbreviations

3DG	3D Printable Graphene
AMSC	Adipose-derived mesenchymal stem cells
BDNF	Brain-derived neurotrophic factor
BMSC	Bone marrow mesenchymal stem cells, bone marrow stromal cells
CNS	Central nervous system
CNTF	Ciliary neurotrophic factor
CVD	Chemical vapor deposition

M.T. Portolés
Department of Biochemistry and Molecular Biology I, Universidad Complutense de Madrid, Ciudad Universitaria s/n, IdISSC, 28040 Madrid, Spain
e-mail: portoles@quim.ucm.es

M.C. Serrano (✉)
Hospital Nacional de Parapléjicos, Servicio de Salud de Castilla-La Mancha, Finca de La Peraleda s/n, 45071 Toledo, Spain
e-mail: mslopezterradas@sescam.jccm.es

G. Gonçalves et al. (eds.), *Graphene-based Materials in Health and Environment*, Carbon Nanostructures, DOI 10.1007/978-3-319-45639-3_6

DMAEMA	Dimethylaminoethyl methacrylate
ELF-EMF	Extremely low frequency electromagnetic fields
ES	Electrical Stimulation
FGF	Fibroblast growth factor
GABA	Gamma-aminobutyric acid
GalC	Galactocerebroside
GAP-43	Growth-associate protein 43
GBM	Graphene-based materials
GDNF	Glial cell line-derived neurotrophic factor
GFAP	Glial fibrillary acidic protein
GO	Graphene oxide
hOR	Human olfactory receptors
IκB	Inhibitor of kappa B
IL-1β	Interleukin 1β
ITO	Indium tin oxide
LPS	Bacterial lipopolysaccharide
Map-2	Microtubule-associated protein 2
MSC	Mesenchymal stem cells
MTT	3-(4,5-Dimethylthiazol-2-yl)-2,5-diphenyltetrazolium bromide
NGC	Nerve guidance channel
NGF	Nerve growth factor
NSC	Neural stem cells
NT-3	Neurotrophin 3
Olig2	Oligodendrocyte transcription factor 2
O4	Oligodendrocyte marker O4
PDGRβ	Platelet-derived growth factor receptor β
PDMS	Poly(dimethylsiloxane)
PEDOT	Poly(3,4-ethylene dioxythiophene)
PLGA	Polylactic-*co*-glycolic acid
PNS	Peripheral nervous system
rGO	Partially reduced graphene oxide
ROS	Reactive oxygen species
RT-PCR	Reverse transcription polymerase chain reaction
sPSC	Spontaneous post-synaptic currents
TCPS	Tissue culture polystyrene
TENG	Tissue engineered nerve graft
TNF-α	Tumor necrosis factor α
USFDA	Food and Drug Administration of the USA
VEGF	Vascular endothelial growth factor

1 Introduction

Graphene, along with carbon nanotubes and carbon nanofibers, has been recognized as an attractive candidate to repair injured nerves and be incorporated into experimental neural prostheses due to its excellent properties [1], some of them acquiring more relevance when referring to their applicability in neural repair. The closely packed honeycomb lattice of graphene, composed of a sheet of sp^2-bonded carbon atoms with the thickness of one atom, displays an extraordinary surface area ($\sim 2600\ m^2\ g^{-1}$) enabling superior adsorptive properties [2]. Graphene oxide (GO), the principal graphene derivative, presents an even stronger ability to absorb molecules as a consequence of its larger amount of oxygen-containing functionalities that serve attracting adhesive moieties, cell media components, and therapeutic drugs with high loading capacity [3]. Moreover, these functional groups confer a higher hydrophilicity to GO molecules, responsible for an increased dispersibility and a lower tendency to form aggregates in a size large enough to induce toxic responses in biological scenarios. Additional features that make graphene-based materials (GBM) attractive for the treatment of injuries in the central nervous system as drug carriers are their ability to cross the blood–brain barrier and their stability in blood circulation after functionalization. In these lines, Yang et al. [4] have demonstrated the potential application of functionalized GO nanosheets as a pirfenidone carrier to treat subarachnoid hemorrhage in a mouse model, preventing secondary bleeding and cerebral infarction. As delivery systems, GBM are also receiving a significant attention in combination with magnetic nanoparticles for magnetic chemotherapy, photothermal therapy, X-ray imaging, and gene therapy [5], as described in other chapters of this book. Other remarkable properties of these materials, such as their charge carrier mobility in excess of $2 \times 10^5\ cm^{-2}\ V^{-1}\ s^{-1}$ [6], show interest for electrical stimulation and recording in neural cells. Moreover, their attractive mechanical properties, with Young modulus values as high as 1 TPa [7], facilitate the preparation of flexible and resistant platforms with mechanical compliance with neural tissues.

In this chapter, we first describe some key features of the central and peripheral nervous systems to take into account before addressing any reparative initiative in these targets. Later on, we present a systematic discussion of the most relevant publications to date focused on the exploration of GBM for neural repair, including regeneration, recording, and stimulation approaches in nervous tissue. Besides work discussed herein, readers are referred to excellent reviews on the topic for further details, including other types of carbon-based materials intended for neural repair [1, 8–13].

2 Lesions in the Central Nervous System

The human nervous system consists of the central nervous system (CNS), including the brain and the spinal cord, and the peripheral nervous system (PNS), which refers to the nerves and ganglia outside the central core. The CNS can suffer different physical injuries (e.g., traumatic lesions) or neurological disorders (e.g., brain stroke, multiple sclerosis, and Parkinson's, Huntington's, and Alzheimer's diseases) involving important structural and functional alterations with loss of both neurons and supporting cells [14]. Only in the United States of America, there are ca. 1.5 million people with traumatic injuries affecting the brain and 265,000 affecting the spinal cord each year [15]. Unfortunately, the treatment of this type of lesions remains a challenge for both clinicians and scientists. The rapid formation of inhibitory fibroglial scars and the reduced intrinsic ability of the CNS to regenerate can be counted within the most crucial events preventing neural tissue and function repair [16, 17]. Moreover, the CNS lacks of the reparative Schwann cells that exist at the PNS, which are able to support axonal regrowth after lesion [10]. For these reasons, regeneration strategies in the CNS strongly differ from those applied in the PNS and generally include combined therapies to moderate reactive gliosis, while promoting axon regrowth and nervous tissue regeneration in a harsher environment [18].

Controlled and local biomolecule delivery has been extensively explored to stimulate endogenous repair mechanisms for CNS regeneration. Drug delivery strategies include the use of polymer micro/nanoparticles, implanted polymer scaffolds, injectable hydrogels, and particle/scaffold composites. The combination of controlled and sustained delivery of both biomolecules/growth factors and cells capable of regenerating damaged tissue as neural stem cells (NSC), with the capacity to differentiate into neurons, astrocytes, and oligodendrocytes, is a promising strategy to enhance tissue regeneration in the injured CNS [19]. In this scenario, a key aspect to support the physiological function of NSC in the implanted site is the use of 3-D scaffolds that mimic the functional organization of the native nervous tissue [20]. Additional parameters when designing implantable materials for the CNS include size and mechanical compliance, as they play a fundamental role in the extent of fibroglial scar formation around the implant. For instance, 50-μm diameter implants lead to less extensive foreign body reaction and greater proximal neuron survival than those with 200 μm in diameter [21].

3 Lesions in Peripheral Nerves

Peripheral nerve injury is a global clinical problem that also affects the quality of patients' lives significantly and causes an enormous socioeconomic burden [22, 23]. When the damage is reduced, most of PNS lesions are capable of spontaneous healing without surgical intervention, but tubular prostheses (i.e., autografts, allografts, and polymeric implants) are required to connect the nerve stumps when the defect does not permit direct nerve repair. Neural tissue engineering has emerged in the last years as a subfield of tissue engineering with the aim of producing biological and artificial nerve grafts (tissue engineered nerve grafts, TENG) in order to supplement and substitute autologous nerve grafts, which still remain as the gold standard [9]. The preparation of these constructs requires the knowledge of important aspects of neurophysiology and neurobiology to determine the requirements that the biomaterial, cells and molecular components of the TENG structure must fulfill. Despite intensive research, TENG implants capable of addressing all requisites for nerve regeneration are not yet a reality due to the numerous factors affecting nerve graft success [8]. Particularly, aspects such as the time elapsed since injury, the type and extent of injury, the vascularity in the area, the orientation of the nerve graft, the length of the graft required, the tension on the repaired nerve, and the age and medical condition of the patient can be cited as some of the most important ones [24]. Since early attempts with a biologically inert silicone elastomer, different classes of biodegradable synthetic polymers are being explored nowadays as scaffolds for peripheral nerve repair, some of them already approved by the USFDA and commercially available [25, 26].

Regarding 3-D configuration, TENG were initially prepared as cylindrical tubes whose single lumen was empty (nerve guidance channel, NGC). More recent approaches describe more complex structures including internal microarchitectures or luminal multiple-component compositions. Thus, different biomaterial-based fillers (e.g., fibers, filaments, gels, or sponges) have been included into the lumen of NGC to provide topological characteristics that might enhance nerve regeneration [27–29]. The incorporation of cells, growth factors, and/or cytokines within TENG has also attracted much interest to improve the outcome of these substitutes. Schwann cells, neural stem cells, embryonic stem cells, and marrow stromal cells have been the most studied supporting cells [30]. Autologous Schwann cells are difficult to obtain in a large quantity and allogeneic Schwann cells can induce immunological rejections. However, bone marrow mesenchymal stem cells (bone marrow stromal cells, BMSC) have become a promising alternative to Schwann cells in this scenario, showing considerable success in experimental studies [31–33]. Alternatively, adipose-derived mesenchymal stem cells (AMSC) are potentially valuable because of their multi-lineage differentiation capability resembling that of BMSC [34, 35]. They are easily accessible from the patient for an autologous cell therapy, thus eliminating the risk of rejection, and can be expanded in a controlled and reproducible manner [36]. Gene-modified stem cells present new options for neural tissue engineering. For instance, BMSC can be genetically engineered to express nerve growth factor (NGF) and then be used for preparing bioactive neural scaffolds [37].

4 Tissue Engineering Criteria Applied to Neural Substitutes

In general terms, the design of nervous tissue substitutes includes the selection of appropriate biomaterials that fulfill physical–chemical (e.g., porosity, permeability, and composition), biomechanical (e.g., flexibility, resistance), and biological (e.g., biocompatibility, biodegradability, immunogenicity) properties. Such features will prompt their successful performance to protect axonal regrowth in the injured nervous tissue, acting also as a carrier for the delivery of biochemical agents [38, 39]. Regarding physical features, scaffold configuration must be conceived in a manner that facilitates cell distribution and growth of new neural structures in 3-D [40]. This structure should act as a substrate for neural cell adhesion, proliferation, migration, and function, eliciting adequate interactions with the native tissue [41]. In this sense, modern nanotechnology techniques are being extensively used to prepare advanced neural scaffolds with nanoscale topography mimicking the natural extracellular matrix architecture [42], thus improving their surface properties and supporting enhanced cell guidance for neural regeneration [43]. Other current technologies such as 3-D bioprinting are also showing a promising future for the development of permissive 3-D substrates with positive in vivo integration in the nervous tissue [20]. Based on composition, natural materials are the most commonly used for the preparation of neural substitutes, with two major categories: (1) autologous non-neural tissues and allogeneic/xenogeneic neural/non-neural tissues that have been decellularized [44] and (2) naturally derived polymers, including extracellular matrix molecules and other polysaccharides and proteins [45]. Beside natural materials, synthetic polymers, metallic-based materials, and even ceramics are being investigated for their use as part of neural scaffold materials. Carbon nanostructures, including nanotubes, nanofibers and graphene, are being under a very intensive research to prepare experimental neural prostheses and guides [1, 46].

The biological requisites for neural substitutes are more complex to attain, mainly because the exact description of those needed is still unclear. As the local presence of growth factors at the injury site is critically hindered after damage, the increased presence of molecules such as neurotrophins (NGF, BDNF, and NT-3) and growth factors with neurotrophic actions (GDNF, CNTF and FGF) is aimed to promote neural regeneration [30, 47]. Diverse strategies have been designed for the controlled and continuous release of growth factors from neural substitutes, including adsorption to the scaffold surface, incorporation during scaffold fabrication, entrapment of growth factor-loaded microspheres into the scaffold, covalent immobilization onto the scaffold, and even implantation of an osmotic mini-pump or injection device [9]. Angiogenesis (i.e., growth of new blood vessels) is also an important requisite for tissue engineered constructs as cell and tissue survival depends on the supply of oxygen and nutrients and the removal of metabolites by a branched blood vessel system [48, 49]. In this sense, vascular endothelial growth factor (VEGF), a fundamental regulator of angiogenesis [50], has demonstrated neurotrophic activity to stimulate axonal outgrowth and to enhance survival and proliferation of Schwann cells [51], as well as capacity to improve intraneural angiogenesis by promoting endothelial growth during peripheral nerve regeneration [52, 53]. For these reasons, VEGF has been incorporated in different scaffolds designed for neural regeneration [54, 55]. Finally, inflammatory responses to implanted biomaterials must be also taken into account as they may induce both positive and negative effects on neural regeneration after nervous tissue damage [56].

Last but not least, electrical signals applied after neural injury improve the amount and accuracy of motor and sensory reinnervations [57–60]. Moreover, there are increasing evidences about the clinical potential that the treatment of damaged neural tissues with electrical fields might have [61], thus underlining the importance of the interaction between neural cells and electrically conductive biomaterials [18]. In this sense, carbon nanomaterials such as carbon nanotubes, nanofibers, and graphene, with extraordinary electrical properties, are being widely investigated as promising candidates for the development of electroactive neural substitutes [8].

5 In Vitro Studies with Graphene-Based Materials for Neural Repair

A broad and diverse repertoire of cell types has been used to explore in vitro the interactions of GBM with biological systems, many of them profoundly discussed in this book. When referring to neural tissue, this variety is clearly diminished including only a narrow spectrum of cells indicated as follows. PC12 cells, derived from a rat adrenal gland pheochromocytoma, are an easy-to-use and readily available neural-like cell line traditionally used for cytocompatibility tests with biomaterials due to their ability to differentiate into neural phenotypes when exposed to NGF. The human neuroblastoma SHSY5Y cell line is also being widely used as a cellular model for neuronal differentiation studies in vitro. These cells evolve to a neuron-like phenotype with elongated neural processes when exposed to retinoic acid [62]. More recently, advances in stem cell research are prompting their use in the field of tissue engineering as promising therapeutic tools. Particularly, NSC are been extensively explored as part of cell therapy approaches for the treatment of the injured spinal cord due to their potential to boost functional neural recovery [63]. Besides MSC and related ones, induced pluripotent stem cells, which are pluripotent cells typically derived from somatic cells by introducing a cocktail of reprogramming factors, can be also mentioned. These cells are able to either maintain an undifferentiated state or differentiate into cells of any of the three germ layers (i.e., endoderm, mesoderm, and ectoderm) and they are being the focus for an increasing number of studies in neural repair. In this section, we describe specific findings reported to date on the in vitro interactions of GBM with these different cell types.

5.1 Primary Neural Cells and Neural Cell Lines

Early studies on the interaction of neural cells with GBM mainly focused on basic biocompatibility features such as viability, morphology, and neurite outgrowth. For instance, work by Li et al. described how graphene films, synthesized by chemical vapor deposition (CVD) and posteriorly coated with poly-L-lysine, enhanced the neurite number at day 2 and the average length per neurite from 2 to 7 days in mouse postnatal hippocampal cells in culture [64]. The expression of GAP-43 protein (growth-associate protein 43) was significantly boosted by the graphene substrate, likely responsible for the observed enhancement in neurite sprouting and outgrowth. CVD-synthesized graphene 2-D substrates were also able to support the growth of rat neuronal cells [65], although the necessity of functionalization was noticed due to its low hydrophilicity. Hippocampal neurons have been also used to explore the ability of linear and nodal-grid patterns on graphene-based substrates to guide neurite growth (Fig. 1) [66], as neural cell alignment is being pursued as a major goal for the development of neural prostheses and interfaces. Kim et al. used

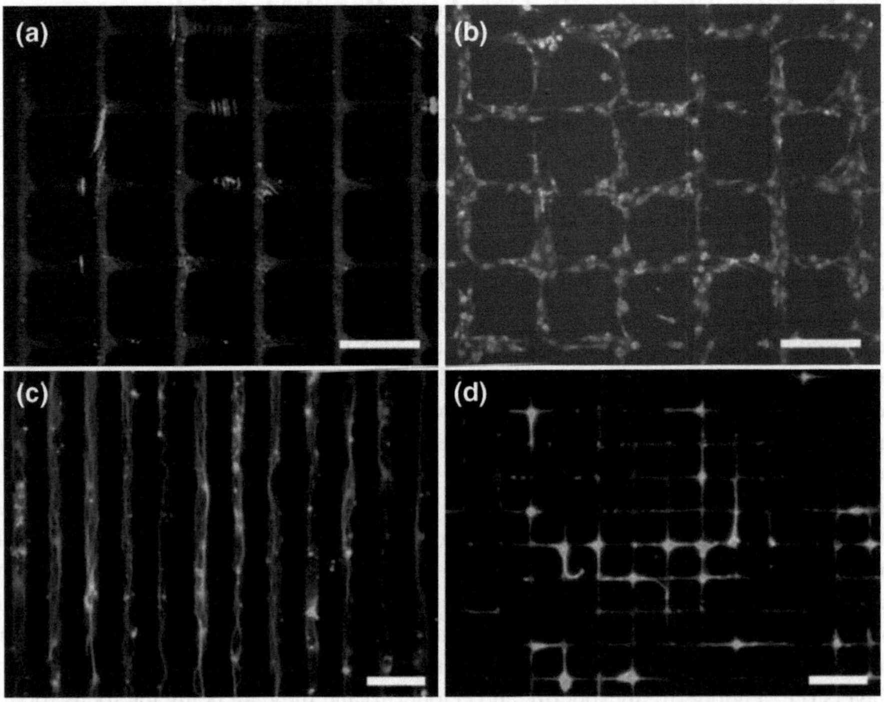

Fig. 1 **a** Confocal laser scanning microscopy image of a poly-D-lysine-FITC-patterned graphene film and **b** fluorescence micrograph of a fibroblast-patterned graphene film. *Scale bars* 200 μm. **c**, **d** Fluorescence micrographs of micropatterns of hippocampal neurons. *Scale bars* 100 μm. Reproduced from [66] with permission of John Wiley & Sons Inc

this same cell type to demonstrate good viability values on partially reduced GO (rGO) multilayers when cultured for up to 30 days [67].

Studies with the human neuroblastoma SHSY5Y cells confirmed the absence of cytotoxicity of GO nanosheets at low concentrations (<80 μg ml^{-1}) for 4 days [68]. Higher concentrations induced a dose- and time-dependent deleterious effect on cell viability, although no apoptosis was detected in any case. When assisted by retinoic acid loaded in the nanosheets, GO enhanced the neural differentiation of this cell line as revealed by analyses of neurite outgrowth and expression of Map-2, a specific protein typically expressed in neuron dendrites and somas. More recent studies by Lee et al. deepened on the molecular mechanisms involved in the stimulation of cellular neurogenesis on SHSY5Y cells exerted by graphene [69]. In this work, monolayer graphene films were deposited by CVD and transferred to glass substrates by a wet transfer technique. The observed enhanced neurite outgrowth, even in the absence of neurogenic factors and in both neurite length and number, was related to the upregulation of genes such as neurofilament light chain, nestin and Map-2. Signaling cascades as those involving focal adhesion kinase and p38 mitogen-activated protein kinase were involved in the favorable response induced by graphene.

Regarding neurite outgrowth and branching, recent studies by Wang and colleagues have demonstrated that these phenomena can be controlled by manipulating the chemical properties of GO [70]. Particularly, carboxylated GO was chemically modified with amino-, poly-*m*-aminobenzene sulfonic acid- or methoxyl-terminated functional groups to change its surface charge. Findings from this work, carried out with primary rat hippocampal neurons cultured on GO films, evidenced that positively charged GO (amino-functionalized) was more beneficial for neurite outgrowth and branching (i.e., maximum neurite length, total outgrowth and cell body area per neuron, and number of branches per neurite and per neuron) than any other surface modification. Negatively charged surfaces seemed ineffective to promote the initiation of growth cones. It is worth noting that viability values were maintained over 96 % in all the cases. In previous work, these authors investigated the ability of biomimetic choline-like GO composites to promote neurite sprouting and outgrowth [71]. In this approach, acetylcholine-like (dimethylaminoethyl methacrylate, DMAEMA) or phosphorylcholine-like (2-methacryloyloxyethyl phosphorylcholine) units were covalently bond to thin GO films on glass coverslips. Both molecules charged the surface of the material positively by the presence of amine groups. When in contact with primary rat hippocampal neurons, cells displayed an enhancement in the number of neurites and branches, cell body area, and neurite length (total, maximum, and averaged) with respect to non-functionalized GO. Interestingly, the filopodia area was also increased. For all the parameters evaluated, GO–DMAEMA displayed the highest performance. These findings were supported by a significant increase in the expression of GAP-43 as evidenced by western blot analyses. In all cases, cell viability was preserved and maintained over 96 % after 7 days in culture.

Besides the topographical and chemical cues of GBM for neural growth and differentiation, studies relaying on the electrical properties of these materials have been also carried out. Highly conductive films composed of graphene nanosheets have been used to enhance the growth and neural differentiation of PC12 cells by means of electrical stimulation (ES) [72]. Differences were found depending on the application of constant or programmed ES (100 mV mm^{-1}), the stimulation time (from 4 to 48 h) and its frequency (1 vs. 10 Hz). Programmed ES at 10 Hz for 6 h seemed to cause the most favorable neural cell responses. SHSY5Y cells have been also used to test the ability of graphene electrodes to control neural cell-to-cell interactions through non-contact electrical field stimulation [73]. It was found that a weak electric field stimulation (pulse duration of 10 s, 4.5 mV mm^{-1} for 32 min) was effective for increasing the number of cells forming new cell-to-cell couplings and cells strengthening existing ones when cultured on a graphene/polyethylene terephthalate film stimulator. In a different study, ES was used to release NGF in a controlled fashion from thiol-functionalized rGO/poly(methacrylic acid) microcapsules arrayed on an indium tin oxide (ITO) flexible substrate [74]. PC12 cells cultured on these electrically active platforms showed a fourfold increase in averaged neurite length, which reached nearly 90 μm in 2 days.

When used as dopants, GBM have also demonstrated an impact on neural cell responses. Particularly, recent studies have revealed the ability of graphene, obtained by chemical reduction from GO and used in combination with polypyrrole, to improve the mechanical and electrical properties of gelatin/chitosan-based porous scaffolds intended for nerve regeneration [75]. These structures were able to support human primary Schwann cells attachment and proliferation for 14 days. In another study, the addition of moderate concentrations (0.3 and 0.5 wt.%) of GO nanosheets (400–600 nm in width, 1.2 nm in thickness) to nanofiber scaffolds composed of polycaprolactone prepared by electrospinning techniques supported adhesion and spreading of PC12-L cells with respect to fibers without GO [76]. Proliferation and viability were not affected and neural differentiation was significantly enhanced on days 6 and 9. Concentrations as high as 1.0 wt% induced a dramatic reduction in cell proliferation and neural marker expression (i.e., β-tubulin III and GFAP). In a different approach, 2-D and 3-D electrospun polycaprolactone nanofibrous scaffolds were functionalized with graphene–heparin/poly-L-lysine by layer-by-layer deposition to confer them both electrical and biological properties [77]. In these studies, graphene nanosheets were obtained by chemical reduction of exfoliated GO. Both 2-D and 3-D substrates were able to support neuron cell adhesion and neurite outgrowth of neural cells isolated from E14 rat embryos cerebral cortices.

5.2 *Stem and Progenitor Cells*

GBM show also a promising future for promoting the differentiation of stem cells towards neural phenotypes. Specifically, Park et al. [78] pioneered the investigation of the interaction of graphene with NSC, demonstrating not only the ability of laminin-coated graphene films to enhance the differentiation of immortalized human

NSC to neurons rather than glial cells but also the ES of these cells when using graphene as a transparent electrode. No harmful effects on cell adhesion, proliferation and viability were reported. During the differentiation process, cells exhibited a stronger attachment to graphene samples than to control glass substrates. Further experiments with microarray techniques evidenced that cells cultured on graphene underwent an upregulation of laminin-related receptors, thus accounting for the promoted adhesion and, indirectly, for the observed enhanced neuronal differentiation. Other genes such as those implicated in calcium signaling pathways were also upregulated. Finally, differentiated cells cultured on a graphene film functioning as an electrode were exposed to a series of voltage pulses (typically, 1–10 of 500 mV monophasic/cathodic voltage pulses with 1–100 ms duration in a second). Increases in the intracellular Ca^{2+} concentrations derived from this ES were detected by using the Fluo-4-AM probe in different neurons in the culture.

Solanki et al. [79] achieved neurite alignment and enhanced neuronal differentiation of NSC when cultured on hybrid structures made of positively charged silica nanoparticles (300 nm in diameter) and GO nanosheets. Before cell culture, all the samples were coated with laminin. Findings revealed that GO was able to promote neurite alignment (labeled specifically for Tuj1, a neuron-specific β-III tubulin), with or without the simultaneous use of silica nanoparticles. Besides enhancing viability in both GO substrates after 3 weeks, neurites extension was further promoted in the presence of silica nanoparticles. Moreover, neuronal differentiation (measured as the expression of specific markers including Tuj1, GAP-43, Map-2, and synapsin) was boosted by these hybrid platforms as a result of the combined effects of silica nanoparticles and GO. Interestingly, the authors demonstrated the feasibility of these structures to be transferred to flexible poly(dimethylsiloxane) (PDMS) films for future utility as implantable samples in the injured neural tissue.

Studies by Akhavan and Ghaderi revealed the ability of nanogrids made of GO nanoribbons to support the differentiation of human NSC into neural networks for their use as photocatalytic stimulators [80]. Nanoribbons are a novel nanostructure of graphene composed of elongated strips of graphene with a high length-to-width ratio and straight edges. These structures, placed on a SiO_2 matrix containing TiO_2 nanoparticles (~40 nm in diameter) at their surface by using a paint-brushing method, were fabricated by the oxidative unzipping of multi-walled carbon nanotubes. Findings from this work revealed that NSC density was higher after 3 days in those nanogrids containing GO rather than rGO, likely due to its higher hydrophilicity and subsequent cytocompatibility. The toxicity of graphene nanoribbons is typically associated with the reactive oxygen species (ROS) generated by graphene at high concentrations and the rupture of cell membranes by direct interaction with sharp edges [81]. The number of differentiated cells on these platforms increased 5.9-fold in comparison to cells on quartz substrates after 3 weeks. When flash photo-stimulated (what caused the injection of photo-excited electrons from TiO_2 nanoparticles onto the nanoribbons through the Ti–C electron channel), the increment raised 26.8-fold with a significant orientation toward the neuronal phenotype ($Tuj1^+$) rather than the glial one ($GFAP^+$), as well as certain degree of cell alignment along the nanogrids. Neural differentiation on nanogrids was also enhanced with respect to that found on sheets, result that the authors ascribed to the physical stress caused by the surface topographic features of the nanogrids. Additionally, effective cell-to-cell and cell-to-nanogrid electrical couplings were demonstrated after 3 weeks by measuring the intensity-voltage curves. The photo-stimulation consisted on the application of flash photo-pulse trains with 10 mW cm^{-2} of intensity, pulse duration of 4 s and frequency of 1 Hz, repeated after 60 s time intervals for 30 min every 12 h by using a camera xenon lamp. Similar findings were obtained when rGO was combined with TiO_2 as heterojunction films to grow human NSC [82]. In more recent work, rGO sheets demonstrated ability to support the proliferation of this same cell type for 3 days and to enhance the differentiation of these cells into neurons rather than glial cells [83]. Specifically, chemically exfoliated GO was reduced with Asian red ginseng, with a reduction level and electrical conductivity comparable to that obtained by hydrazine treatment. This ginseng-rGO was more stable in terms of aggregation in aqueous suspensions. The authors hypothesized that those beneficial effects were caused by the higher electron transfer of these sheets, their higher hydrophilicity and biocompatibility and the attachment of potent antioxidants such as ginsenoside molecules. When the GO nanomeshes containing TiO_2 were reduced with hydrazine to achieve rGO nanomesh semiconductors, the resulting material was also found to enhance human NSC differentiation into neurons after exposure to near infrared laser stimulation [84]. This effect was attributed to the low-energy photo-excited electrons injected from the semiconductors into the cells. This pulsed laser stimulation not only promoted human NSC differentiation into neurons, but also caused the self-organization of a radial neuronal network on the rGO sheets (Fig. 2) [85]. More recently, these authors have developed electrically conductive 3-D scaffolds from rolled GO foam layers with desirable scales and applied them in directional growth of neural fibers by means of neuronal differentiation of human NSC under ES. This stimulation enhanced cell proliferation and accelerated the differentiation to neurons (rather than glial cells) [86].

Functional neural circuits from rat hippocampal NSC have been also obtained on graphene substrates fabricated by CVD [87]. Samples were coated with laminin prior to cell culture. After confirming basic biocompatibility parameters such as cell adhesion and proliferation, the intracellular calcium ion concentration was monitored by using the Fluo-4-AM probe and the graphene substrate as an electrode. Cells exhibited a 30 % increase in fluorescence as a consequence of ES through the graphene layer (a series of 10–100 ms monophasic cathodic pulses at intervals of 5 s, stimulation threshold 0.5–1 μA). Furthermore, almost double amount of cells showed spontaneous Ca^{2+} oscillations when cultured on graphene, being the frequency of these basal oscillations also enhanced with respect to tissue culture polystyrene (control) (Fig. 3). Using single-cell patch clamp techniques, most of neurons cultured on graphene showed spontaneous post-synaptic currents, thus indicating a normal activity of the neural networks, with higher current peak amplitudes and an enhanced frequency. Finally, this boost in the electrical activity of the neural networks formed on graphene films was accompanied by a significant increment in the number of neurons per mm^2.

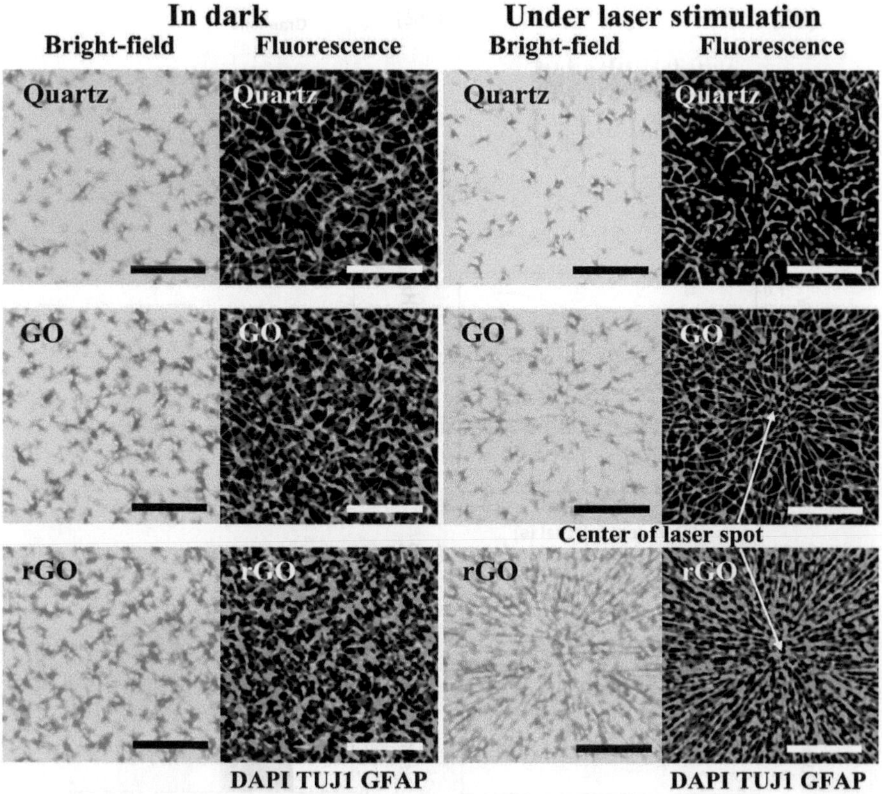

Fig. 2 Bright-field and immunofluorescence images of human NSC differentiated on GO and rGO sheets for 3 weeks in the dark and under pulsed laser stimulation. DAPI was used to stain cell nuclei (*blue*), GFAP for glial cells (*red*), and Tuj1 for neural cells (*green*). *Scale bars* 200 μm. Reproduced from [85] with permission of The Royal Society of Chemistry

Besides neurons, rat hippocampal NSC were also able to predominantly differentiate into oligodendrocytes when cultured on hybrid nanofibrous scaffolds composed of electrospun polycaprolactone nanofibers (200–300 nm in diameter) coated with GO (Fig. 4) [88]. These hybrid substrates were coated with laminin prior to cell culture. Interestingly, the expression of neural markers showed a GO concentration-dependent tendency, being the highest concentration (1 mg ml^{-1}) the one promoting the largest differentiation into mature oligodendrocytes (for which the myelin basic protein, Olig2, O4, and GalC were selected as specific markers). In the absence of polycaprolactone nanofibers, GO itself was found to promote a modest oligodendrocyte differentiation (twofold with respect to tenfold increase in the hybrid structures). This enhancement was related to changes in gene expression that included the overexpression of early and mature oligodendrocyte markers and several key integrin-related intracellular signaling molecules implicated in cytoskeletal remodeling and process extension during oligodendrocyte development. More recent

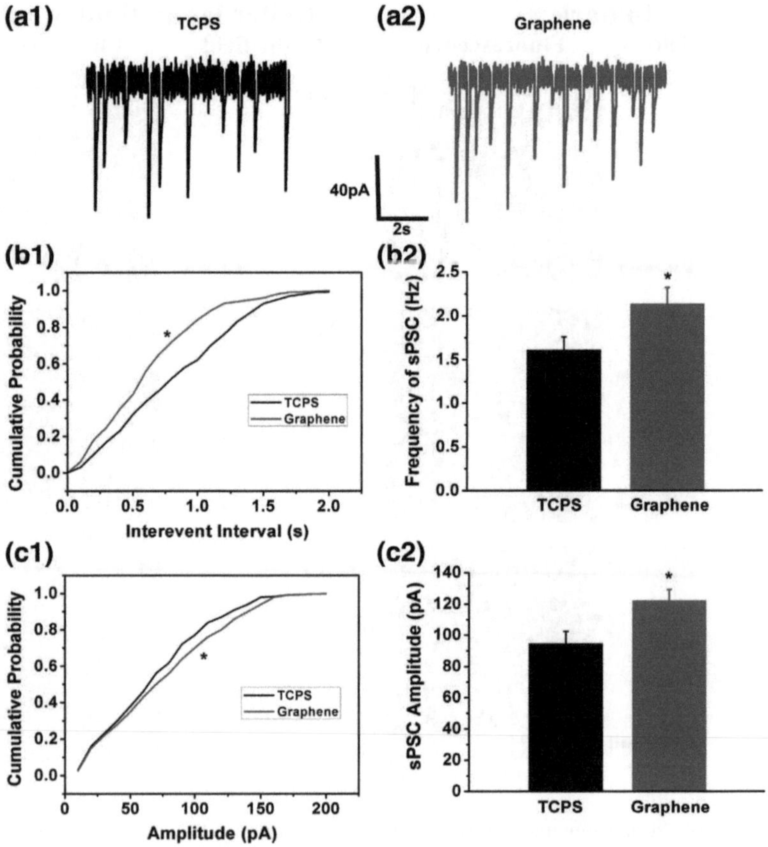

Fig. 3 Graphene substrates increase spontaneous synaptic activity and firing. **a1**, **a2** Representative spontaneous post-synaptic currents (sPSC) in tissue culture polystyrene (TCPS) and graphene. Cumulative probability plots of inter-event interval (**b1**) and amplitude (**c1**) of sPSC in TCPS and graphene. Histogram plots of sPSC frequency (**b2**) and amplitude (**c2**) to identify the difference of selected index. Statistical significance: $*p < 0.05$. Reprinted from [87], Copyright (2013), with permission from Elsevier

work by Weaver et al. [89] have demonstrated the ability of composites of GO and poly(3,4-ethylene dioxythiophene) (PEDOT) to favor the differentiation of NSC to oligodendrocytes when functionalized with platelet-derived growth factor by using carbodiimide chemistry. Functionalization of these platforms with interferon-γ promoted the differentiation of these cells to neurons.

Embryonic neural progenitor cells have been also used as a model in vitro to evaluate the potential of GBM for neural repair. Particularly, the interaction of these cells with compact and flexible microfibers exclusively made of rGO (121 ± 5 μm in diameter, conductivity values of 4.64 ± 0.90 S cm^{-1}) has revealed the ability of these structures to support the growth of mature neural networks, highly interconnected and composed of both neurons and glial cells [90]. Numerous synaptic

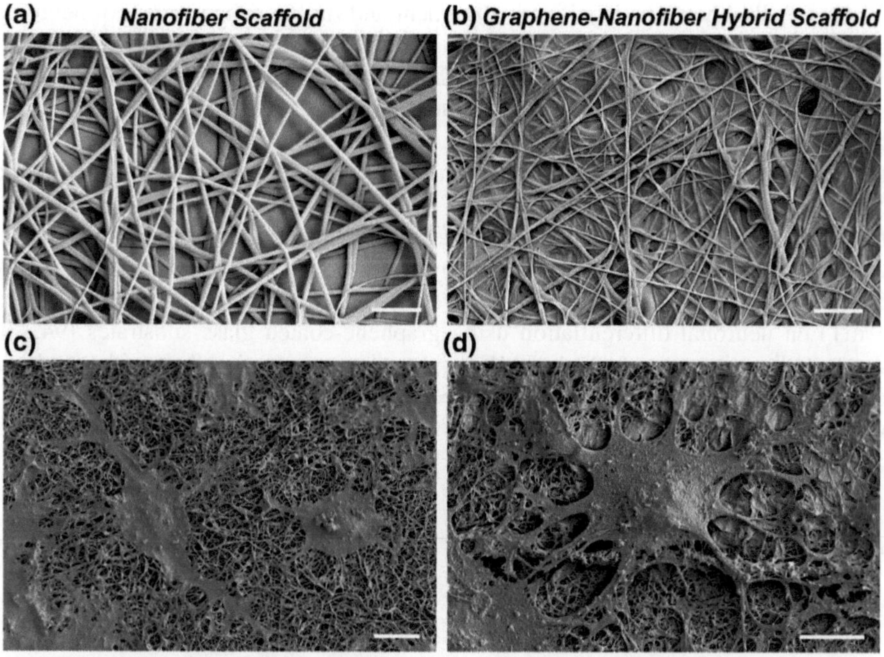

Fig. 4 Morphology of nanofibrous scaffolds and respective cultures of NSC by field-emission scanning electron microscopy. Micrographs of polycaprolactone nanofibers (**a**) and GO-coated polycaprolactone nanofibers (**b**). *Scale bars* 2 μm. Micrographs of differentiated NSC cultured on polycaprolactone nanofiber scaffolds (**c**) and graphene–nanofiber hybrid scaffolds (**d**) after 6 days in culture. *Scale bars* 10 μm. Reproduced from [88] with permission of John Wiley & Sons Inc

contacts were identified in close contact to the microfibers and at their periphery. Moreover, the colonization by meningeal fibroblasts was dramatically hindered when microfibers were coated with N-cadherin. Other stem cells such as MSC have been also targeted to prompt their differentiation into neural cells on GBM. Specifically, Loh and colleagues described the use of these cells for screening the utility of fluorination to functionalize graphene substrates [91]. In the case of pharmaceutical drugs, the introduction of fluorine groups, highly electronegative, and small in size, is known to improve their metabolic stability and increase the strength of the interaction with the targeted proteins [92]. In this particular approach, fluorinated graphene films enhanced the proliferation and polarization of human BMSC after 7 days. These cells were promoted to differentiate into neural phenotypes (higher expression of Tuj1 and Map-2, lower values for nestin and GFAP). This neuronal differentiation was further boosted by the addition of retinoic acid to the culture media. When intentionally patterned (30-μm microchannels defined by PDMS lines of 3 mm in length, 150 μm in width, and 50 μm line spacing), cells showed a significant alignment and further enhancement of neuronal differentiation (Fig. 5). Contrary to these results, Oh et al. [93] reported only 50 % cell viability of human neuroblastoma SHSY5Y cells, besides no significant cell spreading, when cultured on fluorine-treated graphene sheets. In this approach, the graphene sheets were purchased from Samsung Co. (Korea) and the fluorine functionalization achieved by plasma treatment in a C_3F_8 gas environment with a 50 kHz radiofrequency plasma source at 30 W. In this same study, oxygen-functionalized graphene sheets by plasma treatment supported mitochondrial activity values of 138 %, as measured by MTT assay. BMSC were also used by Sung and colleagues to report on the beneficial influence of ELF-EMF (50 Hz, 1 mT) on neuronal differentiation using graphene-coated glass substrates [94]. In these studies, exposure to ELF–EMF displayed a synergistic effect with graphene on neuronal differentiation. Careful inspection of cellular phenomena by RT–PCR, western blot, and microarray analyses demonstrated an upregulation of cell

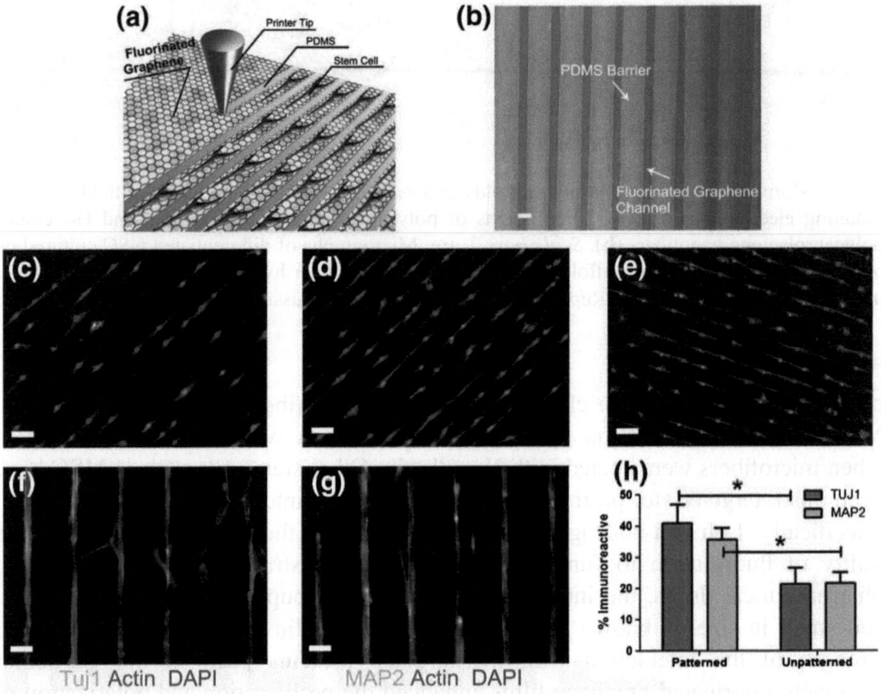

Fig. 5 **a** Schematic drawing of patterning MSC by printing PDMS barriers on graphene films directly. **b** Optical microscope image of printed PDMS on a fluorinated graphene film (*scale bar* 50 μm). **c–e** The aligned growth of stem cell on graphene, partially fluorinated graphene and fluorinated graphene with printed PDMS pattern, respectively (*scale bar* 100 μm). **f, g** MSC preferentially attached on fluorinated graphene strips and their F-actin aligned (*red*) and expressed neural specific markers: Tuj1 and Map-2 (*green*) (*scale bar* 50 μm). **h** Percentage of $Tuj1^+$ and $Map\text{-}2^+$ cells on unpatterned and patterned fluorinated graphene strips ($n = 6$, $*p < 0.05$). Reproduced from [91] with permission of John Wiley & Sons Inc

adhesion, increased intracellular calcium level and activated focal adhesion kinase signaling pathways, as well as a stimulated extracellular matrix production (e.g., collagen Iα2 and fibronectin). AMSC, another type of MSC, were used in recent studies by Lee et al. [95] to investigate the ability of combinatorial patterns of GO to promote their differentiation into ectodermal neurons. The GO grid patterns were fabricated using a microcontact printing technique on gold substrates, mimicking elongated, and interconnected neural networks. The conversion efficiency on the GO patterns was of *ca.* 30 %, with a significantly higher percentage of Tuj1^{+} cells and longer neurites than on control substrates. In a different study, the spontaneous differentiation of induced pluripotent stem cells was promoted by graphene and GO platforms [96]. In these experiments, substrates consisted of glass coverslips coated with either GO or rGO (obtained by direct hydrazine reduction from GO). Interestingly, adhesion and proliferation were enhanced in GO rather than rGO, the latter displaying a similar behavior to control glass. Ectodermal and mesodermal differentiation processes were quite similar between both substrates, while the endodermal one (e.g., hepatocytes, insulin-producing cells, and lung epithelium) was expedited in GO.

Despite the numerous studies of 2-D GBM (e.g., films, coatings) interfacing neural cells, this interaction has been rarely studied in 3-D structures, even though this spatial configuration creates more efficient microenvironments for cell growth. One of the few reports in these lines is work by Li et al. [97], which pioneered the exploration of 3-D porous foams composed of graphene and coated with laminin as substrates for NSC culture. The substrates were fabricated by CVD with a Ni foam template. Findings from this study revealed the ability of these structures to support NSC growth (viability values over 80 %) and maintain them in an actively proliferative state, with upregulation of Ki67 with respect to 2-D graphene films. Further studies confirmed an enhanced differentiation of these cells into neurons (Fig. 6), even illustrating electrical graphene–cell couplings by analyzing intracellular Ca^{2+} movements. Other material configurations such as nanoparticles have been also explored in this scenario. Specifically, the ability of GO nanoparticles to promote the differentiation of mouse embryonic stem cells (CGR8 cell line) into dopamine neurons has been reported [98]. In comparison to carbon nanotubes and graphene (obtained from GO by reduction via L-ascorbic acid), only GO was able to efficiently promote this specific differentiation as confirmed by an increased expression of tyrosine hydroxylase (72 % of increase in GO at 100 μg ml^{-1}).

5.3 *Other Types of Neural Cells*

The specific interaction of GBM with immune cells within the nervous tissue and their ability to initiate pro-inflammatory, anti-inflammatory, or even immunomodulatory responses is still unknown. Pioneer work by Song et al. [99] has brought some clarity into this topic. Specifically, these authors explored the interaction of

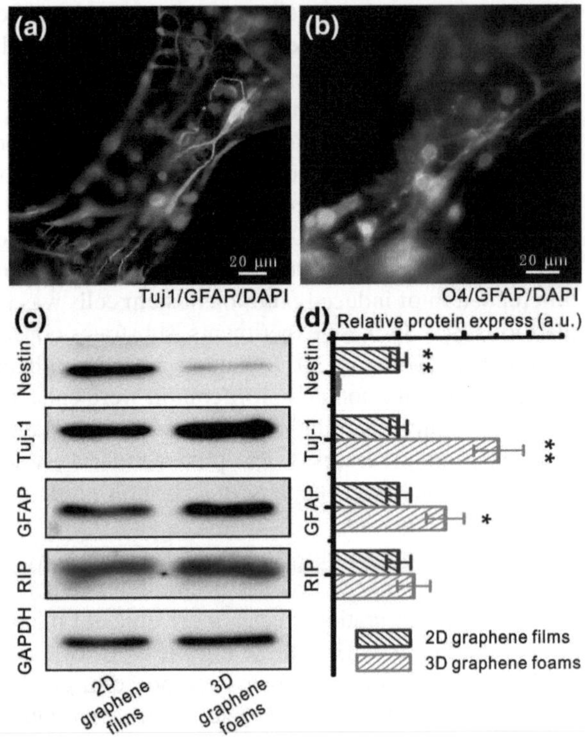

Fig. 6 Differentiation studies of NSC on 3-D graphene foam scaffolds. **a**, **b** Representative fluorescence images of differentiated NSC under differentiation conditions: Tuj1 for neurons (*green*, **a**), GFAP for astrocytes (*red*, **a**, **b**), O4 for oligodendrocytes (*green*, **b**), and DAPI for nuclei (*blue*, **a**, **b**). **c** Western blot analysis of nestin, Tuj1, GFAP and RIP (oligodendrocyte marker) protein expression of differentiated NSC on 2-D graphene films and 3-D graphene foams. **d** Relative optical densities of bands of markers shown in (**c**). Data presented as mean ± SE, *p 0.05; **p 0.01. Reprinted by permission from Macmillan Publishers Ltd [97], copyright (2013)

2-D graphene films and 3-D graphene foams, synthesized by CVD and without any type of biological coating, with the murine BV2 microglial cell line. The rationale behind the selection of this cell type is that microglial cells are the resident macrophages of the brain and spinal cord and they function as the first defense barrier of the innate immune system in the CNS [100]. BV2 cells displayed similar morphologies and high viability, measured by the MTT assay, on graphene materials than those on control substrates. Interestingly, the level of intracellular ROS was found inferior on 3-D graphene foams, with or without stimulation with LPS, while enhanced on 2-D graphene films with respect to the control. When specific pro-inflammatory parameters were investigated (e.g., TNF-α, IL-1β and nitric oxide concentrations, induced nitric oxide synthase and cyclooxygenase-2 levels), results revealed not statistically significant differences between BV2 cells grown on control substrates and 2-D graphene films (after LPS stimulation). On the contrary, all these parameters severely decreased when cells were cultured on 3-D graphene foams, attributed to a partial inhibition of the degradation of IκB proteins in overactivated BV2 cells. Further studies included the exploration of the impact of conditioned media from activated microglial cell cultures on murine NSC and PC12 cells. Conditioned media from 3-D graphene foams seemed to facilitate the growth of both cell types with high viability, while ROS production was not significantly increased. Taking together, these findings outline a potentially protective anti-inflammatory role played by 3-D graphene foams, able to mitigate the activation of microglial cells (by both reducing the production of pro-inflammatory responses after stimulation and limiting their morphological transformations) and alleviate the LPS-induced cytotoxicity of NSC caused by activated microglial cells.

6 In Vivo Studies with Graphene-Based Materials for Neural Repair

Despite the tremendous potential that GBM have extensively demonstrated in vitro with neural cells, their in vivo exploration in neural tissue scenarios is still rare. In the particular context of the CNS, Vicario-Abejón and colleagues recently reported pioneer findings on the ability of rGO to mediate the in vivo generation of new neurons in the adult brain [101]. rGO nanosheets were injected in the brain along with a retroviral vector expressing the green fluorescent protein to label dividing progenitor cells in the core of the adult olfactory bulb. Results demonstrated that the presence of rGO did not alter de novo neurogenesis or neuronal and astrocyte survival. It has to be emphasized that there were no signs of the activation of microglial responses. In a different study, Serrano and colleagues pioneered the investigation of the response of the injured rat spinal cord to porous and flexible rGO scaffolds, encouraged by in vitro positive findings with embryonic neural progenitor cells [102]. At the subacute stage, the substrates were completely colonized by extracellular matrix molecules (e.g., collagen) and cells, mainly positive for markers such as vimentin, platelet-derived growth factor receptor β (PDGRβ) and ED1 [103]. Besides the absence of toxicity in liver, spleen, lungs, and kidneys, the mechanical compliance of these structures with the spinal cord tissue promoted the formation of a soft interface that did not induce additional fibroglial scaring with respect to lesions not carrying scaffolds. Moreover, cell populations were preserved at the perilesional areas and a small fraction of pro-regenerative M2 macrophages was identified both at the interface and inside the scaffold. After 30 days (chronic state), results revealed the existence of a pro-regenerative scenario at the lesion site in the presence of rGO scaffolds characterized by significant immunomodulatory changes (reduction in the total amount of macrophages and presence of those displaying M2 phenotype), enhanced collagen fibers infiltration and angiogenesis, supporting the growth of some new axons inside the scaffold structure [104].

Novel work by Jakus et al. [105] have demonstrated that a solution-based, scalable ink composed of graphene (3–8 atomic layers in thickness, 5–20 μm in length/width) could be 3-D printed into arbitrarily shaped scaffolds (e.g., porous constructs, hollow tubes) with filaments ranging from 100 to 1000 μm in diameter (Fig. 7). Graphene comprised up to 60 vol.% of the resulting scaffolds, being the rest made of polylactic-*co*-glycolic (PLGA). These 3-D structures presented superior properties to similar materials reported to date, including conductivities greater than 800 S cm^{-1}, compressive stress of 600 kPa up to 100 °C and in vitro

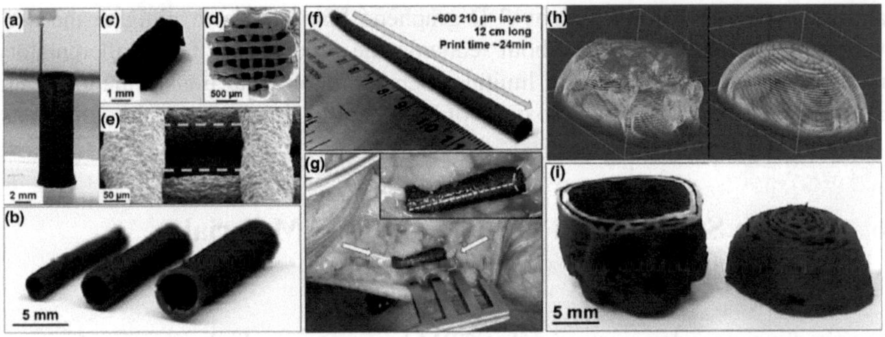

Fig. 7 **a** 3-D printable graphene (3DG) ink can be rapidly 3-D printed into self-supporting tubular structures (140 layers) of **b** various sizes that could serve as custom-sized nerve graft conduits. **c**, **d** Uniaxial, multichannel nerve guides may also be 3-D printed from 3DG inks. **e** Scanning electron microscopy micrograph of multichannel 3DG nerve conduit with every other layer close to contact (*yellow box*), minimizing or eliminating pores orthogonal to the major axis of nerve guide. **f** 3DG can be 3-D printed into structures composed of many hundreds of layers, such as this high aspect ratio (24:1) 5 mm diameter hollow tube, which can be cut to size as needed. **g** Photograph of tubular 3DG nerve conduit cut from **f** implanted into a human cadaver via longitudinal transection and wrapping around the ulnar nerve (*white arrows*). The 3DG nerve conduit was then sutured close along the previously described longitudinal transection (*white dotted line*), as well as to the surrounding epineuro and nerve tissue (*inset*, *yellow circle*). **h** Digitally sliced file of skull and skull cap and **i** photograph of resulting 3-D printed 3DG skull and skull cap. Reprinted with permission from [105]. Copyright (2015) American Chemical Society

biocompatibility and neurogenic ability with human MSC. Subcutaneous implantation in mouse revealed that these structures did not elicit inflammatory responses or fibrous encapsulation up to 30 days, while being actively disassembled by the surrounding cells. Further implantation tests in a human cadaver as a conduit in the ulnar nerve confirmed their exceptional handling characteristics and potential feasibility for fine surgical procedures.

7 Graphene-Based Materials for Neural Stimulation and Recording

An increasing body of work is supporting the idea that mechanically compliant electrodes should comprise the next generation of neural interfaces [106]. Based on this, the outstanding mechanical properties of GBM, alongside with their electrical conductivity, are fueling their exploration in this field. For instance, Chen et al. [107] have developed flexible graphene-based microprobes for the detection of electrophysiological signals not only of action potentials in axons but also QRS complex and T waves in electrocardiograms (Fig. 8). Recordings showed a significantly higher resolution when treated by steam plasma in order to decrease the

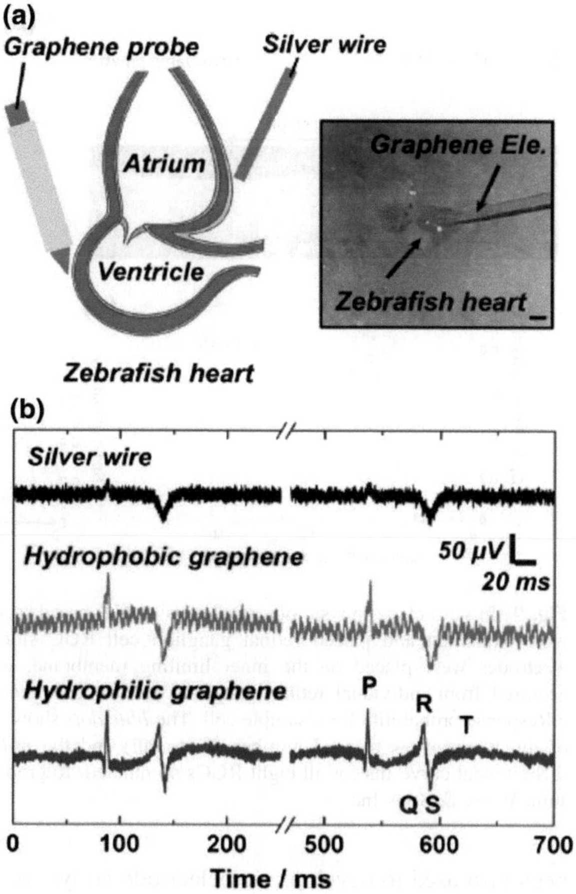

Fig. 8 Electrocardiographic recording: **a** Schematic and actual view of the cardiac recording system for zebrafish (*scale bar* 500 μm). **b** Amplitude of P, QRS complex and T waves recorded using a hydrophilic graphene electrode. Reprinted from [107], Copyright (2013), with permission from Elsevier

interfacial impedance between graphene and the electrolyte. These electrodes, used to record electrical signals in crayfish axons and zebrafish heart, displayed capacitance values of 1.4×10^{-3} mF cm^{-2}, impedance values of 5424 Ω mm^{-2}, action potentials of 102.64 m $V_{p\text{-}p}$, and a noise signal of 4.2 μ V_{rms}. In a more recent approach, strong and conductive microfibers (40–50 μm in diameter) obtained by wet-spinning liquid crystalline dispersions of GO were proposed as a new generation of compliant and free-standing electrodes [108]. After biocompatible insulation with parylene-C, a laser treatment was used to open and expand the end of the fiber (3.5 times) in a "brush" electrode configuration. Electrochemical characterization demonstrated that the resulting electrodes had a high charge injection capacity in the range of tens of mC cm^{-2}. These electrodes were successfully used for in vitro electrophysiology studies, stimulating retinal ganglion cells in live retina and recording at high quality when implanted in the visual cortex after encapsulation within a water-soluble sucrose microneedle (Fig. 9), thus encouraging further work for their implementation in electrode arrays. Graphene obtained by CVD has

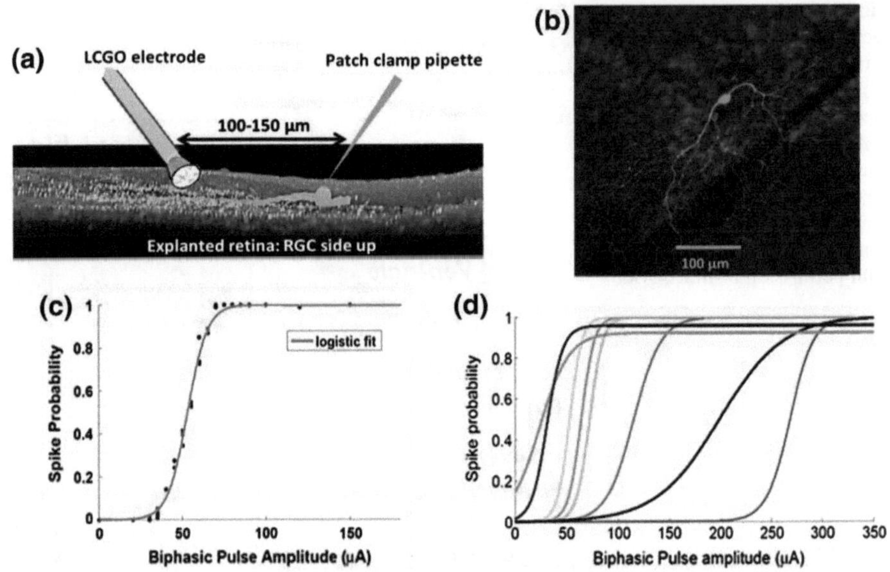

Fig. 9 In vitro electrophysiology studies using GO-based stimulating electrodes. **a** Whole retinas were explanted and placed retinal ganglion cell RGC side up in a perfusion chamber. The electrodes were placed on the inner limiting membrane, while patch clamp recordings were acquired from individual retinal ganglion cells. **b** 3-D reconstruction of an exemplary RGC. **c** Response probability for a sample cell. The *blue dots* show the raw probability (ratio of number of direct responses to total number of stimuli) and the *red line* shows a sigmoidal curve fit. **d** Sigmoidal curve fits for all eight RGCs stimulated. Reproduced from [108] with permission of John Wiley & Sons Inc

been also used to develop microelectrode arrays for neural recording [109]. In this approach, the description of the interface graphene–electrolyte was explained by a modified double-layer theory. These electrodes were able to detect neural activities from dissociated E18 rat cortical neurons with a signal-to-noise ratio of 10.31 ± 1.2.

Novel functionalization strategies of stimulation and recording electrodes tend to promote cell adhesion processes due to the relevance that they play in signal transduction in microelectrode arrays [110]. That is the case of work by Wallace and colleagues, in which polymers composed of polylactic acid or PLGA were coated with a conductive graphene layer for the preparation of unique electrodes [111]. When PC12 cells were cultured in contact with these electrodes, their differentiation towards neural phenotypes was enhanced (increase in the average length of neurites per cell and neurite length) using ES through the graphene layer (100 µs pulses of ± 0.1 mA cm^{-2} with a 20 µs interphase gap and a 3.88 ms rest period in each cycle, applied for 8 h a day during three consecutive days). This ES protocol, although promoting the outgrowth of existing neurites, did not induce the formation of new ones. The interest of these findings is crucial to the field as they enable the use of non-conducting biocompatible materials for electrode fabrication by simply coating their surface with conductive graphene layers. Regarding their performance, electrodes with a 50:50 PLGA composition showed specific capacitance values of *ca.* 7 μF cm^{-2} by cyclic voltammetry. In a different study, Casañ-Pastor and colleagues decorated electrodes for neural recording with nanostructured pristine graphene electrodeposited by means of iridium oxide nanoparticles [112]. In comparison to GO, this electrodeposited graphene displayed charge storage capacity values one order of magnitude larger. These hybrid coatings demonstrated adequate in vitro cytocompatibility with cortical neural cells (isolated from mouse E16 embryos), supporting their growth and release of neurotransmitters such as GABA and glutamate after six divisions. Luo et al. [113] doped PEDOT with GO to fabricate biocompatible electrodes. The resulting PEDOT/GO films supported the growth of neural cells obtained from E18 rat embryos cortical tissue, with significantly longer neurites in the presence of GO. Viability tests at 24 h revealed values over 96 % even in the absence of any biological coating of the material. When p20, a laminin fragment peptide known to potentiate neurite outgrowth, was covalently immobilized into the films, neurons attached to the films were able to sprout significantly longer neurites. Tian et al. [114] has also reported conducting polymer nanocomposites composed of GO-doped PEDOT as useful electrode–tissue interfaces with excellent electrochemical performance and in vitro cytocompatibility.

In the olfactory system, recent work by Jang et al. [115] has presented the first artificial multiplexed superbioelectronic nose mimicking the human olfactory sensory system with high-performance odorant discriminatory ability in mixtures (Fig. 10). These portable noses were fabricated using highly uniform graphene micropatterns (designed by CVD and photolithography) conjugated with different human olfactory receptors (hOR), employed as transducers in a liquid-ion gated field-effect transistor. Specifically, two types of human olfactory receptors (hORs) were selected, being their respective target odorants amyl butylate and helional, and immobilized on the graphene-based substrate by Schiff-base reactions. By introducing alternative sensing materials, the artificial nose provided excellent odorant discrimination in experiments involving various odorants and mixtures, with a minimum detectable level of 0.1 fM.

8 Other Uses of Graphene-Based Materials in Neural Scenarios

An interesting and novel approach by Choi and colleagues described the use of GO for encapsulating gold nanoparticles in a 3-D structure effective for the detection of the differentiation potential of NSC by surface-enhanced Raman spectroscopy [116]. Curiously, undifferentiated cells on these nanoparticles produced signal peaks 3.5 times higher than normal metal structures and distinguishable from differentiated cells. Moreover, the number of C=C bonds (related to the saturation level of cell membranes) and the Raman intensity at 1656 cm^{-1} showed a

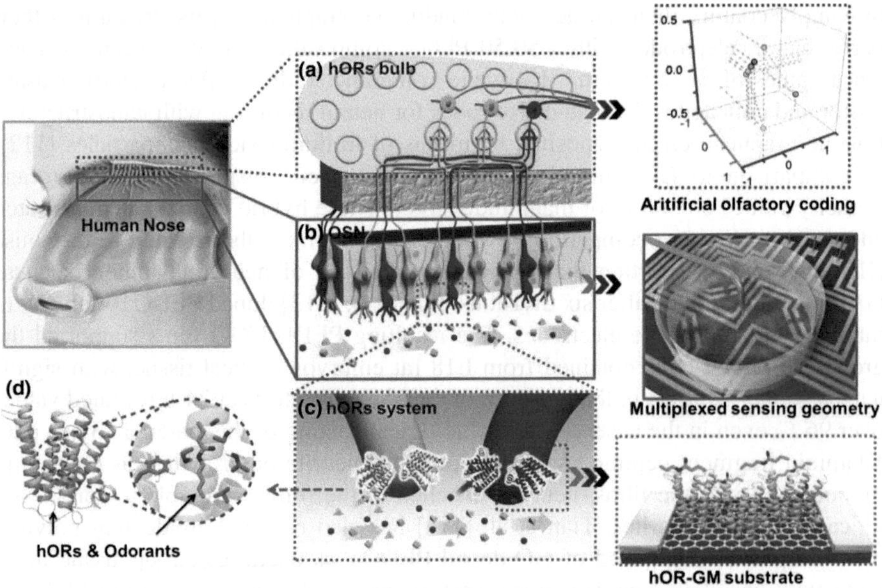

Fig. 10 Schematic diagram of functional anatomy of human olfactory system and components of a multiplexed superbioelectronic nose simulating each functional stages of human nose. **a** Olfactory bulb, where the olfactory signals generated by olfactory sensory neurons (OSN) are combined for the generation of combinatorial olfactory codes, matching with artificial olfactory codes generated by the electronic nose. **b** Olfactory sensory neurons, where olfactory signals triggered by the specific binding of hORs and odorants, matching with graphene micropatterns functionalized with hOR. **c** hORs for the specific recognition of odorants. **d** Illumination of the specific interaction between the hOR and the odorant. Reprinted with permission from [115]. Copyright (2015) American Chemical Society

correlation with the differentiation state of the cells. The versatility of these nanoparticles increases when considering its applicability as a nondestructive in situ monitoring tool for the differentiation stage of other stem cells, not only the neural ones.

More recent studies described the utility of GO nanoparticles decorated with platinum nanoparticles for the modification of ITO plates aiming to the highly sensitive and rapid detection of acetylcholine [117]. Nanoparticles were electrochemically deposited onto the ITO-coated glass plates. The resulting biosensors displayed an improved analytical performance in terms of low working potential (0.2 V), short response time (4 s), low detection limit (5 nM), and high storage stability (4 months) and might also serve for the improvement of other types of biosensors. In this sense, Hossain et al. [118] have reported the development of graphene nanoribbon neuro-sensors for detection and imaging of glycine, an inhibitory neurotransmitter in the CNS that plays critical roles in processing motor and sensory information related to control movement, vision, and audition. The zigzag-edged graphene nanoribbon noticeably changed its density of states, transmission coefficient, and conductance in response to the attachment of glycine molecules. This change is reflected in a significant drop of current flow through the device. Interestingly, the presence of a nitrogen-vacancy center in the device resulted in lowering the power consumption and increasing the current through the device up to 35 %, thus allowing for the concurrent application of these nanoribbons in both biosensing and imaging in in vitro settings. Different graphene-based strategies are also under investigation for the detection of neuron-specific enolase, a lung cancer marker, including nickel hexacyanoferrate nanoparticles assembled on gold nanocrystals electrodes posteriorly coated with gold nanoparticles-functionalized graphene nanosheets [119], graphene nanosheets in combination with magnetic nanoparticles [120], and guanine-decorated graphene nanostructures [121], to cite a few examples. For further details on the use of GBM for biosensing and related applications, readers are referred to further chapters in this book and excellent reviews published before [122].

9 Final Discussion and Future Perspectives

The need of substitutes for damaged and diseased tissues/organs are fueling experts in different fields to design and develop an increasing variety of implantable biomedical devices, many of them already in the market or under investigation in clinical trials [123]. In the particular case of neural implants, some of the most advanced devices include BrainGate™ 2 (Cyberkinetics Neurotechnology Systems Inc.) and Argus® II (Second Sight Medical Products Inc.). The first one is a sensor device to be implanted on the brain motor cortex in patients with impaired motor function and is currently ongoing clinical evaluation [124]. Argus® II, a wireless sensory micro-conduit with a video camera and processing unit to replace the function of light-sensing cells, has received approval by the USFDA and European Union for the treatment of patients with *retinitis pigmentosa* [125]. Although advancing, the field of neural implants is still in its infancy and requires devoted efforts to accomplish the development of more sophisticated, advanced and efficient devices.

In this scenario, GBM appear as promising materials with four major attractive features for interfacing the neural tissue: biocompatibility, electrical activity, feasibility for bio-functionalization, and high surface to volume ratio for drug delivery. Along this chapter, we have extensively revised most of the work involving the use of GBM for neural-related applications, pointing out their potential in the different concrete scenarios. It is worth noting that the interest of GBM for neural regeneration is not only based on their both in vitro and in vivo biocompatibility with neural tissue components, but also on their fascinating interactions with stem cells including NSC and MSC [126]. This aspect opens an even wider perspective for the utility of these materials in regenerative medicine. Nonetheless, we must be prudent because the concrete role played by the different features of graphene and its derivatives in cell differentiation processes is still unknown. Another field that could clearly benefit from the attractive properties of GBM is biosurface engineering [127], with applicability not only on the development of more efficient neural interfaces as mentioned in this chapter, but also in dental implants, joint replacements, and bone substitutes, among others.

Despite this potential and interest, there are still open questions regarding the toxicity of GBM (in both human health and the environment) that boost further investigation to thoroughly clarify so critical aspects. Moreover, contradictory results in some of the published studies also hamper progress in the biomedical use of these materials. In this sense, the youth of these materials is likely one of the principal reasons for the existence of physical–chemical features and properties of GBM that might be dismissed or neglected but playing relevant roles on the applicability of these materials in biomedicine, as recently evidenced by Delacour and colleagues for the crystalline quality of pristine graphene [128]. A cautious reflection must be placed in the biomedical use of GBM, because most of studies have been performed in vitro and, typically, there is not a direct correlation or translation from those to in vivo models. We hope this revision of the state of the art in the use of GBM for neural repair serves to impulse the research community to translate the potentiality of these materials from cell cultures to animal models, a closer scenario for their future safe use in human healthcare.

Acknowledgments MCS thanks the *Instituto de Salud Carlos III* and the *Ministerio de Economía y Competitividad* of Spain for a Miguel Servet I contract (MS13/00060, co-funded by FEDER).

References

1. Tavangarian F, Li Y (2012) Carbon nanostructures as nerve scaffolds for repairing large gaps in severed nerves. Ceram Int 38:6075–6090
2. Li D, Müller MB, Gilje S et al (2008) Processable aqueous dispersions of graphene nanosheets. Nat Nanotechnol 3:101–105
3. Yang X, Zhang X, Liu Z et al (2008) High-efficiency loading and controlled release of doxorubicin hydrochloride on graphene oxide. J Phys Chem C 112:17554–17558
4. Yang L, Wang F, Han H et al (2015) Functionalized graphene oxide as a drug carrier for loading pirfenidone in treatment of subarachnoid hemorrhage. Colloids Surf B 129:21–29
5. Chau NDQ, Ménard-Moyon C, Kostarelos K, Bianco A (2015) Multifunctional carbon nanomaterial hybrids for magnetic manipulation and targeting. Biochem Biophys Res Commun 468:454–462
6. Soldano C, Mahmood A, Dujardin E (2010) Production, properties and potential of graphene. Carbon 48:2127–2150
7. Lee C, Wei X, Kysar JW et al (2008) Measurement of the elastic properties and intrinsic strength of monolayer graphene. Science 321:385–388
8. Fraczek-Szczypta A (2014) Carbon nanomaterials for nerve tissue stimulation and regeneration. Mater Sci Eng C 34:35–49
9. Gu X, Ding F, Williams DF (2014) Neural tissue engineering options for peripheral nerve regeneration. Biomaterials 35:6143–6156
10. Zhang N, Yan H, Wen X (2005) Tissue-engineering approaches for axonal guidance. Brain Res Rev 49:48–64

11. Nakanishi W, Minami K, Shrestha LK, Ji Q, Hill JP, Ariga K (2014) Bioactive nanocarbon assemblies: nanoarchitectonics and applications. Nano Today 9:378–394
12. Fattahi P, Yang G, Kim G, Abidian MR (2014) A review of organic and inorganic biomaterials for neural interfaces. Adv Mater 26:1846–1885
13. John AA, Subramanian AP, Vellayappan MV, Balaji A, Mohandas H, Jaqanathan SK (2015) Carbon nanotubes and graphene as emerging candidates in neuroregeneration and neurodrug delivery. Int J Nanomed 10:4267–4277
14. Li X, Katsanevakis E, Liu X et al (2012) Engineering neural stem cell fates with hydrogel design for central nervous system regeneration. Prog Polym Sci 37:1105–1129
15. Goshmitra S, Diercks D, Mills NC, Hynds A, Ghosh S (2012) Role of engineered nanocarriers for axón regeneration and guidance: current status and future trends. Adv Drug Deliv Rev 64:110–125
16. Chen R, Cohen LG, Hallett M (2002) Nervous system reorganization following injury. Neuroscience 111:761–773
17. Fawcett JW, Asher RA (1999) The glial scar and central nervous system repair. Brain Res Bull 49:377–391
18. Seil JT, Webster TJ (2010) Electrically active nanomaterials as improved neural tissue regeneration scaffolds. WIREs Nanomed Nanobiotechnol 2:635–647
19. Donaghue IE, Tam R, Sefton MV, Shoichet MS (2014) Cell and biomolecule delivery for tissue repair and regeneration in the central nervous system. J Control Rel 190:219–227
20. Hsieh FY, Lin HH, Hsu SH (2015) 3D bioprinting of neural stem cell-laden thermoresponsive biodegradable polyurethane hydrogel and potential in central nervous system repair. Biomaterials 71:48–57
21. Thelin J, Jorntell H, Psouni E et al (2011) Implant size and fixation mode strongly influence tissue reactions in the CNS. PLoS One 6:e16267
22. Robinson LR (2000) Traumatic injury to peripheral nerves. Muscle Nerve 23:863–873
23. Taylor CA, Braza D, Rice JB et al (2008) The incidence of peripheral nerve injury in extremity trauma. Am J Phys Med Rehabil 87:381–385
24. Wolford LM, Stevao ELL (2003) Considerations in nerve repair. BUMC Proc 16:152–156
25. Meek MF, Coert JH (2008) US Food and Drug Administration/Conformit Europe-approved absorbable nerve conduits for clinical repair of peripheral and cranial nerves. Ann Plast Surg 60:466–472
26. Kehoe S, Zhang XF, Boyd D (2012) FDA approved guidance conduits and wraps for peripheral nerve injury: a review of materials and efficacy. Injury 43:553–572
27. Matsumoto K, Ohnishi K, Kiyotani T et al (2000) Peripheral nerve regeneration across an 80-mm gap bridged by a poly-glycolic acid (PGA)-collagen tube filled with laminin-coated collagen fibers: a histological and electrophysiological evaluation of regenerated nerves. Brain Res 868:315–328
28. Cai J, Peng X, Nelson KD et al (2005) Permeable guidance channels containing microfilament scaffolds enhance axon growth and maturation. J Biomed Mater Res A 75:374–386
29. Chew SY, Mi R, Hoke A et al (2007) Aligned protein-polymer composite fibers enhance nerve regeneration: a potential tissue-engineering platform. Adv Funct Mater 17:1288–1296
30. Gu X, Ding F, Yang Y et al (2011) Construction of tissue engineered nerve grafts and their application in peripheral nerve regeneration. Prog Neurobiol 93:204–230
31. Hu N, Wu H, Xue C et al (2013) Long-term outcome of the repair of 50 mm long median nerve defects in rhesus monkeys with marrow mesenchymal stem cells-containing chitosan-based tissue engineered nerve grafts. Biomaterials 34:100–111
32. Ding F, Wu J, Yang Y et al (2010) Use of tissue-engineered nerve grafts consisting of a chitosan/poly (lactic-*co*-glycolic acid)-based scaffold included with bone marrow mesenchymal cells for bridging 50-mm dog sciatic nerve gaps. Tissue Eng Part A 16:3779–3790
33. Yang Y, Yuan X, Ding F et al (2011) Repair of rat sciatic nerve gap by a silk fibroin-based scaffold added with bone marrow mesenchymal stem cells. Tissue Eng Part A 17:2231–2244

34. Kingham PJ, Kalbermatten DF, Mahay D et al (2007) Adipose-derived stem cells differentiate into a Schwann cell phenotype and promote neurite outgrowth in vitro. Exp Neurol 207:267–274
35. Xu Y, Liu L, Li Y et al (2008) Myelin-forming ability of Schwann cell-like cells induced from rat adipose-derived stem cells in vitro. Brain Res 1239:49–55
36. Georgiou M, Golding JP, Loughlin AJ et al (2015) Engineered neural tissue with aligned, differentiated adipose-derived stem cells promotes peripheral nerve regeneration across a critical sized defect in rat sciatic nerve. Biomaterials 37:242–251
37. Rooney GE, Moran C, McMahon SS et al (2008) Gene-modified mesenchymal stem cells express functionally active nerve growth factor on an engineered poly lactic glycolic acid (PLGA) substrate. Tissue Eng Part A 14:681–690
38. Hudson TW, Evans GR, Schmidt CE (2000) Engineering strategies for peripheral nerve repair. Orthop Clin N Am 31:485–498
39. Evans GR (2001) Peripheral nerve injury: a review and approach to tissue engineered constructs. Anat Rec 263:396–404
40. Hopkins AM, DeSimone E, Chwalek K, Kaplan DL (2015) 3D in vitro modeling of the central nervous system. Prog Neurobiol 125:1–25
41. Pancrazio JJ, Wang F, Kelley CA (2007) Enabling tools for tissue engineering. Biosens Bioelectron 22:2803–2811
42. Cao H, Liu T, Chew SY (2009) The application of nanofibrous scaffolds in neural tissue engineering. Adv Drug Deliv Rev 61:1055–1064
43. Spivey EC, Khaing ZZ, Shear JB et al (2012) The fundamental role of sub-cellular topography in peripheral nerve repair therapies. Biomaterials 33:4264–4276
44. Schmidt CE, Leach JB (2003) Neural tissue engineering: strategies for repair and regeneration. Annu Rev Biomed Eng 5:293–347
45. Khaing ZZ, Schmidt CE (2012) Advances in natural biomaterials for nerve tissue repair. Neurosci Lett 519:103–114
46. Jin GZ, Kim M, Shin US et al (2011) Neurite outgrowth of dorsal root ganglia neurons is enhanced on aligned nanofibrous biopolymer scaffold with carbon nanotube coating. Neurosci Lett 501:10–14
47. Jiang X, Lim SH, Mao HQ et al (2010) Current applications and future perspectives of artificial nerve conduits. Exp Neurol 223:86–101
48. Oh SH, Kim JH, Song KS et al (2008) Peripheral nerve regeneration within an asymmetrically porous PLGA/Pluronic F127 nerve guide conduit. Biomaterials 29:1601–1609
49. Novosel EC, Kleinhans C, Kluger PJ (2011) Vascularization is the key challenge in tissue engineering. Adv Drug Deliv Rev 63:300–311
50. Ferrara N (1999) Role of vascular endothelial growth factor in the regulation of angiogenesis. Kidney Int 56:794–814
51. Sondell M, Lundborg G, Kanje M (1999) Vascular endothelial growth factor has neurotrophic activity and stimulates axonal outgrowth, enhancing cell survival and Schwann cell proliferation in the peripheral nervous system. J Neurosci 19:5731–5740
52. Hobson MI, Green CJ, Terenghi G (2000) VEGF enhances intraneural angiogenesis and improves nerve regeneration after axotomy. J Anat 197:591–605
53. Pola R, Aprahamian TR, Bosch-Marcé M et al (2004) Age-dependent VEGF expression and intraneural neovascularization during regeneration of peripheral nerves. Neurobiol Aging 25:1361–1368
54. Aizawa Y, Wylie R, Shoichet M (2010) Endothelial cell guidance in 3D patterned scaffolds. Adv Mater 22:4831–4835
55. Lee YB, Polio S, Lee W et al (2010) Bio-printing of collagen and VEGF-releasing fibrin gel scaffolds for neural stem cell culture. Exp Neurol 223:645–652
56. Benowitz LI, Popovich PG (2011) Inflammation and axon regeneration. Curr Opin Neurol 24:577–583

57. Al-Majed AA, Neumann CM, Brushart TM et al (2000) Brief electrical stimulation promotes the speed and accuracy of motor axonal regeneration. J Neurosci 20:2602–2608
58. Brushart TM, Hoffman PN, Royall RM et al (2002) Electrical stimulation promotes motoneuron regeneration without increasing its speed or conditioning the neuron. J Neurosci 22:6631–6638
59. Brushart TM, Jari R, Verge V et al (2005) Electrical stimulation restores the specificity of sensory axon regeneration. Exp Neurol 194:221–229
60. Geremia NM, Gordon T, Brushart TM et al (2007) Electrical stimulation promotes sensory neuron regeneration and growth-associated gene expression. Exp Neurol 205:347–359
61. McCaig CD, Rajnicek AM, Song B et al (2005) Controlling cell behavior electrically: current views and future potential. Physiol Rev 85:943–978
62. Clagett-Dame M, McNeill EM, Muley PD (2006) Role of all-trans retinoic acid in neurite outgrowth and axonal elongation. J Neurobiol 66:739–756
63. Teng YD, Lavik EB, Qu X et al (2002) Functional recovery following traumatic spinal cord injury mediated by a unique polymer scaffold seeded with neural stem cells. Proc Natl Acad Sci USA 99:3024–3029
64. Li N, Zhang X, Song Q et al (2011) The promotion of neurite sprouting and outgrowth of mouse hippocampal cells in culture by graphene substrates. Biomaterials 32:9374–9382
65. Sahni D, Jea A, Mata JA et al (2013) Biocompatibility of pristine graphene for neuronal interface: laboratory investigation. J Neurosurg Pediatr 11:575–583
66. Hong D, Bae K, Yoo S et al (2013) Generation of cellular micropatterns on single-layered graphene film. Macromol Biosci 14:314–319
67. Kim SM, Joo P, Ahn G et al (2013) Transparent conducting films based on reduced graphene oxide multilayers for biocompatible neuronal interfaces. J Biomed Nanothechnol 9:403–408
68. Lv M, Zhang Y, Liang L et al (2012) Effect of graphene oxide on undifferentiated and retinoic acid-differentiated SHSY5Y cells line. Nanoscale 4:3861–3866
69. Lee JS, Lipatov A, Ha L et al (2015) Gaphene substrate for inducing neurite outgrowth. Biochem Biophys Res Commun 460:267–273
70. Tu Q, Pang L, Chen Y et al (2014) Effects of surface charges of graphene oxide on neuronal outgrowth and branching. Analyst 139:105–115
71. Tu Q, Pang L, Wang L et al (2013) Biomimetic choline-like graphene oxide composites for neurite sprouting and outgrowth. ACS Appl Mater Interfaces 5:13188–13197
72. Meng S (2014) Nerve cell differentiation using constant and programmed electrical stimulation through conductive non-functional graphene nanosheets film. Tissue Eng Reg Med 11:274–283
73. Heo C, Yoo J, Lee S et al (2011) The control of neural cell-to-cell interactions through non-contact electrical field stimulation using graphene electrodes. Biomaterials 32:19–27
74. Liu HW, Huang WC, Chiang CS et al (2014) Arrayed rGO_{SH}/PMA_{SH} microcapsule platform integrating surface topography, chemical cues, and electrical stimulation for three-dimensional neuron-like cells growth and neurite sprouting. Adv Funct Mater 24:3715–3724
75. Baniasadi H, Ramazani A, Mashayelhan S et al (2015) Design, fabrication and characterization of novel porous conductive scaffolds for nerve tissue engineering. Int J Polym Mater Polym Biomater 64:969–977
76. Song J, Gao H, Zhu G et al (2015) The preparation and characterization of polycaprolactone/graphene oxide biocomposite nanofiber scaffolds and their application for directing cell behaviors. Carbon 95:1039–1050
77. Zhou K, Thouas GA, Bernand CC et al (2012) Method to impart electro- and biofunctionality to neural scaffolds using graphene–polyelectrolyte multilayers. ACS Appl Mater Interfaces 4:4524–4531
78. Park SY, Park J, Sim SH et al (2011) Enhanced differentiation of human neural stem cells into neurons on graphene. Adv Mater 23:H263–H267

79. Solanki A, Dean Chueng ST, Yin PT et al (2013) Axonal alignment and enhanced neuronal differentiation of neural stem cells on graphene-nanoparticle hybrid structures. Adv Mater 25:5477–5482
80. Akhavan O, Ghaderi E (2013) Differentiation of human neural stem cells into neural networks on graphene nanogrids. J Mater Chem B 1:6291–6301
81. Akhavan O, Ghaderi E, Emamy H et al (2013) Genotoxicity of graphene nanoribbons in human mesenchymal stem cells. Carbon 54:419–431
82. Akhavan O, Ghaderi E (2013) Flash photo stimulation of human neural stem cells on graphene/TiO_2 heterojunction for differentiation into neurons. Nanoscale 5:10316–10326
83. Akhavan O, Ghaderi E, Abouei E et al (2014) Accelerated differentiation of neural stem cells into neurons on ginseng-reduced graphene oxide sheets. Carbon 66:395–406
84. Akhavan O, Ghaderi E, Shirazian SA (2015) Near infrared laser stimulation of human neural stem cells into neurons on graphene nanomesh semiconductors. Colloids Surf B Biointerfaces 126:313–321
85. Akhavan O, Ghaderi E (2014) The use of graphene in the self-organized differentiation of human neural stem cells into neurons under pulsed laser stimulation. J Mater Chem B 2:5602–5611
86. Akhavan O, Ghaderi E, Shirazian SA et al (2016) Rolled graphene oxide foams as three-dimensional scaffolds for growth of neural fibers using electrical stimulation of stem cells. Carbon 97:71–77
87. Tang M, Song Q, Li N et al (2013) Enhancement of electrical signaling in neural networks on graphene films. Biomaterials 34:6402–6411
88. Shah S, Yin PT, Uehara TM et al (2014) Guiding stem cell differentiation into oligodendrocytes using graphene-nanofiber hybrid scaffolds. Adv Mater 26:3673–3680
89. Weaver CL, Cui XT (2015) Directed neural stem cell differentiation with a functionalized graphene oxide nanocomposite. Adv Healthc Mater 4:1408–1416
90. González-Mayorga A, Gutiérrez MC, Collazos-Castro JE, Ferrer ML, del Monte F, Serrano MC (2016) Graphene oxide microfibers as selective substrates for neural regeneration. J Mater Chem B (under review)
91. Wang Y, Lee WC, Manga KK et al (2012) Fluorinated graphene for promoting neuro-induction of stem cells. Adv Mater 24:4285–4290
92. Müller K, Faeh C, Diederich F (2007) Fluorine in pharmaceuticals: looking beyond intuition. Science 317:1881–1886
93. Oh HG, Nam HG, Kim DH et al (2014) Neuroblastoma cells grown on fluorine or oxygen treated graphene sheets. Mater Lett 131:328–331
94. Lee YJ, Jang W, Im H et al (2015) Extremely low frequency electromagnetic fields enhance neuronal differentiation of human mesenchymal stem cells on graphene-based substrates. Curr Appl Phys 15:S95–S102
95. Kim TH, Shah S, Yang L et al (2015) Controlling differentiation of adipose-derived stem cells using combinatorial graphene hybrid-pattern arrays. ACS Nano 9:3780–3790
96. Chen GY, Pang DWP, Hwang SM et al (2012) A graphene-based platform for induced pluripotent stem cells culture and differentiation. Biomaterials 33:418–427
97. Li N, Zhang Q, Gao S et al (2013) Three-dimensional graphene foam as a biocompatible and conductive scaffold for neural stem cells. Sci Rep 3:1604
98. Yang D, Li T, Xu M et al (2014) Graphene oxide promotes the differentiation of mouse embryonic stem cells to dopamine neurons. Nanomedicine (Lond) 9:2445–2455
99. Song Q, Jiang Z, Li N et al (2014) Anti-inflammatory effects of three-dimensional graphene foams cultured with microglial cells. Biomaterials 35:6930–6940
100. Block ML, Zecca L, Hong JS (2007) Microglia-mediated neurotoxicity: uncovering the molecular mechanisms. Nat Rev Neurosci 8:57–69

101. Defterali C, Verdejo R, Peponi L et al (2016) Thermally reduced graphene is a permissive material for neurons and astrocytes and de novo neurogenesis in the adult olfactory bulb in vivo. Biomaterials 82:84–93
102. Serrano MC, Patiño J, García-Rama C et al (2014) 3D free-standing porous scaffolds made of graphene oxide as substrates for neural cells growth. J Mater Chem B 2:5698–5706
103. López-Dolado E, González-Mayorga A, Portolés MT et al (2015) Subacute tissue response to 3D graphene oxide scaffolds implanted in the injured rat spinal cord. Adv Healthc Mater 4:1861–1868
104. López-Dolado E, González-Mayorga A, Gutiérrez MC, Serrano MC (2016) Immunomodulatory and angiogenic responses induced by graphene oxide scaffolds in chronic spinal hemisected rats. Biomaterials 99:72–81
105. Jakus AE, Secor EB, Rutz AL et al (2015) Three-dimensional printing of high-content graphene scaffolds for electronic and biomedical applications. ACS Nano 9:4636–4648
106. Nguyen JK, Park DJ, Skousen JL et al (2014) Mechanically-compliant intracortical implants reduce the neuroinflammatory response. J Neural Eng 11:056014
107. Chen CH, Lin CT, Hsu WL et al (2013) A flexible hydrophilic modified graphene microprobe for neural and cardiac recording. Nanomed Nanotechnol Biol Med 9:600–604
108. Apollo NV, Maturana MI, Tong W et al (2015) Soft, flexible freestanding neural stimulation and recording electrodes fabricated from reduced graphene oxide. Adv Funct Mater 25:3551–3559
109. Du X, Wu L, Cheng J et al (2015) Graphene microelectrode arrays for neural activity detection. J Biol Phys 41:339–347
110. Blau A (2013) Cell adhesion promotion strategies for signal transduction enhancement in microelectrode array in vitro electrophysiology: an introductory overview and critical discussion. Curr Opin Coll Interf Sci 18:481–492
111. Sherrell PC, Thompson BC, Wassei JK et al (2014) Maintaining cytocompatibility of biopolymers through a graphene layer for electrical stimulation of nerve cells. Adv Funct Mater 24:769–776
112. Pérez E, Lichtenstein MP, Suñol C et al (2015) Coatings of nanostructured pristine graphene-IrOx hybrids for neural electrodes: layered stacking and the role of non-oxygenated graphene. Mater Sci Eng C 55:218–226
113. Luo X, Weaver CL, Tan S et al (2013) Pure graphene oxide doped conducting polymer nanocomposite for bio-interfacing. J Mater Chem B 1:1340–1348
114. Tian HC, Liu JQ, Wei DX et al (2014) Graphene oxide doped conducting polymer nanocomposite film for electrode–tissue interface. Biomaterials 35:2120–2129
115. Kwon OS, Song HS, Park SJ et al (2015) An ultrasensitive, selective, multiplexed superbioelectronic nose that mimics the human sense of smell. Nano Lett 15:6559–6567
116. Kim TH, Lee KB, Choi JW (2013) 3D graphene oxide-encapsulated gold nanoparticles to detect neural stem cell differentiation. Biomaterials 34:8660–8670
117. Chauhan N, Narang J, Jain U (2015) Highly sensitive and rapid detection of acetylcholine using an ITO plate modified with platinum–graphene nanoparticles. Analyst 140:1988–1994
118. Hossain FM, Al-Dirini F, Skafidas E (2014) A graphene nanoribbon neuro-sensor for glycine detection and imaging. J Appl Phys 115:214303
119. Han J, Zhuo Y, Chai YQ, Yuan YL, Yuan R (2012) Novel electrochemical catalysis as signal amplified strategy for label-free detection of neuron-specific enolase. Biosens Bioelectron 31:399–405
120. Li GZ, Tian F (2013) Magnetic graphene nanosheet-based electrochemical immunosensing platform for sensitive monitoring of neuron-specific enolase. Sens Lett 11:1931–1936
121. Li GZ, Tian F (2013) Guanine-decorated graphene nanostructures for sensitive monitoring of neuron-specific enolase based on an enzyme-free electrocatalytic reaction. Anal Sci 29:1195–1201

122. Medina-Sánchez M, Miserere S, Merkoçi A (2012) Nanomaterials and lab-on-a-chip technologies. Lab Chip 12:1932–1943
123. Arsiwala A, Desai P, Patravale V (2014) Recent advances in micro/nanoscale biomedical implants. J Control Rel 189:25–45
124. www.cyberkinetics.com. Last retrieved 03 June 2016
125. www.2-sight.eu/en/patients-families-en. Last retrieved 03 June 2016
126. Gardin C, Piattelli A, Zavan B (2016) Graphene in regenerative medicine: focus on stem cells and neuronal differentiation. Trends Biotechnol. doi:10.1016/j.tibtech.2016.01.006
127. Dong HS, Qi SJ (2015) Realising the potential of graphene-based materials for biosurfaces: a future perspective. Biosurf Biotribol 1:229–248
128. Veliev F, Briançon-Marjollet A, Bouchiat V, Delacour C (2016) Impact of crystalline quality on neuronal affinity of pristine graphene. Biomaterials 86:33–41

Graphene-Based Smart Nanomaterials: Novel Opportunities for Biology and Neuroengineering

Antonina M. Monaco and Michele Giugliano

Abstract In the last three decades, nanotechnologies have so deeply integrated themselves with medicine, that a new term, "nanomedicine," was specifically coined (Freitas in Nanomedicine, volume I: basic capabilities. Landes Bioscience, Georgetown, 1999, [110]) to indicate "*the process of diagnosing, treating, and preventing disease and traumatic injury, relieving pain, and preserving and improving human health, using molecular tools and molecular knowledge of the human body. In short, nanomedicine is the application of nanotechnology to medicine.*"

As Freitas underlined in the same paper [1], though it has been formalized in the late 1980s [2, 3], the concepts themselves of nanotechnology and nanomedicine directly come from the famous visionary talk "There's plenty of room at the bottom" in which the Nobel prize winner Richard Feynman foresaw the great possibilities of the scale-down method: applying this method repeatedly he hypothesized the possibility of the construction of machines able to manipulate atoms and molecules [4].

Since then, so much progress has been made and so many goals have been achieved in several fields that, though not being completely aware of it, we can now consider nanotechnology as an integral part of our everyday life: Titanium dioxide nanoparticles are present in sunscreen lotions and in orthopedic implants; Silver

A.M. Monaco · M. Giugliano (✉)
Theoretical Neurobiology and Neuroengineering Lab, Department of Biomedical Sciences, University of Antwerp, Universiteitsplein 1, 2610 Wilrijk, Belgium
e-mail: Michele.Giugliano@uantwepen.be

M. Giugliano
Brain Mind Institute, Swiss Federal Institute of Technology, Lausanne, Switzerland

M. Giugliano
Department of Computer Science, University of Sheffield, S1 4DP Sheffield, UK

G. Gonçalves et al. (eds.), *Graphene-based Materials in Health and Environment*, Carbon Nanostructures, DOI 10.1007/978-3-319-45639-3_7

nanoparticles are used as antimicrobial agents in textiles and Iron oxide ones to improve the scratch resistance of paints; Carbon fibers make our umbrellas lighter; nanosensors, nanometric drug carriers and *lab-on-chip* are more and more often used for theranostics at the nanoscale.

Nanoparticles and, more in general, nanodevices used in nanomedicine are synthesized from several elements, such as Gold, Silver, Titanium, and Carbon. The latter one, being the key element of life itself, has been intensively studied for biomedical and biological applications, in all its allotropic forms and, among them, researchers identified in nanocrystalline diamond, carbon nanotubes, and graphene extremely good candidates for drug and gene delivery systems, materials for coating electrodes for nervous system and cardiac stimulation, biosensors, and photothermal therapy.

Here we review some of the most important biological applications of graphene and its derivatives, such as graphene oxide (GO) and reduced graphene oxide (rGO), with emphasis on the applications in Neurosciences.

1 Toxicity

Interfacing new materials, regardless their being micro- or nano-, with biological systems requires in-depth biocompatibility evaluations, in terms of cyto- and genotoxity, generation of reactive oxygen species (ROS), interaction of the materials with cells constituents, biological media and organs, and depletion of essential nutrients for cell functions by absorption on surface's materials.

Though toxicity of nanomaterials based on Carbon (C) (i.e. carbon nanotubes, carbon black, nanodiamonds, graphene, and its derivatives) has been extensively investigated, the debate in the scientific community is still quite heated given the controversial results and the remaining open questions about the precise mechanisms of internalization of these materials, as well as their localization once entered in the cytoplasm [5–8]. Furthermore, if on one side the most relevant limitations of CNTs in biological applications, such as the presence of metallic impurities and their asbestos-like shape [9, 10], can be overcome by graphene's morphology and synthesis methods, biocompatibility evaluations of graphene are made more difficult by different physicochemical characteristics of the forms employed, such as single-, few- or multilayer graphene, GO, rGO, nanosheets, nanoplatelets, and nanoflakes.

As reported for other C-based nanomaterials [11–13], surface functionalization of graphene reduces its unfavorable effects [14, 15], which are however due also to the size and shape of graphene itself.

Influence of lateral size of graphene and its derivatives on internalization and cell viability has been studied for several cell lines. Akhavan et al. observed and correlated, for the first time, in human mesenchymal stem cells (hMSc) treated with reduced graphene oxide nanoplateles and nanosheets [16] and graphene nanoribbons [17], increased levels of ROS, reduced viability and chromosomal aberrations with the concentration and the lateral size of the material. A similar study, conducted on human alveolar adenocarcinoma (A549), human colon carcinoma (Caco2), and monkey kidney cells (Vero) cell lines treated with GO at different concentrations and flakes' size, highlighted a loss in viability for A549 cell line and a positive increase of viability for Caco2 and Vero cell line, and relevant genotoxic effects for nanometer-sized GO flakes [18].

Effects of GO on A549 cell line were investigated also by the research group of Wang, who studied viability, morphology, membrane disruption, and mortality of cells treated with three GO samples, different in size [19]. In this study, it was observed that GO slightly affects cell viability even at the highest tested concentrations (200 μg/ml), in correspondence of which almost the 70 % of cells resulted to be healthy regardless the exposure time (24, 48, and 72 h), and that none of the three sample affects morphology and ultrastructure of cells. However, treating A549 cells with GO led to a significant increase of ROS, strictly dose- and size-dependent: the smaller GO tested (s-GO, 170 nm) induced high ROS generation, almost four times the control condition at 200 μg/ml, which was anyway three times less than the ROS level for the positive control. Authors thus conclude that, overall, GO display a good biocompatibility and that is a suitable substrate for cells growth (Fig. 1).

These findings, though in good agreement with previous studies discussing the use of GO as drug delivery system [20, 21], are opposite to those ones of Cui et al., who reported reduced viability of human fibroblast cells exposed to GO for concentrations higher than 50 μg/ml, combined with lower cell adhesion, morphological changes and localization of GO sheets in mitochondria, endoplasm, and lysosomes [22]. In the same study, authors also discuss the effects of GO, administered via tail vein injection, on mice, highlighting that injection dose of GO at 0.1 and 0.25 mg/ml did not affect mice lifespan, while four of the nine mice treated with high dose of GO (0.4 mg/ml) died within 7 days from the injection. Furthermore, histopathological analysis of tissue and organs of mice after 7 and 30 days revealed the formation of granuloma localized in liver, lung and spleen; interestingly, no GO was found in brain tissues, suggesting that GO is probably not able to cross the blood–brain barrier. To explain these observations, authors suggested two possible mechanisms of GO toxicity: once in contact with human cells in vitro, GO triggers processes of signal transduction inside cells and nuclei, which lead to downregulation of genes associated to adhesion proteins, while its presence inside cellular structures leads to alterations of gene transcription and cell energy metabolism. When administrated to mice via vessel injection, GO might be recognized as a foreign body, thus triggering a defensive response. Once entered in lung tissues, GO is attacked by macrophages, this leading to the formation of granulomas. Moreover, because of their particular shape, GO flakes cannot be extruded by kidneys, liver and spleen; these organs seem to be not affected by GO presence at the lowest tested concentration, but the observed mice mortality for the highest dose suggests that this GO heap interferes with organs functionality, thus resulting in mice death.

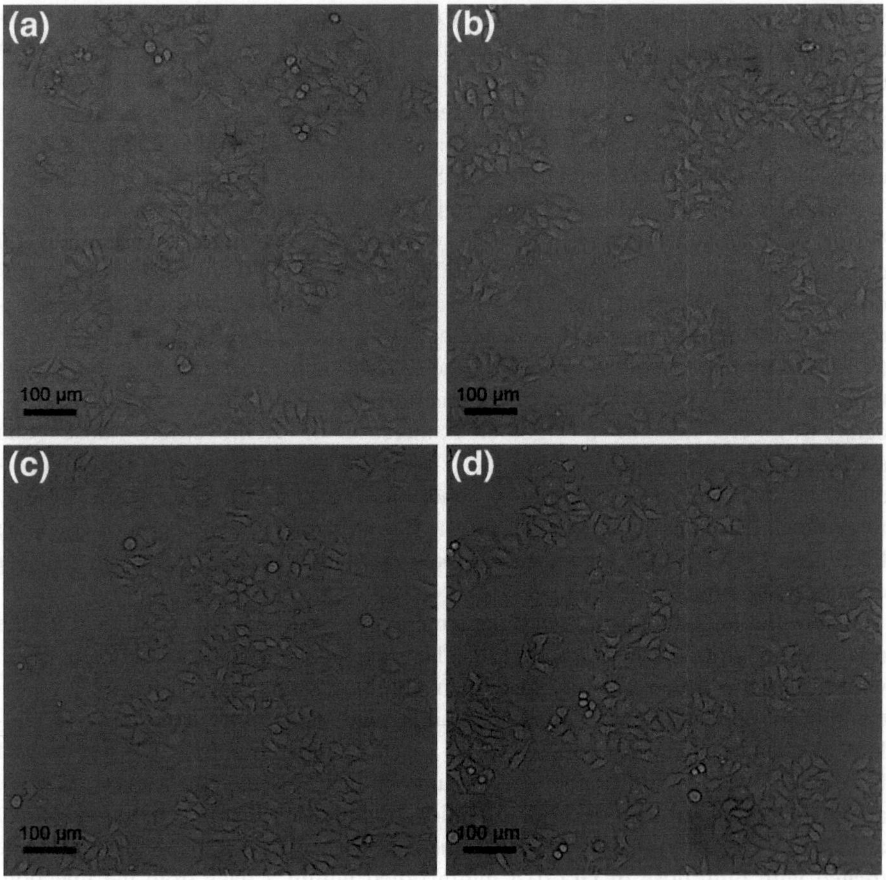

Fig. 1 Optical microscopy images of A549 cells grown on GO (**a–c**) and on control (**d**) substrates. Reproduced with permission from [109]

Another mechanism of GO toxicity, proposed by Hu et al. [23], suggests that physical damages observed in cellular membrane can be due to the electronic interactions taking place between the positively charged lipid layer and the negatively charged groups present on GO surface.

As already mentioned, surface modifications of graphene affect its toxicity; results of several studies show that carboxylated graphene and rGO exert less toxic effects than native graphene and GO [24], that chitosan coating of GO modulates its cytotoxicity [25], and that intravenous administration of amine-functionalized graphene (G-NH_2) to mice do not trigger any macrophages response leading to pulmonary thromboembolisms [26]. These studies also investigated haemocompatibility of graphene and GO [24, 25], a key aspect for drug delivery systems, which requires systemic route of administration. Graphene sheets resulted to be slightly more toxic than GO, and both materials did not affect coagulation pathways, though they induced a dose-dependent haemolysis of red blood cells.

Lack of consensus in cellular viability is reflected also in evaluations of graphene and its derivatives toxicity for bacterial cells; if, on one side, the use of GO, rGO, graphene, and composites GO–Ag can be investigated for designing antimicrobial coatings as pathogens such as *Staphylococcus aureus*, *Escherichia coli*, *Fusarum oxysporum,* and *Aspergillus niger* are inactivated by the presence of these materials [27–31], on the other side this antimicrobial activity has not been observed for both same and different families of bacteria, such as Shewanella [32–34].

In vivo assessments of contingent negative effects of graphene and its derivatives are less numerous than in vitro ones; in addition to the already cited study of Wang et al., the research conducted by Singh et al. on human blood platelets showed for the first time that intravenous injection of thin GO sheets in mice affects the release mechanisms of Ca^{2+} and the activation of Src kinases, this resulting in the formation of aggregates leading to pulmonary thromboembolism. These adverse effects showed a dependence on surface charge distribution of the material, as they were mitigated, albeit not completely removed, administrating rGO thin sheets [35].

Similarly to in vitro evaluations, material's functionalization leads to a modulation of its toxicity, as observed by Sahu et al. [36] studying the effects of PEGylated GO used as component for injectable hydrogels that resulted to be stable after administration to mice, without triggering sever toxic reactions.

Granuloma, pulmonary edema and rise of inflammations are reported for intravenous injection of GO at the dose of 10 mg/kg in mice, while no pathological changes of organs were found at 1 mg/kg. Lungs seem to be the preferred target organ for GO; although this makes it a good candidate for targeted drug delivery, its difficult excretion might lead, in a long-term scenario, to the same adverse effects observed for higher doses because of GO retention in lung tissues [37].

2 Biomedical Applications

The remarkable properties of graphene directly derive from its peculiar chemical structure, in particular its surface easy to functionalize, the ability to adsorb several aromatic biomolecules, and of being processed in aqueous solutions, their having both hydrophilic and hydrophobic groups as well as the fluorescence quenching ability, and make it and its derived nanomaterials extremely interesting for biomedical applications.

This kind of applications can be divided in four main classes: (1) biosensors, (2) substrates, coatings and scaffolds for implants and tissue engineering, (3) biomedical imaging, and (4) drug delivery systems. Here we briefly review some of the most relevant studies.

2.1 Delivery Systems

Graphene's chemical structure and surface area makes it an extremely interesting material as drug carrier because of the possibility of binding pharmacological molecules on both sides of the graphene sheet. However, graphene is insoluble in water and this requires, as first step, the oxidation to its water-soluble form, graphene oxide and the subsequent functionalization with surfactants, mainly PEG, to avoid the clustering of the material once in contact with biological media.

These aspects were investigated by the group of Dai et al. [20, 21], who synthesized PEGylated nano-graphene oxide (NGO) loaded with Doxorubucin and SN-38 (7-ethyl-10-hydroxy-camptothecin), two water-insoluble anticancer drugs, and with an antigen to a specific activated-glycosylated phosphoprotein over expressed in cancer cells in order to target specifically the latters. Stability of these systems exhibited a pH-dependency and the efficiency of the SN-38 loaded NGO, tested on a human colon cancer cell line (HCT-116) resulted to be comparable with the free SN-38 in DMSO but remarkably more potent than a similar drug, camptothecin (CPT-11), incubated with the same cell line. Furthermore, the intrinsic photoluminescence of these NGO was used to image living cells in the near-infrared region (NIR) with very little background.

The pH-dependency highlighted by Dai et al. was investigated by other research groups, who observed that the increased release of drug molecules, related to their improved solubility for lowered pH, might eventually lead to a controlled release of the drugs themselves into lysosomes once the system "drug-carrier" is internalized in cells by endocytosis [38–40].

Binding GO, covalently functionalized with sulfonate groups, to folic acid (FA) allows the specific targeting of human breast cancer cells (MCF-7), as they express FA receptors; exploiting this specificity, Zhang et al. [41] demonstrated that FA–GO loaded with a controlled mix of CPT and Doxorubicin efficiently targets only cells expressing FA receptors and that it is more toxic to MCF-7 cells compared to FA–GO loaded with only a single drug. FA–GO has been also tested as carrier for Ce6, a photosensitizer used in photodynamic therapy; results of in vitro studies demonstrated that incubating human stomach cancer cell lines (MGC803) with FA–GO–Ce6 and then irradiating them significantly affects viability of cells [42].

Thanks to their strong optical adsorption in the NIR region, graphene and its derivatives have also been investigated as agent for photothermal therapy. Hu et al. [43] synthesized and tested in vitro a quantum-dot-tagged rGO nanocomposite (QD-rGO) covalently bonded with FA to be used for both cell/tumor bioimaging and photothermal therapy of MCF-7 cells, observing a selective uptake of QD-rGO in the targeted cells and their consequent death following 4 min of irradiation at 808 nm. Zhang et al. [44] explored the simultaneous use of PEGylated nano-GO as photothermal agent and as Doxorubicin carrier, thus obtaining a nanocomposite able to deliver both a chemotherapic agent and heat. This nanocomposite was tested in vitro on a murine cancer cell line (EMT6) and in vivo on a Xenograft tumor mouse model, and it resulted to be more efficient than the two therapies singularly applied, leading to a complete destruction of tumors without any recurrence.

GO was also investigated for gene delivery and for the combined delivery of drugs and gene; Bao et al. [45] reported the use of chitosan-functionalized GO as a carrier to separately deliver CPT and pDNA into human liver and cervical cancer cell lines (HepG2 and HeLa cells), while Zhang et al. [46] designed polyethylenimine (PEI) functionalized GO loaded with Doxorubicin and short interfering RNA (siRNA) which, inhibiting the protein expression of targeted proteins, might overcome the problem of multiple drug resistance of cancer cells.

2.2 *Bioimaging*

Given their intrinsic photoluminescence in the VIS and in the IR spectral regions, graphene-based materials have been investigated to image living cells and several biomolecules inside living cells in the NIR via fluorescence, magnetic resonance (MRI), and positron emission tomography (PET) imaging.

Hong et al. [47] covalently functionalized nano-GO with a specific monoclonal antibody (TRC105) binding a vascular marker for tumor angiogenesis (CD105) and investigated its tumor targeting efficacy and pharmacokinetics in an in vivo model of murine breast cancer using PET and biodistribution studies. Results showed that nanocomposites are mainly excreted through renal and hepatobiliary pathways and that TRC105-GO effectively targets the tumor, this suggesting the possible use of this nanocomposite as combined agent for photothermal therapy and drug delivery system.

The use of composites of GO and dextran-coated iron oxide nanoparticles (Fe_3O_4-GO) as T_2-weighted contrast agent for MRI has been reported by Chen et al. [48], who also highlighted that these composites exhibit significantly enhanced cellular MRI signal.

Several research groups focused their attention on photoluminescence of graphene quantum dots (GQDs) and on how different preparation methods and surface functionalization can affect it. Peng et al. [49] obtained 1–4 nm sized GQDs by chemical exfoliation and acid treatment of carbon fibers, modulating the color of their photoluminescence by changing the temperature of the reaction. Green-photoluminescent GQDs were tested in vitro on human breast cancer cell lines and obtained results showed that GQDs can be used in high contrast bioimaging applications (Fig. 2).

Blue-fluorescent amino- and carboxy-functionalized GQDs were obtained by hydrothermal cutting of graphene sheets in presence of ammonia and/or water solutions [50, 51]; yellow-fluorescent GQDs were electrochemically synthesized by graphite rods and their uptake and toxicity were tested on three different kinds of stem cells—pancreas and cardiac progenitor cells (PPCs and CPCs), and neurospheres cells (NSCs)—observing that these GQDs can easily be internalized by cells without affecting their viability [52].

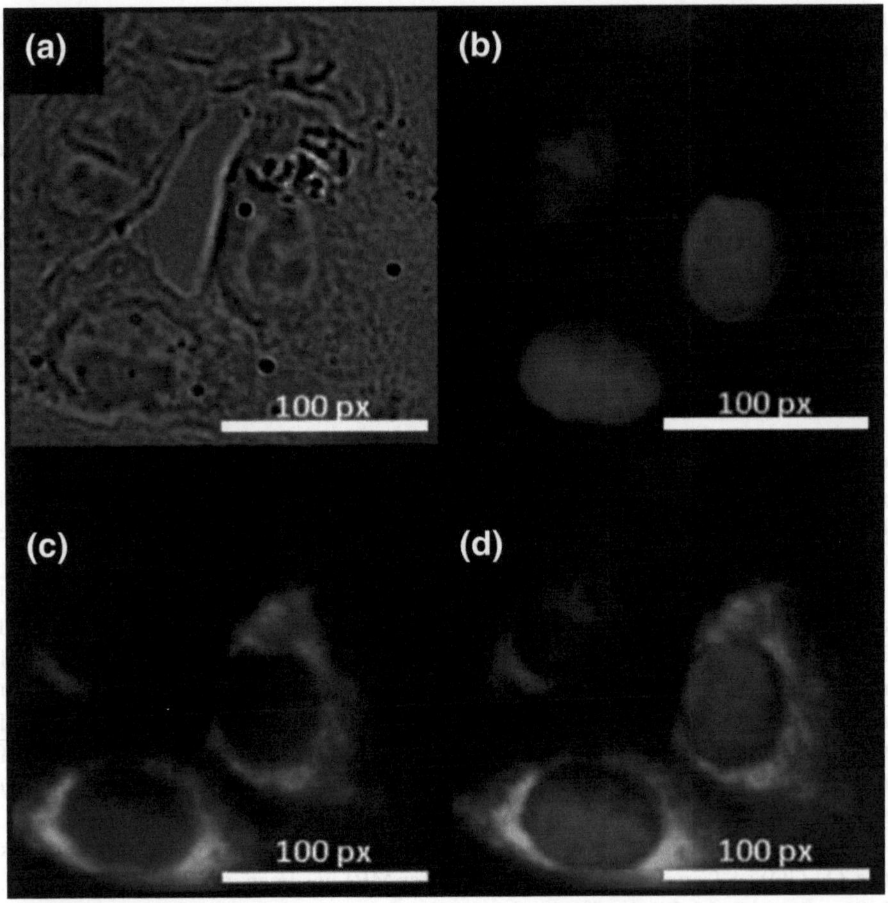

Fig. 2 Phase contrast picture (**a**) and fluorescent images (**b–d**) of human breast cancer cells incubated with green graphene quantum dots (GQDs). Nuclei are stained in *blue* (DAPI) and GQDs have *green fluorescence*; panel **d** shows the overlay of panels **b** and **c**. Reproduced with permission from [49]. © 2012 American Chemical Society

GO nanosheets, combined with DNA/RNA aptamers, were used as sensing platform for simultaneous, selective, and in situ detection of nucleotides involved in the regulation of several biological reactions, such as adenosine-5′-triphospahte (ATP) and guanosine-5′-triphosphate (GTP) [53]. This kind of detection is possible because no hydrolysis of ssDNA by deoxyribonuclease (DNase) has been reported, this suggesting that, once adsorbed onto the surface of GO, ssDNA is protected from enzymatic digestion [54].

2.3 Biosensors

Accordingly with the IUPAC definitions, a chemical sensor is a

> a device that transforms chemical information, ranging from the concentration of a specific sample component to total composition analysis, into an analytically useful signal. Chemical sensors contain usually two basic components connected in series: a chemical (molecular) recognition system (receptor) and a physicochemical transducer. Biosensors are chemical sensors in which the recognition system utilizes a biochemical mechanism. [55]

Graphene, thanks to its excellent electrochemical properties, seems to be a promising material to be used in electrodes for detecting biomolecules.

Given the intrinsic fluorescence of GO from NIR to UV wavelengths [56], this material found use in the fabrication of fluorescence resonance energy transfer (FRET) sensors mainly interfaced with single-strand DNA (ssDNA), as the interaction between GO surface and the exposed bases leads to a strong adsorption of ssDNA to material surface. In this way it is possible to detect and quantify multiple ssDNA, as well as microRNA and double-strand DNA (dsDNA) [57–59]. A device composed by Silver nanoparticles as acceptor and GO, chemically treated with *n*-butylamine, as donor in a FRET sensor was developed and used to optically detect DNA, glutathione, cysteine, and immunoglobulin G [60].

So far, GO FRET biosensors have been used to detect a broad variety of biomolecules, such as insulin [61], proteins [62], and nucleotides [63], as well as metal ions [64, 65].

CVD-grown graphene was exploited in field effect transistor (FET)-based biosensors for detection of nucleic acids, growth factors and proteins [66, 67]; Loh et al. designed a graphene sensor integrated with microfluidic flow cytometry in order to detect red blood cells infected by malaria [19].

Wang et al. [68] developed a graphene modified electrode to be used for selective detection of dopamine, overcoming the limitations of the simultaneous presence of ascorbic acid, whose oxidation potential is quite close to that one of dopamine. A similar graphene/Pt-modified glassy carbon electrode was designed and successfully tested to simultaneously detect dopamine, ascorbic acid and uric acid by Sun et al. [69]. This electrode was compared to only glassy carbon and to only graphene electrodes, and it resulted to provide better measurements of current and potential both using cyclic and differential pulse voltammetry.

2.4 Substrates, Scaffolds and Tissue Engineering

As other nanomaterials [70, 71], graphene and its derivatives have been used as substrates and scaffolds for differentiation of stem cells and antibacterial effects, as well as for culturing primary mammalian cells. Park et al. [72] reported an enhanced neuronal differentiation of human neuronal stem cells (hNSCs) grown on graphene films and measured the neural activity of these cultures using the graphene film itself as a stimulating electrode. A similar result was reported by Wang et al. [73], who observed that bone marrow derived mesenchymal stem cells (MSCs) cultured on fluorinated graphene showed neuron-like morphology with visible neurite protrusions and that they expressed neuronal gene markers.

Chen et al. [74] cultured induced murine pluripotent stem cells (iPSCs) on graphene and GO substrates, observing not only that both materials support iPSCs culture and allow for spontaneous differentiation into ectodermal and mesodermal lineages, but also that different substrates lead to distinct cell proliferation and differentiation characteristics. In particular, iPSCs proliferate and differentiate at a faster rate on GO than the control and the graphene condition.

The reasons of these enhanced growth and differentiation of stem cells on graphene and GO were investigated by Loh et al., who demonstrated that both graphene and GO act as preconcentration platforms for accelerated stem cell growth and differentiation through molecular interactions with growth agent [75].

3 Graphene in the Neurosciences

Their nanoscale dimensions, similar to those ones of the central nervous system (CNS), make nanomaterials ideal candidates for applications in neurosciences, and this drove researchers to investigate them *(a)* for developing both stimulating and sensing technologies to be interfaced with brain tissue and/or nerve cells to repair the brain on its own scale, *(b)* for refining brain imaging, *(c)* as a helpful tool in neurosurgery, and *(d)* for improving noninvasive diagnosis techniques allowing direct access to the CNS.

Graphene and its derivatives, as well as other C-based nanomaterials, has attracted great interest for applications in Neurosciences thanks to their chemical stability and electrically conductive properties. Three are the major applications: *(a)* as substrates and 3-D scaffold for neural growth, *(b)* as material for coating the electrodes of micro elcctrode arrays (MEAs), and *(c)* as material used for field effect transistors (FETs).

3.1 Graphene for Extracellular Stimulations and Recordings of Neuronal Activity: MEAs and FETs

Substrate-integrated microelectrode arrays (MEAs) are devices consisting of metallic electrodes (e.g., made of Pt, Au, and titanium nitrate) embedded in a planar substrate. Such devices allow the study of neuronal physiology, pathology and circuit-connectivity, both in vitro and in vivo, through extracellular recordings of neuronal activity. This methodology, though offering several advantages, such as the simultaneous detection of extracellular field potentials in a completely noninvasive way as well as the possibility of studying the very same neuronal networks over weeks and even months, is not (yet) able to detect synaptic potentials generated by single cells [76], and it has also to meet specific requirements, such as high spatial resolution, large signal-to-noise ratio (*S*/*N*), large charge injection limits, and great biocompatibility. If designing MEAs with smaller electrodes leads to improved spatial resolution, on the other hand this decreases the injected charge limits and worsen the *S*/*N* ratio, as a consequence of the reduced electrode's surface exposed to the electrolyte or to neuronal cell membranes. It is thus clear that the unique electrical properties of graphene make it a very interesting and promising material for the design of a novel class of (micro)electrodes whose use is not merely confined to electronic and material sciences applications [77–81].

An easy and relatively cheap technique to fabricate graphene-MEAs from a CVD-grown graphene films, deposited on quartz substrates, on which Au/Ti electrodes were patterned by lithography was developed by Du et al. [82]; such devices resulted to have good transparence and a *S*/*N* ratio comparable with that one of commercially available MEAs made of other materials and they were used to successfully detect extracellular spontaneous activity of cortical rat neurons from 14 to 40 days in vitro. Interestingly, once cleaned by means of conventional techniques (i.e., via mechanical washing and/or enzymatic digestion), graphene-MEAs exhibited a slightly higher value of impedance with respect to the value before their use, this indicating a long-term stability of the devices.

Another fabrication method has been recently proposed by Koerbitzer et al. [83], who deposited a film of CVD-grown graphene on Gold and on Silicon Dioxide substrates to evaluate how graphene coating can influence the performance of, respectively, conductive yet opaque and not conductive yet transparent electrodes. Results of characterization of these devices showed that, when deposited on Au electrodes, graphene does not significantly modify the electrochemical properties of the electrodes themselves while, when deposited on SiO_2, it improves charge injection capacity so that these electrodes display performances comparable to those of TiN electrodes. These MEAs also showed good cell adhesion properties and biocompatibility, as they were used to culture cryoconserved embryonic cortical rat neurons for several weeks; however, the ability of these devices to detect extracellular signals remains an open question, as authors did not perform recordings or stimulation of the electrical activity in neuronal networks.

The possibility of designing transparent and flexible graphene-MEAs is extremely intriguing, as it might open new horizons in the investigation of electrical properties of populations of neurons such as the simultaneous optical imaging, optogenetic modulation and electrophysiological recordings. Park et al. [84], for instance, developed an implantable graphene-based, carbon-layered electrode array (CLEAR) allowing high-resolution neurophysiological recordings. Characterization of these MEAs by means of cyclic voltammetry and electrical impedance

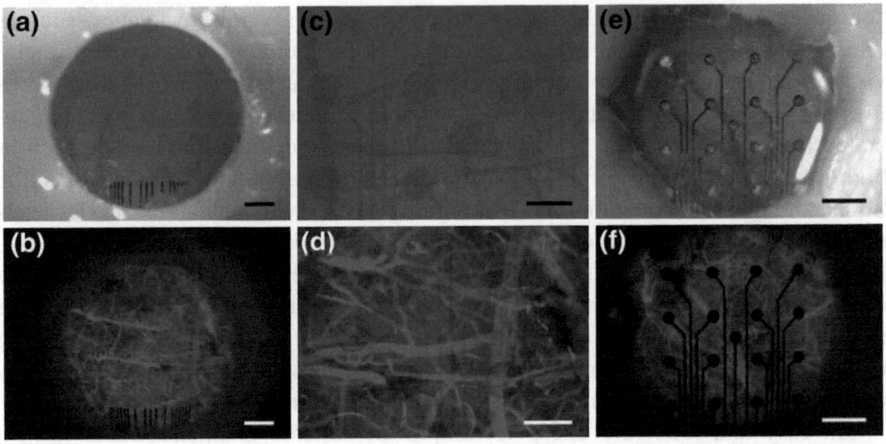

Fig. 3 In vivo cortical vasculature images through CLEAR device. Bright-field (**a**, **c**) and fluorescence (**b**, **d**) images of CLEAR device implanted on the cerebral cortex at different magnifications; *scale bars* 500 nm (**a**, **b**) and 250 nm (**c**, **d**). Bright-field (**e**) and fluorescence (**f**) images of standard micro-ECoG arrays; *scale bars* 750 nm. Reproduced with permission from [84]. © 2014 American Chemical Society

spectroscopy revealed a slightly higher impedance value, compared to conventional Pt microelectrodes arrays, which is thought not to affect recordings of neuronal activity, and similar CV curves were observed for the CLEAR device and Au microelectrodes. Efficiency of these CLEAR MEAs was then tested in vivo, by implanting them in both mice and rats, in comparison to conventional Pt devices; results show that CLEAR MEAs allow to record neuronal signals without difference from Pt MEAs, but with the significant advantage to allow optogenetic stimulation, as well as fluorescence and OCT imaging, directly through the electrode sites, made possible thanks to graphene transparency (Fig. 3).

Similar transparent devices were developed by transferring undoped or nitric acid-doped graphene grown by CVD onto flexible polyimide substrates, previously patterned with Au contacts, and single electrodes (doped and undoped graphene: $50 \times 50\ \mu m^2$; Au: $500 \times 500\ \mu m^2$) were tested for in vitro recordings from brain slices and in vivo electrocorticography recordings. Graphene and doped graphene electrodes showed lower impedances than Au electrodes, especially for frequency lower than 1 kHz, and they allowed in vivo recordings of neural activity with high *S*/*N* ratio, as well as calcium imaging in hippocampal slices by both two-photon and confocal microscopy [85].

Heo et al. [86] had also investigated the use of graphene for in vitro or in vivo stimulator devices; their research led to the design of a graphene/PET film to test the effects of non-contact field stimulations on cell-to-cell coupling. The electrical stimulation delivered through this film, whose biocompatibility and suitability for cell proliferation were demonstrated, affected the regulation of cytoskeleton protein related to cellular mobility, such as actin, this reflecting in morphological changes in cellular edges.

FETs are another kind of arrays used to record electrical activity of tissue and electrogenic cells and, if on one side they offer some advantages with respect to MEAs (i.e., easier fabrication of high-density structures, intrinsic amplification and better *S*/*N* ratio for structures of similar dimensions), on the other side they present the relevant setback of low stability of Silicon (their major component) in aqueous solutions, as well as the sharp edges and poor flexibility of crystalline structures needed in order to achieve a high *S*/*N* ratio; these drawbacks thus set limitations to the use of such devices for in vitro, but especially for in vivo, investigations [87–89].

Graphene, by virtue of its extraordinary electrical and optical properties and chemical stability, has attracted the interest of many researchers working in this field, and this led to the design and development of flexible graphene solution-gated FETs (graphene-SGFETs) with better gate sensitivity than common FETs and interesting *S*/*N* ratio, that were successfully used to record action potentials in electrogenic cells [78, 90, 91].

3.2 Graphene and Neuronal Growth: Neural Stem Cells

As shown in the previous paragraph, several studies have demonstrated that graphene and its derivatives enhance, though in a nonspecific way, cellular growth and the differentiation of different kind of stem cells—such as human neural stem cells (hNSCs) and mesenchymal stem cells (MSCs)—into neurons.

The important role of substrate's surface chemistry in the differentiation of MSCs into neurons has been investigated by Wang et al. [73], who reported stronger polarization and higher proliferation of MSCs seeded on fluorinated graphene substrates. This specific surface functionalization of graphene-induced morphological changes and promoted the differentiation of MSCs into neurons both in presence and in absence of neuron-inductive chemical inducers, such as retinoic acid.

Akhavan et al. [92] compared the contingent effects of GO and rGO, reduced by both conventional hydrazine-based and by an innovative green ginseng-based methods, on hNSCs; they highlighted a better attachment and a higher proliferation for cells grown on GO and ginseng-rGO when compared with both control and hydrazine-rGO condition, probably due to the higher presence of Oxygen group on the surface of ginseng-rGO and GO. Moreover, 3 weeks after the induced differentiation of hNSCs, by means of culture medium lacking growth factors, cells grown on rGO, and especially on ginseng-rGO, displayed significant morphological differences. These results seem to be related to the rGO higher capability for electron transfer and to the already mentioned higher hydrophilicity, and thus to a better biocompatibility, of GO and ginseng-rGO.

The same research group has also investigated the differentiation of hNSCs on GO nanogrids deposited on a substrate made of TiO_2 nanoparticles over a film of SiO_2 [93], on rGO/TiO_2 heterojunction films [94] and on GO and rGO films deposited by drop casting onto quartz substrates [95]. The particular design of these substrates, as well as their post-synthesis treatments, allowed for their use as biocompatible flash photo stimulators for effective differentiation of hNSCs into neurons, which led to a more differentiation of hNSCs into neurons than glia, and to a more pronounced increase in cell growth and alignment along the geometrical pattern of the nanogrids. It has also been observed that, after pulsed laser stimulation, cells grown on rGO-coated substrates exhibit the self-organization of neuronal networks by elongation of the differentiated cells in the radial direction, probably due to the higher thermal conductivity of rGO (with respect to only quartz and GO-coated quartz substrates) that might induce on rGO surface, by thermal gradient, a sort of radial stress originating from the center of the laser spot.

Enhanced neuronal differentiation has been also reported by Solanki et al. [96] (Fig. 4), who designed a substrate composed by positively charged 300 nm Silica nanoparticles, known for promoting axonal growth for neuronal cultures in vitro, coated by GO nanosheets, because of the presence of Oxygen groups on GO surface. hNSCs were seeded on *(a)* glass (control condition), *(b)* only Silica NPs substrates, *(c)* only GO-coated substrates, and *(d)* on GO–silica nanoparticle (SiNP–GO) and the differentiation was induced; while in the first 5 days axons growth was randomly directed in all the substrates, from the sixth day hNSCs on GO and on SiNP–GO displayed an aligned axonal growth not observed in the other two conditions. 14 days after the induced differentiation, cultures on SiNP–GO were characterized by a higher average length of the axons compared to control condition (about 20 %) and to only GO one (about 10 %). Moreover, hNSCs differentiated on SiNP–GO substrates showed the highest expression levels for neuronal markers. These results are due exclusively to the unique chemical structure of GO, as they were not observed for hNSCs grown on Molibdenum disulfide (MoS_2), a two-dimensional material with physical structure similar to GO.

Tang et al. [97] demonstrated that culturing neural stem cells on CVD-grown graphene films leads not only to morphologically healthy, but also developed and active neuronal networks. Using both calcium imaging and whole cell patch clamp recordings, authors observed that cells grown on graphene films *(a)* exhibit higher frequency of Ca^{2+} basal oscillations and *(b)* generate both spontaneous (sPSCs) and miniature postsynaptic currents (mPSCs), hallmark of network's normal activity and of synapse formation, with higher frequency and, only for sPSCs, higher mean current peak amplitude, with respect to the control condition. These results, though not shedding light on the mechanisms responsible for these observed features, suggest that graphene affects synaptic contacts, presynaptic events, and postsynaptic features.

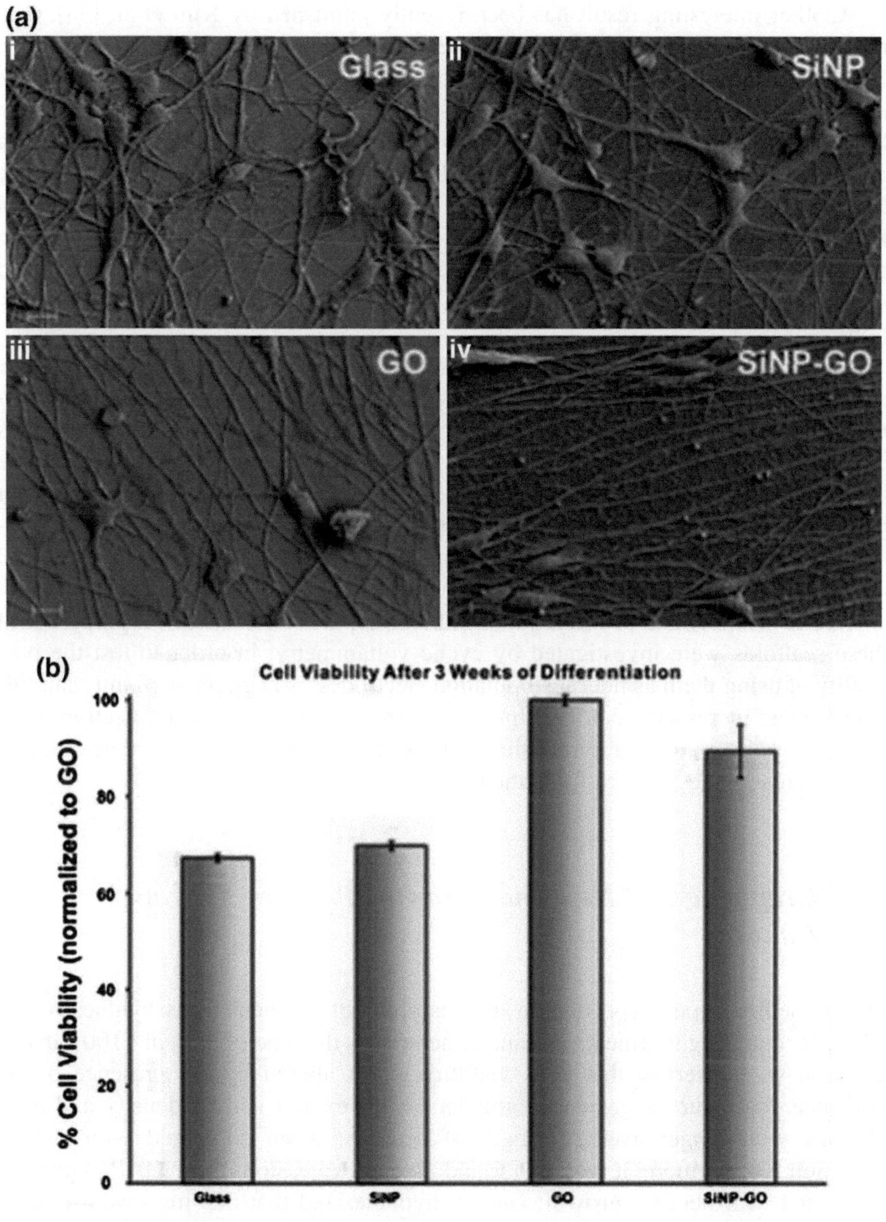

Fig. 4 **a** Scanning electron microscopy images of differentiated hNSCs on GO and SiNP–GO; axons are clearly aligned on GO and SiNP–GO, contrary to what is observed for control condition; scale bar: 10 μm. **b** MTS assay results showing GO and SiNP–GO biocompatibility. Reproduced with permission from [96]

Another interesting result has been recently published by Kim et al. [98], who reported the neurogenesis of hMSCs even in absence of any external neurogenic factors. Authors considered this "spontaneous" neurogenesis to be due to the enhanced formation, on graphene substrates, of three-dimensional clusters of hMSCs that, by mimicking an in vivo-like situation, might promote the secretion of cytokines and chemotaxic factors. This hypothesis seems to be confirmed by the fact that no remarkable differences were found for the body and nuclei shapes of hMSCs grown on glass and graphene-coated substrates, thus suggesting that graphene has a specific, yet still unclear, effect on the formation of these 3D spheroid structures and on the regulation of the growth and the neural differentiation of hMSCs.

The importance of developing innovative methods that, overcoming limitations of conventional cell culturing techniques leading to 2-D networks, allow the formation of 3-D neuronal networks where cells exhibit closer features to the complex in vivo conditions in terms of network morphology and gene expression, is at the basis of the work of Li et al. [99], who designed a 3-D graphene foam scaffold for neural stem cells. Such a scaffold resulted to be not only an extremely good substrate for cell proliferation and adhesion, allowing the formation of 3-D neural networks, but also to be able to up-regulate the expression of a protein, Ki-67, associated with cellular proliferation. Furthermore, the electrochemical properties of these scaffolds were investigated by cyclic voltammetry, in order to test the possibility of using them as neural stimulation electrodes; 3-D graphene foam scaffolds exhibited an increased electrical stimulation via a capacitive charge injection when compared conventional graphene film electrodes, probably due to the larger specific surface area of the 3-D scaffolds themselves.

3.3 *Graphene and Neuronal Growth: Primary Neuronal Cultures*

One of the first papers reporting on successful use of graphene films, synthesized by CVD, for culturing murine hippocampal neurons is that one of Li et al. [100]; in this works it was observed that cells viability is not altered by the presence of the graphene, that neurons grown on graphene films exhibit similar density and morphology with longer average length of neurites, when compared with control condition. Furthermore, it was also found an overexpression of the GAP43 protein, associated with neurites growth; authors hypothesized that this improved neurites sprouting and, consequently, the GAP43 overexpression might be due to both the nanoscale morphology of graphene films and to its high electrical conductivity.

As already highlighted, graphene's unique conductive properties make it one of the best candidates for interfacing with electroactive cells, as several physiological functions involve electrical or charge transfer. This aspect was studied by Zhou et al. [101], who coated poly-ε-caprolactone (PCL) nanofibrous scaffolds with a graphene layer-by-layer self-assembly, in order to obtain electronically conductive tridimensional architectures with specific surface chemistry, that were successfully used as 3-D scaffolds for neuronal growth in vitro.

Sahni et al. [102] investigated the biocompatibility of CVD-grown graphene films interfaced with neuronal cultures, in terms of viability and of neurites outgrowth of cortical neurons on bare, graphene- and poly-D-lysine (PDL)-coated plastic polymer dishes. Remarkable differences were found in neuronal viability, higher on graphene and PDL substrates than in the bare ones, in their adhesion on graphene films, probably due to Van der Waals forces between the material surface and cell membranes, as well as in neuronal morphology, with neurons cultured on graphene displaying more linear dendritic structures compared to the other two conditions.

In order to investigate the properties of smaller isolated neuronal networks, both in terms of cell morphology and electrical properties, neurons can be forced to grow on an ordered pattern, obtained by a variety of techniques and using several materials and/or proteins to be patterned on the substrates. Results obtained by Lorenzoni et al. [103], place themselves in this very context: CVD-grown graphene deposited on glass and on silicon wafers were irradiated by single KrF excimer laser pulses to obtain series of stripes with higher surface roughness than the underneath glass, exposed by the laser, length of 800 μm and width variable from 30 to 60 μm, that were used, after coating with poly-D-lysine, to culture primary hippocampal embryonic neurons. After 7 days in vitro, neurons were found to grow and develop only on the graphene stripes, showing a healthy morphology despite the formation of cell clusters, and stayed healthy up to 3 weeks (Fig. 5).

Another confirmation that graphene can be used as a nontoxic material for interfacing neurons comes from the work of Bendali et al. [104], who successfully cultured retinal ganglion cells (RGC) on glass coverslips, CVD-grown graphene transferred on sapphire substrates and sapphire substrates, either bare or coated with laminin and poly-D-lysin. Interestingly, retinal neurons were found viable, after 6 days in vitro, on both bare and peptide-coated graphene, though a statistically significant reduction in the number of viable cells, as well a slight difference in cells size, were observed for cells grown on bare graphene when compared to peptide-coated substrates. Nevertheless, authors concluded that the observed experimental evidences indicate that RGC can grow and survive on bare graphene, though, in this condition, cells head for aggregation and formation of neurites bundles, as confirmed by the fact that cell processes resulted to be ticker on uncoated sapphire and graphene.

Luo et al. [105] designed and synthesized a biocompatible conducting polymer-based nanocomposite through the electropolymerization of poly (3,4-ethylene dioxythiophene) (PEDOT) in the presence of GO as dopant agent,

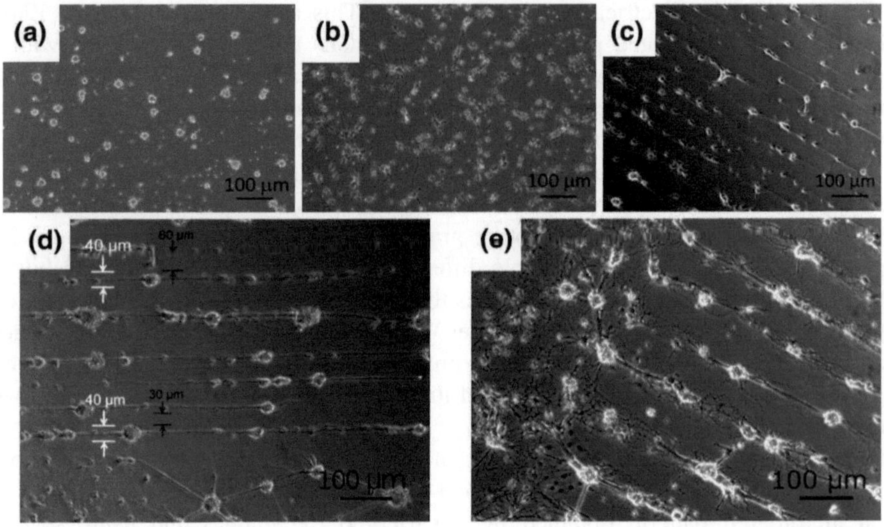

Fig. 5 Wide field transmission images of neurons seeded on different substrates: **a** bare glass/graphene (no neural network development observed); **b** PDL coated glass/graphene substrates (presence of neural networks). In **c**–**e** neural networks oriented along line patterns. Reproduced with permission from [103]

and used these PEDOT/GO films as substrates for neuronal growth, reporting no remarkable toxic effects and the development of neuronal networks with significantly longer neurites than control condition, even in absence of protein commonly used to increase cell adhesion.

As mentioned above, surface functionalization of graphene and C-based nanomaterials in general improves their biocompatibility; when it comes to neurons, this is not the only effect of chemically modifying graphene surfaces, as shown by Tu et al. [106]. In their study, they demonstrated that adhesion and outgrowth of neuronal cells, seeded onto graphene substrates, feel the effects of surface charges; rat hippocampal neurons were grown on carboxylated GO (GO–COOH) as control condition with negative surface charge, and on GO–COOH whose surface had been functionalized with three different functional groups: *(a)* methoxy (–OCH_3), with almost neutral surface charge, *(b)* amino (NH_2), with positively charged surface, and *(c)* poly-*m*-aminobenzene sulfonic acid (–NH_2/–SO_3H, PABS) which resulted to be zwitterionic. After 7 days in vitro, almost the 90 % of neurons were still viable on all the four substrates and neurons cultured on amino functionalized GO substrates showed a greater number of branches per neurite and of neurites per neuron, as well as a longer length of neurites, even without exhibiting relevant differences in cell morphology. However, it is difficult to comment these findings in terms of contingent applications of these substrates as scaffolds and/or as electrodes material for neural stimulation, given the lack of a direct comparison with the conventional control conditions, such as glass or plastic culture substrates.

One of the key questions when novel materials are interfaced with neural cells for future translational applications is whether these materials allow the formation of fully developed and active neural networks. In this framework, we investigated the properties of GO and rGO as substrates for neuronal growth, with a particular attention to their biocompatibility and to the contingent alterations of the electrical properties of neurons and networks; we observed that, though no remarkable differences were found for the percentage of living cells of out the total across the three conditions, the total density of neurons grown on GO was reduced to almost the 35 % of the initial seeding density, while it was almost the 50 % for both control and rGO conditions. We explained this difference taking into account the fact that although GO, being atomically rougher than rGO, should promote neuronal adhesion, its superficial charge is more negative than rGO, and this aspect might have been then predominant, under our culture conditions. We also reported that both passive (i.e., input resistance; membrane capacitance, time constant, and resting potential) and active (i.e., action potential threshold; the peak of AP amplitude) neuronal properties did not significantly differ across the three conditions, with the only exception of the AP width at half amplitude that was slightly, yet significantly larger, on GO and rGO, compared to control; this can be attributed to differences in ionic channels expression, (e.g., KV), as their density and membrane distribution are known to affect AP shape. Furthermore, neurons grown on GO and rGO substrates exhibited a slightly higher spontaneous activity than control conditions, thus suggesting an earlier formation of synaptic connections or a stronger synaptic connectivity; this enhanced activity can be explained in terms of increased length and number of neurites, as previously reported, and in terms of the efficacy of excitatory synaptic connections and their number [107].

Similar results in terms of viability and ability of developing functional neuronal networks have been recently reported by Fabbro et al. [108], who grown hippocampal neurons on graphene substrates obtained by ball milling or liquid phase exfoliation of graphite. Such substrates resulted to be inert neuron-interfacing materials and they supported the development of neuronal networks in absence of any protein or polymer promoting adhesion; however, no perturbation of neuronal network synaptic performances has been observed with respect to control condition. Differences in impact on neuronal activity between graphene and CNTs might be due, according to the authors, to morphological differences between the two materials, especially in terms of their roughness which is, for CNTs, significantly higher (Table 1).

Table 1 Graphene and its derivatives in neurosciences

Material	Note	Model	Observed effects	References
CVD-grown graphene		Primary neuronal cultures	Good cell viability; longer neurites; overexpression of GAP43 protein	[100]
Graphene/PET films	Extracellular stimulation	Primary neuronal cultures	Biocompatibility; morphological modifications	[86]
Fluorinated graphene		MSCs	Morphological changes; promoted differentiation into neurons	[73]
Layer-by-layer graphene on PLC nanofibrous scaffolds	3-D structures	Primary neuronal cultures	Good adhesion and neuronal networks development	[101]
GO nanogrids on SiO_2 films + TiO_2 NPs		hNSCs	Biocompatibility; alignment along the geometrical pattern of the nanogrids	[93]
rGO/TiO_2 heterojunction films		hNSCs	Biocompatibility; enhanced differentiation of hNSCs into neurons than glia	[94]
CVD-grown graphene on Sapphire substrates		Retinal ganglion cells	Reduced viability and neurites bundles on pristine graphene; no remarkable difference with the control when treated with laminin	[104]
CVD-grown graphene	Solution-gated FETs	Retinal ganglion cells	Biocompatibility	[78]
3-D CVD-grown graphene scaffolds	hNSCs		Biocompatibility; upregulation of Ki-67 protein	[99]
Patterned CVD-grown graphene		Primary neuronal cultures	Ordered growth along the patterned stripes	[103]
PEDOT/GO films	Pristine graphene		No toxic effects; longer neurites	[105]
CVD-grown graphene		Primary neuronal cultures	Good cell viability and adhesion	[102]
Silica NPs coated with GO nanosheets		hNSCs	Aligned axonal growth; higher average length of axons; higher level of neuronal markers	[96]
CVD-grown graphene films		NSCs	Healthy and active neuronal networks; higher frequency of sPSCs	[97]
GO and rGO		hNSCs	Effective differentiation into neurons; self-organization of neuronal networks	[95]

(continued)

Table 1 (continued)

Material	Note	Model	Observed effects	References
GO and rGO	Hydrazine-rGO and ginseng-rGO	hNSCs	Higher cell proliferation on GO and ginseng-rGO; morphological changes in ginseng-rGO	[92]
CVD-grown graphene on polymeric substrates	Undoped and Nitric Acid-doped graphene; MEA	Primary neuronal cultures and in vivo	Recordings of neural activity in vitro and in vivo electrocorticography	[85]
Graphene-based carbon layer electrode arrays (CLEAR)	Transparent MEA	In vivo implantation in rodents	Recordings of neural signals; optogenetic stimulation	[84]
GO–COOH + different functional groups leading to different surface charge	GO–COOH; GO–COOH–NH_2; GO–COOH–OCH_3; CO–COOH–PABS	Primary neuronal cultures	Better cell adhesion; more branches per neurites; more and longer neurites per neuron on GO–COOH–NH_2	[106]
CVD-grown graphene on quartz	MEA	Primary neuronal cultures	Detection of extracellular spontaneous activity	[82]
Graphene		hMSCs	Spontaneous neurogenesis without neurogenic factors; 3-D clusters of MSCs	[98]
GO and rGO	PEI treatment necessary	Primary neuronal cultures	Biocompatibility; higher frequency of spontaneous activity on GO and rGO	[107]
BM and LPE graphene	Pristine materials	Primary neuronal cultures	Formation of neuronal networks; no alterations of synaptic activity	[108]
CVD-grown graphene on Au and SiO_2	MEA		More charge injection capacity No recordings from cells	[83]

4 Conclusions

In this review, we highlighted the reasons why graphene and its derivatives have gained more and more importance in the very last few years in the context of biomedical applications; we reviewed their most important applications, with a particular emphasis on the field of neuroscience given their intriguing affinity for neuronal interfacing.

As we showed, interesting and extremely promising results have already been achieved, yet additional research is needed in order to shed light on the mechanism underlying the coupling between these materials and living cells, thus allowing for the design and development of smart engineered materials, specifically tailored for the system they will interact to, to be used in advance applications in the fields of neuroscience and neuroprosthetics.

Acknowledgments Authors are grateful to Drs. C. Bittencourt, J.-F. Colomer and P. De Marco for discussions. Financial support from the European Commission (FP7-PEOPLE-IEF "INCA-NANEP", Contract No. 328214; FP7-NMP "MERIDIAN", Contract No. 280778-02), the Flanders Research Foundation (FWO, Contract No. G088812N), and the Belgian Science Policy Office (BELSPO) is kindly acknowledged.

References

1. Freitas RA Jr (2005) What is nanomedicine? Nanomed Nanotechnol Biol Med 1(1):2–9. doi:10.1016/j.nano.2004.11.003
2. Drexler KE (1986) Engines of creation: the coming era of nanotechnology. Anchor Press, Norwell
3. Drexler KE, Peterson C, Pergamit G (1991) Unbounding the future: the nanotechnology revolution. William Morrow/Quill Books, New York
4. Feynman R (1960) There's plenty of room at the bottom, vol 23. Eng Sci (CalTech), Pasadena
5. Porter AE, Gass M, Muller K, Skepper JN, Midgley PA, Welland M (2007) Direct imaging of single-walled carbon nanotubes in cells. Nat Nanotechnol 2(11):713–717. doi:10.1038/nnano.2007.347
6. Schrand AM, Dai L, Schlager JJ, Hussain SM, Osawa E (2007) Differential biocompatibility of carbon nanotubes and nanodiamonds. Diam Relat Mater 16(12):2118–2123. doi:10.1016/j.diamond.2007.07.020
7. Schrand AM, Lin JB, Hens SC, Hussain SM (2011) Temporal and mechanistic tracking of cellular uptake dynamics with novel surface fluorophore-bound nanodiamonds. Nanoscale 3(2):435–445. doi:10.1039/c0nr00408a
8. Yang M, Flavin K, Kopf I, Radics G, Hearnden CH, McManus GJ, Moran B, Villalta-Cerdas A, Echegoyen LA, Giordani S, Lavelle EC (2013) Functionalization of carbon nanoparticles modulates inflammatory cell recruitment and NLRP3 inflammasome activation. Small 9(24):4194–4206. doi:10.1002/smll.201300481
9. Poland CA, Duffin R, Kinloch I, Maynard A, Wallace WA, Seaton A, Stone V, Brown S, Macnee W, Donaldson K (2008) Carbon nanotubes introduced into the abdominal cavity of mice show asbestos-like pathogenicity in a pilot study. Nat Nanotechnol 3(7):423–428. doi:10.1038/nnano.2008.111

10. Pumera M, Sánchez S, Ichinose I, Tang J (2007) Electrochemical nanobiosensors. Sens Actuators B Chem 123(2):1195–1205. doi:10.1016/j.snb.2006.11.016
11. Dumortier H, Lacotte S, Pastorin G, Marega R, Wu W, Bonifazi D, Briand JP, Prato M, Muller S, Bianco A (2006) Functionalized carbon nanotubes are non-cytotoxic and preserve the functionality of primary immune cells. Nano Lett 6(7):1522–1528. doi:10.1021/nl061160x
12. Vaijayanthimala V, Tzeng YK, Chang HC, Li CL (2009) The biocompatibility of fluorescent nanodiamonds and their mechanism of cellular uptake. Nanotechnology 20(42):425103. doi:10.1088/0957-4484/20/42/425103
13. Yehia HN, Draper RK, Mikoryak C, Walker EK, Bajaj P, Musselman IH, Daigrepont MC, Dieckmann GR, Pantano P (2007) Single-walled carbon nanotube interactions with HeLa cells. J Nanobiotechnol 5:8. doi:10.1186/1477-3155-5-8
14. Lu CH, Zhu CL, Li J, Liu JJ, Chen X, Yang HH (2010) Using graphene to protect DNA from cleavage during cellular delivery. Chem Commun 46(18):3116–3118. doi:10.1039/b926893f
15. Yang K, Wan J, Zhang S, Tian B, Zhang Y, Liu Z (2012) The influence of surface chemistry and size of nanoscale graphene oxide on photothermal therapy of cancer using ultra-low laser power. Biomaterials 33(7):2206–2214. doi:10.1016/j.biomaterials.2011.11.064
16. Akhavan O, Ghaderi E, Akhavan A (2012) Size-dependent genotoxicity of graphene nanoplatelets in human stem cells. Biomaterials 33(32):8017–8025. doi:10.1016/j.biomaterials.2012.07.040
17. Akhavan O, Ghaderi E, Emamy H, Akhavan F (2013) Genotoxicity of graphene nanoribbons in human mesenchymal stem cells. Carbon 54:419–431. doi:10.1016/j.carbon.2012.11.058
18. De Marzi L, Ottaviano L, Perrozzi F, Nardone M, Santucci S, De Lapuente J, Borras M, Treossi E, Palermo V, Poma A (2014) Flake size-dependent cyto and genotoxic evaluation of graphene oxide on in vitro A549, Caco2 and Vero cell lines. J Biol Regul Homestatic Agents 28(2):281–289
19. Ang PK, Li A, Jaiswal M, Wang Y, Hou HW, Thong JT, Lim CT, Loh KP (2011) Flow sensing of single cell by graphene transistor in a microfluidic channel. Nano Lett 11 (12):5240–5246. doi:10.1021/nl202579k
20. Liu Z, Robinson JT, Sun X, Dai H (2008) PEGylated nanographene oxide for delivery of water-insoluble cancer drugs. J Am Chem Soc 130(33):10876–10877. doi:10.1021/ja803688x
21. Sun X, Liu Z, Welsher K, Robinson JT, Goodwin A, Zaric S, Dai H (2008) Nano-graphene oxide for cellular imaging and drug delivery. Nano Res 1(3):203–212. doi:10.1007/s12274-008-8021-8
22. Wang K, Ruan J, Song H, Zhang J, Wo Y, Guo S, Cui D (2010) Biocompatibility of graphene oxide. Nanoscale Res Lett. doi:10.1007/s11671-010-9751-6
23. Hu W, Peng C, Lv M, Li X, Zhang Y, Chen N, Fan C, Huang Q (2011) Protein corona-mediated mitigation of cytotoxicity of graphene oxide. ACS Nano 5(5):3693–3700. doi:10.1021/nn200021j
24. Sasidharan A, Panchakarla LS, Sadanandan AR, Ashokan A, Chandran P, Girish CM, Menon D, Nair SV, Rao CN, Koyakutty M (2012) Hemocompatibility and macrophage response of pristine and functionalized graphene. Small 8(8):1251–1263. doi:10.1002/smll.201102393
25. Liao KH, Lin YS, Macosko CW, Haynes CL (2011) Cytotoxicity of graphene oxide and graphene in human erythrocytes and skin fibroblasts. ACS Appl Mater Interfaces 3(7):2607–2615. doi:10.1021/am200428v
26. Singh SK, Singh MK, Kulkarni PP, Sonkar VK, Gracio JJ, Dash D (2012) Amine-modified graphene: thrombo-protective safer alternative to graphene oxide for biomedical applications. ACS Nano 6(3):2731–2740. doi:10.1021/nn300172t
27. Akhavan O, Ghaderi E (2010) Toxicity of graphene and graphene oxide nanowalls against bacteria. ACS Nano 4(10):5731–5736. doi:10.1021/nn101390x

28. Akhavan O, Ghaderi E, Esfandiar A (2011) Wrapping bacteria by graphene nanosheets for isolation from environment, reactivation by sonication, and inactivation by near-infrared irradiation. J Phys Chem B 115(19):6279–6288. doi:10.1021/jp200686k
29. Hu W, Peng C, Luo W, Lv M, Li X, Li D, Huang Q, Fan C (2010) Graphene-based antibacterial paper. ACS Nano 4(7):4317–4323. doi:10.1021/nn101097v
30. Liu L, Liu J, Wang Y, Yan X, Sun DD (2011) Facile synthesis of monodispersed silver nanoparticles on graphene oxide sheets with enhanced antibacterial activity. New J Chem 35 (7):1418. doi:10.1039/c1nj20076c
31. Sawangphruk M, Srimuk P, Chiochan P, Sangsri T, Siwayaprahm P (2012) Synthesis and antifungal activity of reduced graphene oxide nanosheets. Carbon 50(14):5156–5161. doi:10.1016/j.carbon.2012.06.056
32. Dellieu L, Lawarée E, Reckinger N, Didembourg C, Letesson JJ, Sarrazin M, Deparis O, Matroule JY, Colomer JF (2015) Do CVD grown graphene films have antibacterial activity on metallic substrates? Carbon 84:310–316. doi:10.1016/j.carbon.2014.12.025
33. Ruiz ON, Fernando KA, Wang B, Brown NA, Luo PG, McNamara ND, Vangsness M, Sun YP, Bunker CE (2011) Graphene oxide: a nonspecific enhancer of cellular growth. ACS Nano 5(10):8100–8107. doi:10.1021/nn202699t
34. Wang G, Qian F, Saltikov CW, Jiao Y, Li Y (2011) Microbial reduction of graphene oxide by Shewanella. Nano Res 4(6):563–570. doi:10.1007/s12274-011-0112-2
35. Singh SK, Singh MK, Nayak MK, Kumari S, Shrivastava S, Gracio JJ, Dash D (2011) Thrombus inducing property of atomically thin graphene oxide sheets. ACS Nano 5 (6):4987–4996. doi:10.1021/nn201092p
36. Sahu A, Choi WI, Tae G (2012) A stimuli-sensitive injectable graphene oxide composite hydrogel. Chem Commun 48(47):5820–5822. doi:10.1039/c2cc31862h
37. Zhang X, Yin J, Peng C, Hu W, Zhu Z, Li W, Fan C, Huang Q (2011) Distribution and biocompatibility studies of graphene oxide in mice after intravenous administration. Carbon 49(3):986–995. doi:10.1016/j.carbon.2010.11.005
38. Depan D, Shah J, Misra RDK (2011) Controlled release of drug from folate-decorated and graphene mediated drug delivery system: synthesis, loading efficiency, and drug release response. Mater Sci Eng C 31(7):1305–1312. doi:10.1016/j.msec.2011.04.010
39. Yang K, Wan J, Zhang S, Zhang Y, Lee ST, Liu Z (2011) In vivo pharmacokinetics, long-term biodistribution, and toxicology of PEGylated graphene in mice. ACS Nano 5 (1):516–522. doi:10.1021/nn1024303
40. Yang XY, Zhang XY, Liu ZF, Ma YF, Huang Y, Chen Y (2008) High-efficiency loading and controlled release of doxorubicin hydrochloride on graphene oxide. J Phys Chem C 112 (45):17554–17558. doi:10.1021/Jp806751k
41. Zhang L, Xia J, Zhao Q, Liu L, Zhang Z (2010) Functional graphene oxide as a nanocarrier for controlled loading and targeted delivery of mixed anticancer drugs. Small 6(4):537–544. doi:10.1002/smll.200901680
42. Huang P, Xu C, Lin J, Wang C, Wang X, Zhang C, Zhou X, Guo S, Cui D (2011) Folic acid-conjugated graphene oxide loaded with photosensitizers for targeting photodynamic therapy. Theranostics. doi:10.7150/thno/v01p0240
43. Hu SH, Chen YW, Hung WT, Chen IW, Chen SY (2012) Quantum-dot-tagged reduced graphene oxide nanocomposites for bright fluorescence bioimaging and photothermal therapy monitored in situ. Adv Mater 24(13):1748–1754. doi:10.1002/adma.201104070
44. Zhang W, Guo Z, Huang D, Liu Z, Guo X, Zhong H (2011) Synergistic effect of chemo-photothermal therapy using PEGylated graphene oxide. Biomaterials 32(33):8555–8561. doi:10.1016/j.biomaterials.2011.07.071
45. Bao H, Pan Y, Ping Y, Sahoo NG, Wu T, Li L, Li J, Gan LH (2011) Chitosan-functionalized graphene oxide as a nanocarrier for drug and gene delivery. Small 7(11):1569–1578. doi:10.1002/smll.201100191
46. Zhang L, Lu Z, Zhao Q, Huang J, Shen H, Zhang Z (2011) Enhanced chemotherapy efficacy by sequential delivery of siRNA and anticancer drugs using PEI-grafted graphene oxide. Small 7(4):460–464. doi:10.1002/smll.201001522

47. Hong H, Yang K, Zhang Y, Engle JW, Feng L, Yang Y, Nayak TR, Goel S, Bean J, Theuer CP, Barnhart TE, Liu Z, Cai W (2012) In vivo targeting and imaging of tumor vasculature with radiolabeled, antibody-conjugated nanographene. ACS Nano 6(3):2361–2370. doi:10.1021/nn204625e
48. Chen W, Yi P, Zhang Y, Zhang L, Deng Z, Zhang Z (2011) Composites of aminodextran-coated Fe_3O_4 nanoparticles and graphene oxide for cellular magnetic resonance imaging. ACS Appl Mater Interfaces 3(10):4085–4091. doi:10.1021/am2009647
49. Peng J, Gao W, Gupta BK, Liu Z, Romero-Aburto R, Ge L, Song L, Alemany LB, Zhan X, Gao G, Vithayathil SA, Kaipparettu BA, Marti AA, Hayashi T, Zhu JJ, Ajayan PM (2012) Graphene quantum dots derived from carbon fibers. Nano Lett 12(2):844–849. doi:10.1021/nl2038979
50. Pan D, Zhang J, Li Z, Wu M (2010) Hydrothermal route for cutting graphene sheets into blue-luminescent graphene quantum dots. Adv Mater 22(6):734–738. doi:10.1002/adma.200902825
51. Tetsuka H, Asahi R, Nagoya A, Okamoto K, Tajima I, Ohta R, Okamoto A (2012) Optically tunable amino-functionalized graphene quantum dots. Adv Mater 24(39):5333–5338. doi:10.1002/adma.201201930
52. Zhang M, Bai L, Shang W, Xie W, Ma H, Fu Y, Fang D, Sun H, Fan L, Han M, Liu C, Yang S (2012) Facile synthesis of water-soluble, highly fluorescent graphene quantum dots as a robust biological label for stem cells. J Mater Chem 22(15):7461. doi:10.1039/c2jm16835a
53. Wang Y, Li Z, Weber TJ, Hu D, Lin CT, Li J, Lin Y (2013) In situ live cell sensing of multiple nucleotides exploiting DNA/RNA aptamers and graphene oxide nanosheets. Anal Chem 85(14):6775–6782. doi:10.1021/ac400858g
54. Tang Z, Wu H, Cort JR, Buchko GW, Zhang Y, Shao Y, Aksay IA, Liu J, Lin Y (2010) Constraint of DNA on functionalized graphene improves its biostability and specificity. Small 6(11):1205–1209. doi:10.1002/smll.201000024
55. Thevenot DR, Tóth K, Durst RA, Wilson GS (1999) Electrochemical biosensors: recommended definitions and classification. Pure Appl Chem. doi:10.1351/pac199971122333
56. Loh KP, Bao Q, Eda G, Chhowalla M (2010) Graphene oxide as a chemically tunable platform for optical applications. Nat Chem 2(12):1015–1024. doi:10.1038/nchem.907
57. Balapanuru J, Yang JX, Xiao S, Bao Q, Jahan M, Polavarapu L, Wei J, Xu QH, Loh KP (2010) A graphene oxide-organic dye ionic complex with DNA-sensing and optical-limiting properties. Angew Chem Int Ed Engl 49(37):6549–6553. doi:10.1002/anie.201001004
58. He S, Song B, Li D, Zhu C, Qi W, Wen Y, Wang L, Song S, Fang H, Fan C (2010) A graphene nanoprobe for rapid, sensitive, and multicolor fluorescent DNA analysis. Adv Funct Mater 20(3):453–459. doi:10.1002/adfm.200901639
59. Lu CH, Yang HH, Zhu CL, Chen X, Chen GN (2009) A graphene platform for sensing biomolecules. Angew Chem Int Ed Engl 48(26):4785–4787. doi:10.1002/anie.200901479
60. Mei Q, Zhang Z (2012) Photoluminescent graphene oxide ink to print sensors onto microporous membranes for versatile visualization bioassays. Angew Chem Int Ed Engl 51 (23):5602–5606. doi:10.1002/anie.201201389
61. Pu Y, Zhu Z, Han D, Liu H, Liu J, Liao J, Zhang K, Tan W (2011) Insulin-binding aptamer-conjugated graphene oxide for insulin detection. The Analyst 136(20):4138–4140. doi:10.1039/c1an15407a
62. Chang H, Tang L, Wang Y, Jiang J, Li J (2010) Graphene fluorescence resonance energy transfer aptasensor for the thrombin detection. Anal Chem 82(6):2341–2346. doi:10.1021/ac9025384
63. He Y, Wang ZG, Tang HW, Pang DW (2011) Low background signal platform for the detection of ATP: when a molecular aptamer beacon meets graphene oxide. Biosens Bioelectron 29(1):76–81. doi:10.1016/j.bios.2011.07.069

64. Huang WT, Shi Y, Xie WY, Luo HQ, Li NB (2011) A reversible fluorescence nanoswitch based on bifunctional reduced graphene oxide: use for detection of Hg^{2+} and molecular logic gate operation. Chem Commun 47(27):7800–7802. doi:10.1039/c1cc11631b
65. Zhao XH, Kong RM, Zhang XB, Meng HM, Liu WN, Tan W, Shen GL, Yu RQ (2011) Graphene-DNAzyme based biosensor for amplified fluorescence "turn-on" detection of Pb^{2+} with a high selectivity. Anal Chem 83(13):5062–5066. doi:10.1021/ac200843x
66. Kwon OS, Park SJ, Hong JY, Han AR, Lee JS, Lee JS, Oh JH, Jang J (2012) Flexible FET-type VEGF aptasensor based on nitrogen-doped graphene converted from conducting polymer. ACS Nano 6(2):1486–1493. doi:10.1021/nn204395n
67. Ohno Y, Maehashi K, Matsumoto K (2010) Label free biosensors based on aptamer-modified graphene field-effect transistors. J Am Chem Soc 132(51):18012–18013. doi:10.1021/ja108127r
68. Wang Y, Li Y, Tang L, Lu J, Li J (2009) Application of graphene-modified electrode for selective detection of dopamine. Electrochem Commun 11(4):889–892. doi:10.1016/j.elecom.2009.02.013
69. Sun CL, Lee HH, Yang JM, Wu CC (2011) The simultaneous electrochemical detection of ascorbic acid, dopamine, and uric acid using graphene/size-selected Pt nanocomposites. Biosens Bioelectron 26(8):3450–3455. doi:10.1016/j.bios.2011.01.023
70. Kim NJ, Lee SJ, Atala A (2013) Biomedical nanomaterials in tissue engineering. Elsevier, Amsterdam. doi:10.1533/9780857097231.1
71. Stout DA, Durmus NG, Webster TJ (2013) Synthesis of carbon based nanomaterials for tissue engineering applications. doi:10.1533/9780857097231.1.119
72. Park SY, Park J, Sim SH, Sung MG, Kim KS, Hong BH, Hong S (2011) Enhanced differentiation of human neural stem cells into neurons on graphene. Adv Mater 23(36): H263–H267. doi:10.1002/adma.201101503
73. Wang Y, Lee WC, Manga KK, Ang PK, Lu J, Liu YP, Lim CT, Loh KP (2012) Fluorinated graphene for promoting neuro-induction of stem cells. Adv Mater 24(31):4285–4290. doi:10.1002/adma.201200846
74. Chen GY, Pang DW, Hwang SM, Tuan HY, Hu YC (2012) A graphene-based platform for induced pluripotent stem cells culture and differentiation. Biomaterials 33(2):418–427. doi:10.1016/j.biomaterials.2011.09.071
75. Lee WC, Lim CH, Shi H, Tang LA, Wang Y, Lim CT, Loh KP (2011) Origin of enhanced stem cell growth and differentiation on graphene and graphene oxide. ACS Nano 5(9):7334–7341. doi:10.1021/nn202190c
76. Spira ME, Hai A (2013) Multi-electrode array technologies for neuroscience and cardiology. Nat Nanotechnol 8(2):83–94. doi:10.1038/nnano.2012.265
77. Cohen-Karni T, Qing Q, Li Q, Fang Y, Lieber CM (2010) Graphene and nanowire transistors for cellular interfaces and electrical recording. Nano Lett 10(3):1098–1102. doi:10.1021/nl1002608
78. Hess LH, Seifert M, Garrido JA (2013) Graphene transistors for bioelectronics. Proc IEEE 101(7):1780–1792. doi:10.1109/jproc.2013.2261031
79. Li F, Xue M, Ma X, Zhang M, Cao T (2011) Facile patterning of reduced graphene oxide film into microelectrode array for highly sensitive sensing. Anal Chem 83(16):6426–6430. doi:10.1021/ac200939g
80. Li X, Zhu Y, Cai W, Borysiak M, Han B, Chen D, Piner RD, Colombo L, Ruoff RS (2009) Transfer of large-area graphene films for high-performance transparent conductive electrodes. Nano Lett 9(12):4359–4363. doi:10.1021/nl902623y
81. Zhao J, Pei S, Ren W, Gao L, Cheng HM (2010) Efficient preparation of large-area graphene oxide sheets for transparent conductive films. ACS Nano 4(9):5245–5252. doi:10.1021/nn1015506
82. Du X, Wu L, Cheng J, Huang S, Cai Q, Jin Q, Zhao J (2015) Graphene microelectrode arrays for neural activity detection. J Biol Phys 41(4):339–347. doi:10.1007/s10867-015-9382-3

83. Koerbitzer B, Krauss P, Nick C, Yadav S, Schneider JJ, Thielemann C (2016) Graphene electrodes for stimulation of neuronal cells. 2D Mater 3(2):024004. doi:10.1088/2053-1583/3/2/024004
84. Park DW, Schendel AA, Mikael S, Brodnick SK, Richner TJ, Ness JP, Hayat MR, Atry F, Frye ST, Pashaie R, Thongpang S, Ma Z, Williams JC (2014) Graphene-based carbon-layered electrode array technology for neural imaging and optogenetic applications. Nat Commun 5:5258. doi:10.1038/ncomms6258
85. Kuzum D, Takano H, Shim E, Reed JC, Juul H, Richardson AG, de Vries J, Bink H, Dichter MA, Lucas TH, Coulter DA, Cubukcu E, Litt B (2014) Transparent and flexible low noise graphene electrodes for simultaneous electrophysiology and neuroimaging. Nat Commun 5:5259. doi:10.1038/ncomms6259
86. Heo C, Yoo J, Lee S, Jo A, Jung S, Yoo H, Lee YH, Suh M (2011) The control of neural cell-to-cell interactions through non-contact electrical field stimulation using graphene electrodes. Biomaterials 32(1):19–27. doi:10.1016/j.biomaterials.2010.08.095
87. Basham E, Parent D (2012) Design optimization of transistors used for neural recording. Act Passive Electron Compon 2012:1–10. doi:10.1155/2012/472306
88. Cabruja E, Merlos A, Cané C, Lozano M, Bausells J, Esteve J (1991) Influence of the degradation on the surface states and electrical characteristics of EOS structures. Surf Sci 251–252:364–368. doi:10.1016/0039-6028(91)91015-p
89. Offenhäusser A, Sprössler C, Matsuzawa M, Knoll W (1997) Field-effect transistor array for monitoring electrical activity from mammalian neurons in culture. Biosens Bioelectron 12 (8):819–826. doi:10.1016/s0956-5663(97)00047-x
90. Blaschke BM, Lottner M, Drieschner S, Calia AB, Stoiber K, Rousseau L, Lissourges G, Garrido JA (2016) Flexible graphene transistors for recording cell action potentials. 2D Mater 3(2):025007. doi:10.1088/2053-1583/3/2/025007
91. Hess LH, Jansen M, Maybeck V, Hauf MV, Seifert M, Stutzmann M, Sharp ID, Offenhausser A, Garrido JA (2011) Graphene transistor arrays for recording action potentials from electrogenic cells. Adv Mater 23(43):5045–5049, 4968. doi:10.1002/adma.201102990
92. Akhavan O, Ghaderi E, Abouei E, Hatamie S, Ghasemi E (2014) Accelerated differentiation of neural stem cells into neurons on ginseng-reduced graphene oxide sheets. Carbon 66:395–406. doi:10.1016/j.carbon.2013.09.015
93. Akhavan O, Ghaderi E (2013) Differentiation of human neural stem cells into neural networks on graphene nanogrids. J Mater Chem B 1(45):6291. doi:10.1039/c3tb21085e
94. Akhavan O, Ghaderi E (2013) Flash photo stimulation of human neural stem cells on graphene/TiO_2 heterojunction for differentiation into neurons. Nanoscale 5(21):10316–10326. doi:10.1039/c3nr02161k
95. Akhavan O, Ghaderi E (2014) The use of graphene in the self-organized differentiation of human neural stem cells into neurons under pulsed laser stimulation. J Mater Chem B 2 (34):5602. doi:10.1039/c4tb00668b
96. Solanki A, Chueng ST, Yin PT, Kappera R, Chhowalla M, Lee KB (2013) Axonal alignment and enhanced neuronal differentiation of neural stem cells on graphene-nanoparticle hybrid structures. Adv Mater 25(38):5477–5482. doi:10.1002/adma.201302219
97. Tang M, Song Q, Li N, Jiang Z, Huang R, Cheng G (2013) Enhancement of electrical signaling in neural networks on graphene films. Biomaterials 34(27):6402–6411. doi:10.1016/j.biomaterials.2013.05.024
98. Kim J, Park S, Kim YJ, Jeon CS, Lim KT, Seonwoo H, Cho S-P, Chung TD, Choung P-H, Choung Y-H, Hong BH, Chung JH (2015) Monolayer graphene-directed growth and neuronal differentiation of mesenchymal stem cells. J Biomed Nanotechnol 11(11):2024–2033. doi:10.1166/jbn.2015.2137
99. Li N, Zhang Q, Gao S, Song Q, Huang R, Wang L, Liu L, Dai J, Tang M, Cheng G (2013) Three-dimensional graphene foam as a biocompatible and conductive scaffold for neural stem cells. Sci Rep 3:1604. doi:10.1038/srep01604

100. Li N, Zhang X, Song Q, Su R, Zhang Q, Kong T, Liu L, Jin G, Tang M, Cheng G (2011) The promotion of neurite sprouting and outgrowth of mouse hippocampal cells in culture by graphene substrates. Biomaterials 32(35):9374–9382. doi:10.1016/j.biomaterials.2011.08.065
101. Zhou K, Thouas GA, Bernard CC, Nisbet DR, Finkelstein DI, Li D, Forsythe JS (2012) Method to impart electro- and biofunctionality to neural scaffolds using graphene–polyelectrolyte multilayers. ACS Appl Mater Interfaces 4(9):4524–4531. doi:10.1021/am3007565
102. Sahni D, Jea A, Mata JA, Marcano DC, Sivaganesan A, Berlin JM, Tatsui CE, Sun Z, Luerssen TG, Meng S, Kent TA, Tour JM (2013) Biocompatibility of pristine graphene for neuronal interface. J Neurosurg Pediatr 11(5):575–583. doi:10.3171/2013.1.PEDS12374
103. Lorenzoni M, Brandi F, Dante S, Giugni A, Torre B (2013) Simple and effective graphene laser processing for neuron patterning application. Sci Rep 3:1954. doi:10.1038/srep01954
104. Bendali A, Hess LH, Seifert M, Forster V, Stephan AF, Garrido JA, Picaud S (2013) Purified neurons can survive on peptide-free graphene layers. Adv Healthc Mater 2(7):929–933. doi:10.1002/adhm.201200347
105. Luo X, Weaver CL, Tan S, Cui XT (2013) Pure graphene oxide doped conducting polymer nanocomposite for bio-interfacing. J Mater Chem B 1(9):1340. doi:10.1039/c3tb00006k
106. Tu Q, Pang L, Chen Y, Zhang Y, Zhang R, Lu B, Wang J (2014) Effects of surface charges of graphene oxide on neuronal outgrowth and branching. The Analyst 139(1):105–115. doi:10.1039/c3an01796f
107. Monaco AM, Moskalyuk A, Motylewski J, Vahidpour F, Ng AMH, Ping Loh K, Nesladek M, Giugliano M (2015) Coupling (reduced) graphene oxide to mammalian primary cortical neurons in vitro. AIMS Mater Sci 2(3):217–229. doi:10.3934/matersci.2015.3.217
108. Fabbro A, Scaini D, Leon V, Vazquez E, Cellot G, Privitera G, Lombardi L, Torrisi F, Tomarchio F, Bonaccorso F, Bosi S, Ferrari AC, Ballerini L, Prato M (2016) Graphene-based interfaces do not alter target nerve cells. ACS Nano 10(1):615–623. doi:10.1021/acsnano.5b05647
109. Chang Y, Yang ST, Liu JH, Dong E, Wang Y, Cao A, Liu Y, Wang H (2011) In vitro toxicity evaluation of graphene oxide on A549 cells. Toxicol Lett 200(3):201–210. doi:10.1016/j.toxlet.2010.11.016
110. Freitas RAJ (1999) Nanomedicine, volume I: basic capabilities. Landes Bioscience, Georgetown

Stimulus Responsive Graphene Scaffolds for Tissue Engineering

Sofia S. Almeida, André F. Girão, Gil Gonçalves, António Completo and P.A.A.P. Marques

Abstract Tissue engineering (TE) is an emerging area that aims to repair damaged tissues and organs by combining different scaffold materials with living cells. Recently, scientists started to engineer a new generation of nanocomposite scaffolds able to mimic biochemical and biophysical mechanisms to modulate the cellular responses promoting the restoration of tissue structure or function. Due to its unique electrical, topographical and chemical properties, graphene is a material that holds a great potential for TE, being already considered as one of the best candidates for accelerating and guiding stem cell differentiations. Although this is a promising field there are still some challenges to overcome, such as the efficient control of the differentiation of the stem cells, especially in graphene-based microenvironments. Hence, this chapter will review the existing research related to the ability of graphene and its derivatives (graphene oxide and reduced graphene oxide) to induce stem cell differentiation into diverse lineages when under the influence of electrical, mechanical, optical and topographic stimulations.

Keywords Differentiation · Graphene · Scaffold · Stem cells · Stimulation · Tissue engineering

1 Introduction

Recent developments of medicine and technology have led to an outburst of research on regenerative medicine allowing the emergence of new and enhanced biomaterials with the potential to fully heal damaged tissues and organs. Primary studies were focused on skin equivalents for treating burns but with the increasing

S.S. Almeida (✉)
Faculty of Engineering, University of Porto, Porto, Portugal
e-mail: sofialmeida22@gmail.com

A.F. Girão · G. Gonçalves · A. Completo · P.A.A.P.Marques (✉)
TEMA, Mechanical Engineering Department, University of Aveiro, Aveiro, Portugal
e-mail: paulam@ua.pt

G. Gonçalves et al. (eds.), *Graphene-based Materials in Health and Environment*, Carbon Nanostructures, DOI 10.1007/978-3-319-45639-3_8

needs of medicine, different tissue types started to be engineering such as bone, blood vessels, liver, muscle and nerve conduits [1]. Today, research is focused more significantly in stem cell technologies. The capacity of these cells to self-renew and generate differentiated progenies gives them the ability to treat numerous diseases and injuries all over the body [2]. For improved results in this field it is essential to use substrates that enable cell attachment, proliferation and differentiation [3, 4]. Indeed, materials that are able to initiate, stimulate and sustain the series of complex events that lead to cell differentiation produce more favourable results when implanted in the human body [2, 5]. The design of scaffolds is therefore vital for cells to proliferate, differentiate and eventually generate the desired tissue [6]. According to the literature, the choice of the material and consequently scaffold's mechanical properties has a direct impact in tissue formation by inducing cell differentiation into the desired phenotype through mechanotransduction [6, 7]. For that reason, it is important to use materials that possess intrinsic characteristics for sustaining cell growth and inducing differentiation which may possibly lead to important progresses in stem cell research.

Graphene is a single-atom thick sheet of hexagonally bonded carbon atoms arranged in a two-dimensional (2D) honeycomb structure with unique electronic, physical, chemical and mechanical properties, including high electrical conductivity, elasticity and high molecule absorption [8, 9]. Presently considered the thinnest, strongest and stiffest material [10], graphene is leading to numerous promising advances in fundamental science, especially in nanobiology and nanomedicine [11]. Moreover, this material has a great potential to improve the performance of a broad range of biomedical devices since it is highly biocompatible, has low toxicity [12, 13] and exhibits low inflammatory responses [14]. Furthermore, graphene and its derivative, graphene oxide (GO) have the capability of being easily functionalized [15], which has made them popular in several fields of biomedical engineering including anti-bacterial [16], viral [17] and parasitical applications [18] as well as bioassays [19], biosensors [20], cancer cell targeting, imaging and therapy [21–25] and stem cell-based tissue engineering (TE) [25–27]. As a result, graphene-based materials are now receiving increasing attention in the field of TE [26]. Many researchers have already proposed graphene as one of the best candidates for accelerating and guiding stem cell differentiations into specific lineages [28]. In fact, its exceptional properties led graphene to be considered as a powerful platform that not only allows stem cell attachment and growth but also enhances the differentiation of stem cells into specialized cells [29]. It has already been demonstrated that graphene is an effective substrate to promote the adherence of human osteoblasts and mesenchymal stromal cells [30], to stimulate osteogenic [31], myogenic [32] and neuronal [33] differentiation of human mesenchymal stem cells (hMSCs) and to induce the differentiation of neural stem cells (NSCs) in three-dimensional (3D) porous structures [34].

However, despite of the significant developments in applications using graphene-based materials, the stem cell-based TE using graphene is still a field of science that is not entirely mastered, in particular the efficient control of the differentiation of stem cells into specific cell types of interest, especially in graphene-based microenvironments [35]. Additionally, because it is possible to use stimuli to control stem cells' physiological activities, such as its viability, division, migration and differentiation [36–40], graphene has recently been investigated as a template and/or electrode for inducing stem cell differentiation under various stimulation types.

Therefore, in the following, the capacity of graphene-based materials to induce stem cell differentiation when under the influence of various stimulations (including electrical, mechanical and chemical) will be revised.

2 Graphene Substrates as Platforms for Directing Cellular Differentiation: From 2D to 3D

Scaffolds play a very important role in the success of TE, since they are able to not only provide cell support and attachment, by acting as a biomimetic structure, but also to modify cellular responses over time and to supply growth factors and drugs [41]. For the fabrication of an ideal scaffold several parameters should be taken in account, such as the mechanical strength so that it can support the predesigned tissue structure, porosity, absence of toxicity, ability to transport oxygen, provision of attachment sites for cells, provision for nutrients and biodegradability [42]. Although the use of 3D scaffolds is the most common stem cell culture method in TE [43], the use of 2D constructs can also be very advantageous, since, for example, the smart layering of 2D sheets can facilitate the creation of a more organized structure allowing a more appropriated cell placement.

The fabrication of 2D graphene templates is usually accomplished by two main methods [29]. First, by the chemical vapour deposition (CVD) method [44], in which thin graphene sheets are produced. For allowing the enhancement of graphene's excellent features, including its electrical conductivity and mechanical properties, the fabrication of 2D graphene scaffolds by this method led researchers to start to see graphene as a potential material for TE applications. Alternatively, 2D graphene constructs can be fabricated through the chemical exfoliation of graphite [45], which allows the production of GO and reduced graphene oxide (rGO) sheets. In this case, it is essential to perform the exfoliation appropriately once it was already demonstrated in previous works [28] that the surface chemistry of graphene determines the surface adhesion properties, which can directly lead to the increasing or decreasing of cells' proliferation and differentiation. In fact the lack of standardization of graphene-based materials remains a significant problem for the reproducibility of the results, ion since the several methods reported for the production of graphene sheets can have, as final products, materials with different number of layers and/or chemical groups, affecting the behaviour of cultured stem cells in a different manner [46]. Many researchers have already demonstrated that the use of graphene as a 2D scaffold has a great potential to enhance the proliferation and differentiation of a vast number of cells [32, 34, 47–49]. However, for a successful in vitro implementation of stem cell differentiation systems, the use of 2D scaffolds may not be always sufficient, since effective differentiation requires appropriated features such as a suitable topography with cell growth channels as well as the natural synergistic effects of cell-to-cell and cell-to-extracellular matrix (ECM) interactions that can only be assured by 3D systems [43]. In addition, their high heterogeneity, limited scalability and low reproducibility are making 2D scaffolds a more less attractive system when compared with 3D ones since they are not entirely compatible with the development of in vitro models that accurately simulate the native stem cell niche [34].

Concerning the importance of the topographic cues for cells behaviour reported above, the effects of graphene-based polymer composites in the form of 2D and 3D scaffolds on the biological response of osteoblasts were investigated by Kumar et al. [50]. As shown schematically in Fig. 1, first, the authors fabricated the scaffolds by incorporating 1 % (wt/wt) of GO and rGO in polycaprolactone (PCL) nanocomposites, being that the 2D substrates were circular discs (5 mm diameter and 30.5 mm height) prepared by compression moulding, whereas the 3D macroporous scaffolds were prepared by the salt leaching method. Then the cells were cultured on both 2D and 3D substrates and biological studies were performed in order to observe cells' behaviour. They demonstrated that there is in fact a difference in the cellular responses obtained by the two different substrates. They found that the cells tend to organize into aggregates in 3D scaffolds in contrast to 2D ones where cells tend to spread and become randomly distributed.

As a result, the use of 3D scaffolds in TE is becoming a more popular method for stem cell culture, since it not only allows a substantial improvement of stem cell viability and function but also because it is a method that has higher efficiency, consistency and predictability [43]. Although graphene has proved to have a great potential in this field it is still a very recent material, meaning that there is a limited number of studies related with 3D graphene structures. However, its exceptional

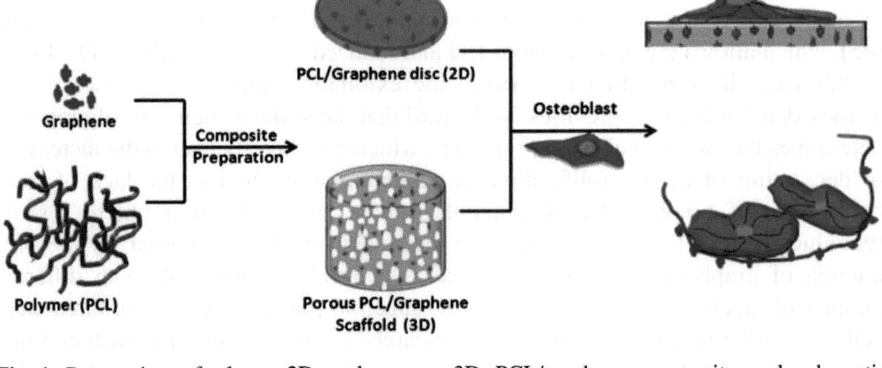

Fig. 1 Preparation of planar 2D and porous 3D PCL/graphene composite and schematic representation of in vitro osteoblast response on 2D and 3D scaffolds. Reproduced with permission [50]. © 2015 WILEY PERIODICALS, INC

properties are exciting more and more the scientific community to develop further methods in order to provide 3D graphene scaffolds with several topographies.

Primary attempts for the fabrication of 3D graphene structures were reported by Chen et al. [51] in which they performed the direct synthesis of 3D foam-like graphene macrostructures using the CVD method, allowing the production of thin films named by them as graphene foams (GFs) (Fig. 2).

This was an important step in graphene's research since it excited the development of more studies using 3D graphene scaffolds in the field of TE, in particular, the investigation of the ability of this graphene structures to regulate cell behaviour and induce stem cell differentiation. Consequently, Crowder et al. [31] investigated the use of 3D GFs as cell culture substrates for the evaluation of its capacity to promote osteogenic differentiation of hMSCs. The GFs were produced by growing graphene on 3D Ni scaffolds, in which Ni was posteriorly removed by $FeCl_3$ etching. Their results indicated that 3D GFs are capable of supporting the attachment and viability of hMSCs and also of inducing the spontaneous osteogenic differentiation of hMSCs without the need of extrinsic biochemical manipulation, demonstrating once more the potential of graphene in biomedical applications.

Later on, the capacity of 3D GF to act as a biocompatible and conductive scaffold for NSCs was evaluated. In this study Li et al. [52] reported the first use of a 3D graphene porous foam, as a novel scaffold for inducing the proliferation and differentiation of NSCs in vitro. The synthesis of the 3D GFs was made by the CVD

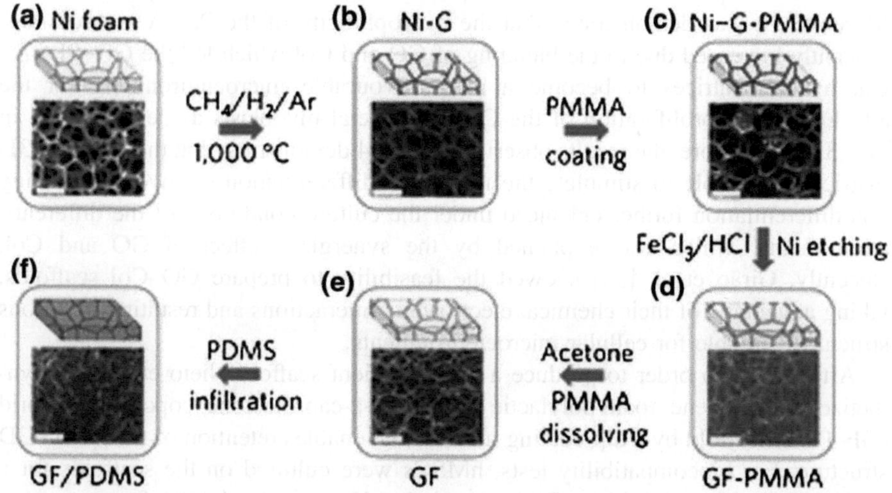

Fig. 2 Synthesis of a GF and integration with polydimethylsiloxane (PDMS). **a**, **b** CVD growth of graphene films (Ni–G) using a nickel foam as a 3D scaffold template. **c** An as-grown graphene film after coating a thin poly(methyl methacrylate) (PMMA) supporting layer (Ni–G–PMMA). **d** A GF coated with PMMA (GF–PMMA) after etching the nickel foam with hot HCl (or $FeCl_3$/HCl) solution. **e** A free-standing GF after dissolving the PMMA layer with acetone. **f** A GF/PDMS composite after infiltration of PDMS into a GF. All the *scale bars* are 500 μm. Adapted by permission from Macmillan Publishers Ltd: Nature Materials [51], © 2011

method using a Ni foam as a template, in which the Ni foams were removed afterwards by appropriated chemical solvents. The results presented in this study indicate that these graphene structures are able to effectively support NSC growth as well as induce more cells' proliferation when compared with the traditional CVD-grown 2D templates, since it was verified an upregulation of Ki67 protein expression (known as a cellular marker for proliferation). Additionally, it was also demonstrated by phenotypic analysis that 3D GFs tend to enhance the NSCs differentiation towards astrocytes and neurons in particular.

Another study concerning the use of 3D GFs to promote the repair of neural injuries was presented by Serrano et al. [53]. In this paper, it was reported a new method for the fabrication of the 3D porous structures, in which scaffolds were produced by a biocompatible freezer-casting procedure. The possible utility in neural tissue regeneration was assessed using embryonic neural progenitor cells which allowed the study of the cellular adhesion, morphology and viability as well as the neuronal/glial differentiation. The results proved that highly viable and interconnected neural networks were formed on these scaffolds having presented both neurons and glial cells and synaptic connections.

Later on, Shin et al. [54] reported the fabrication of GO-impregnated biomimetic matrices composed of poly(lactic-co-glycolic acid) (PLGA) and collagen (GO–PLGA–Col) by performing electrospinning for the enhancement of the myoblast differentiation. After analysing the physicochemical and mechanical properties of the hybrid scaffold as well as its biocompatibility and biofunctionality, they found out that GO dispersed and blended adequately with collagen in the hybrid matrices. Moreover, it was demonstrated that the hydrophilicity of the PLGA matrices significantly increased due to the blending of GO and Col which led the GO–PLGA–Col hybrid matrices to become a more favourable microenvironment for the attachment and proliferation of the C2C12 skeletal myoblasts as demonstrated in Fig. 3. Furthermore, the results observed in Fig. 4 demonstrate that these hybrid 3D matrices were able to stimulate the myogenic differentiation of myoblasts, being this differentiation further enhanced under the culture conditions of the differentiation media, which was explained by the synergistic effect of GO and Col. Recently, Girão et al. [55] showed the feasibility to prepare GO–Col scaffolds, taking advantage of their chemical electrostatic interactions and resulting in porous structures suitable for cellular microenvironments.

Afterwards, in order to produce a more efficient scaffold Nieto et al. [56] synthetized a graphene foam/polylactic acid–poly-ε-caprolactone copolymer hybrid (GF–PLC) scaffold by a dip-coating method that enables retention of the porous 3D structure. For biocompatibility tests, hMSCs were cultured on the scaffolds for a period of 28 days. Hybrid scaffolds revealed to be a good substrate for supporting chondrogenesis. Therefore, this study demonstrated that 3D GFs combined with other materials are also advantageous for TE applications, in particular for this case, it is suited for musculoskeletal applications, such as the growth of new cartilage in order to replace damaged one.

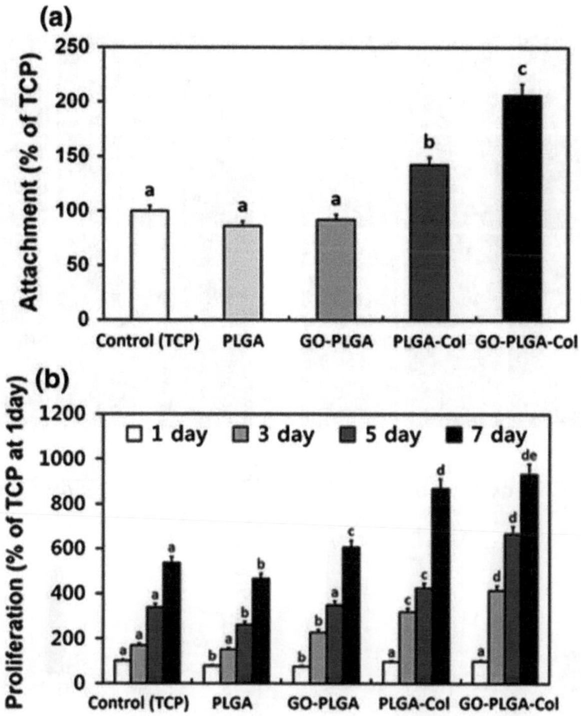

Fig. 3 Initial attachment and proliferation of C2C12 skeletal myoblasts. **a** Initial attachment of C2C12 skeletal myoblasts on tissue culture plastic (TCP), PLGA matrices, GO–PLGA matrices, PLGA–Col matrices, and GO–PLGA–Col matrices were measured using a CCK-8 assay at 6 h after seeding. **b** Proliferation of C2C12 skeletal myoblasts was measured using CCK-8 assay on 1, 3, 5 and 7 days after incubation. The *different letters* in **a** denote the significant differences between the control and experimental groups, $p < 0.05$. The *different letters* in **b** denote the significant differences between the control and experimental groups at the same time point, $p < 0.05$. If two groups have the same single letter (*a*, *b*, *c*, etc.), there is no significant difference between them. If a group is marked with a dual letter (e.g. *de*), it has a significant difference from the control and other groups marked with '*a*', '*b*' or '*c*', but does not from another group marked with '*d*'. Reproduced with permission from [54]. © 2015 BioMed Central

3 Graphene-Based Stimulus Responsive Scaffolds

Previous works have already demonstrated that by stimulating stem cells it is possible to control several physiological activities of the cells, including its differentiation [36–40]. In addition, graphene has already proved to be a good structure for inducing stem cells' differentiation when several stimuli are applied [57–59]. As a result, in the last few years, the possibility of controlling stem cell fates by applying different stimulations to graphene scaffolds has attracted an increasing attention of many researchers in the field of stem cell-based therapies.

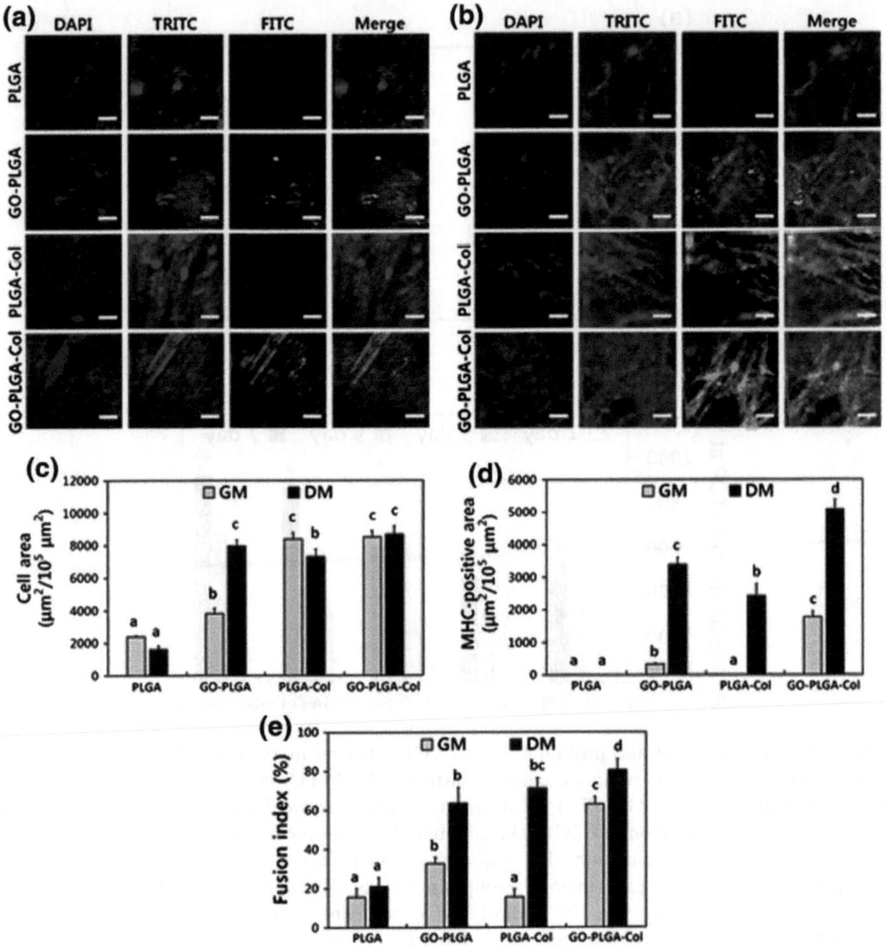

Fig. 4 Myogenic differentiation analysis with immunofluorescence staining. Two-photon excitation fluorescence images of C2C12 skeletal myoblasts in **a** growth media (GM) and **b** differentiation media (DM). The cells were cultured in GM for 2 days and then cultured in GM or DM for additional 5 days. The cell nuclei were counterstained with DAPI (*blue*), the F-actins were stained with TRITC-labelled phalloidin (*red*) and the myosin heavy chains (MHCs) were stained with FITC-labelled anti-MHC antibody (*green*). The *scale bars* are 50 μm. Quantification of **c** the cell area, **d** MHC-positive area, and **e** fusion index. The fusion index was calculated as a percentage of the nuclei number in multinucleate myotubes with more than two nuclei to the total number of nuclei. Quantitative analysis was performed using ImageJ Software. The *different letters* in **c** and **d** denote the significant differences between each experimental group, $p < 0.05$. The *different letters* in **e** denote the significant differences between each experimental group, $p < 0.05$. If two groups have the same single letter (*a*, *b*, *c*, etc.), there is no significant difference between them. If a group is marked with a dual letter (e.g. *bc*), it has a significant difference from the control and other groups marked with ‘*a*’, but does not from another group marked with ‘*b*’. Reproduced with permission from [54]. © 2015 BioMed Central

3.1 Electrical Stimulation

Electrical stimulation plays an important role in the guidance of stem cell differentiation [60, 61] and because of its particularity of being a non-invasive method it is now being used as a promising therapy, in particular for neurological diseases [62, 63]. In addition it is considered one of the most advantageous stimulations once it is a more controllable, quantifiable and reliable technique that allow the precise delivery of the stimulus through the use of electrodes [64].

Previous works [37, 65] have already demonstrated that the use of electrical stimulation indeed influences cells' behaviour being already established that it effectively controls various stem cell's physiological activities such as cell viability, division, migration and more importantly cells' differentiation. Due to its favourable electrical conductivity and high transmittance [10] graphene is considered a promising material for modifying the behaviour of electrically sensitive cells. In fact, since neural cells and their functions are based on electrical activities, the unique electrical properties of graphene suggest that it could be a successful platform for neuronal stimulation and treatment [66]. Moreover, graphene not only has adequate chemical characteristics for the effective attachment of neural tissues [67], but also is susceptible to have its electronic properties tailored in order to match the charge transport that is necessary for electrical cellular interfacing [68, 69].

The first work that reported the use of graphene combined with electrical stimulation was presented by Heo et al. [70]. The authors proposed the fabrication of an enhanced electrical stimulator supported by a substrate that combines the superior properties of graphene with the non-cytotoxic properties of the polymer polyethylene terephthalate (PET) [71] in order to minimize cellular damage after electrical stimulation. Thus, non-contact electrical field stimulation, produced by charge-balanced biphasic stimuli, was applied through the flexible, transparent and non-cytotoxic graphene-PET film electrodes in order to stimulate neural cells that were cultured in a dish and to further examine its effects on cell-to-cell coupling. Their results showed a superior effectiveness in the shape of cell-to-cell interaction of human neuroblastoma cells when weak electric field stimulation of 4.5 mV/mm with pulse duration of 10 s during 32 min was applied. Additionally, for this stimulation they also observed a significant increase in the strength of the existing cell-to-cell couplings as well as an increment in the number of cells which enabled the formation of new cell-to-cell couplings (Fig. 5). The observed altered cellular interactions were explained based on possible changes on the regulation of the endogenous cytoskeletal proteins fibronectin, actin and vinculin. Hence, this work allowed, for the first time, to demonstrate the benefits of using graphene in electrical stimulation therapies, opening tremendous opportunities for graphene-based materials in the field of TE.

Meantime, in the study of Park et al. [34], already mentioned above, a transparent graphene electrode was used for the first time for the electrical stimulation of already differentiated cells of hNSCs (neurons and glia cells) in order to confirm their neuronal activity. Therefore, a series of voltage pulses typically $1 \approx 10$ of

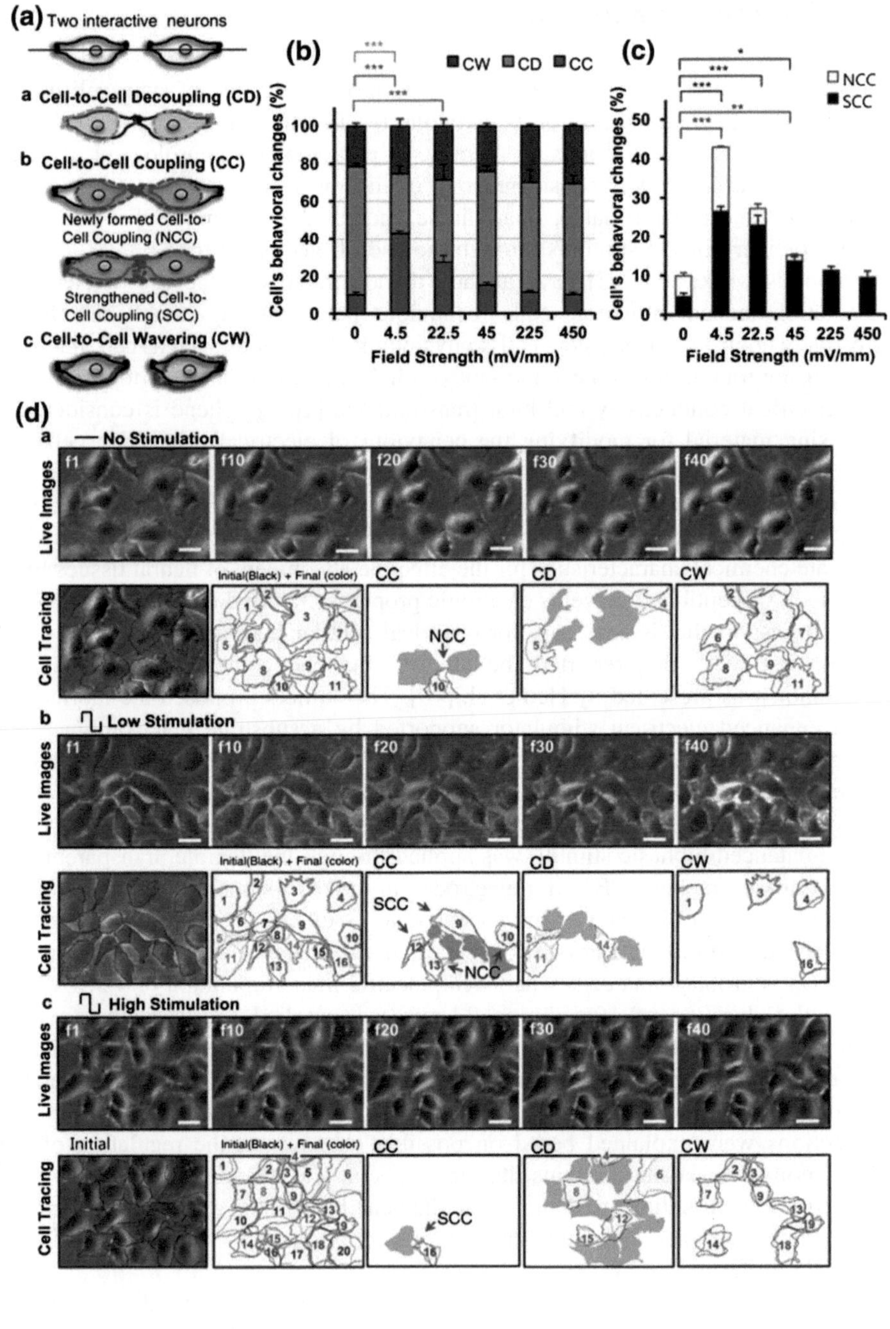
(a) Two interactive neurons
a Cell-to-Cell Decoupling (CD)
b Cell-to-Cell Coupling (CC)
Newly formed Cell-to-Cell Coupling (NCC)
Strengthened Cell-to-Cell Coupling (SCC)
c Cell-to-Cell Wavering (CW)
(b)
CW CD CC
Cell's behavioral changes (%)
Field Strength (mV/mm)
0 4.5 22.5 45 225 450
(c)
NCC
SCC
Cell's behavioral changes (%)
Field Strength (mV/mm)
0 4.5 22.5 45 225 450
(d)
a No Stimulation
b Low Stimulation
c High Stimulation
Live Images
Cell Tracing
f1 f10 f20 f30 f40
Initial(Black) + Final (color)
CC CD CW
NCC
SCC
Initial

◀ **Fig. 5** Cell-to-cell interactive reactions to electric field stimulation. **a** Schematic illustrations of cell-to-cell interactive reactions between two separated cells under electric field stimulation. **a** Cell-to-cell decoupling (CD). Cells belonging to the CD group separated from each other after stimulation. **b** Cell-to-cell coupling (CC). The CC group was further classified into two groups: The newly formed cell-to-cell coupling (NCC) group and the strengthened cell-to-cell coupling (SCC) group. The NCC represents a group of cells that respond to electric field stimulation by forming new contacts between cells. The SCC represents a group of cells strengthening existing contacts between cells after electric field stimulation. **c** Cell-to-cell wavering (CW). Cells belonging to the CW group exhibit a wavering behaviour following electric field stimulation. **b** A bar graph categorizing behavioural reactions to electric field strengths. The percentage of cells categorized as CC (*red*) is the highest at 4.5 mV/mm and the second highest is at 22.5 mV/mm. On the contrary, the percentage of cells categorized as CD (*green*) is the lowest at 4.5 mV/mm and the second lowest is at 22.5 mV/mm. There was no effect of electric field on the CW (*blue*) group. **c** The categorization of CC cells. When we further categorized CC into two groups, there was a clear effect of electric field on NCC. In particular, at 4.5 mV/mm, the lowest electric field stimulation, the percentage of cells belonging to NCC was the highest among other electric fields. Under the weak electric fields, cells exhibited a trend of increased percentages of SCC. **d** A typical example of a cellular reaction to the electric field. **a** Live images and tracing of cellular changes with no electric field stimulation. "f1" indicates the 1st image and "f40" indicates the 40th image taken from the optical microscope during stimulation experiment. The initial shape of the cell is represented as a *black tracing line* and the final shape of the cell is represented with colour coding (CC-*red*, CD-*green* and CW-*blue*). With no stimulation, the majority of cells were categorized into the CW group. Few NCC and SCC were observed with no stimulation. **b** Live images and tracing of cellular changes with 4.5 mV/mm. At 4.5 mV/mm, the largest percentages of cells were categorized as CC and we observed clear NCC and SCC. **c** Live images and tracing of cellular changes at 450 mV/mm. At 450 mV/mm, the majority of cells reacted to the electric field and were categorized into CD and CW groups. The *scale bar* represents 30 μm. Values from the bar graphs are mean (n = 6 for 22.5, 45, and 225 mV/mm, and n = 10 for 0, 4.5 and 450 mV/mm). Significantly different $p < 0.05$, 0.01 and 0.001 are represented by *, ** and *** symbols. Reproduced from Heo et al. [70], © 2011, with permission from Elsevier

500 mV monophasic/cathodic voltage pulses with 1 $\approx$ 100 ms duration in a second were applied to the differentiated cells from hNSCs using graphene electrodes. The fluorescence intensity of the cells which indicated calcium levels inside the cell was continuously monitored with a fluorescence microscope. By observing the results before and after electrical stimuli, they found out that cells exhibited increased calcium levels, since its fluorescence intensity increased significantly after stimulation. In addition, by plotting the fluorescence intensity versus the stimulation time period they observed that electrical stimulation caused an increase of fluorescence intensity between 60 and 70 %. Thus, their findings clearly suggested that the differentiated cells from hNSCs on graphene films were operating as a neuron, and more importantly that graphene films can indeed be used as neural stimulation electrodes.

The response of NSCs cultured on graphene substrates to electrical stimulation has been also studied by Tang et al. [72]. Herein, a series of 10–100 ms monophasic cathodic pulses with intervals of 5 s were applied in the cultured graphene substrates. After several tests, they established a stimulation threshold current of 0.5–1 μA. Furthermore, their findings confirmed not only that graphene is able to improve neural performance and electrical signalling in the network but also that it can work as a 2D material for conducting electrical current to neurons, confirming, hence, its potential to act as an effective neural interface for TE applications.

The effects of electrical stimulation in 3D graphene porous structures were also investigated in the work presented by Li et al. [52] that first proved that these scaffolds were able to support NSCs growth and induce cells' proliferation, as mentioned in the previous section. Here, 3D GFs were used as a cell stimulation electrode by being subjected to a series of monophasic cathodic pulse produced by a function generator, in which it was demonstrated that the stimulation threshold was 20–30 μA. To evaluate the influence of electrical stimulation, the differentiated NSCs cultured on the 3D structures were stained with Fluo-4 AM dye before stimulation in order to monitor the change of intracellular calcium ion concentrations, once it is known by previous works that submitting a neuron under voltage pulse stimuli may favour the opening of calcium ion channels, consequently increasing calcium ion concentrations in the cell, which results in a superior fluorescence intensity of the dye present in the cells [73]. The results presented in this study show an increase of 50–60 % in fluorescence intensity after electrical stimulation, clearly demonstrating that 3D GFs also have the capacity to operate as an effective conductive platform for electrically stimulate cells, which validates that 3D graphene structures exhibit a great potential for advanced strategies in several areas of regenerative medicine.

Despite of graphene effectiveness in acting as a viable material for regulating cells' behaviour it may not always produce the most desirable results when submitted to electrical stimulation. For that reason, researchers started to study and fabricating enhanced graphene-based structures combined with other materials in order to act as a platform for applications based on electrical stimulation. Concerning this, Zhang et al. [74] studied the performance of a new amphiphilic rGO template with an enhanced charge injection capacity in the electrical stimulation of neural cells to assess its capacity to act as a viable interface for the passage of electrical current. To obtain the rGO sheets and further fabricate this template, GO sheets were simultaneous submitted to covalent functionalization and thermal reduction treatments. The covalent functionalization was performed by applying methoxy poly(ethylene glycol) (mPEG) chains on the surface of the rGO sheets, allowing the enhancement of this material as it not only allows the material to have a high dispersibility in various solvents, which enables several post-treatment processes, but also increases its charging capacitance. By performing calcium imaging tests in the PC12 neural cells cultured on the amphiphilic mPEG–rGO films and hydrazine–rGO sheets, the authors found out that the electrical stimulation, executed by applying a series of 1–100 ms monophasic anodic with durations of 10 s and potentials lower than 0.6 V, significantly increased the percentage of cells with higher action potentials cultured on the mPEG–rGO films due to their higher charge injection. These findings not only demonstrated that this new amphiphilic mPEG–rGO material is capable of being used for neural prostheses applications in a safer and efficacious manner but also that graphene can be combined with several other materials in order to effectively modify electrically sensitive cells.

In another work Berit et al. [75] investigated the suitability of graphene to work as a material for electrical stimulation purposes when combined with other materials, by analysing the electrochemical properties of the fabricated graphene-based electrodes. As it was already recognized that graphene had the ability to improve the electrical interface between neuronal cells and electrodes, the authors proposed the use of this material as a biocompatible coating for commonly used electrode materials, more specifically, for gold and silica (SiO_2). After performing electrochemical and Raman characterization on both gold and SiO_2 electrodes coated with graphene, they found out that graphene on SiO_2 substrate is a more promising material combination for the fabrication of superior stimulation electrodes.

More recently, the ability of graphene-based materials to serve as viable platforms for the regeneration of optical nerves using electrical stimulation was studied by Yan et al. [76]. Optical nerves are usually damaged due to atrophy, apoptosis or death of retinal ganglion cells (RGCs) and could cause serious problems, such as the permanent visual field loss and irreversible ocular diseases. The use of electrical stimulation to heal injured neurons in vitro and to improve the survival rates and axon growth of central nervous system in vivo has already produced reasonable results. Despite this, the use of electrical stimulation-based therapies still remains a challenge, mostly due to electrode's characteristics, which have to include high safe charge injection limit (Qinj, i.e. electrochemical capacitance), high performance for long periods of stimulation and high biocompatibility. Graphene has a low Qinj, which forces researchers to combine graphene with other materials to achieve success. Hence, in this work, the authors proposed a new graphene-based electrode for the regeneration of optic nerve via electrical stimulation. The fabrication of the electrode was performed by co-electrospinning biocompatible PLGA with the polypyrrole-functionalized graphene (PPy-G)/ethanol dispersion. To synthetize the PPy-G, the polymerization-enhanced ball milling method was utilized (Fig. 6a), which enables the efficient introduction of electron-acceptor nitrogen increasing the material's capacitance and also allowing it to remain a conductive platform for charge transportation. Moreover, randomly oriented PPy-G/PLGA nanofibers and aligned PPy-G/PLGA nanofibers with different percentages of (w/w) PPy-G were fabricated under the same electrospinning conditions, as shown in Fig. 6.

In order to investigate the biocompatibility of the proposed substrate, first, its cytotoxicity was evaluated using the CCK-8 assay. As observed in Fig. 7e, the RGC cell viability was higher than 80 % in all the substrates, demonstrating that the as-synthesized PPy-G/PLGA nanofibers are highly biocompatible substrates. Moreover, the authors demonstrated that the PPy/G–PLGA nanofibers containing 6 % (w/w) PPy-G are the best substrate structure for cell attachment and growth since it showed a cell viability of approximately 100 % (similar to the control substrate). Then electrical stimulation was applied to the PPy-G/PLGA-based aligned nanofibers, with an optimized step potential of ±700 mV/cm. For effects of comparison, PPy/PLGA aligned and random nanofibers were also submitted to the same electrical stimulation as well as random PPy-G/PLGA nanofibers. The results

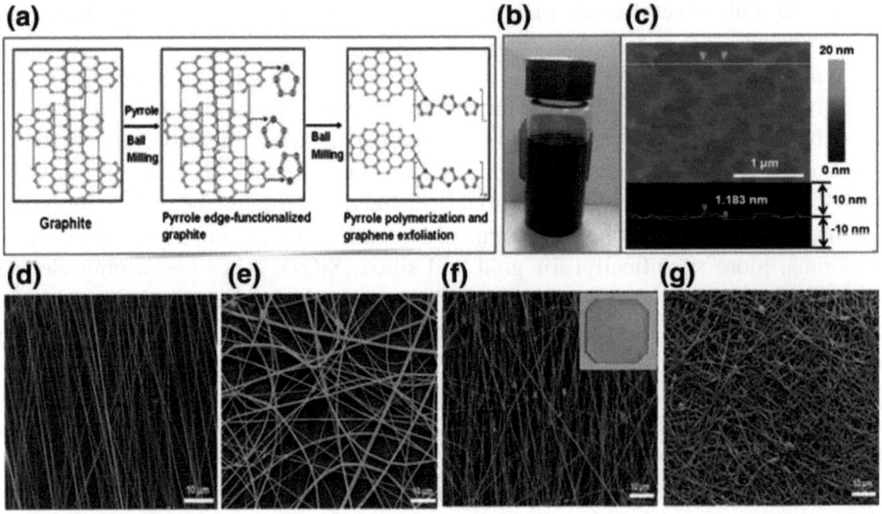

Fig. 6 **a** Schematic synthesis of the PPy-G via the polymer polymerization enhanced ball milling method. **b** Digital photo of the PPy-G well dispersed in ethanol. **c** AFM micrograph of the PPy-G. **d** SEM micrograph of the aligned nanofibers from 1 % (w/w) PPy-G/PLGA dispersion compared to (**e**) that of the random nanofibers prepared at the identical conditions. **f** SEM micrograph of the aligned nanofibers from 6 % (w/w) PPy-G/PLGA dispersion compared to (**g**) that of the random nanofibers synthesized at the identical conditions. *Inset* **f** is the digital photo of the nanofibrous scaffold used for electrical stimulation. Reproduced with permission from Yan et al. [76]. © 2016 American Chemical Society

showed that electrical stimulation led to a clear enhancement in the viability, neurite outgrowth and antiaging ability of RGCs. The influence of aligned or random nanofibers was also evaluated by analysing scanning electron microscopy (SEM) images of RGCs that were seeded on these both types of substrates (Fig. 7a, a′, b, b′, c, c′).

They found out that although electrical stimulation has promoted cell growth and density in both types of nanofibers, the aligned nanofibers were capable of allowing a longer cell length (137 % improvement in cell length after electrical stimulation), as shown in Fig. 7d, which not only demonstrates that electrical stimulation enhances the regeneration of RGCs but also that aligned fibrous substrates are more favourable to do this.

The impact of electroactive graphene composite scaffolds on neural TE was also studied by Zhou et al. [77], who have successfully functionalized 2D and 3D PCL structures with graphene-polyelectrolyte multilayers. Briefly, graphene-heparin / poly-L-lysine polyelectrolytes were assembled via layer-by-layer deposition onto 2D PCL surfaces and 3D PCL electrospun fibers in order to fabricate alternative biocompatible microenvironments with tunable electrical conductance properties capable of enhancing both neuron cell adhesion and neurite outgrowth.

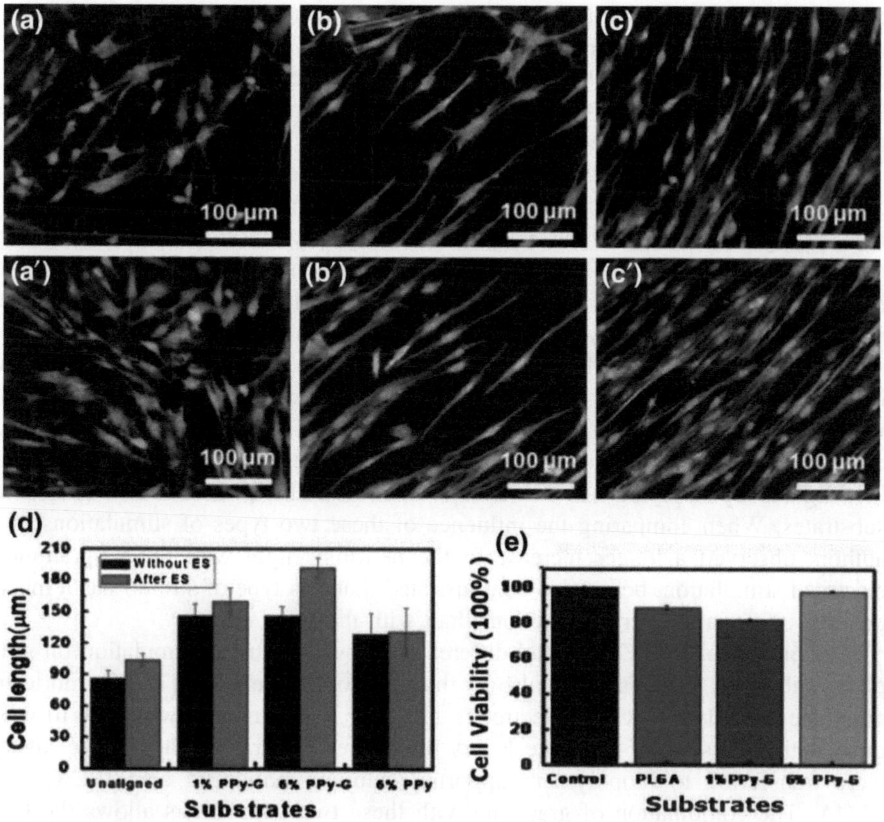

Fig. 7 Confocal microscopy images of RGC cells seeded on **a** the random PPy-G/PLGA nanofibers without ES and **a′** after ES; **b** the aligned PPy-G/PLGA nanofibers with 1 % (w/w) PPy-G without ES and **b′** after ES; **c** the aligned PPy/G–PLGA nanofibers containing 6 % (w/w) PPy-G without ES and **c′** after ES. **d** The average cell length of RGCs without and after ES. **e** Cell viability of RGCs cultured on the different substrates. ES conditions: Step potential was pulsed between −700 and 700 mV/cm. ES was performed 1 h everyday and lasted for 3 days. Reproduced with permission from Yan et al. [76]. © 2016 American Chemical Society

At this point, graphene had already proved that it could be a potential material for inducing stem cell differentiation via electrical stimulation; however, despite the studies mentioned above, there were still few investigations that could confirm this possibility. Thus, to address this challenge and evaluate, for the first time, the influence of electrical stimulation on the differentiation of muscle cells on graphene-based materials, Ahadian et al. [58] presented a study where C2C12 myoblasts were cultured not only on ultrathin thermally reduced graphene (TRG) films but also on GO and glass slide substrates. After finding out that TRG films are more favourable for cell adhesion and spreading, they also demonstrated that myoblast cell differentiation was significantly increased after electrical stimulation (8 V, 1 Hz and 10 ms for 2 continuous days). As a result, this work enabled the confirmation that graphene substrates are able to electrically regulate cells' behaviour, including its differentiation, leading to the development of numerous possible applications in the field of TE.

Meanwhile, several other studies have been exploring the capacity of graphene-based materials to control cells' behaviour via electrical stimulation. Concerning this, Meng et al. [59] used a large size non functional graphene nano-film (NGNF) as a substrate for the electrical stimulation and consequent behaviour modification of rat pheochromocytoma PC-12 cells. The graphene nano-film was fabricated by spray coating several high conducting graphene sheets on top of the polyurethane film. After performing numerous trials, they defined the intensity of 100 mV/mm as the optimized constant electrical stimulation condition, which revealed a significant enhance of PC-12 cells differentiation as well as neurite extension and growth. The effects of a programmed electrical stimulation were also investigated by applying 100 mV/mm at 1 and 10 Hz to the cultured graphene substrates. When comparing the influence of these two types of stimulations, the authors observed a longer increase in the neuritis length with the programmed electrical stimulation, being also demonstrated that this type of stimuli has a much positive effect in the nerve behaviour than with the other one.

The promotion of PC-12 cells' differentiation via electrical stimulation on graphene substrates was confirmed also by the study of Sherrel et al. [78]. To fabricate graphene–biopolymer (GBP) electrodes, first, the CVD method was used to produce highly crystalline graphene layers on copper substrates. Then, these layers were transferred to biopolymer supports, including polylactic acid (PLA) and PLGA. The combination of graphene with these two copolymers allows the formation of a unique electrode structure with several benefits, including superior surface properties and the possibility of electrical communication with excitable cells. Electrical stimulation was applied by submitting the cells that were seeded in the GBP structures to a biphasic current waveform with 250 Hz, which consisted in 100 μs pulses of ± 0.1 mA/cm^2 with a 20 μs interphase gap, and a 3.88 ms rest period for each cycle, during 8 h per day for 3 days. The results presented in this study corroborated that electrical stimulation enhances the differentiation of PC-12 cells when cultured on graphene conductive layers, since it was visible a great increase in the neurite length and neurites' connectivity when compared to unstimulated cells. The authors also demonstrated that graphene can be used as a coating for non-conducting, flexible biopolymer surfaces without hampering its biocompatibility, which is favourable for the creation of viable electrodes from non-conducting materials with optimized cell compatibility with graphene, which provides the suitable electrical properties for electrically stimulate cells.

Meantime, Guo et al. [79] suggested that an electrical stimulation-assisted culture process should enhance the ability of engineered conductive scaffolds to regulate the differentiation of MSCs towards the neural lineage. To obtain a self-powered electrical stimulation system for the enhancement of neural differentiation of MSCs on graphene substrates, an highly effective triboelectric nanogenerator (TENG), used to supply pulsed electric simulation signals, was combined with poly(3,4-ethylenedioxythiophene) (PEDOT)–rGO hybrid microfiber (80 μm in diameter), which allowed the enhance of the scaffold's electrical conductivity. First, they found out that the conductive rGO–PEDOT hybrid microfiber exhibited enhanced MSCs proliferation as well as good neural differentiation tendency. Then, graphene microfibers were submitted to electrical stimulation using the TENG, triggered by human walking steps, to apply 3000 pulses/day during 21 days with a frequency of about 120 times/min, resulting in a significant increase in the neural differentiation of MSCs. Their findings illustrate the potential of rGO–PEDOT hybrid microfiber scaffolds for neural TE applications and also demonstrate the viability of a human motion-driven self-powered TENG to electrically stimulate stem cells cultured on graphene substrates.

In a more recent study reported by Akhavan et al. [80] GO foam layers were fabricated by precipitation of chemically exfoliated GO sheets in an aqueous suspension under ultraviolet (UV) irradiation. This is a very versatile method where 2D graphene layers can be manipulated in order to produce 3D scaffolds with the desirable shape and size. Hence, rolled laminin-functionalized GFs were developed as electrically conductive 3D scaffolds and applied in directional growth of neural fibres, demonstrating that these structures allow the differentiation of hNSCs into neurons under an electrical stimulation. It was established that under low voltages (<5 V), the electrical sheet resistance of GFs is low enough ($\sim$170 Ω/sq) to produce the electrical stimulation currents ($\sim$20 mA) necessary to allow the differentiation of the neural cells. Moreover, it was demonstrated that the rolled shape of the GFs was also an important factor since it led to the formation of cross sections with superhydrophilic characteristics which induced the effective proliferation and differentiation of the hNSCs through the pores and scaffold's interfaces. Finally, it was investigated the influence of the electrical stimulation in the differentiation of the hNSCs cultured on the GFs. A series of 100 ms cathodic voltage pulses were applied on the two ends of the graphene roll in intervals with duration of 10 s. This stimulation resulted in an increased cells' proliferation and in an accelerated differentiation of the hNSCs into neurons (rather than glia) through the pores of the foam. Therefore, the findings presented in this study validate the capacity of 3D GFs to act as a flexible and conductive scaffold for the regeneration of nervous systems and TE. This work excited further studies as the one presented by Ahadian et al. [81] in which they demonstrated the influence of electrical stimulation on the differentiation of cardiac stem cells that are seeded on graphene substrates. Here, graphene nanosheets were embedded into a 3D structure of mouse embryoid bodies (EBs) using the hanging drop technique, in a ratio of 0.2 mg per mL graphene in the EBs, being posteriorly confirmed that this inclusion did not hamper stem cells' viability. Moreover, results showed that the inclusion of graphene in the EBs led to a decrease in the stem cell proliferation, which can be explained by the accelerated cell differentiation caused by graphene. In addition, it was demonstrated that the inclusion of graphene not only allowed the enhancement of the mechanical and electrical properties of the EBs, but also allowed to significantly increase the cardiac differentiation on the EB–graphene substrates. To assess the influence of electrical stimuli on the cells seeded on this structure, an electrical current with a voltage of 4 V, frequency of 1 Hz and a duration of 10 ms during 2 continuous days was applied. The results presented in this study reveal an enhanced cardiac differentiation of the EBs after electrical stimulation, which demonstrates once more that graphene is a promising material for the differentiation of several types of stem cells via electrical stimulation, revealing its potential for inumerous stem cell-based therapies.

3.2 Optical Stimulation

3.2.1 Laser Stimulation

For many years, the use of electrical current to stimulate neurons was the most appellative technique [64]. However, several limitations such as the necessity of having direct contact between the stimulating electrode and the tissue [82], the possible toxicity associated with the electrode's material [83, 84] and the influence of tissue impedance and coupling on the effective stimulation [85] led researchers to explore other viable alternatives to effectively stimulate the differentiation of stem cells. Pulsed laser stimulation is already being used as an alternative to electrical stimulation for evoking neural activity in motor and sensory systems [82, 86, 87]. In fact, the use of lasers for inducing neural responses is becoming a more appealing method as it is not only able to stimulate the desired tissue without the necessity of having direct contact but also offers an improved spatial resolution of stimulation [88]. Regarding its influence in the behaviour of stem cells, as laser irradiation has the capacity of providing intensive electric fields to the culture media, it is expected that it can open the calcium ion channels of the cells and that its magnetic field component can control the released calcium ions, which in theory would result in an increased calcium ion concentration inside the neurons, therefore enhancing neural regeneration [28]. Previous works have already demonstrated the efficacy of lasers for modifying the behaviour of stem cells [89, 90]; however, until very recently there were no reported studies that investigated the influence of pulsed laser irradiation on the stem cells' behaviour when these are seeded on graphene sheets. In this regard, Akhavan et al. [57] reported for the first time the use of graphene in the self-organized differentiation of hNSCs into neurons under pulsed laser stimulation. After the preparation of the films, by drop-casting a GO suspension onto quartz substrates and reducing some of the GO films with hydrazine, hNSCs were seeded on the synthetized GO and rGO films to assess, first, the biocompatibility of these templates in the proliferation of the neural cells. The results revealed that GO films are able to induce a higher proliferation of the cells which was explained by GO's superior hydrophilicity. On the other hand, it was demonstrated that the better electrical conductivity of rGO sheets enable them to induce more differentiation of the stem cells into neurons. To promote hNSCs differentiation into neurons and to rearrange their orientation towards the centre of the laser spot, GO and rGO templates seeded with hNSCs were submitted to nanosecond pulsed laser stimulation using a ND–YAG nanosecond pulsed laser system (wavelength of 532 nm, pulse duration of 5 ns, repetition rate of 10 Hz, power of 0.3 W, maximum pulse energy of 300 mJ and spot size of $\approx 0.5\ cm^2$) which was used at its second harmonic oscillation for 30 s at each hour to stimulate the cells through a photoexciting method. The results presented in this study clearly show that rGO sheets, as a 2D biocompatible scaffold, when stimulated by pulsed laser irradiation, not only provide an accelerated differentiation of hNSCs into neurons (rather than glia) but also cause the self-organization of a neuronal network on its surface by elongating the differentiated cells in the radial direction, which can be explained by the radial stress that is induced by the surface thermal gradient. Contrariwise, a decelerated differentiation was observed on the quartz substrates, which was assigned to the extra local heating produced by the irradiation on the quartz substrate. When comparing the efficacy of GO and rGO substrates it was visible that rGO films produced better differentiation results mainly due to its higher thermal and electrical conductivity, which allow it to transfer the extra heat generated by the laser irradiation outward from the laser spot. Hence, their findings not only demonstrate the excellent ability of chemically exfoliated rGO sheets for neural regeneration and repairing, but also encourage further investigations using graphene as the substrate for the promotion of stem cells' differentiation by applying pulsed laser irradiation.

3.2.2 Flash-Photo Stimulation

In order to reduce the possibility of causing cellular damages, which are usually induced by electric field stimulation, flash photo stimulation also started to be investigated. The use of this type of stimulation in graphene-based scaffolds is advantageous when these are combined with metal oxide semiconducting materials, such as titanium oxide (TiO_2) or zinc oxide (ZnO), once graphene and GO have the capacity of trapping extra electrons and the metals are able to effectively generate photoexcited electron–hole pairs with those trapped electrons. When the conditions are favourable, appropriate chemical bonds, such as Ti–C and/or Ti–O–C, are formed at the existing interface between graphene and the metal oxide, leading to the relocation of the electrons present in the pairs to graphene materials. The increasing number of photoexcited electrons present on the graphene sheets ultimately induces local electric fields. The induction of this local electrical field is crucial once it allows the modification of cell's behaviour, such as its differentiation, by affecting cells' calcium ion channels. Therefore, graphene-based materials are a potential viable substrate for the photocatalytic-based stimulation of cells [28, 91–93]. In this regard, Akhavan et al. [66] presented a pioneer work in which the influence of flash photo stimulation in the differentiation of hNSCs into neurons on graphene–TiO_2 heterojunctions was investigated. Herein, for purposes of comparison the differentiation behaviour of hNSCs was evaluated with and without flash photo stimulation on three different substrates (TiO_2, GO/TiO_2 and rGO/TiO_2). The hNSCs cultured on the graphene-based substrates were submitted to a flash photo stimulation by applying flash photo pulse trains (produced by Xenon lamp of a Canon camera) with pulse duration of 4 s and a frequency of 1 Hz (4 flashes/s) that would repeat after each 60 s in a total time of 30 min at each 12 h. Moreover, in order to find the optimal conditions of stimulation, the concentration of the ascorbic acid in the cultured media was varied as well as the flash intensity, which was varied from 1 to 100 mW/cm^2 by changing the distance between the flash and sample. As shown in Fig. 8, the results presented in this study show that flash photo stimulation induced different responses in the three substrates, since on the rGO/TiO_2 substrate the number of cell nuclei was increased by a factor of ~ 1.5, whereas on the GO/TiO_2 substrate and on the TiO_2 substrate a smaller increase was observed with rates of 48 and 24 %, respectively. This increase observed under flash photo stimulation in all the materials was explained by the ability of this stimulation to prevent cell death in the absence of cell growth factors. Furthermore, when the influence of flash photo stimulation on the differentiation of hNSCs was analysed, it was observed that the rGO/TiO_2 film was the substrate that presented better results. In fact, an 88 % decrease in the ratio of glial cells and an 81 % increase in the ratio of neuronal cells were observed, contrary to the other substrates (GO/TiO_2 and TiO_2) where only a 25 and a 15 % increase in the neuronal cells rate was, respectively, observed, revealing that although photo flash stimulation enhances the differentiation of hNSCs it is also dependent on the chemical state of graphene sheets as well as the composition formed between the graphene sheets and the beneath TiO_2 layer. The enhanced differentiation verified on the rGO/TiO_2 films was assigned to the fact that the photoexcited TiO_2 films enable the injection of electrons from this substrate into the cells that are cultured on the rGO sheets through the Ti–C and Ti–O–C bonds formed at the interface of graphene and TiO_2 films.

Finally, the optimal conditions for an efficient flash photo stimulation of hNSCs cultured on graphene-based substrates were assessed (Fig. 9), being revealed that with a concentration of 15.0 $mmol/dm^3$ of ascorbic acid in the cell culture medium and with an optimum flash intensity of 10 mW/cm^2 it is possible to not only

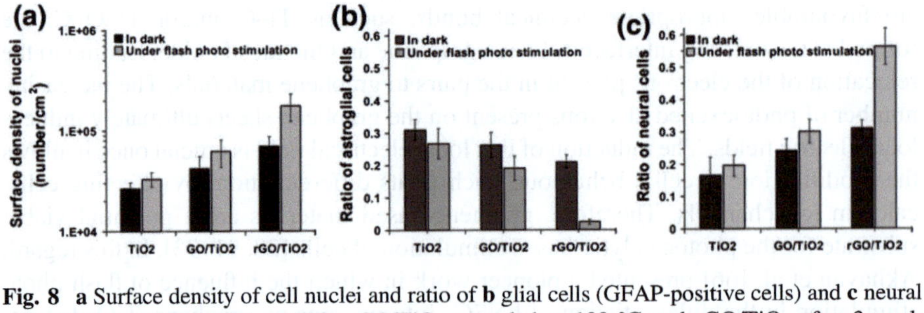

Fig. 8 **a** Surface density of cell nuclei and ratio of **b** glial cells (GFAP-positive cells) and **c** neural cells (TUJ1-positive cells) on TiO_2, GO/TiO_2 annealed at 100 °C and rGO/TiO_2 after 3-week differentiation in dark and under flash photo stimulation ($n = 5$, $p < 0.01$). Reproduced from Akhavan et al. [66] with permission from the Royal Society of Chemistry

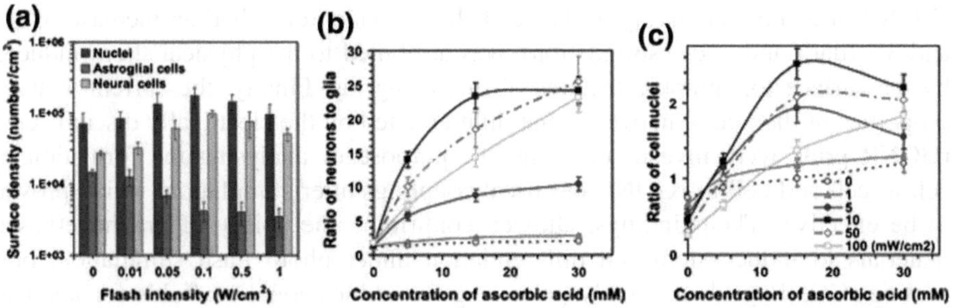

Fig. 9 **a** Surface density of cell nuclei, glia and neurons ($n = 3$, $p < 0.05$), **b**) ratio of neurons to glia ($n = 3$, $p < 0.05$) and **c** ratio of cell nuclei ($n = 5$, $p < 0.05$) on rGO/TiO_2 film after 3-week differentiation in a culture medium containing 15.0 mM ascorbic acid (**a**) and different concentrations of ascorbic acid (**b**, **c**), under flash photo stimulation with various intensities. Reproduced from Akhavan et al. [66] with permission from the Royal Society of Chemistry

increase the number of cell nuclei and differentiated neurons on rGO/TiO_2 (by factors of ~2.5 and 3.6), but also to decrease the number of glial cells (by a factor of ~0.28). These findings allowed introducing for the first time rGO/TiO_2 heterojunction film as a biocompatible flash photo stimulator for the effective differentiation of hNSCs into neurons, exciting further researches related with the photo stimulation of stem cells.

Later on, in another work presented by Akhavan et al. [94] flash photo stimulation was used to induce the differentiation of hNSCs into 2D neuronal networks, which contained cell-to-cell and cell-to-graphene electrical connections, on graphene nanogrids. To fabricate the nanogrids, graphene nanoribbons were synthesized by the oxidative unzipping of multiwalled CNTs and deposited on a SiO_2 matrix containing TiO_2 nanoparticles (NPs), being this structure proposed as a photocatalytic stimulator for the accelerated differentiation of hNSCs. Then, the hNSCs were stimulated to differentiate by applying, flash photo pulse trains (generated by a xenon lamp of a Canon camera) with 10 mW/cm^2 of intensity, 4 s of pulse duration and 1 Hz of frequency (4 flashes/s), repeated after 60 s time intervals for 30 min at every 12 h to the cultured substrates. The influence of flash photo stimulation on the differentiation of the cells was studied on both rGO nanoribbon (rGONR) grid/TiO_2 NPs/SiO_2 and quartz and rGO/TiO_2 substrates, in order to not only assess its effects but also to study the influence of the surface topography of the nanogrids on the differentiation of hNSCs. Thus, the results presented in this study show that the number of cell nuclei differentiated on rGONR grid/TiO_2 NPs/SiO_2 films increased 5.9- and 26.8-fold when compared with the number of cells on quartz substrates, in the dark and under photo stimulation, respectively. In addition, the stimulation of the cells cultured on the rGONR grids, originated by the injection of photoexcited electrons from the TiO_2 NPs into the cells on the nanogrids, resulted in higher neural differentiation than differentiation of glial cells (1.8 and 0.17, respectively). The enhanced differentiation on the rGONR grids comparatively to the rGO sheets, which only had an increase of 1.6 and 3.1 (dark and under stimulation), was attributed to the physical stress induced by the surface topographic features of the nanogrids. Finally, the current–voltage properties of the neural networks that differentiated on the electrically disconnected rGONR grids were investigated with the purpose of analysing the formation of cell-to-cell and cell-to-rGONR electrical couplings after stimulation, which proved to be effective. Their findings, allowed confirming the ability of graphene-based materials to induce stem cell differentiation under photo flash stimulation, presenting rGONR grids as highly promising two-dimensional scaffolds for applications in regenerative medicine and stem cell-based therapies, exciting further researches in the field of TE.

3.2.3 Near Infrared (NIR) Stimulation

Although flash photo stimulation has demonstrated to be an efficient method for the effective differentiation of hNSCs into neurons on graphene-based substrates [66], this is a method where the presence of a semiconducting photocatalyst material is mandatory, meaning that this stimulation will not produce favourable results when applied to substrates composed only by pristine graphene, once it acts like a zero band gap semiconductor. Despite that, it has already been proved that is possible to induce band gap opening and modulation of graphene not only by the fabrication of N-doped graphene sheets [95], but also by graphene nanoribbons [96] and graphene nanomeshes [28, 68, 69]. Although these structures have already been investigated for NIR photothermal therapies [23, 24, 97], up to now, there has been only one work that reported the study of the applicability of these graphene-based semiconductors in neural stimulation and differentiation. Therefore, in a more recent research presented by Akhavan et al. [98], the effects of applying NIR stimulation on hNSCs cultured on graphene nanomesh semiconductors were assessed. To fabricate GO nanomeshes (GONMs), GO sheets were, first, photocatalytic degraded using TiO_2 NPs embedded on the surface of a SiO_2 layer. For purpose of biological applications, several GONM films were posteriorly reduced by hydrazine vapour, obtaining thereby rGO nanomeshes (rGONMs), which act as p-type semiconductors with a band energy of approximately 1 eV. The biocompatibility of the graphene-based structures was first evaluated by analysing the adhesion and proliferation of stem cells in those substrates, being demonstrated that rGONMs have a similar biocompatibility in growth of hNSCs to GO sheets. Additionally, the number of cell nuclei and neural cells per unit area of the substrates (Quartz, GO, GONMs and rGONMs) studied was assessed, being established that GONMs are the most beneficial substrates for the proliferation of the hNSCs. These results were assigned to the excess of oxygen functional groups formed on edge defects of GONMs substrates, resulting in an enhancement of the surface's hydrophilicity. Afterwards, graphene-based nanomeshes were applied as semiconductor templates in NIR laser stimulation to induce the differentiation of hNSCs into neurons. In this work an 808 NIR laser with a spot size of $\sim$0.5 cm^2 and a power density of

0.1 W/cm^2 was applied to the graphene-based nanomeshes for 10 min at each hour. In addition, the temperature of the samples was continuously monitored using a thermocouple that was positioned inside the medium, since that for extended periods (10 min) of NIR irradiation the temperature of the incubated medium tends to increase to over than 45 °C, which can result in undesirable cell damages. Their observations demonstrated that under NIR laser stimulation rGONMs (with better chemical stability and higher electrical conductivity than GONMs) showed a higher level of cell differentiation, including elongation of the cells and increased differentiation into neurons than glia, when compared to quartz and rGO substrates. The higher biological activity of hNSCs cultured on the nonzero band gap rGONM semiconductor than on the zero band gap rGO sheets was attributed to the response of the low-energy (<0.5 eV) photoexcited electrons, which are injected from the rGONM semiconductors into the neural cells, to the NIR laser stimulation applied. In addition to that, the high-energy ($\sim$1.5 eV) photoelectrons present in the stimulated rGO could possibly inhibit cell proliferation and/or even cause cell damages. Finally, since the use of this laser irradiation produces thermal heating, the authors also investigated the influence of conventional heating of the culture media up to $\sim$43 °C (the temperature typically reached under the irradiation) in order to assess the real influence of the NIR laser stimulation. Interestingly, they found out that no significant differentiation was induced under these conditions, which further confirmed that photoelectrons play an important role in the promotion of the hNSCs differentiation. Their findings not only suggest that rGONMs are promising scaffolds for neural regeneration and repairing, but also excite more researches related with the induction of stem cells' differentiation under NIR laser stimulation.

3.3 Mechanical Stimulation

Although it was already demonstrated that the behaviour of stem cells is influenced by the microenvironment in which they reside, the understanding of the mechanisms that regulate the fate decisions of stem cells is a subject that still attracts the attentions of many researchers [99]. In fact, it was just recently discovered that physical and mechanical factors have a significant influence in the modulation of cells' behaviour, being already demonstrated that the mechanical properties of the extracellular environment and the application of mechanical forces on cells can, indeed, trigger several cellular responses, including cells' self-renewal and its lineage specification [100–105]. As a result, the application of mechanical stimuli for inducing stem cell differentiation is becoming an increasingly recognized tool within the scientific community [99]. Since cells have the ability to sense forces, and transduce them into biochemical signals, when a perturbation on the cellular microenvironment occurs, including a change in the ECM stiffness, topography or composition, the cell immediately responds to those cues, adopting different shapes, generating traction stress or even producing other mechanical forces that can be transmitted to the neighbouring cells [102, 105–107]. Therefore, when a mechanical stimulus, such as fluid shear stress, is applied to the cell surface several mechanosensitive ion channels, protein kinases and other membrane-associated signal-transduction molecules are activated, which triggers downstream signalling cascades that lead to changes in gene expression [108].

Graphene is a very versatile material that has the potential to act as an effective platform for inducing stem cell differentiation under mechanical stimulation. In a recent study, Kang et al. [109] reported the enhanced osteogenic differentiation of hMSCs by covalently conjugating the mechanical stiffness of GO flakes to 3D collagen scaffolds. The higher levels of osteogenic differentiation observed on the stiffer scaffolds were said to be mediated by MSCs mechanosensing, since the molecules that were involved in cell adhesion to stiff substrates were either upregulated or activated. This study confirmed the importance of scaffold's mechanical properties and therefore, mechanical stimulation for promoting stem cell differentiation.

Furthermore, several theoretical and experimental studies have already suggest that mechanical strain influences the vibrational and electronic band structure of 2D graphene [110] and graphene nanoribbons (GNRs) [111]. In fact, in the work presented by Chen et al. [112], the effects of uniaxial strain in individual GNRs were for the first time successfully investigated. After applying uniaxial strains (0–6 %) to individual GNRs with highly smooth edges by atomic force microscopy (AFM) manipulation, they were able to demonstrate that strain engineering GNRs allows to tune the band gap of graphene in a non-monotonic manner, which could be beneficial for the modulation of cells' behaviour.

Despite of the unique properties of graphene which allow it to be a very advantageous platform for inducing stem cell differentiation, to this date there are not yet any reported investigations that demonstrate the effects of applying several mechanical strains on stem cells cultured on graphene-based substrates.

3.4 Chemical Stimulation

Due to its interesting chemical diversity, graphene-based materials have also been studied as substrates for controlling the differentiation of stem cells via chemical stimulation. Concerning this, Wang et al. [113] studied the effects of using modified graphene sheets on the MSCs neuronal commitment. Since cell growth is influenced by the surface chemistry of graphene templates [28], the authors produced a scaffold for stem cell growth composed by fluorinated graphene sheets. First, the CVD method was used to prepare graphene sheets and then, to perform the fluorination of graphene several samples were exposed to a fluorinating agent. After studying the influence of cellular adhesion, morphology, gene expression and differentiation in terms of the surface chemistry, topography and mechanical properties of the substrate, they observed that fluorinated graphene substrates were able to induce a higher level of proliferation and also stronger polarization of MSCs when compared to substrates composed by graphene. To study the influence of chemical stimulation in the neuronal differentiation of these cells, the authors examined the expression of neuronal gene markers after the use of retinoic acid. When comparing the results of the cells cultured with and without the addition of the retinoic acid, it was observed that neural differentiation can be significantly enhanced with the addition of the retinoic acid, which demonstrates that it can act as a neuron-inductive chemical agent.

The influence of the surface chemistry on the differentiation of stem cells cultured on graphene structures was also investigated by Akhavan et al. [114] which explored an alternative way to produce improved graphene sheets using Asian red ginseng for the green reduction of chemically exfoliated GO into rGO. Concerning this, they analysed the influence of GO, hydrazine–rGO and ginseng–rGO films in the differentiation of NSCs into neurons. The results obtained show that ginseng–rGO films presented a better stability against aggregation when compared to the hydrazine–rGO ones in aqueous suspensions. Moreover, hydrazine–rGO films, which are hydrophobic, exhibited no toxicity against hNSCs, whereas the hydrophilic GO and ginseng–rGO films proved to be more biocompatible since it allowed stem cells' proliferation after only 3 days. When evaluating the influence of these three substrates in the neural differentiation, it was demonstrated that ginseng–rGO films were the substrates that allowed higher differentiation of hNSCs into neurons (rather than glia), followed by hydrazine–rGO substrates, being this explained by fact that rGO films have higher capability for electron transfer when compared to GO films. In the work already mentioned above, besides studying the influence of electrical stimulation, Tang et al. [72] also investigated the influence of chemical stimulation on the neural response of NSCs cultured on graphene films. Hence, high K^+ stimulation was applied to the cells by adding 50 mmol/dm^3 KCl to the culture media. To assess the cell response the intracellular Ca^{2+} changes upon stimulation were evaluated. The results presented show that K^+ stimulation was able to induce a superior response (65 % of fluorescence intensity) to electrical stimulation (30 %) in the intracellular Ca^{2+} change, by allowing the depolarization of the cells and activating voltage-operated calcium channels which resulted in an increased extracellular Ca^{2+} influx. Furthermore, the authors also investigated the effects of this stimulation in the activation of C-jun, which is an inducible transcription factor that has the capacity to regulate other gene expression in response to extracellular stimuli [115]. Their examinations allowed to the conclude that K^+ stimulation induced the activation of C-jun in the neural networks on graphene substrates, once it was demonstrated that under stimulation the number of cells positive for phosphorylated C-jun was only 14 %, whereas with stimulation the number significantly increased to 43 %.

Meantime, in another investigation, Akhavan et al. [116] also tested the influence of using chemical inducers on the osteogenic differentiation of hMSCs cultured on graphene nanogrids. After demonstrating that the proposed hydrophilic graphene nanogrids were biocompatible, the authors investigated the influence of chemical stimulation on this type of substrate by adding known osteogenic inducers, such as dexamethasone and ascorbic acid. The results presented in this study reveal that the chemical stimulation of the rGONR grid induced an enhanced osteogenic differentiation of the hMSCs, revealing that the amount of osteogenesis in the patterned substrates was $\sim$2.2-folds greater than the differentiation on the rGO sheets, which were uniform substrates. When comparing to the substrates that were not chemically stimulated, the authors observed lower levels of osteogenic differentiation on the patterned substrates and no differentiation on the graphene sheets. Hence, besides confirming the potential of chemical stimulation in the promotion of stem cell differentiation, this work also opened the path for further researches related with the surface topographic features since it was demonstrated that patterned substrates showed enhanced differentiation.

Later on, in the work reported by Weaver and Cui [117], a new nanocomposite composed by of conducting polymer PEDOT and GO nanosheets was fabricated and its NSC scaffolding performance evaluated. They found out that the biomolecules interferon-γ (IFNγ) and platelet-derived growth factor (PDGF) are able to selectively stimulate neuronal or oligodendrocyte lineage differentiation, by covalently cross-linking to the surface of the GO or PEDOT nanocomposite, respectively, via carboxylic acid functional groups provided by GO. Additionally, it was demonstrated that when the surfaces are stimulated with FNγ they have the capacity to support a larger population of neurons whereas when they are stimulated with PDGF they support a larger population of oligodendrocytes.

3.5 Topographical Stimulation

The microenvironment present on the cell imposes specific biophysical cues, including cells' topographical and mechanical properties, which can act through cell–substrates interactions and lead ultimately to the modulation of cellular behaviours such as differentiation [35]. In addition, since stem cells have the ability to sense the physical characteristics of their environment, specific biomechanical signals can be transmitted to them via various substrate topographical features such as pillars, grooves or pits [118]. Therefore, in the last years, researchers have started to further investigate the possibility of controlling stem cell fates by modulating some biophysical characteristics of the scaffolds, including the design of micro/nano-patterns, elasticity and scaffold's porosity [119–121].

The first work that studied the influence of morphological stimulation using graphene-based materials on the differentiation of stem cells was presented by Wang et al. [113], which investigated the possibility of inducing neural differentiation of MSCs by confining them into microchannels of fluorinated graphene surrounded by polydimethylsiloxane (PDMS) parallel lines in the absence of chemical inducers. To design the PDMS lines, these were ink-jet printed onto the fluorinated graphene substrates with the length and the width of, respectively, 3 mm and 150 μm and a line spacing of $\sim$50 μm, which resulted in a 30-μm-wide fluorinated graphene microchannel in between the PDMS lines. The results

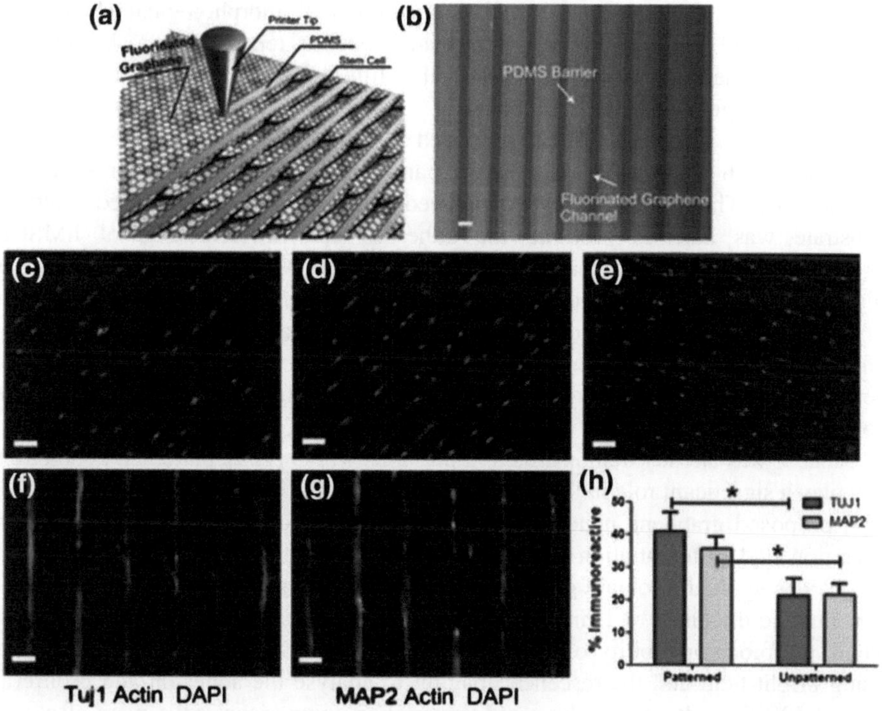

Fig. 10 **a** Schematic drawing of patterning MSCs by printing PDMS barriers on graphene films directly. **b** Optical microscope image of printed PDMS on fluorinated graphene film (*scale bar* 50 μm). **c–e** The aligned growth of stem cell on graphene, PFG and FG with printed PDMS pattern, respectively (*scale bar* 100 μm). **f**, **g** MSCs preferentially attached on the FG strips and their F-actin aligned (*red*) and expressed neural-specific markers—Tuj1 and MAP2 (*green*) (*scale bar* 50 μm). **h** Percentage of immunoreactive cells for Tuj1 and MAP2 on unpatterned and patterned FG strips. Note that the patterned FG strips induce higher expression of Tuj1 and MAP2 in the absence of retinoic acid ($n = 6$, $p < 0.05$). Reproduced with permission [113] © 2012 WILEY-VCH Verlag GmbH & Co. KGaA, Weinheim

presented in this study reveal that there is a correlation between the density of aligned MSCs and the coverage of fluorine on the graphene film and that the MSCs that were randomly seeded attached preferentially onto the fluorinated graphene microchannels, assuming a more elongated morphology as illustrated in Fig. 10.

Moreover, they found out that not only the MSCs showed a preferential adhesion to the fluorinated graphene strips but also that their cytoskeleton aligned along the length of the pattern. To assess the effects on the neuronal differentiation, the expression of neuronal markers of the cells that were cultured on the micropatterns was analysed. When comparing with the results of the cells cultured under non-patterned fluorinated graphene, they observed that the cells that were cultured under patterned fluorinated graphene demonstrated enhanced expression of the neuronal markers Tuj1 and MAP2 even without the presence of a neuron-inductive agent (Fig. 10). This work allowed confirming that morphological changes in graphene-based substrates that induce variations in terms of cytoskeletal and nuclear alignment can, indeed, promote the differentiation of MSCs towards the neuronal lineage.

Afterwards, Kim et al. [122] proposed a simple method that uses graphene combined with chitosan to fabricate a nanotopographic substrate for stem cell engineering. The results presented showed that graphene-incorporated chitosan substrate was able to promote the adhesion and differentiation of hMSCs. Moreover, it was proved that the differentiation on rGO–chitosan substrate was higher than the ones obtained on the chitosan substrate and polystyrene regardless of the use of osteogenic of osteogenic induction media. Finally, they proved that nanotopographic cues of the substrate play a significant role in the enhancement of cell-to-cell and cell-to-material interactions, which results in an improvement of cells' functions.

Since it was already demonstrated in previous works [123], patterned geometries can play a significant role in the guidance of stem cell differentiation. Akhavan et al. [94] proposed graphene nanogrids as a viable 2D-patterned template for the proliferation and differentiation of hNSCs. The oxidative unzipping of multiwalled CNTs was applied to obtain graphene nanoribbons (length of 10 μm, width of 50–200 nm and thickness of 1 nm) and to posteriorly fabricate the graphene nanogrids. Then, the biocompatibility of rGONR grid/TiO_2 NPs/SiO_2 films was investigated using bright-field and fluorescence imaging to analyse the adhesion and proliferation of hNSCs cultured on those substrates. For comparison, cells were also cultured on quartz, TiO_2 NPs/SiO_2, and GO nanoribbon (GONR) TiO_2 NPs/SiO_2 substrates. After analysing the fluorescence images, they observed that the hNSC were immunopositive for nestin (a green-coloured marker used for staining NSCs). This observation allowed researchers to conclude that not only stem cells were able to proliferate on the films, but also that they exhibited the neural stem cell property as they induced positive results for nestin. Moreover, the hydrophilic graphene nanogrid-coated films, in particular the GONR, demonstrated to have better cells' attachment as well as a increased proliferation of the hNSCs in the two vertically preferred orientations, coincident with the pattern of the nanogrids, when comparing with the quartz substrate and the TiO_2 NPs/SiO_2 films, which exhibited lower cell growths with random orientations. Finally, they found out that elongated patterns of rGO nanogrids were able to demonstrate increased differentiation of hNSCs into neurons in the absence of any chemical inducer when compared with the SiO_2 substrate.

Meantime, Solanki et al. [124] reported the fabrication of a new substrate consisting of nanoparticle-based nanotopographical features modified with graphene. This substrate was studied for the enhancement of neuronal differentiation and growth. The arrays of graphene–silica NPs hybrid structures (SiNP-GO) were produced by coating GO sheets on the surface of positively charged silica NPs followed by packing on the surface of glass substrates. Since GO contains high concentration of oxygen functional groups attached to its basal plane (high electronegativity), they have a higher capacity to readily attach positively charged molecules or surfaces, which in this case was the surface of 300 nm silica nanoparticles (SiNPs). The substrates were then treated with the ECM protein laminin (10 μg/mL for 4 h) in order to provide further attachment and growth of the hNSCs on the GO. To investigate the influence of the proposed substrate (SiNP–GO) in the neuronal differentiation, other three substrates were used as control: a glass substrate with a positively charged surface; a glass substrate with a monolayer of positively charged NPs and a glass substrate with a positively charge surface coated with GO. Then the hNSCs were seeded in the four substrates and the influence of SiNPs, GO and SiNP–GO on the neuronal differentiation was analysed by performing immunocytochemistry and quantitative PCR on the differentiated cells after 14 days. Their findings revealed that in the first days of differentiation axons tended to be aligned only on the GO and SiNP–GO substrates and not on glass and SiNPs substrates. In fact, after 14 days of differentiation, these substrates exhibited very well-aligned and well-extended axons, in opposing with the other substrates (glass and SiNPs) where although they demonstrated to have extended axons, they did not show any cell alignment. After analysing the images, the variation in the angle of orientation of the axons extending from differentiated hNSCs on substrates containing GO was quantified and compared with the orientation of the axons present on the control SiNPs and glass substrates. Their results allowed the authors to confirm the observations made previously, since the angle of orientation of the axons from differentiated hNSCs on the GO and SiNP–GO substrates was $\pm 17.8°$ and $\pm 9.16°$ respectively, whereas in the other two substrates the axons extended randomly having a much higher variation of $\pm 42°$ and $\pm 46.11°$, respectively. In addition to this, they also investigated the influence of nanotopographical features on the length of the axons extending from hNSCs that were differentiated on the different substrates on the day 14. It was confirmed that the length of axons present on the SiNPs was 20.76 % more than the average length of those cultured on glass and 11–3 % more than those cultured on GO, confirming that the alignment of the axons was exclusively induced by the presence of GO in the interior of the ECM, while the SiNPs monolayer can induce an increase in the average length of the axons from hNSCs differentiated on SiNP–GO. Afterwards, to determine whether the alignment of the axons present on GO and SiNP–GO was due to the crowding of hNSCs and by consequence dependent on the seeding density, the authors reduced the cell density by 50 % and after observing the cells' behaviour during 2 weeks, they confirmed that the axonal alignment of the differentiating hNSCs on the SiNP–GO substrates is not dependent of the cellular density, as the cells performed similarly. The effects of SiNP–GO on the neuronal differentiation of hNSCs were investigated by analysing the expression of immature and mature neuronal markers in the differentiated cells after 2 weeks. After performing immunostaining tests, it was demonstrated that a majority part of the aligned axons present in that substrate were characterized by the expression of several neuronal markers, including TuJ1, MAP2 and synapsin, and the axonal marker, GAP43. Moreover, it was demonstrated that the hNSCs differentiated on SiNP–GO substrates showed the highest expression levels for all neuronal markers. Next, for the first time the authors compared the axonal alignment of the differentiated cells on GO and CVD-grown graphene sheets. Although they observed axonal alignment on pristine graphene similar to the alignment observed on GO, they concluded that this was not sufficient for effective differentiation, being the GO substrates more advantageous for coating SiNPs, assembling ECM proteins and aligning the axons from differentiating hNSCs. Then, they further investigated the influence of the chemical structure of graphene (hexagonal lattice carbon-based structure) in the axonal alignment. In this regard, a nanomaterial, molybdenum disulphide (MoS_2), with a physical structure similar to graphene was selected and deposited on glass substrates, which were already coated by laminin. Their observations allowed to conclude that the unique chemical structure of graphene caused the axons to align, once although the hNSCs have grown well and differentiated on these MoS_2 substrates, they did not reveal to have any axonal alignment. Hence, this work demonstrated the importance of morphological stimulation on the differentiation of stem cells, once it proved that is possible to enhance neuronal differentiation as well as axonal alignment by engineering specific microenvironments, which in this case consisted in nanotopographical features modified with GO.

In another work, Shah et al. [125] reported the use of graphene–nanofiber hybrid scaffolds for the guidance of NSCs differentiation into oligondendrocytes. In order to fabricate these hybrid scaffolds, GO was directly deposited on the surface of oxygen plasma-treated PCL electrospun nanofibers and then coated with the ECM protein laminin. They found out that GO–PCL nanofiber scaffolds have a higher capacity to promote the growth of more elongated cells when comparing to the simple PCL substrates. In addition, they demonstrated that the GO–PCL nanofiber scaffolds were able to induce a more selective differentiation of the NSCs into oligodendrocytes, which was confirmed by the presence of several oligodendrocyte markers, including CNP, PDGFR, Olig1 and Olig2, as well as mature oligodendrocytes markers, such as PLP, MBP, MAG and MOG. Moreover, it was observed that the use increased amounts of GO on the electrospun nanofibers further promotes the differentiation of NSCs into mature oligodendrocytes. Finally, they also demonstrated that GO contents of the nanofiber scaffolds have a significant influence on the overexpression of several key integrin-related intracellular signalling proteins, including focal adhesion kinase (FAK), Akt, integrin-linked kinase (ILK) and Fyn kinase (Fyn) proteins, known by promoting the oligodendrocyte differentiation.

In a more recent work, Kim et al. [35] hypothesize that the fabrication of combinatorial patterns of graphene-based nanomaterials could be a more advantageous approach to have a precise control of stem cell differentiation, in particular for the differentiation of human adipose-derived mesenchymal stem cells (hADMSCs). To generate the desired patterns over large surface areas the microcontact printing (MCP) method was applied. They found out that GO line patterns were effective in the modulation of cells' morphology, which resulted in an enhanced differentiation of hADMSCs into osteoblasts. It was also confirmed that

GO grid patterns are a promising method for the neuronal differentiation, once it allowed the highly efficient conversion of mesodermal stem cells into ectodermal neuronal cells, explained by its ability to mimic interconnected/elongated neuronal networks.

4 Conclusion

Graphene and its derivatives show to be emerging materials that hold great potential for biomedical applications, in particular on the development of stimulus responsive platforms for TE. Due to its unique structure and diversity of electrical, mechanical, chemical and morphological characteristics, graphene and its derivatives (GO and rGO) are now receiving increasing attention within the scientific community. In fact, its exceptional characteristics have already led many researchers to consider graphene as a promising platform for stem cell-based therapies, since graphene has already demonstrated to play an important role in the adhesion, proliferation and differentiation of several stem cells, demonstrating its potential not only for neuronal differentiation, but also for osteogenic and cardiac differentiation.

Moreover, besides of being an effective substrate for inducing spontaneous stem cell differentiation, graphene has already proved to be appropriated for inducing stem cell differentiation under several types of stimulation, including electric, NIR, flash photo, chemical, mechanical and morphological. Although the research with graphene for tissue regeneration is still in the early stages of development, this carbon nanomaterial may have a bright future in different clinical scenarios.

Acknowledgments Sofia S. Almeida thanks the Mobility Program: Programme Almeida Garret —Studies during the Academic Year 2015/2016. PAAP Marques thank the Portuguese Foundation for Science and Technology (FCT) for the investigator grant (IF/00917/2013) and for the Project IF/00917/2013/CP1162/CT0016 which supports the researcher grant of André Girão. Gil Gonçalves thank FCT for respectively the grants SFRH/BDP/84419/2012. This work was partially supported by the funding of Program COMPETE-FEDER, Programa Operacional Competitividade e Internacionalização through the Project POCI-01-0145-FEDER-016574 and by FCT, IP through the Project PTDC/EMS-TEC/3263/2014.

References

1. Persidis A (1999) Tissue engineering Nat Biotech 17(5):508–510. doi:10.1038/8700
2. van der Kooy D, Weiss S (2000) Why stem cells? Science 287(5457):1439–1441
3. Bae H, Chu H, Edalat F, Cha JM, Sant S, Kashyap A, Ahari AF, Kwon CH, Nichol JW, Manoucheri S, Zamanian B, Wang Y, Khademhosseini A (2014) Development of functional biomaterials with micro- and nanoscale technologies for tissue engineering and drug delivery applications. J Tissue Eng Regen Med 8(1):1–14. doi:10.1002/term.1494
4. Rosa V, Bona AD, Cavalcanti BN, Nör JE (2012) Tissue engineering: from research to dental clinics. Dent Mater 28(4):341–348

5. Dawson E, Mapili G, Erickson K, Taqvi S, Roy K (2008) Biomaterials for stem cell differentiation. Adv Drug Deliv Rev 60(2):215–228. doi:10.1016/j.addr.2007.08.037
6. Kemppainen JM, Hollister SJ (2010) Tailoring the mechanical properties of 3D-designed poly(glycerol sebacate) scaffolds for cartilage applications. J Biomed Mater Res A 94(1):9–18. doi:10.1002/jbm.a.32653
7. Leong NL, Jiang J, Lu HH (2006) Polymer-ceramic composite scaffold induces osteogenic differentiation of human mesenchymal stem cells. Conf Proc IEEE Eng Med Biol Soc 1:2651–2654. doi:10.1109/iembs.2006.259459
8. Rao CN, Sood AK, Subrahmanyam KS, Govindaraj A (2009) Graphene: the new two-dimensional nanomaterial. Angew Chem Int Ed Engl 48(42):7752–7777. doi:10.1002/anie.200901678
9. Novoselov KS, Geim AK, Morozov SV, Jiang D, Zhang Y, Dubonos SV, Grigorieva IV, Firsov AA (2004) Electric field effect in atomically thin carbon films. Science 306 (5696):666–669. doi:10.1126/science.1102896
10. Geim AK, Novoselov KS (2007) The rise of graphene. Nat Mater 6(3):183–191
11. Peng R, Yao X, Ding J (2011) Effect of cell anisotropy on differentiation of stem cells on micropatterned surfaces through the controlled single cell adhesion. Biomaterials 32 (32):8048–8057. doi:10.1016/j.biomaterials.2011.07.035
12. La WG, Park S, Yoon HH, Jeong GJ, Lee TJ, Bhang SH, Han JY, Char K, Kim BS (2013) Delivery of a therapeutic protein for bone regeneration from a substrate coated with graphene oxide. Small 9(23):4051–4060. doi:10.1002/smll.201300571
13. Wang K, Ruan J, Song H, Zhang J, Wo Y, Guo S, Cui D (2011) Biocompatibility of graphene oxide. Nanoscale Res Lett 6(1):8. doi:10.1007/s11671-010-9751-6
14. Sanchez VC, Jachak A, Hurt RH, Kane AB (2012) Biological interactions of graphene-family nanomaterials: an interdisciplinary review. Chem Res Toxicol 25(1):15–34. doi:10.1021/tx200339h
15. Chen GY, Pang DWP, Hwang SM, Tuan HY, Hu YC (2012) A graphene-based platform for induced pluripotent stem cells culture and differentiation. Biomaterials 33(2):418–427. doi:10.1016/j.biomaterials.2011.09.071
16. Ma J, Zhang J, Xiong Z, Yong Y, Zhao XS (2011) Preparation, characterization and antibacterial properties of silver-modified graphene oxide. J Mater Chem 21(10):3350–3352. doi:10.1039/c0jm02806a
17. Akhavan O, Choobtashani M, Ghaderi E (2012) Protein degradation and RNA efflux of viruses photocatalyzed by graphene-tungsten oxide composite under visible light irradiation. J Phys Chem C 116(17):9653–9659. doi:10.1021/jp301707m
18. Akhavan O, Ghaderi E, Rahimi K (2012) Adverse effects of graphene incorporated in TiO_2 photocatalyst on minuscule animals under solar light irradiation. J Mater Chem 22 (43):23260–23266. doi:10.1039/c2jm35228a
19. Cai Y, Li H, Du B, Yang M, Li Y, Wu D, Zhao Y, Dai Y, Wei Q (2011) Ultrasensitive electrochemical immunoassay for BRCA1 using BMIM·BF4-coated SBA-15 as labels and functionalized graphene as enhancer. Biomaterials 32(8):2117–2123. doi:10.1016/j.biomaterials.2010.11.058
20. Feng L, Chen Y, Ren J, Qu X (2011) A graphene functionalized electrochemical aptasensor for selective label-free detection of cancer cells. Biomaterials 32(11):2930–2937. doi:10.1016/j.biomaterials.2011.01.002
21. Markovic ZM, Harhaji-Trajkovic LM, Todorovic-Markovic BM, Kepic DP, Arsikin KM, Jovanovic SP, Pantovic AC, Dramicanin MD, Trajkovic VS (2011) In vitro comparison of the photothermal anticancer activity of graphene nanoparticles and carbon nanotubes. Biomaterials 32(4):1121–1129. doi:10.1016/j.biomaterials.2010.10.030
22. Yang K, Wan J, Zhang S, Tian B, Zhang Y, Liu Z (2012) The influence of surface chemistry and size of nanoscale graphene oxide on photothermal therapy of cancer using ultra-low laser power. Biomaterials 33(7):2206–2214. doi:10.1016/j.biomaterials.2011.11.064

23. Akhavan O, Ghaderi E, Aghayee S, Fereydooni Y, Talebi A (2012) The use of a glucose-reduced graphene oxide suspension for photothermal cancer therapy. J Mater Chem 22(27):13773–13781. doi:10.1039/c2jm31396k
24. Yang K, Hu L, Ma X, Ye S, Cheng L, Shi X, Li C, Li Y, Liu Z (2012) Multimodal imaging guided photothermal therapy using functionalized graphene nanosheets anchored with magnetic nanoparticles. Adv Mater 24(14):1868–1872. doi:10.1002/adma.201104964
25. Ryu S, Kim B-S (2013) Culture of neural cells and stem cells on graphene. Tissue Eng Regen Med 10(2):39–46. doi:10.1007/s13770-013-0384-6
26. Menaa F, Abdelghani A, Menaa B (2015) Graphene nanomaterials as biocompatible and conductive scaffolds for stem cells: impact for tissue engineering and regenerative medicine. J Tissue Eng Regen Med 9(12):1321–1338. doi:10.1002/term.1910
27. Kim T-H, Lee T, El-Said W, Choi J-W (2015) Graphene-based materials for stem cell applications. Materials 8(12):5481
28. Akhavan O (2016) Graphene scaffolds in progressive nanotechnology/stem cell-based tissue engineering of nervous systems. J Mater Chem B. doi:10.1039/c6tb00152a
29. Dubey N, Bentini R, Islam I, Cao T, Castro Neto AH, Rosa V (2015) Graphene: a versatile carbon-based material for bone tissue engineering. Stem Cells Int 2015:12. doi:10.1155/2015/804213
30. Kalbacova M, Broz A, Kong J, Kalbac M (2010) Graphene substrates promote adherence of human osteoblasts and mesenchymal stromal cells. Carbon 48(15):4323–4329. doi:10.1016/j.carbon.2010.07.045
31. Crowder SW, Prasai D, Rath R, Balikov DA, Bae H, Bolotin KI, Sung HJ (2013) Three-dimensional graphene foams promote osteogenic differentiation of human mesenchymal stem cells. Nanoscale 5(10):4171–4176. doi:10.1039/c3nr00803g
32. Ku SH, Park CB (2013) Myoblast differentiation on graphene oxide. Biomaterials 34 (8):2017–2023. doi:10.1016/j.biomaterials.2012.11.052
33. Kim J, Park S, Kim YJ, Jeon CS, Lim KT, Seonwoo H, Cho SP, Chung TD, Choung PH, Choung YH, Hong BH, Chung JH (2015) Monolayer graphene-directed growth and neuronal differentiation of mesenchymal stem cells. J Biomed Nanotechnol 11(11):2024–2033
34. Park SY, Park J, Sim SH, Sung MG, Kim KS, Hong BH, Hong S (2011) Enhanced differentiation of human neural stem cells into neurons on graphene. Adv Mater 23(36): H263–H267. doi:10.1002/adma.201101503
35. Kim T-H, Shah S, Yang L, Yin PT, Hossain MK, Conley B, Choi J-W, Lee K-B (2015) Controlling differentiation of adipose-derived stem cells using combinatorial graphene hybrid-pattern arrays. ACS Nano 9(4):3780–3790. doi:10.1021/nn5066028
36. Hilpert F, Heiser A, Wieckhorst W, Arnold N, Kabelitz D, Jonat W, Pfisterer J (2005) The impact of electrical charge on the viability and physiology of dendritic cells. Scand J Immunol 62(4):399–406. doi:10.1111/j.1365-3083.2005.01677.x
37. Woo DG, Shim MS, Park JS, Yang HN, Lee DR, Park KH (2009) The effect of electrical stimulation on the differentiation of hESCs adhered onto fibronectin-coated gold nanoparticles. Biomaterials 30(29):5631–5638. doi:10.1016/j.biomaterials.2009.07.026
38. Liu J, Zhao Z, Li J, Zou L, Shuler C, Zou Y, Huang X, Li M, Wang J (2009) Hydrostatic pressures promote initial osteodifferentiation with ERK1/2 not p38 MAPK signaling involved. J Cell Biochem 107(2):224–232. doi:10.1002/jcb.22118
39. Maul TM, Chew DW, Nieponice A, Vorp DA (2011) Mechanical stimuli differentially control stem cell behavior: morphology, proliferation, and differentiation. Biomech Model Mechanobiol 10(6):939–953. doi:10.1007/s10237-010-0285-8
40. Pires F, Ferreira Q, Rodrigues CA, Morgado J (1850) Ferreira FC (2015) Neural stem cell differentiation by electrical stimulation using a cross-linked PEDOT substrate: Expanding the use of biocompatible conjugated conductive polymers for neural tissue engineering. Biochim Biophys Acta 6:1158–1168. doi:10.1016/j.bbagen.2015.01.020

41. Chan BP, Leong KW (2008) Scaffolding in tissue engineering: general approaches and tissue-specific considerations. Eur Spine J 17(Suppl 4):467–479. doi:10.1007/s00586-008-0745-3
42. Ma PX (2004) Scaffolds for tissue fabrication. Mater Today 7(5):30–40. doi:10.1016/S1369-7021(04)00233-0
43. Meng X, Leslie P, Zhang Y, Dong J (2014) Stem cells in a three-dimensional scaffold environment. SpringerPlus 3:80. doi:10.1186/2193-1801-3-80
44. Juang Z-Y, Wu C-Y, Lu A-Y, Su C-Y, Leou K-C, Chen F-R, Tsai C-H (2010) Graphene synthesis by chemical vapor deposition and transfer by a roll-to-roll process. Carbon 48 (11):3169–3174
45. Chung C, Kim Y-K, Shin D, Ryoo S-R, Hong BH, Min D-H (2013) Biomedical applications of graphene and graphene oxide. Acc Chem Res 46(10):2211–2224. doi:10.1021/ar300159f
46. Wick P, Louw-Gaume AE, Kucki M, Krug HF, Kostarelos K, Fadeel B, Dawson KA, Salvati A, Vázquez E, Ballerini L, Tretiach M, Benfenati F, Flahaut E, Gauthier L, Prato M, Bianco A (2014) Classification framework for graphene-based materials. Angew Chem Int Ed 53(30):7714–7718. doi:10.1002/anie.201403335
47. Defteralı Ç, Verdejo R, Peponi L, Martín ED, Martínez-Murillo R, López-Manchado MÁ, Vicario-Abejón C (2016) Thermally reduced graphene is a permissive material for neurons and astrocytes and de novo neurogenesis in the adult olfactory bulb in vivo. Biomaterials 82:84–93. doi:10.1016/j.biomaterials.2015.12.010
48. Park J, Park S, Ryu S, Bhang SH, Kim J, Yoon J-K, Park YH, Cho S-P, Lee S, Hong BH, Kim B-S (2014) Graphene-regulated cardiomyogenic differentiation process of mesenchymal stem cells by enhancing the expression of extracellular matrix proteins and cell signaling molecules. Adv Healthc Mater 3(2):176–181. doi:10.1002/adhm.201300177
49. Park J, Kim YS, Ryu S, Kang WS, Park S, Han J, Jeong HC, Hong BH, Ahn Y, Kim B-S (2015) Graphene potentiates the myocardial repair efficacy of mesenchymal stem cells by stimulating the expression of angiogenic growth factors and gap junction protein. Adv Funct Mater 25(17):2590–2600. doi:10.1002/adfm.201500365
50. Kumar S, Azam D, Raj S, Kolanthai E, Vasu KS, Sood AK, Chatterjee K (2016) 3D scaffold alters cellular response to graphene in a polymer composite for orthopedic applications. J Biomed Mater Res B Appl Biomater 104(4):732–749. doi:10.1002/jbm.b.33549
51. Chen Z, Ren W, Gao L, Liu B, Pei S, Cheng H-M (2011) Three-dimensional flexible and conductive interconnected graphene networks grown by chemical vapour deposition. Nat Mater 10 (6):424–428. doi:http://www.nature.com/nmat/journal/v10/n6/abs/nmat3001.html#supplementary-information
52. Li N, Zhang Q, Gao S, Song Q, Huang R, Wang L, Liu L, Dai J, Tang M, Cheng G (2013) Three-dimensional graphene foam as a biocompatible and conductive scaffold for neural stem cells. Sci Rep 3:1604. doi:10.1038/srep01604. http://www.nature.com/articles/srep01604#supplementary-information
53. Serrano MC, Patino J, Garcia-Rama C, Ferrer ML, Fierro JLG, Tamayo A, Collazos-Castro JE, del Monte F, Gutierrez MC (2014) 3D free-standing porous scaffolds made of graphene oxide as substrates for neural cell growth. J Mater Chem B 2(34):5698–5706. doi:10.1039/c4tb00652f
54. Shin YC, Lee JH, Jin L, Kim MJ, Kim Y-J, Hyun JK, Jung T-G, Hong SW, Han D-W (2015) Stimulated myoblast differentiation on graphene oxide-impregnated PLGA-collagen hybrid fibre matrices. J Nanobiotechnol 13(1):1–11. doi:10.1186/s12951-015-0081-9
55. Girao AF, Goncalves G, Bhangra KS, Phillips JB, Knowles J, Irurueta G, Singh MK, Bdkin I, Completo A, Marques PAAP (2016) Electrostatic self-assembled graphene oxide-collagen scaffolds towards a three-dimensional microenvironment for biomimetic applications. RSC Adv 6(54):49039–49051. doi:10.1039/c6ra10213a
56. Nieto A, Dua R, Zhang C, Boesl B, Ramaswamy S, Agarwal A (2015) Three dimensional graphene foam/polymer hybrid as a high strength biocompatible scaffold. Adv Funct Mater 25(25):3916–3924. doi:10.1002/adfm.201500876

57. Akhavan O, Ghaderi E (2014) The use of graphene in the self-organized differentiation of human neural stem cells into neurons under pulsed laser stimulation. J Mater Chem B 2 (34):5602–5611. doi:10.1039/c4tb00668b
58. Ahadian S, Ramon-Azcon J, Chang H, Liang X, Kaji H, Shiku H, Nakajima K, Ramalingam M, Wu H, Matsue T, Khademhosseini A (2014) Electrically regulated differentiation of skeletal muscle cells on ultrathin graphene-based films. RSC Advances 4 (19):9534–9541. doi:10.1039/c3ra46218h
59. Meng S (2014) Nerve cell differentiation using constant and programmed electrical stimulation through conductive non-functional graphene nanosheets film. Tissue Eng Regen Med 11(4):274–283. doi:10.1007/s13770-014-0011-1
60. Serena E, Figallo E, Tandon N, Cannizzaro C, Gerecht S, Elvassore N, Vunjak-Novakovic G (2009) Electrical stimulation of human embryonic stem cells: cardiac differentiation and the generation of reactive oxygen species. Exp Cell Res 315(20):3611–3619. doi:10.1016/j.yexcr.2009.08.015
61. Akanji OO, Lee DA, Bader DA (2008) The effects of direct current stimulation on isolated chondrocytes seeded in 3D agarose constructs. Biorheology 45(3–4):229–243
62. Sisken BF, Walker J, Orgel M (1993) Prospects on clinical applications of electrical stimulation for nerve regeneration. J Cell Biochem 51(4):404–409
63. Shapiro S, Borgens R, Pascuzzi R, Roos K, Groff M, Purvines S, Rodgers RB, Hagy S, Nelson P (2005) Oscillating field stimulation for complete spinal cord injury in humans: a phase 1 trial. J Neurosurg Spine 2(1):3–10. doi:10.3171/spi.2005.2.1.0003
64. Wells J, Konrad P, Kao C, Jansen ED, Mahadevan-Jansen A (2007) Pulsed laser versus electrical energy for peripheral nerve stimulation. J Neurosci Methods 163(2):326–337. doi:10.1016/j.jneumeth.2007.03.016
65. Yao L, Shanley L, McCaig C, Zhao M (2008) Small applied electric fields guide migration of hippocampal neurons. J Cell Physiol 216(2):527–535. doi:10.1002/jcp.21431
66. Akhavan O, Ghaderi E (2013) Flash photo stimulation of human neural stem cells on graphene/TiO_2 heterojunction for differentiation into neurons. Nanoscale 5(21):10316–10326. doi:10.1039/c3nr02161k
67. Kotov NA, Winter JO, Clements IP, Jan E, Timko BP, Campidelli S, Pathak S, Mazzatenta A, Lieber CM, Prato M, Bellamkonda RV, Silva GA, Kam NWS, Patolsky F, Ballerini L (2009) Nanomaterials for neural interfaces. Adv Mater 21(40):3970–4004. doi:10.1002/adma.200801984
68. Bai J, Zhong X, Jiang S, Huang Y, Duan X (2010) Graphene nanomesh. Nat Nano 5 (3):190–194. http://www.nature.com/nnano/journal/v5/n3/suppinfo/nnano.2010.8_S1.html
69. Akhavan O (2010) Graphene nanomesh by ZnO nanorod photocatalysts. ACS Nano 4 (7):4174–4180. doi:10.1021/nn1007429
70. Heo C, Yoo J, Lee S, Jo A, Jung S, Yoo H, Lee YH, Suh M (2011) The control of neural cell-to-cell interactions through non-contact electrical field stimulation using graphene electrodes. Biomaterials 32(1):19–27. doi:10.1016/j.biomaterials.2010.08.095
71. Kawakami O, Miyamoto S, Hatano T, Yamada K, Hashimoto N, Tabata Y (2005) Accelerated embolization healing of aneurysms by polyethylene terephthalate coils seeded with autologous fibroblasts. Neurosurgery 56(5):1075–1081 (discussion 1075–1081)
72. Tang M, Song Q, Li N, Jiang Z, Huang R, Cheng G (2013) Enhancement of electrical signaling in neural networks on graphene films. Biomaterials 34(27):6402–6411. doi:10.1016/j.biomaterials.2013.05.024
73. Wang K, Fishman HA, Dai H, Harris JS (2006) Neural stimulation with a carbon nanotube microelectrode array. Nano Lett 6(9):2043–2048. doi:10.1021/nl061241t
74. Zhang Q, Xu J, Song Q, Li N, Zhang Z, Li K, Du Y, Wu L, Tang M, Liu L, Cheng G, Liu J (2014) Synthesis of amphiphilic reduced graphene oxide with an enhanced charge injection capacity for electrical stimulation of neural cells. J Mater Chem B 2(27):4331–4337. doi:10.1039/c4tb00279b
75. Berit K, Peter K, Christoph N, Sandeep Y, Joerg JS, Christiane T (2016) Graphene electrodes for stimulation of neuronal cells. 2D Mater 3(2):024004

76. Yan L, Zhao B, Liu X, Li X, Zeng C, Shi H, Xu X, Lin T, Dai L, Liu Y (2016) Aligned nanofibers from polypyrrole/graphene as electrodes for regeneration of optic nerve via electrical stimulation. ACS Appl Mater Interfaces 8(11):6834–6840. doi:10.1021/acsami.5b12843
77. Zhou K, Thouas GA, Bernard CC, Nisbet, DR, Finkelstein DI, Li D, Forsythe JS (2012) Method to impart electro- and biofunctionality to neural scaffolds using graphene–polyelectrolyte multilayers. ACS Appl Mater Interfaces 4(9):4524–4531. doi:10.1021/am3007565
78. Sherrell PC, Thompson BC, Wassei JK, Gelmi AA, Higgins MJ, Kaner RB, Wallace GG (2014) Maintaining cytocompatibility of biopolymers through a graphene layer for electrical stimulation of nerve cells. Adv Funct Mater 24(6):769–776. doi:10.1002/adfm.201301760
79. Guo W, Zhang X, Yu X, Wang S, Qiu J, Tang W, Li L, Liu H, Wang ZL (2016) Self-powered electrical stimulation for enhancing neural differentiation of mesenchymal stem cells on graphene–poly(3,4-ethylenedioxythiophene) hybrid microfibers. ACS Nano. doi:10.1021/acsnano.6b00200
80. Akhavan O, Ghaderi E, Shirazian SA, Rahighi R (2016) Rolled graphene oxide foams as three-dimensional scaffolds for growth of neural fibers using electrical stimulation of stem cells. Carbon 97:71–77. doi:10.1016/j.carbon.2015.06.079
81. Ahadian S, Zhou Y, Yamada S, Estili M, Liang X, Nakajima K, Shiku H, Matsue T (2016) Graphene induces spontaneous cardiac differentiation in embryoid bodies. Nanoscale 8 (13):7075–7084. doi:10.1039/c5nr07059g
82. Izzo AD, Richter CP, Jansen ED, Walsh JT Jr (2006) Laser stimulation of the auditory nerve. Lasers Surg Med 38(8):745–753. doi:10.1002/lsm.20358
83. Geddes LA (2004) Accuracy limitations of chronaxie values. IEEE Trans Biomed Eng 51 (1):176–181. doi:10.1109/tbme.2003.820340
84. Geddes LA, Roeder R (2003) Criteria for the selection of materials for implanted electrodes. Ann Biomed Eng 31(7):879–890
85. Ragheb T, Geddes LA (1990) Electrical properties of metallic electrodes. Med Biol Eng Comput 28(2):182–186
86. Wells J, Kao C, Jansen ED, Konrad P, Mahadevan-Jansen A (2005) Application of infrared light for in vivo neural stimulation. J Biomed Opt 10(6):064003. doi:10.1117/1.2121772
87. Wells J, Kao C, Mariappan K, Albea J, Jansen ED, Konrad P, Mahadevan-Jansen A (2005) Optical stimulation of neural tissue in vivo. Opt Lett 30(5):504–506
88. Izzo AD, Walsh JT, Ralph H, Webb J, Bendett M, Wells J, Richter C-P (2008) Laser stimulation of auditory neurons: effect of shorter pulse duration and penetration depth. Biophys J 94(8):3159–3166. doi:10.1529/biophysj.107.117150
89. Ginani F, Soares DM, Barreto MP, Barboza CA (2015) Effect of low-level laser therapy on mesenchymal stem cell proliferation: a systematic review. Lasers Med Sci 30(8):2189–2194. doi:10.1007/s10103-015-1730-9
90. de Souza SC, Munin E, Alves LP, Salgado MAC, Pacheco MTT (2005) Low power laser radiation at 685 nm stimulates stem-cell proliferation rate in *Dugesia tigrina* during regeneration. J Photochem Photobiol, B 80(3):203–207. doi:10.1016/j.jphotobiol.2005.05.002
91. Bukowski B, Deskins NA (2015) The interactions between TiO_2 and graphene with surface inhomogeneity determined using density functional theory. Phys Chem Chem Phys 17 (44):29734–29746. doi:10.1039/c5cp04073f
92. Štengl V, Bakardjieva S, Grygar TM, Bludská J, Kormunda M (2013) TiO_2-graphene oxide nanocomposite as advanced photocatalytic materials. Chem Cent J 7(1):1–12. doi:10.1186/1752-153x-7-41
93. Stengl V, Popelkova D, Vlacil P (2011) TiO_2-graphene nanocomposite as high performance photocatalysts. J Phys Chem C. doi:10.1021/jp207515z
94. Akhavan O, Ghaderi E (2013) Differentiation of human neural stem cells into neural networks on graphene nanogrids. J Mater Chem B 1(45):6291–6301. doi:10.1039/c3tb21085e

95. Wei D, Liu Y, Wang Y, Zhang H, Huang L, Yu G (2009) Synthesis of N-doped graphene by chemical vapor deposition and its electrical properties. Nano Lett 9(5):1752–1758. doi:10.1021/nl803279t
96. Barone V, Hod O, Scuseria GE (2006) Electronic structure and stability of semiconducting graphene nanoribbons. Nano Lett 6(12):2748–2754. doi:10.1021/nl0617033
97. Akhavan O, Ghaderi E (2013) Graphene nanomesh promises extremely efficient in vivo photothermal therapy. Small 9(21):3593–3601. doi:10.1002/smll.201203106
98. Akhavan O, Ghaderi E, Shirazian SA (2015) Near infrared laser stimulation of human neural stem cells into neurons on graphene nanomesh semiconductors. Colloids Surf B 126:313–321. doi:10.1016/j.colsurfb.2014.12.027
99. Li D, Zhou J, Chowdhury F, Cheng J, Wang N, Wang F (2011) Role of mechanical factors in fate decisions of stem cells. Regen Med 6(2):229–240. doi:10.2217/rme.11.2
100. Engler AJ, Sen S, Sweeney HL, Discher DE (2006) Matrix elasticity directs stem cell lineage specification. Cell 126(4):677–689. doi:10.1016/j.cell.2006.06.044
101. Chen CS, Mrksich M, Huang S, Whitesides GM, Ingber DE (1997) Geometric control of cell life and death. Science 276(5317):1425–1428. doi:10.1126/science.276.5317.1425
102. Engler AJ, Griffin MA, Sen S, Bonnemann CG, Sweeney HL, Discher DE (2004) Myotubes differentiate optimally on substrates with tissue-like stiffness: pathological implications for soft or stiff microenvironments. J Cell Biol 166(6):877–887. doi:10.1083/jcb.200405004
103. McBeath R, Pirone DM, Nelson CM, Bhadriraju K, Chen CS (2004) Cell shape, cytoskeletal tension, and RhoA regulate stem cell lineage commitment. Dev Cell 6(4):483–495
104. Discher DE, Janmey P, Wang YL (2005) Tissue cells feel and respond to the stiffness of their substrate. Science 310(5751):1139–1143. doi:10.1126/science.1116995
105. Wang N, Tytell JD, Ingber DE (2009) Mechanotransduction at a distance: mechanically coupling the extracellular matrix with the nucleus. Nat Rev Mol Cell Biol 10 (1):75–82. doi: http://www.nature.com/nrm/journal/v10/n1/suppinfo/nrm2594_S1.html
106. Yeung T, Georges PC, Flanagan LA, Marg B, Ortiz M, Funaki M, Zahir N, Ming W, Weaver V, Janmey PA (2005) Effects of substrate stiffness on cell morphology, cytoskeletal structure, and adhesion. Cell Motil Cytoskelet 60(1):24–34. doi:10.1002/cm.20041
107. Janmey PA, McCulloch CA (2007) Cell mechanics: integrating cell responses to mechanical stimuli. Annu Rev Biomed Eng 9:1–34. doi:10.1146/annurev.bioeng.9.060906.151927
108. Hahn C, Schwartz MA (2009) Mechanotransduction in vascular physiology and atherogenesis. Nat Rev Mol Cell Biol 10(1):53–62. doi:10.1038/nrm2596
109. Kang S, Park JB, Lee T-J, Ryu S, Bhang SH, La W-G, Noh M-K, Hong BH, Kim B-S (2015) Covalent conjugation of mechanically stiff graphene oxide flakes to three-dimensional collagen scaffolds for osteogenic differentiation of human mesenchymal stem cells. Carbon 83:162–172. doi:10.1016/j.carbon.2014.11.029
110. Ni ZH, Wang HM, Ma Y, Kasim J, Wu YH, Shen ZX (2008) Tunable stress and controlled thickness modification in graphene by annealing. ACS Nano 2(5):1033–1039. doi:10.1021/nn800031m
111. Rasuli R, Rafii-Tabar H, Zad AI (2010) Strain effect on quantum conductance of graphene nanoribbons from maximally localized Wannier functions. Phys Rev B 81(12):125409
112. Chen C, Wu JZ, Lam KT, Hong G, Gong M, Zhang B, Lu Y, Antaris AL, Diao S, Guo J, Dai H (2015) Graphene nanoribbons under mechanical strain. Adv Mater 27(2):303–309. doi:10.1002/adma.201403750
113. Wang Y, Lee WC, Manga KK, Ang PK, Lu J, Liu YP, Lim CT, Loh KP (2012) Fluorinated graphene for promoting neuro-induction of stem cells. Adv Mater 24(31):4285–4290. doi:10.1002/adma.201200846
114. Akhavan O, Ghaderi E, Abouei E, Hatamie S, Ghasemi E (2014) Accelerated differentiation of neural stem cells into neurons on ginseng-reduced graphene oxide sheets. Carbon 66:395–406. doi:10.1016/j.carbon.2013.09.015
115. Pulverer BJ, Kyriakis JM, Avruch J, Nikolakaki E, Woodgett JR (1991) Phosphorylation of c-jun mediated by MAP kinases. Nature 353(6345):670–674. doi:10.1038/353670a0

116. Akhavan O, Ghaderi E, Shahsavar M (2013) Graphene nanogrids for selective and fast osteogenic differentiation of human mesenchymal stem cells. Carbon 59:200–211. doi:10.1016/j.carbon.2013.03.010
117. Weaver CL, Cui XT (2015) Directed neural stem cell differentiation with a functionalized graphene oxide nanocomposite. Adv Healthc Mater 4(9):1408–1416. doi:10.1002/adhm.201500056
118. Flemming RG, Murphy CJ, Abrams GA, Goodman SL, Nealey PF (1999) Effects of synthetic micro- and nano-structured surfaces on cell behavior. Biomaterials 20(6):573–588
119. Solanki A, Shah S, Memoli KA, Park SY, Hong S, Lee K-B (2010) Controlling differentiation of neural stem cells using extracellular matrix protein patterns. Small 6 (22):2509–2513. doi:10.1002/smll.201001341
120. Kshitiz Park J, Kim P, Helen W, Engler AJ, Levchenko A, Kim DH (2012) Control of stem cell fate and function by engineering physical microenvironments. Integr Biol (Camb) 4 (9):1008–1018
121. Yao X, Peng R, Ding J (2013) Cell-material interactions revealed via material techniques of surface patterning. Adv Mater 25(37):5257–5286. doi:10.1002/adma.201301762
122. Kim J, Kim Y-R, Kim Y, Lim KT, Seonwoo H, Park S, Cho S-P, Hong BH, Choung P-H, Chung TD, Choung Y-H, Chung JH (2013) Graphene-incorporated chitosan substrata for adhesion and differentiation of human mesenchymal stem cells. J Mater Chem B 1(7):933–938. doi:10.1039/c2tb00274d
123. Kang K, Choi SE, Jang HS, Cho WK, Nam Y, Choi IS, Lee JS (2012) In vitro developmental acceleration of hippocampal neurons on nanostructures of self-assembled silica beads in filopodium-size ranges. Angew Chem Int Ed Engl 51(12):2855–2858. doi:10.1002/anie.201106271
124. Solanki A, Chueng S-TD, Yin PT, Kappera R, Chhowalla M, Lee K-B (2013) Axonal alignment and enhanced neuronal differentiation of neural stem cells on graphene-nanoparticle hybrid structures. Adv Mater 25(38):5477–5482. doi:10.1002/adma.201302219
125. Shah S, Yin PT, Uehara TM, Chueng S-TD, Yang L, Lee K-B (2014) Guiding stem cell differentiation into oligodendrocytes using graphene-nanofiber hybrid scaffolds. Adv Mater 26(22):3673–3680. doi:10.1002/adma.201400523

Part IV
Environment

Graphene Hybrid Architectures for Chemical Sensors

Parikshit Sahatiya and Sushmee Badhulika

Abstract Graphene, one atom thick allotrope of carbon, has enabled researchers to a new era of exploration due to its unique properties. Graphene is considered to be mother of all carbon materials with excellent electrical, mechanical, optical, and thermal properties that made its use for various engineering applications. Graphene and graphene hybrids have proved over the last decade to be promising material for chemical sensors. High surface-to-volume ratio coupled with high conductivity enabled graphene-based sensors to perform well with high accuracy, high sensitivity and selectivity, low detection limits and long-term stability. To further enhance the properties of graphene, graphene-based hybrids have been synthesized for its use as transducing element in various chemical sensors such as gas and biosensors. These hybrids exhibit the synergistic benefit for both the material for fabrication of efficient sensors with enhanced performance. This chapter focuses on synthesis, characterization and applications of various graphene hybrids in chemical sensors.

Keywords Graphene · Graphene hybrids · Functionalized graphene · Chemical sensors · Electrochemical sensors · Biosensors

1 Introduction

Chemical and biosensors are increasingly becoming indispensable part of our society with applications across various fields such as biomedical, environmental, pharmaceutical, clinical, chemical processing, and military [1–4]. This had led researchers to focus on developing various chemical sensors with low cost process while still exhibiting high sensitivity. Chemical sensors may be defined as a device that responds with selectivity to a particular chemical substance (analyte) present in desired medium to produce a measurable signal at any given analyte concentration.

P. Sahatiya · S. Badhulika (✉)
Department of Electrical Engineering, Indian Institute of Technology Hyderabad, Hyderabad 502285, India
e-mail: sbadh@iith.ac.in

G. Gonçalves et al. (eds.), *Graphene-based Materials in Health and Environment*, Carbon Nanostructures, DOI 10.1007/978-3-319-45639-3_9

Chemical sensors in reality are complex devices fabricated and optimized for a particular application. While each application has its own unique requirement, the ultimate goal of the chemical sensor is to achieve high sensitivity at which single atom or molecule can be resolved. In general there are two main components of the chemical sensors namely (i) a recognition element which is selective to the analyte to be detected and (ii) transduction element that is responsible for generating a measurable signal which can be further processed for qualitative and quantitative analysis. For this purpose, various materials which include ceramics, metals, semiconductors, etc., [5–7] have been synthesized over ages to fulfill the growing demand of the chemical sensors from both environment and healthcare perspective.

In recent years, graphene, densely packed sp^2 bonded carbon atoms having one atom planar thickness have attained significant attention to be used as a next generation electronic material, due to its outstanding properties such as high current density, ballistic transport, high thermal conductivity, optical transmittance, hydrophobicity, and chemical inertness at nanometer scale [8]. The first ever graphene was extracted by mechanical exfoliation of graphite using scotch tape method [9] which has further led to enormous experimental activities. Electron transport experiments on graphene-based devices have shown apart from the above mentioned properties, very high carrier density and exceptionally high electron mobilities (100,000 $cm^2/V\text{-}s$) [10]. These outstanding properties arise due to the unique band structure of graphene, which has zero band gap with conduction and valence band exhibiting near-linear dispersion that touches the brillouin zone corner [11]. Graphene is highly sensitive to changes in the chemical environment which is due to the fact that suspended graphene has extremely high mobility and that the electron transport remains ballistic up to 0.3 μm at room temperature [12]. Moreover, every carbon atom in graphene provides large surface to volume ratio, as a result, electron transport in graphene is highly sensitive to adsorbed molecular species. Finally, due to its high crystallinity graphene has inherently low electrical noise. The above properties make graphene an ideal choice for fabricating ultrahigh sensitive chemical sensors for detection of various analytes which includes various gas and biomolecules.

However, the above properties of graphene emerge only in 2D direction thereby limiting its scope and applications. To further enhance graphene properties and expands the scope of applications that graphene offers, various graphene hybrids have been synthesized, wherein graphene acts as a platform for support, scaffold or a 2D planar substrate for anchoring other nanomaterials. It is always desirable to harness the useful properties of both graphene and its derivatives with various kinds of functional materials in composites. In this regard, hybrids of graphene with various metal oxides, metal nanoparticles [13], semiconducting nanoparticles [14], conducting polymers [15] and other carbon allotropes such as carbon nanotubes (CNTs)/carbon nanofibers (CNFs), etc. [16, 17] have been developed over the past decade. For instance, incorporating graphene with metal oxide in various forms increases the electrical conductivity as compared to pristine metal oxide thereby expanding the scope of metal oxide usage in various forms of chemical and biosensors. Moreover, CNTs, which is a 1D material can be functionalized onto the surface of 2D graphene, forming a hybrid material whose properties emerge in all three dimensions, thereby increasing the active surface area and faster electron kinetics. Similarly, pristine metal oxide/metal nanoparticles have the tendency to aggregate [18] thereby reducing the stability. Graphene with metal oxides/metal nanoparticles serves to overcome the drawback wherein nanoparticles can grow or decorate on graphene sheets or where graphene is wrapped around the nanoparticle. These hybrid structures serves for various exciting applications such as electronics, chemical and biosensors, energy storage, optical, thermal, photodetectors, etc. [19].

In this chapter, we will systematically discuss graphene hybrids architectures in terms of their synthesis, characterization, properties and their applications in various chemical sensors. Potential sensing mechanisms are thoroughly analyzed in an effort to understand the sensing properties which both graphene and its derivatives offer. Furthermore, summary on recent promising progress of graphene hybrids in the area of chemical sensors is made, highlighting, sensing performances such as selectivity, sensitivity, response time, and limit of detection. Finally, wherever applicable, limitations of present approaches and future research perspective have been discussed.

2 Functionalized Graphene Materials

Pristine graphene is highly hydrophobic material and is insoluble in almost all solvents [20]. Therefore, foremost processing of graphene composites begins with solubilization of graphene. To increase the solubility of graphene various functional groups have been attached to the carbon atoms by covalent, non-covalent functionalization, and chemical modification [21, 22].

It is not possible to directly disperse hydrophobic graphene sheets or graphite flakes in water directly without the assistance of dispersing agents. Reduced graphene oxide (rGO) can form stable aqueous solutions due to the presence of oxygen containing groups, but the degree of solubility depends on the reduction of graphene oxide. However, the resulting solubility of rGO is limited with a value of 0.5 mg/mL [23]. There are various other organic solvents such as acetone, ethanol, dimethyl formamide (DMF), dimethyl sulfoxide (DMSO), and tetrahydrofuron (THF) which can be used to form stable suspensions with rGO [24]. Besides that, metal ions and functionalized groups have been known to modify rGO sheets for enhanced solubility. GO sheets mixed with KOH in water and then with hydrazine has shown to produce K-modified graphene [25] which can lead to homogeneous suspensions. In addition, sulfonated groups could be attached to rGO by the use of *p*-phenyl-SO_3H groups which has shown excellent dispersion in water (2 mg/mL) [26].

Covalent functionalization is based on binding of organic functionalities like free radicals and dienophiles on pristine graphene and chemistry of attachment of oxygen groups from Graphene oxide (GO), that is, synthesis method starting from GO. The covalent functionalization reactions mainly includes two methods (i) formation of covalent bonds between free radicals and dienophiles and C=C bond of pristine graphene and (ii) formation of covalent bond between oxygen group of GO and organic functional group. In the work done by Sinitskii et al. [27] they showed upon heating of diazonium salt, a highly reactive free radical is produced, which reacts with the carbon atoms of graphene sheet thus forming covalent bond. The graphene sheets in this work were synthesized by unzipping of carbon nanotubes (CNTs).

The ratio of carbon atoms with sp^2 and sp^3 hybridization in the graphitic lattice is an indication of degree of oxidation or covalent functionalization reaction. As the functional groups increases there is increase in the defects sites which is estimated using Raman spectroscopy as I_D/I_G ratio, I_D represents the defects peak which occur at ~ 1350 cm^{-1} which corresponds to sp^3 carbon atoms and I_G is the graphitic peak which occur at ~ 1580 cm^{-1} which corresponds to sp^2 carbon atoms. As discussed, graphene consists of sp^2 carbon atoms and as such coexistence of sp^3 carbon atoms which are termed as defects. As shown in Fig. 1. I_D/I_G, ratio increases with addition of free radicals.

Apart from free radicals, dienophiles are also known to functionalize the sp^2 carbon atoms of graphene by covalent method. In work done by Georgakilas et al. [29] they functionalized graphene sheets which were produced by graphite dispersed in organic solvents by pyrrolidine rings. The hydroxyl group introduced due to the above reaction increased its dispersibility in organic solvents such as ethanol and DMF. Also, ratio of I_D/I_G remarkably increased which indicated the functionalization of graphene and increase in the sp^3 planar carbon atoms. He et al. [30] developed a facile method for covalent functionalization of graphene sheets with functional groups and or polymeric chains via nitride cycloaddition method. The resulting functionalized graphene sheets are highly conductive and show excellent dispersibility in solvents.

Covalent functionalization can also be achieved by forming a covalent bond with oxygen atoms of GO with the functional group. GO can be defined by randomly

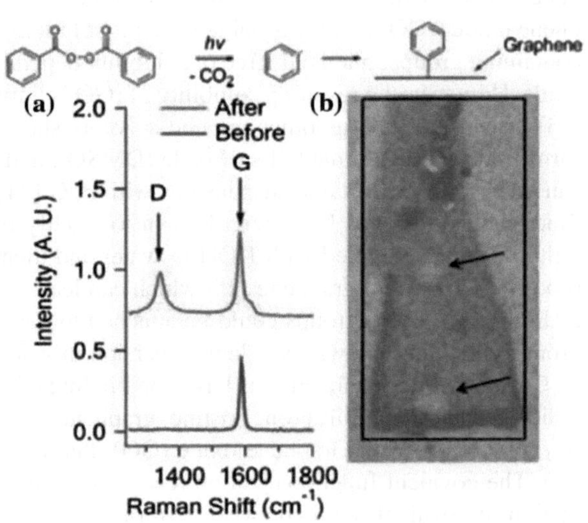

Fig. 1 Schematic of radical addition to single layer graphene **a** Raman spectra of graphene before and after functionalization, **b** optical image of functionalized graphene layer. Reprinted with permission from Ref. [28], © 2009 American chemical society

distributed aromatic sp^2 carbon atoms and oxygenated aliphatic sp^3 carbon atoms containing various functional groups such as epoxy, hydroxyl, carbonyl, and carboxyl [31, 32]. GO forms unstable dispersion in water and organic solvents such as DMF, ethylene glycol and THF, since the exfoliated GO tend to aggregate and forms larger particles of graphite oxide. For stabilization of GO, several stabilizing groups are added such as polystyrene, elastomeric silicon foams, octadecylamine, etc., [33]. However, the use of these stabilizing agents restricts the use of graphene in technological applications. Conceptually, reduction of GO should rehybridize the effected sp^3 carbon atoms to sp^2 carbon atoms, thereby making it defect free. However, experimental evidence of GO reduction has shown that large number of oxygen groups remains which increases the defect sites in graphene. These defects alters the properties of rGO, mainly its conductivity, and hence the reduction method is always specified such as thermally reduced graphene, chemically reduced graphene, chemically converted graphene to distinguish it from pristine graphene. Apart from above mentioned methods, there are several other methods such as addition of chromophores, polymers, etc., [34, 35] for covalent functionalization of graphene to tune its properties and increase the solubility in the common organic solvents.

Non covalent functionalization is another way of functionalizing graphene with other organic compounds by π-interactions. It is an attractive synthetic method, because it offers functionalization of graphene with functional group without altering the electronic network. These π interactions are important in context of fabrication of nanosensors, because subtle changes in the electronic properties of the π systems can lead to dramatic effects in structure and properties of the nano system. In the past couple of decades, extensive literature is available to understand the π complexes which include H–π interaction, nonpolar gas–π interaction, π–π

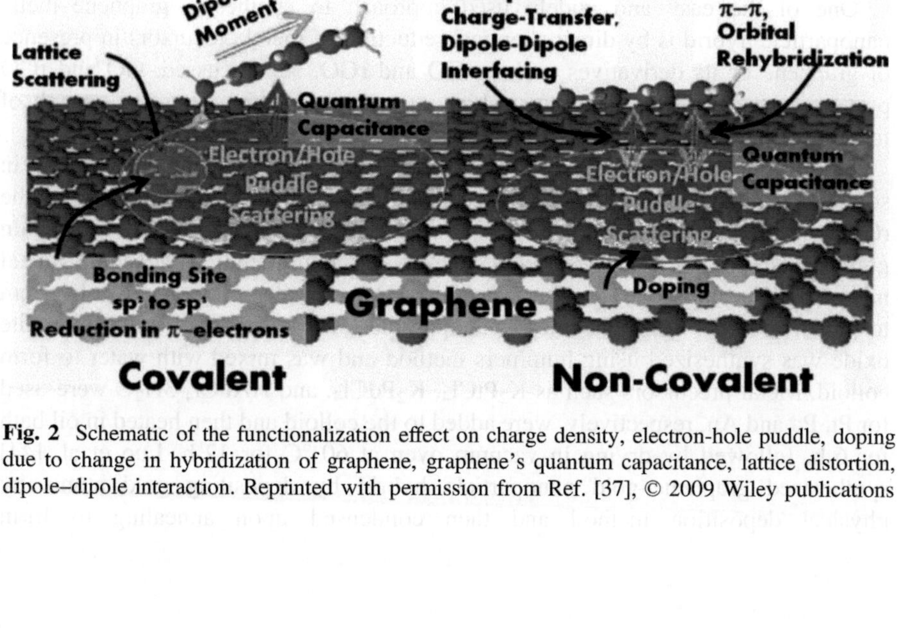

Fig. 2 Schematic of the functionalization effect on charge density, electron-hole puddle, doping due to change in hybridization of graphene, graphene's quantum capacitance, lattice distortion, dipole–dipole interaction. Reprinted with permission from Ref. [37], © 2009 Wiley publications

interaction, cation–π interaction and anion–π interaction. Since describing each of them is beyond the scope of this chapter we refer to review by Georgakilas et al. [36] wherein all the interactions are explained with the relevant literature (Fig. 2).

3 Graphene–Metal Nanoparticle Hybrid

Combination of graphene with nanoparticles thus forming graphene nanoparticles hybrid offers unique physiochemical properties that are highly desirable for both chemical and biosensors, when compared to the use of either material alone. These graphene–nanoparticle hybrid are promising and alluring due to the fact that they not only display the excellent individual properties of the nanoparticles such as optical, electronic, magnetic, and structural [38, 39] that are not observed in bulk material, but also exhibit additional advantage and offer synergistic properties that promise their use in various chemical and biosensors. In general graphene–nanoparticle hybrids can be categorized into two kinds (i) nanoparticles anchored or decorated or grown on graphene sheets and (ii) graphene encapsulated nanoparticle wherein the nanoparticle surface is wrapped around graphene.

There have been many reports of graphene hybrids with metal nanoparticles such as Au, Pd, Ag, Pt, Ni, Cu, etc., [40–42]. Depending upon the type of nanoparticles decorated or grown on graphene, they have been applied in various forms of sensors such as gas sensors for detection for toxic gases and volatile organic compounds, biosensors for detection of cardiac biomarkers, malaria, glucose by different techniques such as electrochemistry, chemiresistive, etc. There are various methods of synthesis of graphene–metal nanoparticles such as chemical reduction, microwave-assisted synthesis, thermal evaporation, photochemical synthesis, and electroless metallization.

One of the easy and widely used approach to synthesize graphene–metal nanoparticle hybrid is by direct chemical reduction of metal precursors in presence of graphene or its derivatives such as GO and rGO. As discussed, GO and rGO possess oxygen functional groups which act as a nucleation sites for growth of metal nanoparticles.

Usually the precursors for nanoparticles are metal salts, which are reduced in solvent that contains GO and rGO. Because of the van der Walls interactions, the rGO tend to aggregate and even stack to form graphite. Hence, in order to obtain graphene as individual sheets and reduce its agglomeration attaching metal nanoparticle is a possible solution. Xu et al. [43] used water–ethylene glycol system to prepare various graphene–metal nanoparticle (Au, Pt, and Pd). First, Graphite oxide was synthesized using hummers method and was mixed with water to form colloid. Metal precursors such as K_2PtCl_4, K_2PdCl_4, and $HAuCl_4 \cdot 3H_2O$ were used for Pt, Pd and Au, respectively, were added to the colloid and then heated in oil bath for 6 h, followed by drying in vacuum oven at 60 °C for 12 h. Luo et al. [44] synthesized graphene–gold nanoparticle hybrid by depositing gold atoms by physical deposition method and then condensed upon annealing to form nanoparticles. The deposition of gold nanoparticles on GO nanoplattes has also been achieved by direct reduction of $AuCl_4$ by $NaBH_4$ in GO THF solution. In order to disperse GO in organic solvent it was first functionalized with octadecyl amine forming GO-ODA. UV–Vis spectroscopy confirmed the formation of Au nanoparticles where the peak at 550 nm was observed. Size was measured by the determining the concentration of GO-ODA in the solution. Oleic acid and oleylamine mixture is strong reducing agent for formation of Au, Ag and Cu nanoparticles in presence of GO [45]. The method showed narrow size distribution as well as fine dispersion of the produced nanoparticles on the GO surface. Phan et al. [46] synthesized Pt/Pd core-shells hybrid on a graphene support by two step chemical (seed-mediated and polymer-assisted) process. At first, colloidal solution of Pd nanocubes were synthesized as core followed by Pt coating as shells. In the second step, graphene was decorated onto graphene by the use of hydrazine in a facile one step reaction. Recently, affinity of Au nanoparticles with graphene has been compared with Pd and silver nanoparticles. The study concluded that palladium has the highest affinity to graphene which was supported by theoretical studies suggesting that higher affinity of Pd with graphene was due to the partial covalent nature of binding between Pd and graphene [47].

Another general method for synthesizing graphene–metal nanoparticle is the reduction of GO by microwave assisted method. Hassan et al. [48] utilized microwaves to reduce aqueous solutions of GO in presence of various reducing agents like ammonium hydroxide, hydrazine dihydrate, or ethylenediamine. Metal salts such as copper nitrate, palladium nitrate, or a mixture is present in liquid phase together with GO and the corresponding nanoparticles were formed and deposited on rGO nanoplatelets. Thermogravimetric analysis showed a significant weight loss up to 750 °C with this method, which indicates that this method was effective in removing the oxygen functionalities from rGO.

Photochemical method is another method of synthesis of graphene–metal nanoparticles hybrids. Gu et al. [49] synthesized various noble metal nanoparticles such as Au, Pt, Pd, and Ag via UV-assisted photocatalytic reduction by the use of ZnO nanorods. First, GO served as the two dimensional support to disperse ZnO nanorods by hydrothermal method. The ZnO nanorods were self-assembled on GO forming ZnO/graphene sheet (GS) composite. This ZnO/GS then served as supporting material for dispersion of metal nanoparticles. The mixture was then irradiated with UV light. Photogenerated electrons from ZnO reduced the metal ions to metal nanoparticles at the location distinct from ZnO nanorods thereby forming graphene–metal nanoparticle–ZnO hybrid. Figure 3 shows the TEM images of graphene—Ag, Au, and Pd nanoparticles hybrids structure.

Du et al. [50] developed a green and facile method for synthesizing graphene–metal nanoparticle hybrid by the use of electroless deposition. They constructed a galvanic cell by using nickel (Ni) foam as a substrate. Various nanoparticles with redox potential higher than that of Ni were deposited which included Au, Pt, Pd and Cu. Figure 4 shows the schematic illustration of the deposition of metal nanoparticles on graphene hydrogel @ nickel foam via electroless deposition.

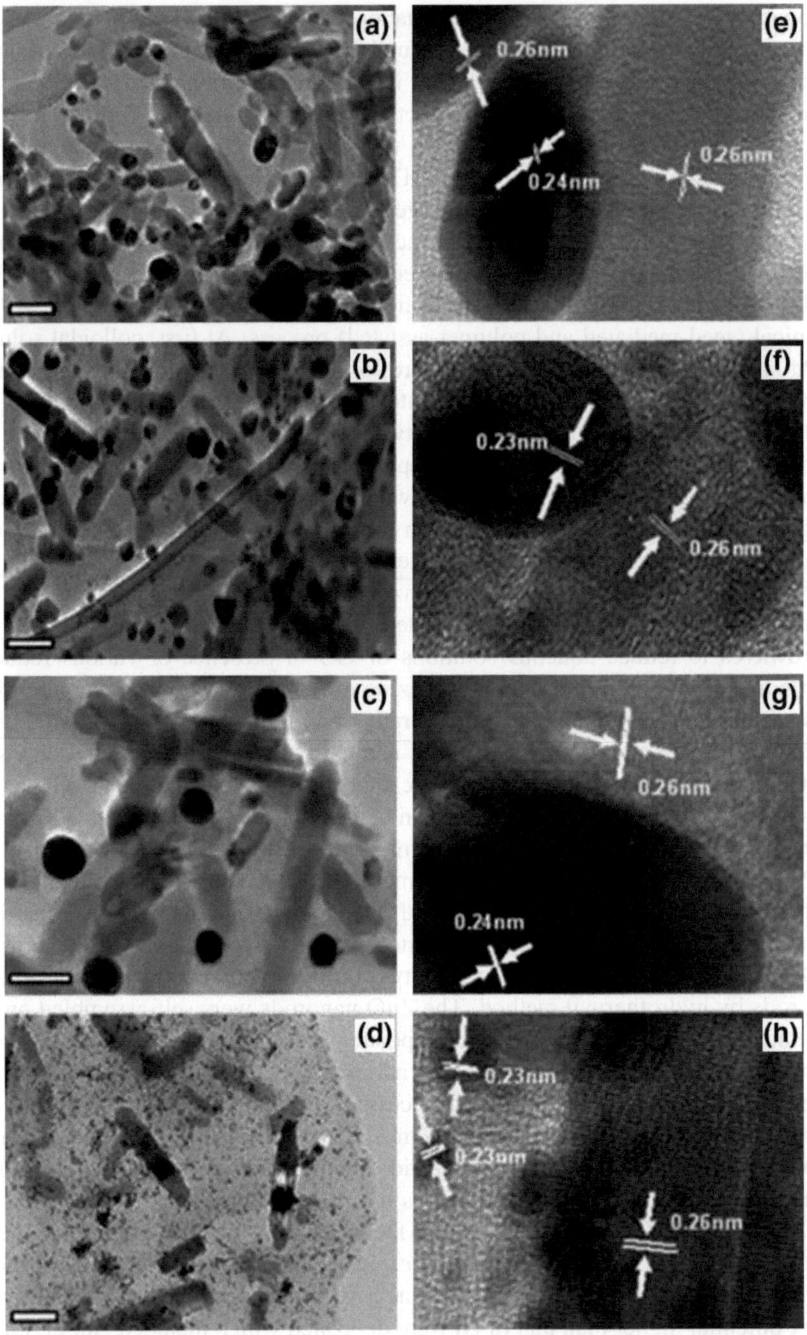

Fig. 3 TEM images of **a** GS/ZnO @Ag, **b** GS/ZnO @Pd, **c** GS/ZnO@Au, **d** GS/ZnO@Pt and HRTEM images of the interface of **e** Ag nanoparticles, **f** Pd nanoparticles, **g** Au nanoparticles and **h** Pt nanoparticles and ZnO nanorods. *Scale* 50 nm. Reprinted with permission from Ref. [49]. © American Chemical Society

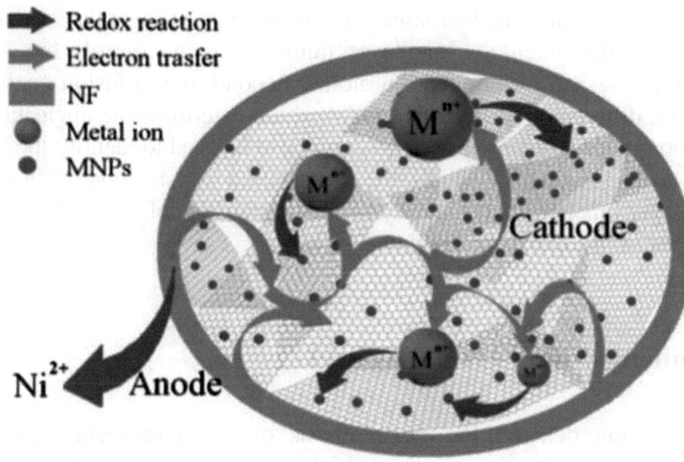

Fig. 4 Schematic illustration of deposition of metal nanoparticles on graphene hybrid @ nickel foam with substrate enhanced electroless deposition. Reprinted with permission from Ref. [50]. © Royal Society of Chemistry

It is worth noting that GO sheets, compared to rGO facilitate the growth of metal nanoparticles which can be due to the larger number of functional groups present in GO as compared to rGO. Several reports regarding this phenomenon including that of Ag and Ni nanoparticles have been published [51, 52]. Also, surface chemistry of chemically modified graphene can affect the nucleation density of as grown metal nanoparticle, whereas the lattice structure of rGO can be useful in patterning of several nanoparticles. Apart from this, number of graphene layer has shown direct relation to particle size and density to the thermally evaporated Au nanoparticles [53]. It was observed that particle size decreases and density increases as the

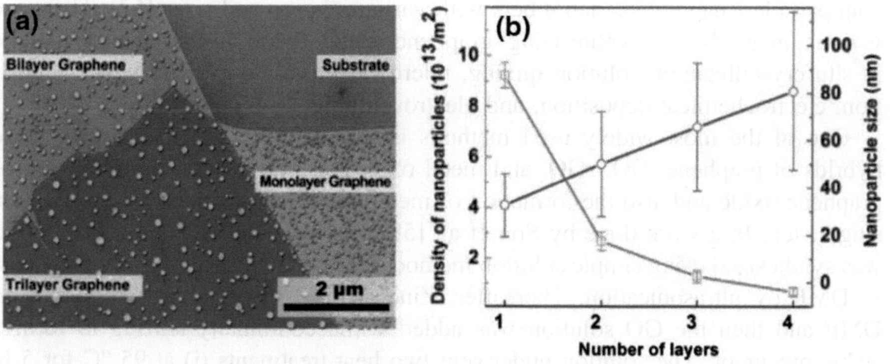

Fig. 5 **a** Distribution of gold nanoparticles on monolayer, bilayer and trilayer graphene and **b** statistics of the size and density of gold nanoparticles on *n*-layers graphene. Reprinted with permission from Ref. [53]. © American Chemical Society

number of layer of graphene increases. This is attributed to fact that surface energy of graphene is dependent on the layer number which controls the interaction between graphene and evaporated Au atoms. Second, the diffusion coefficients of Au atoms are different on different surfaces which determine the nucleation growth of Au nanoparticles. The above method could also be used to identify the number of graphene layers. Figure 5a shows the distribution Au nanoparticles on single, bilayer, trilayer graphene and Fig. 5b shows the statistics of size and density of Au nanoparticles on *n*-number of graphene layers.

4 Graphene–Metal Oxide Hybrid

There is growing demand for the synthesis of graphene–metal oxide hybrids because of their excellent properties which promise applications in electronics, chemical and biosensors, solar cells, Li-ion batteries, and supercapacitors. Till date, there are various kinds of metal oxides synthesized which include TiO_2, ZnO, CdS, NiO, MnO_2, Co_3O_4, Fe_3O_4, Fe_2O_3, Cu_2O, RuO_2, SnO_2, and CdSe [54–56]. The fact that metal oxide has a large bandgap, restricts its direct use in various sensing and electronic device applications because of the low electronic conductivity. To enhance and expand the scope of pristine metal oxides, graphene has been proved to be an efficient dopant which not only increases the conductivity but also increases the overall surface to volume ratio. Moreover, due to the fact the conduction band of the most of the metal oxide is greater than the work function of the graphene, it allows for easy electron transfer from metal oxide to graphene. Graphene, due to its high mobility, increases the conductivity and directs the electrons to the metal contacts as most of the metal contacts used have work function less than that of graphene. Unlike metal nanoparticles, which mostly have the 0D structure with only few reports on 1D metal nanowires, metal oxide have been synthesized in various dimensional structures such as quantum dots, nanoparticles, nanowires, nanofibers with various shapes and sizes [57]. There are various methods for synthesizing graphene–metal oxide hybrid which include in situ crystallization, solution mixing, microwave-assisted growth, vapor deposition, electrochemical deposition, and electrospinning.

One of the most widely used methods is in situ crystallization to synthesize hybrids of graphene, GO, rGO, and metal oxide. In this process, the reduction of graphene oxide and also the formation of metal oxide nanoparticles take place in a single step. In a work done by Son et al. [58] graphene ZnO quantum dot hybrid was synthesized using simple solution method. Firstly, GO was uniformly dispersed in DMF by ultrasonication. Thereafter, Zinc acetate dihydrate was dissolved in DMF and then the GO solution was added while continually stirring to form a stable precursor. The solution underwent two heat treatments (i) at 95 °C for 5 h. Followed by washing with ethanol by centrifugation and finally with water and (ii) at 55 °C for 2 h. The complete schematic of the synthesis process is as shown in Fig. 6. Cao et al. [59] reported facile synthesis of graphene–CdS hybrid by mixing

GO and $Cd(CH_3COO)_2$ in DMSO, followed by heating in autoclave for 12 h at 180 °C. During the process, hydrothermal reaction results in simultaneous formation of CdS nanoparticle and reduction of GO to rGO. The in situ crystallization method is in general applied to other semiconducting metal oxides on graphene platforms such as TiO_2 rods, SnO_2 nanoparticles, and MnO_2 nanoneedles [60, 61]. Mostly, in situ crystallization refers to the hydrothermal method, but alternate to it is in situ microwave assisted method of synthesis to obtain graphene–metal oxide hybrid. Yan et al. [62] GO, synthesized by Hummers method, was reduced to graphene by the use of hydrazine. Thereafter, 100 mL of graphene water suspension was subjected to ultrasonic vibration for 1 h. Potassium permanganate ($KmNO_4$) was then added to the graphene solution and stirred for 10 min. Subsequently, the resulting mixture was heated using a household microwave oven for 5 min and then cooled to room temperature. Finally, the product was washed, filtered, and dried at 100 °C for 5 h to obtain graphene–MnO_2 hybrid.

Another efficient easy and direct means to synthesize graphene metal oxide hybrid is by simple solution mixing approach. Feng et al. [63] pre-synthesized CdS nanoparticle were functionalized with benzyl mercaptan and then mixed with rGO sheets. Through π–π interactions functionalized CdS were absorbed on rGO sheets. Sellapan et al. [64] synthesized graphene/TiO_2 hybrid by the use of chemical vapor deposition and RF magnetron sputtering techniques. In this method, graphene was synthesized by three methods, namely, CVD, catalyst free CVD and solution processed graphene. TiO_2 was synthesized by RF magnetron sputtering with thickness of 50 nm.

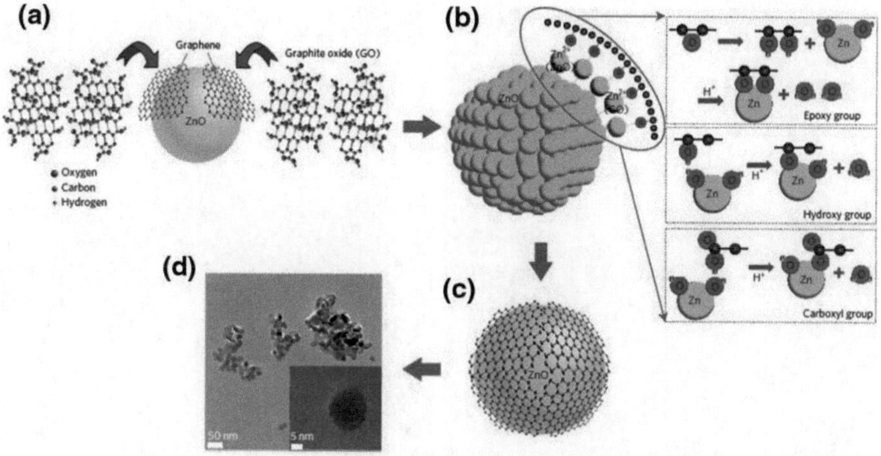

Fig. 6 Schematic of **a** chemical exfoliation of graphene sheets from GO with the aid of ZnO, **b** synthesis of Gr–ZnO quantum dot from GO and zinc acetate dihydrate, **c** graphene covered ZnO quantum dot and **d** TEM image of graphene–ZnO quantum dot. *Inset* HRTEM image of Graphene–ZnO quantum dot. Reprint with permission from Ref. [58]. © Nature publishing group

Various thin films of metal oxide on graphene and its derivatives have been synthesized by electrochemical deposition. This method has the ability to deposit various metal oxides in various shapes, size, and thickness depending upon the applied potential. These methods overcomes the drawbacks of traditional methods like drop casting, spin coating, etc., which in ideal case will have device to device variation. Yin et al. [65] deposited monocrystalline ZnO with high donor concentration on rGO on quartz. As the thickness of the rGO film increases the morphology of ZnO deposited changes from porous like structure to ZnO nanorods which have tremendous applications in electrochemical sensing. Figure 7 shows the FESEM of 4 samples with different thickness of rGO films. It can be clearly seen that ZnO deposited changes its morphology from porous structure to nanoparticle to nanorods structure. Several other metal oxides such as CuO, NiO, and TiO_2 have been deposited using the electrochemical deposition.

Apart from the conventional methods discussed above, new methods such as growing metal oxides on graphene foams, flexible substrates such as polyimide, PET, PDMS, etc., are gaining interest due to recent progress in wearable and

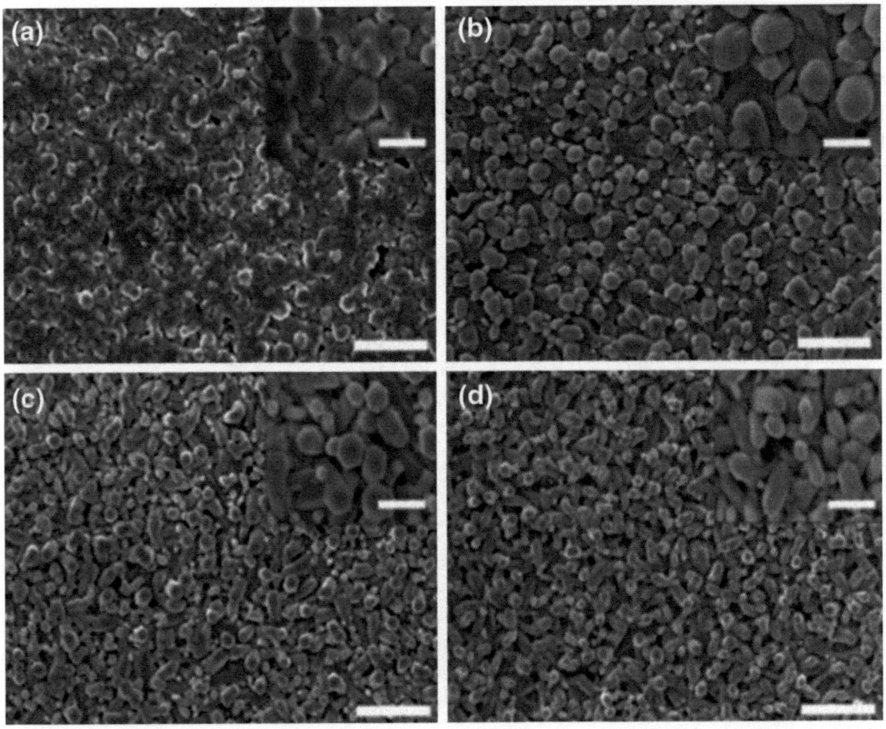

Fig. 7 SEM images of electrochemically deposited ZnO on rGO films of thickness **a** 5 nm showing particle shaped structure, **b** 9 nm showing transition state which is rod like without the typical hexagonal shape, **c** 13 nm, **d** 13 nm exhibiting ZnO hexagonal nanorods. Reprinted with permission from Ref. [65]. © Royal Society of Chemistry

flexible electronics. Boruah et al. [66] demonstrated the growth of ZnO nanowires on graphene foam. Graphene foam was synthesized by CVD on the surface of Ni foam as catalyst, wherein the nickel was etched by dipping the sample in 3 M HCl for 12 h. After nickel etching, the sample was washed with DI and chloroform and then ZnO nanorods were grown using resistive thermal evaporation. Hwang et al. [67] developed method to grow ZnO nanowires on graphene/PDMS substrate. Graphene films were synthesized using chemical exfoliation from natural graphite and subsequently growing ZnO by hydrothermal method.

Other than the methods discussed above, electrospinning is another method to synthesize 1D structures of graphene–metal oxide hybrid. Electrospinning process involves preparation of the polymer solution, and then feeding it into syringe and varying the electrospinning parameters such as applied voltage, flow rate and distance between the syringe and collector. Recently, graphene–ZnO nanofibers were in situ synthesized and align between the gold electrodes by our group [68]. Herein, the electrospinning time was optimized such that to get a single aligned graphene–ZnO nanofiber between pre-patterned gold electrodes. The structural characterization of the synthesized hybrid revealed that graphene exists not only on the surface of ZnO but also inside the ZnO. Varying the parameters of the electrospinning process and the viscosity of the electrospinning solution, it is possible to synthesize the hybrids in various shapes such as nanorods and nanoparticles [69].

5 Graphene-Conducting Polymer Hybrids

Conducting polymers or so called synthetic metals such as Polyaniline (PANi), polypyrole (PPy), polythiophene, are the class of polymers containing large number of sp^2 carbon atoms that permits delocalization of charge carriers. Owing to the unique electrical, chemical, mechanical and structural properties of both conducting polymers and graphene, these hybrids have generated interest in field of chemical and biosensors. Conducting polymers act as a conducting agent that interface with graphene or play the role of sensitizing agents to enhance the properties of a sensor such as sensitivity, lower detection of limit, increase response time and selectivity.

The most widely used method for synthesis of graphene-conducting polymer is the in situ polymerization in a solution containing monomer and graphene or GO, wherein in case of GO post reduction of GO to rGO is necessary. Since both the molecules have conjugated configuration there is a strong affinity between graphene and conducting polymer by π–π stacking. In a typical process, monomer is added into organic solution of graphene or GO dispersion. An initiator which is generally an oxidative species such as $FeCl_3$, H_2O_2, etc., is needed for polymerization reaction. Wang et al. [70] synthesized graphene/PANi hybrid by in situ polymerization of aniline. Three electrode set up was used wherein graphene was used as a working electrode. The electrolyte used was 0.5 M H_2SO_4 and 0.05 M aniline. PANi was in situ electropolymerized on graphene at a constant potential of 0.75 V for different periods. Thereafter, the hybrid was washed with distilled water and dried. Asadin et al. [71] synthesized graphene nanoribbon/PANi hybrid by in situ electropolymerization method, wherein graphene nanoribbon was synthesized via solvothermal route through unzipping of CNTs. For hybrid, 5 μL of graphene/aniline was casted on glassy carbon electrode (GCE) and the potential was scanned between −1.3 and 0.8 V (v/s Ag/AgCl) at scan rate of 50 mV/s for a total of 10 cycles. Figure 8 shows the cyclic voltagram of the electro-polymerization process.

Another method to synthesize graphene-conducting hybrid is by liquid/liquid interfacial polymerization in oil/water system. In work done by Hao et al. [72] water phase containing sulfonated graphene with H_2O_2 and $FeCl_3$, and the oil phase is chloroform containing aniline. The oil/water interface provides a good template for the polymerization of PANi. The sulfonic group acts as acid dopant necessary for the formation of PANi/sulfonated graphene hybrid.

Synthesis of graphene-conducting polymer can also be done using photochemical reaction. Tien et al. [73] developed a facile and one step redox reaction under UV illumination employing GO reduction and pyrrole polymerization via electron transfer at room temperature. GO was synthesized using Hummers method in the above synthesis. Recently, inkjet printing has gained attention due to its ease of fabricating device. Sriprachuabwong et al. [74] used Graphene–PEDOT: PSS solution as ink for inkjet printing by the use of commercial inkjet printer. One to eight layers of graphene–PEDOT: PSS were printed over an area of 3 mm × 5 mm.

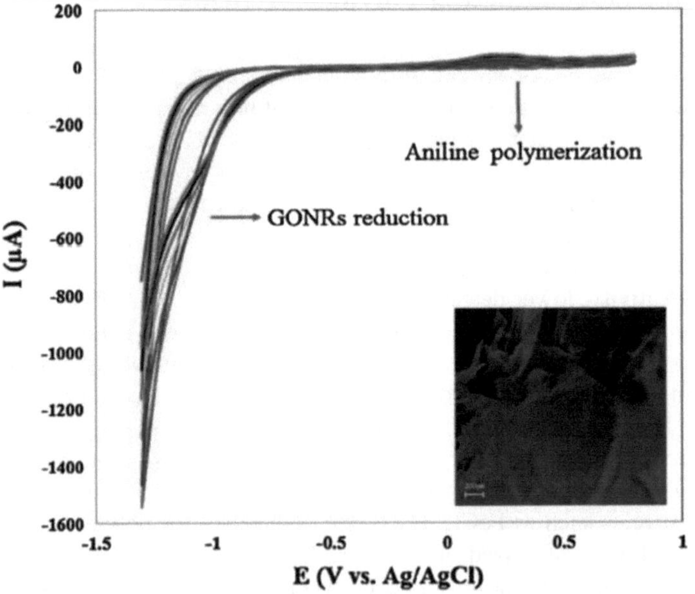

Fig. 8 CV of the electro-polymerization process in 1 M H_2SO_4 at scan rate of 50 mV/s. *Inset* SEM image of graphene/PANi on the surface of GCE. Reprinted with permission from Ref. [71]. © Elsevier

6 Graphene–Carbon Nanomaterials Hybrids

Graphene properties exists in 2D planar direction of graphene structure thereby its scope in applications wherein it is necessary to utilize properties in all axial directions. In this regards, 1D CNTs can be functionalized onto the surface of graphene combining the properties in all directions allowing for high surface area as compared to graphene alone. Badhulika et al. [75] synthesized graphene–CNTs hybrid in a single step CVD method via in situ vapor–liquid–solid and surface catalyzed mechanisms. In this approach, 1 nm Fe nanoparticles were deposited on copper substrates via electron beam evaporation. C_2H_2 gas was used as carbon source in argon and hydrogen atmosphere. Growth of graphene and CNTs take place simultaneously on copper foil and Fe nanoparticle, respectively. Thereafter, copper was etched in $FeCl_3$ solution and residues of Fe were removed by the use of HCl. Further, platinum nanoflowers were electrodeposited on graphene–CNT hybrid by the use of three electrode system using chloroplatinic acid hexahydrate as a precursor for platinum.

Apart from CNTs, carbon nanofibers are also one dimensional structure which has unique and excellent properties in axial direction. Dong et al. [76] synthesized graphene–CNFs hybrid by the use of electrospinning method. Polyacrylonitrile (PAN) polymer was dissolved in DMF under continual stirring to obtain homogenous polymer solution. Secondly, GO was dispersed in water. Thereafter, fibrous polymer was obtained by the use of electrospinning wherein the voltage

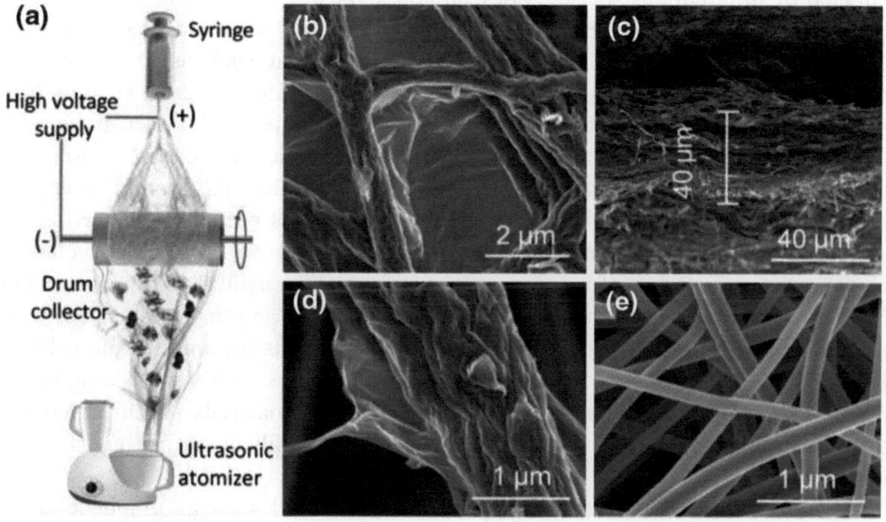

Fig. 9 **a** Schematic of electrospinning process using ultrasonic atomizer, **b** top view and **c** side view of the as prepared graphene–CNF hybrid, **d** high magnification image of a single graphene–CNF hybrid and **e** SEM image of pristine CNF. Reprinted with permission from Ref. [76]. © Elsevier

applied was 22 kV, distance from needle to collector was kept at 15 cm at a flow rate of 1 mL/hr. Meanwhile, spray containing GO was incorporated using ultrasonic atomizer. The complete schematic of the procedure and the FESEM images of the nanofiber are as shown in Fig. 9.

7 Application of Graphene Hybrids in Chemical Sensors

7.1 Gas and Vapor Sensors

One of the most interesting application of graphene and its derivative is gas sensors. The principle of operation of graphene-based gas sensors is based on the change in the graphene conductivity due to adsorption of gas molecule on the surface of graphene. This change in electrical conductivity due to the adsorption of gas molecule can be attributed to the change in the carrier concentration of graphene acting as donors or acceptors. Furthermore, interesting properties of graphene aid to detect single atom or molecular level thereby increasing the sensitivity. Graphene, being a 2D material, all the sp^2 carbon atoms are exposed to the analyte of interest. Moreover, graphene is highly conductive with low Johnson noise (electrical noise generated by thermal agitation of charge carriers) and hence little change in the carrier concentration can cause notable variation in electrical conductivity. Finally, it has very few crystal defects and hence noise caused by thermal switching is low.

Graphene is very sensitive to chemical environment and hence it is easily affected by different gas species. For example, exposure to NO_2 causes an increase in conductance of graphene. However, similar increase in conductance would be observed when exposed with different oxidizing agent such as O_2. Moreover, mixture containing oxidizing agent such as NO_2 and reducing agent such as NH_3 will not create any net change in graphene conductivity. Another issue with graphene-based gas sensor is reversibility. Activation energy needed for molecular desorption for gases such as NH_3 and NO_2 is higher than the thermal energy at room temperature. Hence high temperature desorption is needed in an inert ambient. Most of the graphene-based gas sensor is based on conductance change. Electrical conductivity of graphene is very much sensitive to environmental changes such as moisture, temperature, contaminations, and this creates additional difficulties for reliable sensors. The above problems calls for an alternate solution for gas sensing, which can be functionalizing nanofibers with various conducting polymers/metal oxides/metal nanoparticles/carbon nanomaterials which can control the binding energy of the target molecule to the nanofibers surface and provide better selectivity.

There are several reports on graphene hybrid-based such as graphene–metal nanoparticles/metal oxide/conducting polymers/carbon nanomaterials which have shown excellent sensitivity and selectivity to environmental hazardous gases such as NO_2, NH_3, benzene, hydrogen, and hydrogen sulfide. Li et al. [77] sensed H_2 gas with the Pt/graphene hybrid with the lowest detection of 0.06 %, operating temperature being high as close to 100 °C. Hydrogen gas has usually been detected using noble metal/graphene hybrid at room temperature like Kaniyoor et al. [78] reported lowest detection limit of 30 ppm using Pd/graphene hybrid. Shafiei et al. [79] synthesized Pt/graphene with the lowest detection limit of 50 ppm. Chu et al. [80] reported the detection range of 6–1000 ppm with lowest detection limit of 6 ppm by use of Pd nanocube/RGO hybrid. Toxic gas such as NH_3 and NO_2 has been reported with Ag/RGO composite with the lowest detection limit of 15 ppm and 0.5 ppm by Phan et al. [46] and Tran et al. [81], respectively. Figure 10 shows the response of rGO/Ag nanowire hybrid for NH_3 gas at various levels of injection.

Song et al. [82] reported the detection limit of 1 ppm of NO_2 gas with ZnO/graphene composite. Jiang et al. [83] sensed benzene at 260 °C using composite of SnO_2/graphene with the lowest detection value of 5 ppb. Likewise, Li et al. [84] synthesized CuO_2/graphene composite and detected H_2S at room

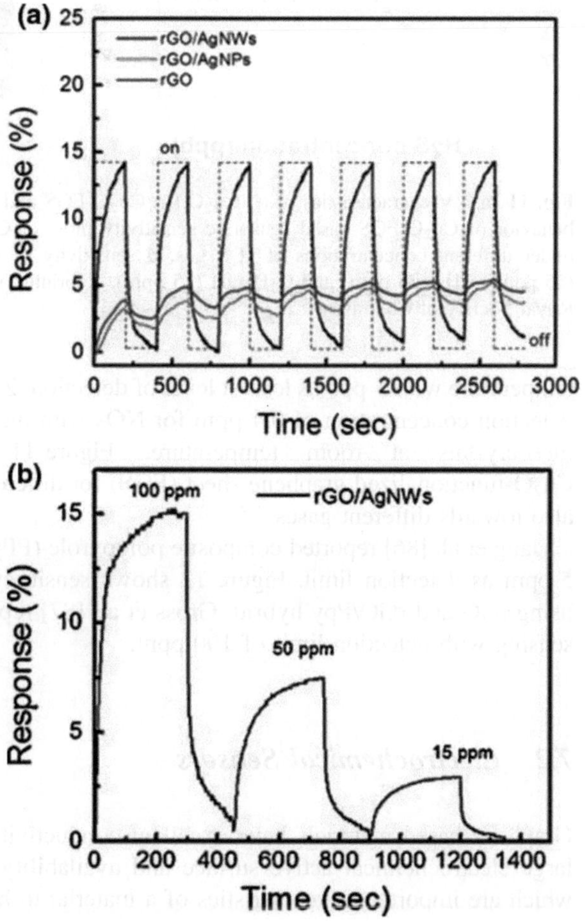

Fig. 10 Response of a rGO, rGO/Ag nanoparticles, rGO/Ag nanowires to 100 ppm NH_3 gas in nitrogen environment at room temperature, **b** rGO/Ag nanowires at different concentrations of NH_3 gas in nitrogen environment at room temperature. Reprinted with permission from Ref. [81]. © Elsevier

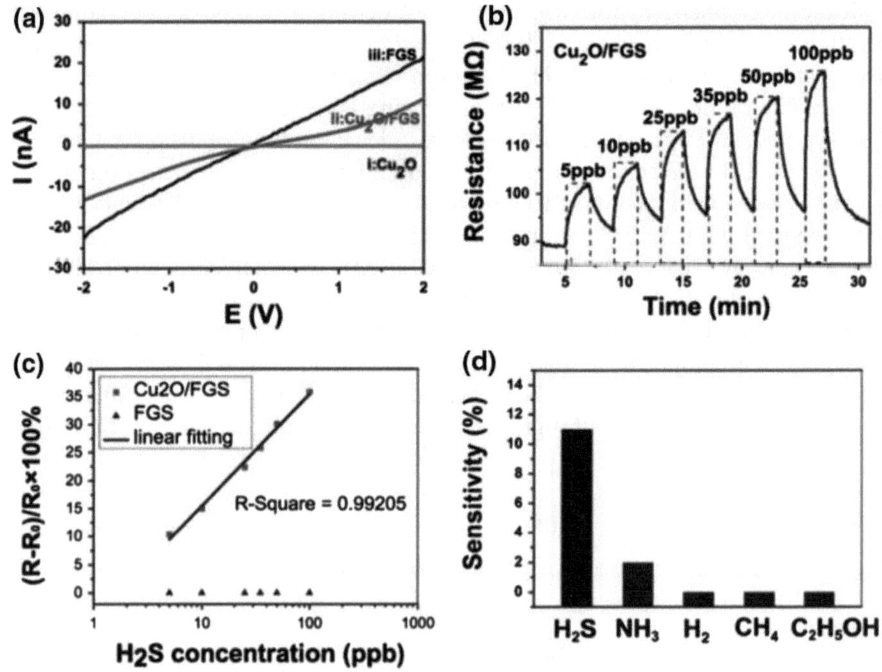

Fig. 11 **a** I–V characteristics of (*i*) Cu_2O (*ii*) Cu_2O-FGS and (*iii*) FGS, **b** dynamic H_2S sensing behavior of Cu_2O-FGS-based sensor, **c** sensitivity plots of Cu_2O-FGS and FGS-based sensors under different concentrations of H_2S gas, **d** sensitivity of different gases H_2S (5 ppb), NH_3 (25 ppm), CH_4 (25 ppm) and C_2H_5OH (25 ppm). Reprinted with permission from Ref. [85]. © Royal Society of Chemistry

temperature with 5 ppb as lowest level of detection. Zhou et al. [85] reported lowest detection concentration of 0.4 ppm for NO_2 with the use of RGO/Cu_2O nanowire mesocrystals at room temperature. Figure 11 shows the response of Cu_2O-functionalized graphene sheet (FGS) for different concentrations of H_2S and also towards different gases.

Jang et al. [86] reported composite polypyrole (PPy)/RGO for sensing NH_3 with 5 ppm as detection limit. Figure 12 shows sensitivity graph of sensor fabricated using rGO and rGO/Ppy hybrid. Gross et al. [87] reported PANI/GO for methanol sensing with detection limit of 100 ppm.

7.2 *Electrochemical Sensors*

Graphene-based materials have excellent conductivity, large specific surface area, large electrochemical active surface and availability for surface functionalization which are important characteristics of a material to be utilized for electrochemical

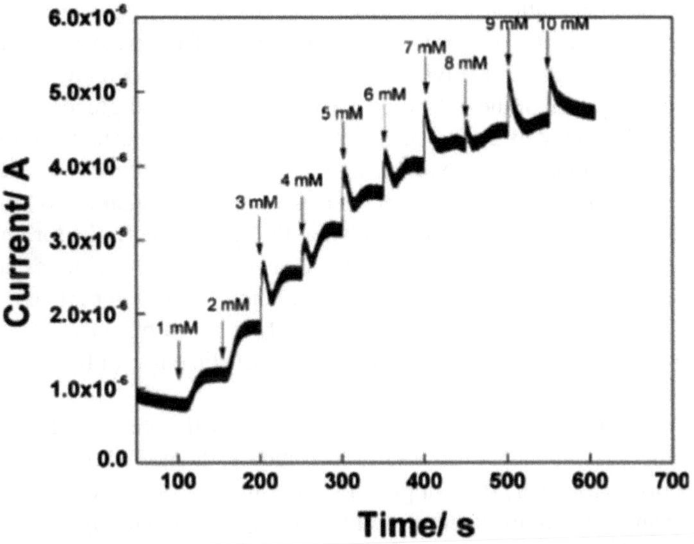

Fig. 12 Real-time amperometric response of graphene/MWCNTs/Pt nanoflowers towards successive addition of glucose at 0.4 V v/s Ag/AgCl in 0.1 M phosphate buffer saline (PBS) with pH 7.4. Reprinted with permission from Ref. [75]. © Wiley

sensors. Graphene has been known to perform superior than other carbon nanomaterials such as CNTs due to presence of more sp^2-like planes and edge defects in graphene. Till date, there are numerous reports on graphene and its derivatives-based electrochemical sensors such as detection of dopamine, H_2O_2, ascorbic acid, volatile organic compounds, DNA, etc.

In order to use graphene hybrids as working electrode of the electrochemical sensor, graphene hybrids have been used to modify conventional glassy carbon electrode, platinum or gold electrode. There are various methods of functionalization of graphene hybrids onto the surface of working electrodes such as the use of APTES when functionalizing with rGO, or by drop casting, or by the use of paraffin film, etc. The surface modification of gold electrode is realized by thiol terminated molecules because of the excellent affinity of thiol molecule towards gold.

In order to take good catalytic activity that metal nanoparticles possess, graphene–metal nanoparticles hybrids have been used for electrochemical sensors. Shan et al. [88] fabricated enzymatic glucose sensors based on graphene–Au nanoparticles–chitosan hybrid. The sensor also exhibited good electrocatalytic activity towards H_2O_2 and O_2. The sensors showed good linear response to glucose from 2 to 10 mM at −0.2 V. Although amperometric glucose sensor based on glucose oxidase exhibit high sensitivity and selectivity, their application is restricted due to the intrinsic nature of enzyme which needs specific temperature, pH, and ionic detergents. Moreover, immobilization step are highly complicated. Hence, non-enzymatic technique seems to be attractive without any of the above-mentioned drawback. Badhulika et al. [75] synthesizes graphene/CNT/Pt nanoflowers hybrid for developing non-enzymatic glucose sensor. Direct oxidation of glucose was observed due to the electrocatalytic property of Pt nanoflowers. The linear range of detection was 1–7 mM with the sensitivity of 11.06μA/mM/cm^2. Figure 12 shows the amperometric response of the hybrid on successive addition of glucose. Li et al. [89] developed facile one step electrochemical method for synthesizing graphene/NiO hybrid as non-enzymatic glucose and methanol sensor. Electrochemical experiments indicated that composite had large surface area compared to its pristine counterpart. Guo et al. [90] used graphene–Pt hybrid electrode for the electrochemical sensing of H_2O_2 and trinitrotoluene. The hybrid was synthesized rapidly using one step microwave assisted heating method. In this report, the enhanced performance of graphene–metal nanoparticle hybrid arises from excellent electrical conductivity and high specific area as well as synergistic effect of both graphene and metal nanoparticle.

Some electrochemical sensors have been developed based on the catalysis of biomacromolecules such as horseradish peroxidase, hemoglobin and myosin for H_2O_2 detection. However, their applications are limited due to the factors such as critical demand on environmental condition, high cost and instability. Hence, research toward non-enzymatic sensors for detection of H_2O_2 is gaining interest. Recent reports have shown that electrodes modified with noble metal nanoparticles, metal oxides, CNTs have shown good catalytic activity for direct detection of H_2O_2. Jia et al. [91] synthesized graphene–Au nanoparticle hybrid by direct electrochemical deposition of Au on graphene/chitosan modified electrode for non-enzymatic detection of H_2O_2. Figure 13 shows the real-time amperometric response of graphene/Au nanoparticle modified electrode for H_2O_2 detection. Devasenathipathy et al. [92] synthesized graphene bismuth hybrid for the detection

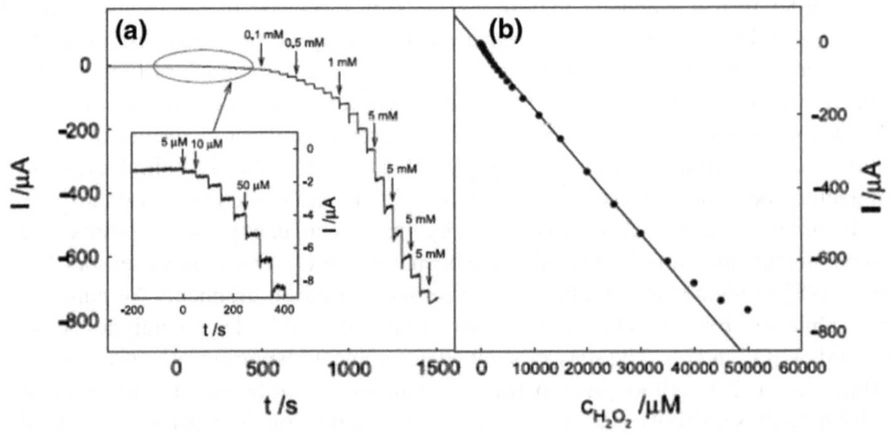

Fig. 13 **a** Amperometric response of graphene/chitosan/Au nanoparticle modified electrode for successive addition of H_2O_2 into N_2 saturated 0.1 M pH 7.2 with an applied potential of −0.4 V under constant stirring, **b** reduction current v/s concentration of H_2O_2. Reprinted with permission from Ref. [91]. © Elsevier

of hydrazine by real time amperometric method. The linear range was 0.02–280 μM with limit of detection (LOD) of 0.005 μM.

7.3 Biosensors

Biosensor is a chemical sensing device that uses biochemical reactions mediated by immunosystems, enzymes and tissues to detect chemical compounds by electrical, optical or thermal signal. A typical biosensor comprises of two parts (i) bioreceptor and (ii) transducer. The bioreceptor is a molecule which detects the target analyte and transducer converts the binding or recognition event into a measurable signal which can even be quantified.

Proper immobilization of enzyme onto the modified electrode surface is of prime importance in development of a biosensor. There are various immobilization techniques that depend on nature of the enzyme and associated detection mode which includes adsorption, covalent binding, affinity to complementary biomolecules, cross linking, and entrapment. The perfect choice of the immobilization method will yield higher sensitivity and hence it is of utmost importance to choose proper technique depending on the enzyme and the detection technique. Graphene hybrids have high surface area, large active sites for functionalization which makes them suitable for enzyme immobilization, thus expanding the range of applications graphene hybrids can offer. Moreover, due to the high electrical conductivity, it can enhance direct electron transfer between enzymes and electrodes. Enzyme are immobilized onto graphene hybrid modified electrode via covalent and non-covalent interactions. Functional groups present in the graphene hybrids aid in covalent functionalization with the enzyme whereas non covalent functionalization happens via weak van der Walls forces, electrostatic interaction, hydrogen bonding or π–π stacking.

Cao et al. [93] developed electrochemical sensor for the detection of cholesterol using Pt–Pd–chitosan–graphene hybrid electrode. Pt–Pd were electrochemically deposited on graphene–chitosan hybrid. The hybrid not only accelerated the electron transfer kinetics but also helps in increased enzyme immobilization. The linear range for the fabricated cholesterol sensor was found to be 2.2–5.2 μM with LOD of 0.75 μM. The response time was less than 7 s. The interference studies with Uric acid, ascorbic acid and glucose displayed high selectivity of the as fabricated sensor towards cholesterol. Kumar et al. [94] reported label free electrochemical detection of cardiac biomarker myoglobin using graphene–CNT hybrid. Here, high sensitivity was achieved due direct electron transfer between the myoglobin and the composite. The linear range was found to be 1 ng/mL to 4 μg/mL with LOD of ~0.34 ng/mL.

Immunosensors are biosensors which are based on antigen–antibody interaction. Immobilization of antibody onto the material of interest is an important step in the development of Immunosensors as antibody are very specific to antigen, which acts a recognition site and thus promote antibody–antigen interaction. Graphene hybrid-based Immunosensors are fabricated by covalent immobilization of antibody onto graphene hybrid. The above method requires graphene to have functional groups where antibody can be immobilized. Next, the carboxyl group on the surface is activated for coupling with amine group, as every antibody has a protein which has a specific amine group. Anitbody with the amine group binds to functional group on graphene hybrids and forms a peptide bond.

Prostate-specific antigen (PSA) is a tumor biomarker for prostate cancer. Su et al. [95] synthesized Gr/MnO_2 hybrid material with the deposition of Au nanoparticles for development of PSA biosensor. Au nanoparticles were deposited on to the Gr/MnO_2 hybrid by the use of poly (diallyldimethylammonium chloride) via simple sonication induced method. Au nanoparticles were used as binding sites for antibody immobilization. The linear range was found to be 0.005–10 ng/mL with LOD of 2.5 pg/mL. Clenbutonal (CLB) is known to reduce stress symptoms and asthma. This had led its wide spread use in treatment of human depression and pulmonary disease.

8 Conclusion and Outlook

Graphene exhibits unique electronic, chemical, mechanical, optical, and thermal properties which have paved way for various interesting applications in the field of electronics, sensors, biomedical and mechanical. The derivatives of graphene, i.e., graphene oxide and reduced graphene oxide are synthesized in various forms to meet the ever growing requirements of graphene as a material for thin film processing, composites, sensors and devices. To further enhance the properties and expand the scope of applications that graphene offers, various hybrids have been synthesized which include graphene–metal nanoparticles/metal oxides/conducting polymers/CNTs/CNFs, etc., and applied in various domains of engineering and technology. The properties of these hybrids do not only depend on the individual component, but also depends upon the synergy between the graphene and the hybrid material. Therefore, it becomes important to control the kind of chemical bonding, distribution density and 3D arrangement of graphene hybrids which would not increase the stability but also enhance the sensing behavior of the hybrid material. Despite the excellent properties of graphene and its hybrids, it still has a long way to go before it could totally replace silicon technology due to the fact that its reproducibility and large-scale production with precise number of layer is still a challenge.

Graphene hybrids have been utilized in developing numerous chemical sensors by varying the hybrid materials, synthesis procedure, optimization of the hybrid material to be doped, device level fabrication and lately developing flexible sensors for wearable devices. The idea is to benefit from the synergistic contribution of both materials to produce an output which not only is selective and sensitive but also repeatable. Next step would be a step ahead in mass fabrication of sensors with low cost which can be commercialized. This could be achieved by suitable microfabrication technique for patterning of graphene. Current approaches of graphene-based FET/microfabrication rely on e-beam evaporation technique whose throughput is very low and is not suitable for mass production. As a future research direction, it would be useful to directly pattern graphene on substrates for precise and rapid device integration and also to modify graphene properties by preparing molecular patterns.

References

1. Chung C, Kim YK, Shin D, Ryoo SR, Hong BH, Min DH (2013) Biomedical applications of graphene and graphene oxide. Acc Chem Res 46(10):2211–2224
2. Chang H, Wu H (2013) Graphene-based nanocomposites: preparation, functionalization, and energy and environmental applications. Energy Environ Sci 6(12):3483–3507
3. Liu J, Cui L, Losic D (2013) Graphene and graphene oxide as new nanocarriers for drug delivery applications. Acta Biomater 9(12):9243–9257
4. Singh K, Ohlan A, Pham VH, Balasubramaniyan R, Varshney S, Jang J, Kong BS (2013) Nanostructured graphene/Fe_3O_4 incorporated polyaniline as a high performance shield against electromagnetic pollution. Nanoscale 5(6):2411–2420
5. Traversa E (1995) Ceramic sensors for humidity detection: the state-of-the-art and future developments. Sens Actuators B Chem 23(2):135–156
6. Wohltjen H, Snow AW (1998) Colloidal metal-insulator-metal ensemble chemiresistor sensor. Anal Chem 70(14):2856–2859
7. Shimizu Y, Egashira M (1999) Basic aspects and challenges of semiconductor gas sensors. MRS Bull 24(06):18–24
8. Allen MJ, Tung VC, Kaner RB (2009) Honeycomb carbon: a review of graphene. Chem Rev 110(1):132–145
9. Novoselov KS, Geim AK, Morozov SV, Jiang D, Zhang Y, Dubonos SA, Firsov AA (2004) Electric field effect in atomically thin carbon films. Science 306(5696):666–669
10. Bolotin KI, Sikes KJ, Jiang Z, Klima M, Fudenberg G, Hone J, Stormer HL (2008) Ultrahigh electron mobility in suspended graphene. Solid State Commun 146(9):351–355
11. Neto AC, Guinea F, Peres NMR, Novoselov KS, Geim AK (2009) The electronic properties of graphene. Rev Mod Phys 81(1):109
12. Mayorov AS, Gorbachev RV, Morozov SV, Britnell L, Jalil R, Ponomarenko LA, Geim AK (2011) Micrometer-scale ballistic transport in encapsulated graphene at room temperature. Nano Lett 11(6):2396–2399
13. Kamat PV (2009) Graphene-based nanoarchitectures. Anchoring semiconductor and metal nanoparticles on a two-dimensional carbon support. J Phys Chem Lett 1(2):520–527
14. Wu ZS, Zhou G, Yin LC, Ren W, Li F, Cheng HM (2012) Graphene/metal oxide composite electrode materials for energy storage. Nano Energy 1(1):107–131
15. Al-Saleh MH, Sundararaj U (2009) A review of vapor grown carbon nanofiber/polymer conductive composites. Carbon 47(1):2–22
16. Yu D, Dai L (2009) Self-assembled graphene/carbon nanotube hybrid films for supercapacitors. J Phys Chem Lett 1(2):467–470
17. Kim CH, Kim BH, Yang KS (2012) TiO_2 nanoparticles loaded on graphene/carbon composite nanofibers by electrospinning for increased photocatalysis. Carbon 50(7):2472–2481
18. Keller AA, Wang H, Zhou D, Lenihan HS, Cherr G, Cardinale BJ, Ji Z (2010) Stability and aggregation of metal oxide nanoparticles in natural aqueous matrices. Environ Sci Technol 44 (6):1962–1967

19. Lee JM, Pyun YB, Yi J, Choung JW, Park WI (2009) ZnO nanorod–graphene hybrid architectures for multifunctional conductors. J Phys Chem C 113(44):19134–19138
20. Li X, Zhang G, Bai X, Sun X, Wang X, Wang E, Dai H (2008) Highly conducting graphene sheets and Langmuir-Blodgett films. Nat Nanotechnol 3(9):538–542
21. Fang M, Wang K, Lu H, Yang Y, Nutt S (2009) Covalent polymer functionalization of graphene nanosheets and mechanical properties of composites. J Mater Chem 19(38):7098–7105
22. Bai H, Xu Y, Zhao L, Li C, Shi G (2009) Non-covalent functionalization of graphene sheets by sulfonated polyaniline. Chem Commun 13:1667–1669
23. Si Y, Samulski ET (2008) Synthesis of water soluble graphene. Nano Lett 8(6):1679–1682
24. Cai D, Song M (2007) Preparation of fully exfoliated graphite oxide nanoplatelets in organic solvents. J Mater Chem 17(35):3678–3680
25. Yan L, Lin M, Zeng C, Chen Z, Zhang S, Zhao X, Guo M (2012) Electroactive and biocompatible hydroxyl-functionalized graphene by ball milling. J Mater Chem 22(17):8367–8371
26. Jeon JH, Cheedarala RK, Kee CD, Oh IK (2013) Dry-type artificial muscles based on pendent sulfonated chitosan and functionalized graphene oxide for greatly enhanced ionic interactions and mechanical stiffness. Adv Funct Mater 23(48):6007–6018
27. Sinitskii A, Dimiev A, Corley DA, Fursina AA, Kosynkin DV, Tour JM (2010) Kinetics of diazonium functionalization of chemically converted graphene nanoribbons. ACS Nano 4 (4):1949–1954
28. Liu H, Ryu S, Chen Z, Steigerwald ML, Nuckolls C, Brus LE (2009) Photochemical reactivity of graphene. J Am Chem Soc 131(47):17099–17101
29. Georgakilas V, Bourlinos AB, Zboril R, Steriotis TA, Dallas P, Stubos AK, Trapalis C (2010) Organic functionalisation of graphenes. Chem Commun 46(10):1766–1768
30. He H, Gao C (2010) General approach to individually dispersed, highly soluble, and conductive graphene nanosheets functionalized by nitrene chemistry. Chem Mater 22 (17):5054–5064
31. Li X, Wang H, Robinson JT, Sanchez H, Diankov G, Dai H (2009) Simultaneous nitrogen doping and reduction of graphene oxide. J Am Chem Soc 131(43):15939–15944
32. Zhang L, Kiny VU, Peng H, Zhu J, Lobo RF, Margrave JL, Khabashesku VN (2004) Sidewall functionalization of single-walled carbon nanotubes with hydroxyl group-terminated moieties. Chem Mater 6(11):2055–2061
33. Paredes JI, Villar-Rodil S, Martinez-Alonso A, Tascon JMD (2008) Graphene oxide dispersions in organic solvents. Langmuir 24(19):10560–10564
34. Yu D, Yang Y, Durstock M, Baek JB, Dai L (2010) Soluble P3HT-grafted graphene for efficient bilayer–heterojunction photovoltaic devices. ACS Nano 4(10):5633–5640
35. Cao Y, Feng J, Wu P (2010) Alkyl-functionalized graphene nanosheets with improved lipophilicity. Carbon 48(5):1683–1685
36. Georgakilas V, Otyepka M, Bourlinos AB, Chandra V, Kim N, Kemp KC, Kim KS (2012) Functionalization of graphene: covalent and non-covalent approaches, derivatives and applications. Chem Rev 112(11):6156–6214
37. Sreeprasad TS, Berry V (2013) How do the electrical properties of graphene change with its functionalization? Small 9(3):341–350
38. Subrahmanyam KS, Manna AK, Pati SK, Rao CNR (2010) A study of graphene decorated with metal nanoparticles. Chem Phys Lett 497(1):70–75
39. Tjoa V, Jun W, Dravid V, Mhaisalkar S, Mathews N (2011) Hybrid graphene–metal nanoparticle systems: electronic properties and gas interaction. J Mater Chem 21(39): 15593–15599
40. Hong W, Bai H, Xu Y, Yao Z, Gu Z, Shi G (2010) Preparation of gold nanoparticle/graphene composites with controlled weight contents and their application in biosensors. J Phys Chem C 114(4):1822–1826
41. Lu LM, Li HB, Qu F, Zhang XB, Shen GL, Yu RQ (2011) In situ synthesis of palladium nanoparticle–graphene nanohybrids and their application in nonenzymatic glucose biosensors. Biosens Bioelectron 26(8):3500–3504

42. Bao Q, Zhang D, Qi P (2011) Synthesis and characterization of silver nanoparticle and graphene oxide nanosheet composites as a bactericidal agent for water disinfection. J Colloid Interface Sci 360(2):463–470
43. Xu C, Wang X, Zhu J (2008) Graphene–metal particle nanocomposites. J Phys Chem C 112 (50):19841–19845
44. Luo Z, Somers LA, Dan Y, Ly T, Kybert NJ, Mele EJ, Johnson AC (2010) Size-selective nanoparticle growth on few-layer graphene films. Nano Lett 10(3):777–781
45. Xu Z, Shen C, Hou Y, Gao H, Sun S (2009) Oleylamine as both reducing agent and stabilizer in a facile synthesis of magnetite nanoparticles. Chem Mater 21(9):1778–1780
46. Phan DT, Uddin AI, Chung GS (2015) A large detectable-range, high-response and fast-response resistivity hydrogen sensor based on Pt/Pd core–shell hybrid with graphene. Sens Actuators B Chem 220:962–967
47. Granatier J, Lazar P, Prucek R, Šafářová K, Zbořil R, Otyepka M, Hobza P (2012) Interaction of graphene and arenes with noble metals. J Phys Chem C 116(26):14151–14162
48. Hassan HM, Abdelsayed V, Abd El Rahman SK, AbouZeid KM, Terner J, El-Shall MS, El-Azhary AA (2009) Microwave synthesis of graphene sheets supporting metal nanocrystals in aqueous and organic media. J Mater Chem 19(23):3832–3837
49. Gu H, Yang Y, Tian J, Shi G (2013) Photochemical synthesis of noble metal (Ag, Pd, Au, Pt) on graphene/ZnO multihybrid nanoarchitectures as electrocatalysis for H_2O_2 reduction. ACS Appl Mater Interfaces 5(14):6762–6768
50. Du C, Yao Z, Chen Y, Bai H, Li L (2014) Synthesis of metal nanoparticle@graphene hydrogel composites by substrate-enhanced electroless deposition and their application in electrochemical sensors. RSC Adv 4(18):9133–9138
51. Zhou X, Huang X, Qi X, Wu S, Xue C, Boey FY, Zhang H (2009) In situ synthesis of metal nanoparticles on single-layer graphene oxide and reduced graphene oxide surfaces. J Phys Chem C 113(25):10842–10846
52. Ji Z, Shen X, Zhu G, Zhou H, Yuan A (2012) Reduced graphene oxide/nickel nanocomposites: facile synthesis, magnetic and catalytic properties. J Mater Chem 22 (8):3471–3477
53. Zhou H, Qiu C, Liu Z, Yang H, Hu L, Liu J, Sun L (2009) Thickness-dependent morphologies of gold on N-layer graphenes. J Am Chem Soc 132(3):944–946
54. Liang Y, Wang H, Casalongue HS, Chen Z, Dai H (2010) TiO_2 nanocrystals grown on graphene as advanced photocatalytic hybrid materials. Nano Res 3(10):701–705
55. Yi J, Lee JM, Park WI (2011) Vertically aligned ZnO nanorods and graphene hybrid architectures for high-sensitive flexible gas sensors. Sens Actuators B Chem 155(1):264–269
56. Peng L, Peng X, Liu B, Wu C, Xie Y, Yu G (2013) Ultrathin two-dimensional MnO_2/graphene hybrid nanostructures for high-performance, flexible planar supercapacitors. Nano Lett 13(5):2151–2157
57. Jana NR, Chen Y, Peng X (2004) Size-and shape-controlled magnetic (Cr, Mn, Fe Co, Ni) oxide nanocrystals via a simple and general approach. Chem Mater 16(20):3931–3935
58. Son DI, Kwon BW, Park DH, Seo WS, Yi Y, Angadi B, Choi WK (2012) Emissive ZnO-graphene quantum dots for white-light-emitting diodes. Nat Nanotechnol 7(7):465–471
59. Cao A, Liu Z, Chu S, Wu M, Ye Z, Cai Z, Liu Y (2010) A facile one-step method to produce graphene–CdS quantum dot nanocomposites as promising optoelectronic materials. Adv Mater 22(1):103–106
60. Tu W, Zhou Y, Liu Q, Yan S, Bao S, Wang X, Zou Z (2013) An in situ simultaneous reduction-hydrolysis technique for fabrication of TiO_2-graphene 2D sandwich-like hybrid nanosheets: graphene-promoted selectivity of photocatalytic-driven hydrogenation and coupling of CO_2 into methane and ethane. Adv Funct Mater 23(14):1743–1749
61. Li Y, Zhao N, Shi C, Liu E, He C (2012) Improve the supercapacity performance of MnO_2-decorated graphene by controlling the oxidization extent of graphene. J Phys Chem C 116 (48):25226–25232
62. Yan J, Fan Z, Wei T, Qian W, Zhang M, Wei F (2010) Fast and reversible surface redox reaction of graphene–MnO_2 composites as supercapacitor electrodes. Carbon 48(13):3825–3833

63. Feng M, Sun R, Zhan H, Chen Y (2010) Lossless synthesis of graphene nanosheets decorated with tiny cadmium sulfide quantum dots with excellent nonlinear optical properties. Nanotechnology 21(7):075601
64. Sellappan R, Sun J, Galeckas A, Lindvall N, Yurgens A, Kuznetsov AY, Chakarov D (2013) Influence of graphene synthesizing techniques on the photocatalytic performance of graphene–TiO_2 nanocomposites. Phys Chem Chem Phys 15(37):15528–15537
65. Yin Z, Wu S, Zhou X, Huang X, Zhang Q, Boey F, Zhang H (2010) Electrochemical deposition of ZnO nanorods on transparent reduced graphene oxide electrodes for hybrid solar cells. Small 6(2):307–312
66. Boruah BD, Mukherjee A, Sridhar S, Misra A (2015) Highly dense ZnO nanowires grown on graphene foam for ultraviolet photodetection. ACS Appl Mater Interfaces 7(19):10606–10611
67. Hwang JO, Lee DH, Kim JY, Han TH, Kim BH, Park M, Kim SO (2011) Vertical ZnO nanowires/graphene hybrids for transparent and flexible field emission. J Mater Chem 21 (10):3432–3437
68. Sahatiya P, Badhulika S (2015) One-step in situ synthesis of single aligned graphene–ZnO nanofiber for UV sensing. RSC Adv 5(100):82481–82487
69. Haridas AK, Sharma CS, Sritharan V, Rao TN (2014) Fabrication and surface functionalization of electrospun polystyrene submicron fibers with controllable surface roughness. RSC Adv 4(24):12188–12197
70. Wang DW, Li F, Zhao J, Ren W, Chen ZG, Tan J, Cheng HM (2009) Fabrication of graphene/polyaniline composite paper via in situ anodic electropolymerization for high-performance flexible electrode. ACS Nano 3(7):1745–1752
71. Asadian E, Shahrokhian S, Jokar E (2014) In-situ electro-polymerization of graphene nanoribbon/polyaniline composite film: Application to sensitive electrochemical detection of dobutamine. Sens Actuators B Chem 196:582–588
72. Hao Q, Wang H, Yang X, Lu L, Wang X (2011) Morphology-controlled fabrication of sulfonated graphene/polyaniline nanocomposites by liquid/liquid interfacial polymerization and investigation of their electrochemical properties. Nano Res 4(4):323–333
73. Tien HN, Hur SH (2012) One-step synthesis of a highly conductive graphene–polypyrrole nanofiber composite using a redox reaction and its use in gas sensors. Phys Status Solidi (RRL) 6(9–10):379–381
74. Sriprachuabwong C, Karuwan C, Wisitsorrat A, Phokharatkul D, Lomas T, Sritongkham P, Tuantranont A (2012) Inkjet-printed graphene–PEDOT: PSS modified screen printed carbon electrode for biochemical sensing. J Mater Chem 22(12):5478–5485
75. Badhulika S, Paul RK, Terse T, Mulchandani A (2014) Nonenzymatic glucose sensor based on platinum nanoflowers decorated multiwalled carbon nanotubes-graphene hybrid electrode. Electroanalysis 26(1):103–108
76. Dong Q, Wang G, Hu H, Yang J, Qian B, Ling Z, Qiu J (2013) Ultrasound-assisted preparation of electrospun carbon nanofiber/graphene composite electrode for supercapacitors. J Power Sources 243:350–353
77. Li Y, Li X, Tang Z, Tang Z, Yu J, Wang J (2015) Hydrogen sensing of the mixed-potential-type MnWO 4/YSZ/Pt sensor. Sens Actuators B Chem 206:176–180
78. Kaniyoor A, Jafri RI, Arockiadoss T, Ramaprabhu S (2009) Nanostructured Pt decorated graphene and multi walled carbon nanotube based room temperature hydrogen gas sensor. Nanoscale 1(3):382–386
79. Shafiei M, Arsat R, Yu J, Kalantar-Zadeh K, Wlodarski W, Dubin S, Kaner RB (2009) Pt/graphene nano-sheet based hydrogen gas sensor. In: IEEE sensors. IEEE, pp. 295–298
80. Chu BH, Nicolosi J, Lo CF, Strupinski W, Pearton SJ, Ren F (2011) Effect of coated platinum thickness on hydrogen detection sensitivity of graphene-based sensors. Electrochem Solid-State Lett 14(7):K43–K45
81. Tran QT, Hoa HTM, Yoo DH, Cuong TV, Hur SH, Chung JS, Kohl PA (2014) Reduced graphene oxide as an over-coating layer on silver nanostructures for detecting NH_3 gas at room temperature. Sens Actuators B Chem 194:45–50

82. Song H, Zhang L, He C, Qu Y, Tian Y, Lv Y (2011) Graphene sheets decorated with SnO_2 nanoparticles: in situ synthesis and highly efficient materials for cataluminescence gas sensors. J Mater Chem 21(16):5972–5977
83. Jiang L, Sun G, Zhou Z, Sun S, Wang Q, Yan S, Xin Q (2005) Size-controllable synthesis of monodispersed SnO_2 nanoparticles and application in electrocatalysts. J Phys Chem B 109 (18):8774–8778
84. Li K, Luo Y, Yu Z, Deng M, Li D, Meng Q (2009) Low temperature fabrication of efficient porous carbon counter electrode for dye-sensitized solar cells. Electrochem Commun 11 (7):1346–1349
85. Zhou L, Shen F, Tian X, Wang D, Zhang T, Chen W (2013) Stable Cu_2O nanocrystals grown on functionalized graphene sheets and room temperature H_2S gas sensing with ultrahigh sensitivity. Nanoscale 5(4):1564–1569
86. Jang WK, Yun J, Kim HI, Lee YS (2013) Improvement of ammonia sensing properties of polypyrrole by nanocomposite with graphitic materials. Colloid Polym Sci 291(5):1095–1103
87. Gross MA, Sales MJ, Soler MA, Pereira-da-Silva MA, da Silva MF, Paterno LG (2014) Reduced graphene oxide multilayers for gas and liquid phases chemical sensing. RSC Adv 4 (34):17917–17924
88. Shan C, Yang H, Han D, Zhang Q, Ivaska A, Niu L (2010) Graphene/AuNPs/chitosan nanocomposites film for glucose biosensing. Biosens Bioelectron 25(5):1070–1074
89. Li SJ, Xia N, Lv XL, Zhao MM, Yuan BQ, Pang H (2014) A facile one-step electrochemical synthesis of graphene/NiO nanocomposites as efficient electrocatalyst for glucose and methanol. Sens Actuators B Chem 190:809–817
90. Guo S, Wen D, Zhai Y, Dong S, Wang E (2010) Platinum nanoparticle ensemble-on-graphene hybrid nanosheet: one-pot, rapid synthesis, and used as new electrode material for electrochemical sensing. ACS Nano 4(7):3959–3968
91. Jia N, Huang B, Chen L, Tan L, Yao S (2014) A simple non-enzymatic hydrogen peroxide sensor using gold nanoparticles-graphene–chitosan modified electrode. Sens Actuators B Chem 195:165–170
92. Devasenathipathy R, Mani V, Chen SM (2014) Highly selective amperometric sensor for the trace level detection of hydrazine at bismuth nanoparticles decorated graphene nanosheets modified electrode. Talanta 124:43–51
93. Cao S, Zhang L, Chai Y, Yuan R (2013) Electrochemistry of cholesterol biosensor based on a novel Pt–Pd bimetallic nanoparticle decorated graphene catalyst. Talanta 109:167–172
94. Kumar V, Shorie M, Ganguli AK, Sabherwal P (2015) Graphene–CNT nanohybrid aptasensor for label free detection of cardiac biomarker myoglobin. Biosens Bioelectron 72:56–60
95. Su M, Zhang Y, Song X, Ge S, Yan M, Yu J, Huang J (2013) Three-dimensional nanoflower-like MnO_2 functionalized graphene as catalytically promoted nanolabels for ultrasensitive electrochemiluminescence immunoassay. Electrochim Acta 97:333–340

Antimicrobial Properties of Graphene Nanomaterials: Mechanisms and Applications

Adel Soroush, Douglas Rice, Md Saifur Rahaman and François Perreault

Abstract Nanotechnology opens new possibilities for the development of antimicrobial materials. Of particular interest are graphene-based nanomaterials, which possess unique antimicrobial properties and offer multiple routes for functionalization into advanced nanocomposite materials. In this chapter, we review the current state of knowledge regarding the fundamental aspects of the antimicrobial interactions of graphene and graphene-based materials. Then, an overview of the multiple graphene-based composite materials developed for antimicrobial applications is provided, with an analysis of the different chemical functionalization routes used to modify graphene and graphene oxide with biocidal compounds. An analysis of the potential of graphene-based nanomaterials in the development of novel antimicrobial surfaces and coatings is also conducted, with an emphasis on the field of membrane processes, where significant developments have been made. Finally, promising avenues for material development are identified and critical questions surrounding graphene-based nanomaterials are discussed, providing a guide for future development and application of antimicrobial graphene-based materials.

Adel Soroush and Douglas Rice have contributed equally to this work.

A. Soroush · Md S. Rahaman
Department of Building, Civil and Environmental Engineering,
Concordia University, 1455 de Maisonneuve Blvd. West, EV-6.139,
Montreal, QC H3G 1M8, Canada

D. Rice · F. Perreault (✉)
School of Sustainable Engineering and the Built Environment,
Arizona State University, Tempe, AZ 85287-3005, USA
e-mail: francois.perreault@asu.edu

D. Rice · F. Perreault
Nanosystems Engineering Research Center for Nanotechnology-Enabled
Water Treatment, Arizona State University, Tempe, AZ 85287-3005, USA

G. Gonçalves et al. (eds.), *Graphene-based Materials in Health and Environment*,
Carbon Nanostructures, DOI 10.1007/978-3-319-45639-3_10

1 Introduction

While microorganisms are an essential component of our man-made world, performing countless beneficial services to humanity, under many situations a careful control of their proliferation is required. When microorganisms represent a potential source of pathogens, cause product spoilage, or hinder the performance of industrial processes, control strategies must be established to limit the deleterious effects of microbial growth.

Antimicrobial compounds, from silverware to plant extracts, have been used throughout history to control microbial proliferation [1]. Modern antibiotics, first extracted from molds and then obtained from synthetic sources, have risen in the 20th century as an extremely efficient approach to fight bacterial infection [1]. However, their widespread use triggered the emergence of resistant strains that are able to cope with these potent antibiotics [2]. The emergence of these antibiotic-resistant bacteria is due to the strong selective pressure of antibiotics, which act on a few selective sites in the cell. Consequently, new antimicrobial compounds that target multiple sites or exploit different inhibitory pathways [1] are being investigated as a response to these drug-resistant strains.

Outside of clinical settings, control of microbial growth is important in any engineered system exposed to aqueous environments, as the development of surface microbial communities, or biofilms, can lead to impaired system performance. Several of the most preeminent examples of biofilms' deleterious effects include biofouling of ships' hulls, heat-exchangers, water treatment membranes, and metallic surfaces susceptible to corrosion [3–6]. The negative impacts of biofouling make it a costly problem, as it leads to material degradation and increased energy and chemicals consumption. In the desalination industry alone, Azis et al. [7] estimated a global cost of 15 billion USD yearly due solely to biofouling.

In environmental systems, antibiotics are not desirable due to their high cost, narrow range of action, and the risk of spreading drug-resistance. Because of this, less specific biocides such as oxidizers, copper, tributyltin, and pesticides are used to prevent the growth of biomass on surfaces. However, environmental concerns over the persistence of biocidal compounds in the environment have triggered a search for alternative compounds capable of imparting broad-range antimicrobial properties on surfaces prone to fouling.

In both biomedical and environmental settings, nanotechnology provides a path to smarter and more efficient antimicrobial material design. Antimicrobial nanomaterials (NMs), summarized in Table 1, can be very diverse in nature, ranging from metallic nanoparticles (NPs), organic nanostructures, photocatalytic materials, and carbon-based compounds. These new materials represent a shift to a new paradigm of microbial control, where surface-active materials are used to provide localized antimicrobial activity [8]. Each material possesses unique advantages and

Table 1 Main types of antimicrobial nanomaterials and their antimicrobial mechanisms

Nanomaterial	Proposed mechanisms	Comments	Ref.
Ag	–Silver ions release	–No particle-specific effect identified –Direct cell contact increases toxicity by enhancing particle dissolution	[9–11]
CuO	–Copper ions release –Particle-mediated oxidative stress	–Particle-specific effect identified –Shape, size, and polymer coating affect particle-mediated toxicity	[12–14]
Dendrimers	–Membrane disruption by polycations binding to lipopolysaccharides	–Antimicrobial activity influenced by dendrimer type and generation –Also used as a carrier for other antimicrobial compounds	[15–17]
ZnO	–Zinc ions release –Radicals formation	–Light-mediated toxicity –Direct contact increases toxicity	[18, 19]
TiO_2	–Radicals formation	–Light-mediated toxicity –Depends on crystal structure –Direct contact increases toxicity	[20, 21]
C_{60}	–Photocatalytic activity –Oxidative stress	–Oxidative stress is caused by direct oxidation of biomolecules by C_{60}	[22, 23]
Vanadium oxide	–Haloperoxidases-like activity	–Generates oxidizers (HOBr and $^1O_2$) in the presence of Br^- and H_2O_2	[24]
Carbon nanotubes	–Membrane disruption –Oxidative stress	–Requires direct contact for toxicity –Size, surface functionalization, and electronic-structure modulate toxicity	[25, 26]
Graphene	–Membrane disruption –Oxidative stress	–Requires direct contact for toxicity –Size, surface functionalization, and sheet orientation modulate toxicity	[27–30]

specific limitations that guide how they should be used for different applications in biomedical or environmental settings. This chapter will focus on graphene-based nanomaterials (GNMs), providing an overview of the most recent understanding of their antimicrobial properties, and highlights the most promising applications and challenges to overcome in the use of GNMs for antimicrobial applications.

1.1 Interactions of Graphene Nanomaterials with Bacterial Systems

1.1.1 Insights from In Silico and In Vitro Studies

The most fundamental insights into the mechanisms of bacterial inactivation of GNMs are obtained from computational studies using molecular dynamics (MD) simulations. These in silico systems allow for a nanoscale resolution of the interactions that can occur when GNMs come into contact with biological membranes. From computational studies, the mechanisms of membrane penetration and membrane disruption by GNMs were investigated.

Penetration of GNMs into the membrane was found to occur by two mechanisms according to the size and oxidation level. Small sheets with low levels of oxidation insert in the membrane and rotate under the influence of Brownian motion and transbilayer lateral pressure until they reach the center of the bilayer, at which point they align their hydrophobic planes with the phospholipid tails [31, 32]. As the sheet size increases to a dimension greater than the membrane thickness, sheets will be oriented at an angle to maximize the amount of their surface embedded the membrane [33]. This motion triggers the formation of lipid vesicles and facilitate GNMs' internalization [31, 34]. The relationship between the internalization pathway and the physicochemical properties of GNMs is shown in Fig. 1a.

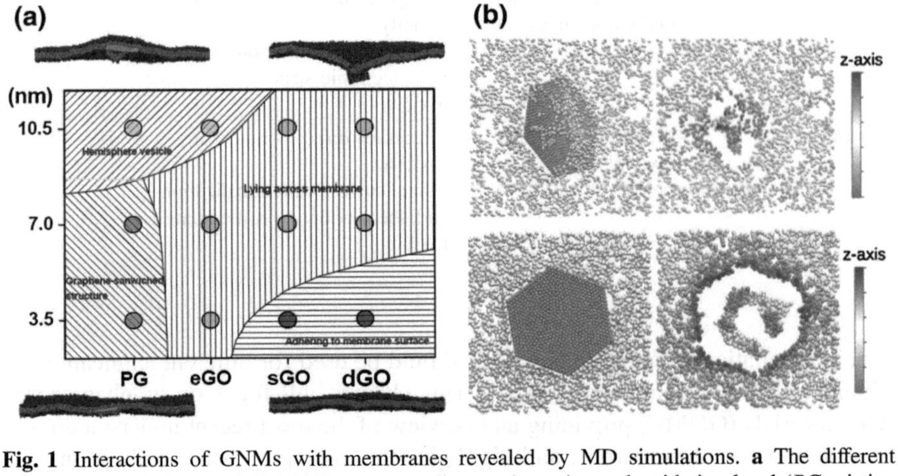

Fig. 1 Interactions of GNMs with membranes revealed by MD simulations. **a** The different interactions of GNMs with lipid bilayers according to sheet size and oxidation level (*PG* pristine graphene, *eGO* edge-oxidized GO, *sGO* sparsely- oxidized GO, *dGO* densely oxidized GO). Adapted from Mao et al. [31] with permission of Elsevier Ltd, © 2014. **b** Disruption of the lipid bilayer integrity when a sheet penetrates the lipid bilayer (*upper panels*) compared to when the sheet adheres on the membrane surface. Adapted from Dallavalle et al. [33] with permission from the American Chemical Society, © 2015

Due to their two-dimensional structure, GNMs may cross biological membranes more readily than other NMs. Indeed, MD studies show that sheet entry is initiated by a piercing of the lipid bilayer by the sharp edges of graphene, which lowers the energy barrier for membrane penetration [33, 35]. Since GNMs always feature sharp edges at the nanoscale, they can be expected to have higher membrane penetration rates than other NMs [35].

Sheets that are unable to enter the lipid bilayer will instead adhere to the surface of the membrane (Fig. 1a). Two levels of surface interaction have been noted from MD studies: a state where sheets are deposited on the membrane without membrane deformation, and a state where GNMS adhere strongly to the membrane [31, 33]. The latter is favored in larger and more oxidized GNMs and leads to significant alteration of the membrane. When sheets are strongly bound to the membrane, phospholipids will move out of their arrangement following the motion of the graphene sheets. As the phospholipid molecule is moved out of its bilayer, the hydrophobic tail become free to reorient and will have a strong affinity to the hydrophobic regions of GNMs. This result in a reversed membrane orientation in the area covered by the sheet (Fig. 1b). Such drastic membrane reorientation can lead to a disruption of the membrane functions and impair the cell metabolism [33].

Membrane disruption by the extraction of phospholipids from the membrane was also observed by Tu et al. [36]. These authors identified three stages of membrane disruption by GNMs: an initial membrane structure (stage I), a weakened membrane stage (stage II), and a final, disrupted membrane where cell integrity is compromised (stage III). From MD simulations, the weakening and disruption of the membrane was attributed to the capacity of GNMs to extract phospholipids from the bilayer onto the hydrophobic planes of the sheet. Phospholipids extraction was triggered by short-range van der Waals forces between the GNMs and lipid molecules [36]. Then hydrophobic forces will allow lipids to accumulate on GNMs, weakening locally the cell membranes. This mechanism was observed to occur in both graphene and GO, due to the unoxidized hydrophobic regions found in GO sheets [36].

Experimental data supporting these simulation findings were obtained with artificial lipid bilayers. When a dioleoylphosphatidylcholine:dipalmitoylphosphatidylcholine (DOPC:DPPC) lipid bilayer was exposed to GO, the interaction between GO and the bilayer promoted an expansion of the lipid structure and a detachment of the lipid layer from its mica substrate [37]. This effect is thought to be governed by electrostatic interactions [38, 39]. Hydration forces were also shown to influence the adhesion kinetics of GO sheets on lipid bilayers [39]. When these forces are overcome, however, irreversible binding and vesicle disruption occurs, demonstrating the potential membrane disrupting properties of GO [39].

1.1.2 Bacterial Inactivation by Graphene Nanomaterials

Model systems are often limited in their capabilities to predict the interactions of GNMs with bacteria due to structural differences between lipid bilayers and actual cell membranes. Bacteria possess a complex cell wall architecture, with lipopolysaccharides and peptidoglycans covering the phospholipid layer [40]. This important difference was previously shown to lead to different types of cellular interactions when the nanoscale interactions of GO were probed by atomic force spectroscopy [41]. Therefore, cellular studies are the most reliable approach to identify which interaction pathway can be associated with antimicrobial activity.

Exposure of bacteria to GNMs leads to several different cellular effects able to generate cell inactivation. These mechanisms, which range from physical disruption of the membrane to chemical oxidative pathways, are described in Fig. 2. The relative importance of these mechanisms in the overall antimicrobial effect has been a subject of debate, with different studies often emphasizing the importance of one mechanism over the other.

A commonly observed effect of bacterial exposure to GNMs is a disruption of the cell integrity. Permeation of the cell membrane and distorted cells have been demonstrated using fluorescent assays, leakage of cellular metabolites, and microscopy analysis [27, 43–45]. This disruption effect was often linked with the properties of GNMs to perturb membrane integrity observed in in vitro or in silico systems [36, 46]. However, membrane disruption can be induced by multiple cellular stresses, and the hypothesis that direct physical alteration of the membrane by GNMs is the main cause for cell disruption remains to be clearly demonstrated.

Recent results have instead emphasized the primary role of oxidative stress in bacterial inactivation by GNMs [27, 47]. For instance, when cells exposed to GO are enriched with α-tocopherol, a lipid-soluble antioxidant used to mitigate cellular oxidative stress [48, 49], cell disruption, and bacterial inactivation were significantly reduced [27]. These results demonstrate that oxidative interactions are a major element of the antimicrobial activity of GO.

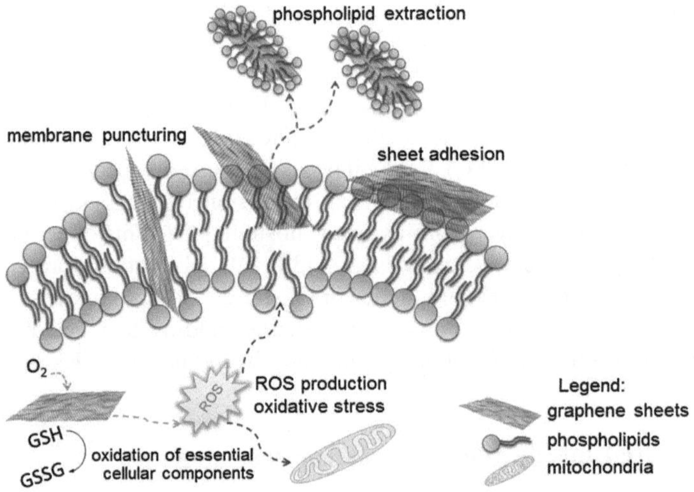

Fig. 2 Overview of the different interactions of GNMs with membranes. Reproduced from Perreault et al. [42] with permission from The Royal Society of Chemistry, © 2015

Despite the evidence supporting the central role of oxidative interactions in bacterial inactivation by GNMs, the mechanisms leading to oxidative stress remain unclear. GNMs have been shown to be able to directly oxidize biomolecules in in vitro conditions [27, 50, 51]. The intrinsic oxidative potential may play a role in the oxidative stress induced by GNMs. However, oxidative stress is a general cellular response; physical stress and alteration of respiratory pathways can also trigger oxidative bursts in bacteria [52]. Due to the complex nature of GNMs, it is most likely that oxidative stress and cell inactivation proceed through multiple pathways.

In addition to these two pathways, which both lead to membrane damage, GNMs can adhere to the cell surface and completely wrap them in an impermeable layer of GNMs [28]. This mechanism is specific to cells exposed to GNMs in suspension and is not observed for GNMs-coated surfaces [27]. Cell wrapping by GNMs does not result in a disruption of the cell integrity, as it is reversible and at least three different groups have reported complete bacterial recovery when the sheets were detached from the cells by mild bath sonication [27, 28, 53].

1.1.3 Physicochemical Determinants of Antimicrobial Activity in Graphene

One of the main paradigms of nanotechnology is that the behavior of NMs is dictated by their physicochemical properties. This statement is also true for antimicrobial NMs, where properties such as size, shape, crystal structure, and surface chemistry were found to be key for their bactericidal properties [54–56]. The determining properties will be dependent on the specific nature of the material. For example, the antimicrobial activity of metal NPs will change according to their solubility [9, 10], while crystal structure and morphology have major impacts on the antimicrobial activity of photocatalytic NMs [20, 57]. While there is limited data on the antimicrobial determinants for GNMs, properties such as size, oxidation level, and sheet orientation were highlighted as determining factors.

Sheet size is an important parameter that modulate the antimicrobial effect of GO. However, how size affects bacterial inactivation was found to be dependent on the exposure conditions. In suspension assays, both Liu et al. and Perreault et al. showed a clear relationship between GO sheet size and the number of viable *Escherichia coli* colonies obtained on agar plates [27, 28]. This increased antimicrobial effect of larger GO sheets was attributed to the previously discussed cell wrapping mechanism, where larger sheets can completely wrap around cells and isolate them from the medium [27, 28, 53]. On the other hand, when bacteria were exposed to a GO-coated surface, cell inactivation increased with *decreasing* sheet size, revealing a completely different mechanism of inactivation. For GO-coated surfaces, the antimicrobial activity was higher for small GO sheets due to the higher concentration of defects and density of edges [27]. However, the exact nature of these reactive sites remains unidentified.

Surface chemistry can influence the antimicrobial activity of NMs [26, 58]. For GNMs, the most significant change in surface functionalization is the oxidation level, which changes the material from the conductive and hydrophobic graphene to its insulating and hydrophilic oxidized counterpart, GO [42, 59]. Multiple studies have compared the antimicrobial potential of GO with graphene or rGO in order to understand how oxidation affects antimicrobial properties. However, there is a lack of agreement between the different studies available. Akhavan et al. [60] compared the toxicity of GO- and rGO-coated stainless steel and found higher antimicrobial activity for rGO, an effect that was attributed to its higher conductivity and stiffness compared to GO. On the other hand, Musico et al. [43] found that a PVK–GO nanocomposite performed better than a PVK–graphene of the same loading. Similar conclusions were obtained in suspension assays, with some reports identifying GO as the more antimicrobial material [29], while for other studies, rGO is the more potent form [61].

This discrepancy highlights the need for additional studies in order to pinpoint the role of the oxidation level on their antimicrobial activity of GNMs. A systematic approach, where a range of partially reduced GO is evaluated, could bring in-depth understanding of the relationship between oxygen content and bacterial inactivation. Such analysis was performed for mammalian cells, where a linear correlation between oxygen content and toxicity was found using 8 GO materials with different oxygen content [62]. The nature of the oxygen functional groups was also found to influence the toxicity of oxidized GNMs to lung epithelial cells, with C=O functionalities providing higher toxicity [63]. Similar in-depth studies of oxygen content and functionalities are still lacking for bacterial toxicity evaluation.

The edges of GNM sheets are very reactive sites compared to the planes [64] and their abundance and availability have been correlated with the antimicrobial properties in both GO and graphene [27, 46]. In a recent study, Pham et al. [46] demonstrated the role of the sheet edges in bacterial inactivation by comparing a smooth highly ordered graphite surface with graphene-based surfaces of different roughness. In these surfaces, increased roughness resulted in a high density of sheet edges exposed for bacterial interactions. Bacterial inactivation was higher for the rougher graphene surface compared to the smoother one, while bacterial viability was virtually not affected for the graphite surface (Fig. 3). However, it should be noted that it is the edge density and not the specific orientation of the sheets that has a high impact on bacterial inactivation. Indeed, when using Langmuir-Blodgett films to expose bacteria to GO sheets on a completely flat surface, high bacterial inactivation was still observed by Mangadlao et al. [30] revealing that orthogonal contact between the sheet and the cell is not needed.

These findings reveal that the antimicrobial activity of GNMs can be tuned by changing their physicochemical properties. Additional research is needed to gather a complete understanding on the relationships between GNMs properties and bacterial inactivation. Beyond size, orientation and oxidation level, there are multiple surface functionalization that can have a strong influence on the antimicrobial effects of GNMs that remains unexplored. Identifying the physicochemical

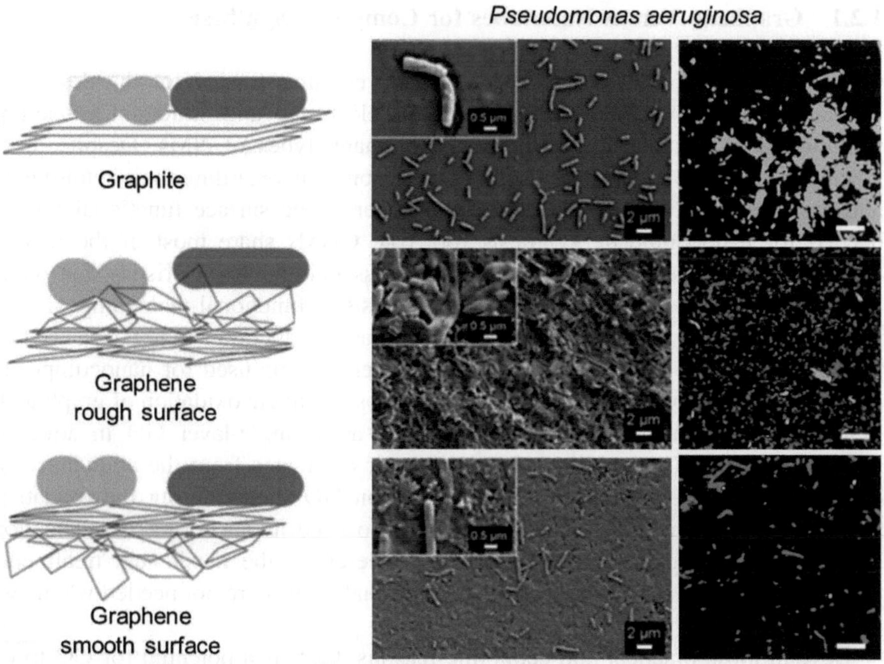

Fig. 3 Effect of sheet edge density on the antimicrobial activity of GNMs. Surfaces of highly ordered graphite are compared with graphene-coated surfaces with either a rough or a smooth surface. Cell integrity is visualized by scanning electron microscopy. Cell viability is quantified using a live/dead fluorescent staining. Adapted with permission from Pham et al. [46] © 2015, American Chemical Society

determinants of the antimicrobial activity of GNMs is critical to ensure optimal antimicrobial performances by graphene-based antimicrobial compounds.

1.2 Graphene-Based Antimicrobial Nanocomposites

Combining different nanostructures can lead to the development of advanced antimicrobial NMs, presenting multifunctional properties, synergistic antimicrobial effects, or offer a more amenable structure to be integrated into material design. For all these objectives, GNMs have shown the potential to be used as a platform for the synthesis of antimicrobial nanocomposites. In this section, the different approaches used to design GNMs-based nanocomposites, along with the different routes for their synthesis, are presented.

1.2.1 Graphene Versus Nanotubes for Composite Synthesis

Carbon nanotubes (CNTs) have been extensively investigated for the design of advanced functional composites [65]. Their electronic and structural properties were found to enhance the performance of many types of NMs. Despite these advantages, the use of CNTs is limited due to concerns regarding their cytotoxicity and the harsh acid treatments required to generate the surface functional groups needed for nanocomposite synthesis [66, 67]. GNMs share most of the unique properties of CNTs; however, they also possess unique characteristics that make them more attractive than CNTs for the synthesis of functional nanocomposites.

From a practical point of view, the ease of mass production and low costs of GO compared to CNTs make it a more viable material to be used for nanocomposite synthesis. GO can be produced through a low-cost chemical oxidation of graphite to graphite oxide, which can then be dispersed as a single-layer GO in aqueous suspensions [42]. The stability of GO in water originates from the abundance of oxygen-containing functional groups formed on GO sheets during the oxidation process. These same functional groups are also used as anchoring or nucleation sites for the synthesis of nanocomposites. Therefore, the harsh acid treatments performed on CNTs to generate oxygen functional groups are not needed when GO is used.

Beyond these practical and economic reasons, there is a potential for GO to be more effective for the synthesis of nanocomposites, as GNMs have a higher surface area and functional groups density than CNTs, which make them better candidates to be used as a support for growing metal NPs [42, 68]. Graphene and GO can be used as substrate, stabilizer, and reducing agent for the development of nanocomposites [69]. They can be used to facilitate the functionalization of materials or to enhance their biocidal properties.

1.2.2 Main Types of Graphene-Based Antimicrobial Nanocomposites

The unique features of GNMs, when combined with NMs, were found to improve their antimicrobial effect. Multiple phenomena are involved in the synergistic interactions of GNMs with other biocides. First, by using GO as a support, aggregation of NPs is prevented; additionally, the nanocomposite exhibits a higher active surface area and increased biocidal activity. Second, for NPs inactivating bacteria by leaching ions, GO also can control the ion release from NPs. In thin film composite (TFC) membranes modified by GO–AgNPs, the presence of GO as a sublayer for AgNPs increases their release durability by increasing the initial silver loading and by providing better interaction between the NPs and the substrate [70]. Third, GO was found to be able to enhance the contact between bacteria and the nanocomposite due to its flexible structure [19]. This direct contact enhances solubilization of metal NPs, enhancing their biocidal potential [11]. In addition, GNMs were hypothesized to damage bacterial membranes and allow a more rapid uptake of biocides [71]. Finally, for NMs inactivating bacteria through photocatalytic or photothermal effects, the electron and thermal conductivity of graphene enhance the activity of the nanocomposite [72]. These synergistic interactions arise as the main appeal for using GNMs in antimicrobial nanocomposites.

Of the many antimicrobial compounds used in conjunction with GNMs, AgNPs are the most common due to their excellent biocidal properties [73]. Silver possesses a broad-spectrum antimicrobial activity, which allows its use in biomedical applications, water and air purification, clothing, and household products [74, 75]. Although the mechanisms of bacterial inactivation by AgNPs have been debated, the most widely recognized is the release of Ag^+ to generate oxidative stress and alter cell membrane function. In this mechanism, a close contact between the cell and AgNPs can enhance their antimicrobial activity by increasing the rate of Ag^+ release from AgNPs [11]. Multiple reports have indicated that the antimicrobial potential of AgNPs can be enhanced by combining them with GNMs [70, 76–78].

Copper is another popular biocide that can be used at a fraction of the cost of silver. Copper NPs, copper oxide NPs, and Cu^{2+} have all demonstrated antimicrobial properties against various bacterial strains [79, 80]. Dissolution of copper-based NPs, as for AgNPs, plays a role in their antimicrobial activity; however, unlike AgNPs, particle-specific effects involving oxidation reactions have also been shown to be an important component of the biological interactions of copper-based NMs [14]. Recently, Ouyang et al. [81] showed that the antimicrobial activity of copper NMs can also be increased by combining them with GNMs.

Photocatalytic antimicrobial NMs are also enhanced in combination with GNMs [82–84]. TiO_2 NPs generate oxidative radicals when illuminated by a light with energy higher than its bandgap energy. When excited, electron–hole pairs diffuse out to the surface of the NPs and the combination of those electrons with water generate hydroxyl radicals. Combining these photocatalytic NMs with GNMs were found to improve the overall photocatalytic performance of the nanocomposite by reducing the rate of electron-hole pair recombination, a phenomenon attributed to the high electron conductivity of the reduced forms of GNMs.

There are other types of metallic NPs that have been used for the decoration of GNMs but that are not as prominent. This group includes, but is not limited to, SnO_2 [85], CdSe [86, 87], MnO_2 [88], and ZnO [19, 72]; when combined with GNMs, each of these compounds showed improved antimicrobial performance. However, their use is mostly limited by the lower costs of the alternative materials. Iron oxide was also used to impart some antimicrobial effects [89]. The main interest in iron oxide is not its antimicrobial properties, which are weak, but rather its magnetic properties, which allow the recycling of NMs after usage [89–92].

GNMs can also be functionalized with organic materials to improve their antibacterial properties. Cai et al. [93] synthesized a brilliant blue/rGO/tetradecyl triphenylphosphonium bromide composite that exhibits high antibacterial activity, water solubility, and mild cytotoxicity. Carpio et al. [94] developed a GO material functionalized with ethylenediamine triacetic acid for both metal removal and bacterial inactivation. Maktedar et al. [95] functionalized GO with 6-aminoindazole through sonochemical nucleophilic substitution. Some et al. [96] synthesized a graphene-poly (L-lysine) (PLL) composites using electrostatic interactions and

Table 2 Summary of the chemical synthesis routes, investigated organisms, and applications for different graphene-based antibacterial nanocomposites

Decorative	Synthesis route	Precursor/reducing agent	Organism	Comments	Ref.
Ag	–Ex situ –Electrostatic	$AgNO_3/NaBH_4$	–*E. coli* –*B. subtilis*	–Synergistic interaction by size control	[98]
Ag	–In situ –Electrostatic	$AgNO_3$/TETA (NaOH)	–*E. coli*	–GO served as sublayer, stabilizer, and reducing agent	[69]
Ag	–In situ	Hydroquinone in citrate buffer	–*E. coli* –*S. aureus*	–Paper-like materials and biocidal coating	[67]
Ag	–In situ	$AgNO_3$/GO	–*C. albicans* –*C. tropical*	–Synergistic interaction by slower release –Antifungal materials	[78]
Ag	–In situ	$AgNO_3$/sodium citrate	–*P. aeruginosa*	–Coating for biofilm control	[99]
Ag	–In situ	$AgNO_3/NaBH_4$ trisodium citrate	–*E. coli* –*P. aeruginosa*	–Antimicrobial activity increased by Ag%	[100]
Ag	–Ex situ –Covalent	$AgNO_3$/gelatin	–*E. coli*	–Synergistic interaction	[117]
Ag	–In situ	Hydrazine monohydrate	–*E. coli* –*S. aureus*	–Synergetic interaction by higher stability and slower release	[71]
Ag	–In situ	$NaBH_4$ trisodium citrate	–*S.aureus* –*B. subtilis*	–Synergetic interaction by better distribution	[105]
Ag	–In situ	$Ag(NH_3)_2^+$/starch	–*P. aeruginosa.*	–Green (microwave), Synergetic interaction	[68]
Ag	–In situ	$AgNO_3$/sodium citrate	–*Colibacillus* –*C. albicans*	–Green	[108]
Ag	–In situ	Ag(hfa)/H_2	–*E. coli* –*L. anguillarum* –*B cereus* –*S. aureus*	–Green (Supercritical CO_2)	[113]

(continued)

Table 2 (continued)

Decorative	Synthesis route	Precursor/reducing agent	Organism	Comments	Ref.
Ag	–In situ	$AgNO_3/NaBH_4$ Sodium citrate	–*E. coli*	–pH affects the stability of the composite –Cell remained intact after contact with GO but not with the composite	[106]
Ag	–In situ –Electrostatic	$AgNO_3$/*colocasia esculenta* leaf	–*E. coli* –*S. aureus* –*C. albicans*	–Green, biocompatible –Biomedical science	[109]
IONP-Ag	–In situ	$AgNO_3$/sodium citrate	–*S. aureus* –*E. coli*	–Recyclable by magnetic separation	[89]
bioAg	–Ex situ –H-bonding	*F. oxysporum*	–*S. typhimurium*	–Green synthesis –The use of protein as capping agent results in small AgNPs	[118]
PDA–Ag	–In situ –Electrostatic	$AgNO_3$/DA	–*E. coli* –*B. subtilis*	–Green, bioinspired materials –Clinical and environmental applications	[107]
PDA–Ag	–In situ –Electrostatic	$AgNO_3$/DA	–*E. coli* –*B. subtilis*	–Green, bioinspired materials	[111]
TTP	–π–π stacking	None	–*E. coli* –*S. aureus*	–Synergetic interaction, targeting capability, mild cytotoxicity	[93]
EDTA	–Silanization –Covalent	None	–*C. metallidurans* *B. subtilis*	–Synergetic interaction –Also used for metal adsorption	[94]
IONPs	–Ex situ –Electrostatic	Mixing ferric acid and ferrous solutions	–*E. coli*	–Synergetic interaction –Recyclable	[92]
Au–TiO_2	–In situ –Electrostatic	Hydrothermal reduction of Au	–*E. coli* –*Rhodopseudanon*	–Biomedical application	[84]

(continued)

Table 2 (continued)

Decorative	Synthesis route	Precursor/reducing agent	Organism	Comments	Ref.
TiO_2	–In situ –Electrostatic	$TiCl_3$	–*E. coli* –*S. aureus* –*C. albicons*	–Electroconductive, self-cleaning, antibacterial textiles	[82]
ZnO	–In situ –Electrostatic	Thermal decomposition of zinc benzoate dihydrazinate	–*E. coli*	–Biosensors, photocatalysts, disinfectants, solar cells	[72]
SnO_2	–Ex situ –Electrostatic	Mixing of GO and SnO_2	–*P. aeruginosa* –*S. aureus*	–Synergetic interaction –Environmental and clinical applications	[85]
CdSe	–Ex situ –Electrostatic	Hydrothermal reduction	–*S. aureus*	–Water purification	[86]
QD	–In situ –Electrostatic	Graphite rods/NaOH/ethanol	–*S. aureus* –*E. coli*	–Synergetic effect in the presence of light	[87]
PLL–Cu	–In situ –Electrostatic	Copper nitrate/hydrazine	–*S. aureus* –*E. coli*	–Synergetic interaction by longer release	[81]
ZnO	–In situ –Electrostatic	Zinc acetate/LiOH	–*E. coli*	–Synergetic interaction –Less toxicity than ZnONPs	[19]
Aminoindazole	–In situ –Covalent	None	–*E. coli* –*S. aureus* –*C. albicans*	–Suitable for antibacterial application	[95]
Fe_3O_4	–In situ –Electrostatic	$FeCl_3 \cdot 6H_2O$ solvothermal	–*E. coli*	–Synergistic interaction –Adsorbent and antibacterial material	[91]
DS-PLL	–Covalent	None	–*E. coli*	–Synergetic interaction –High biocompatibility with human cells	[96]
Lysozyme	–Covalent	lysozyme	–*B. subtilis*	–Controlled release materials	[97]

(continued)

Table 2 (continued)

Decorative	Synthesis route	Precursor/reducing agent	Organism	Comments	Ref.
chlorophenyl	–In situ	Aryl radical	–*S. aureus* –*E. coli*	–Green, synergetic interaction –Food packaging and paper industries	[119]
Ramizol	–Electrostatic	None	–*S. aureus*	–Wound dressing materials –Graphene functions as a drug carrier and to improve mechanical properties	[120]

covalent bonding between a graphene derivative and PLL. Yuan et al. studied the preparation of an antibacterial GO-lysozyme composite. Self-assembly of the sheets endows the film with a multilayered ordered structure with proteins intercalated between the different layers [97]. The summary of different types of GNMs-based antibacterial composites is provided in Table 2.

The different synthesis routes used to achieve functionalization of GNMs can be categorized in two groups: in situ and ex situ formation. In in situ formation, NPs are formed mostly through a nucleation-growth mechanism at the surface of graphene-based materials. On the contrary, ex situ formation involves covalent bonding of bare or functionalized as-prepared NPs and bare or functionalized GNMs. Based on different types of interaction, procedures and formation conditions, the final nanocomposites entail different properties. In the lines that follow, each of in situ and ex situ formation routes will be discussed in detail.

1.3 In Situ Formation of Graphene-Based Nanocomposites

Although the synthesis pathway varies for different types of nanocomposites, the most common approach is through the chemical in situ reduction of metal ions on GNMs. Electrostatic interactions between metal ions and GNMs promote their adhesion to the sheet. The in situ reduction can be done under wet conditions or use thermal reduction pathways.

Wet chemical reduction of metal precursors on GNMs using reducing agents is a common technique for synthesizing nanocomposites [81, 98–100]. GNMs are first exfoliated in a solvent by ultrasonication to form a stable colloidal dispersion. A solution of precursor and a reducing agents are added to initiate the chemical reduction of the precursor salt on GNMs. The most common reducing agents are sodium borohydride and hydrazine monohydrate. Since hydrazine is an extremely strong reducing agent, it can reduce metal precursor to NPs and GO to rGO. The material then becomes decorated with reduced metal NPs. In this process, heterogeneities in the nucleation sites can result in a heterogeneous nucleation of NPs on GO. This was demonstrated by Faria et al. [99], who found that oxidation debris have a crucial role in the nucleation and growth of AgNPs on GO [99]. The schematic picture of the process of GO decoration by AgNPs through in situ wet chemical reduction is presented in Fig. 4.

The main mechanism of NP formation is nucleation and growth [101–103]. On GNMs, this is initiated by the electrostatic adsorption of positive metal ions to the functional groups on GNMs [87]. The oxygenated moieties on GNMs also act as the nucleation and stabilization sites for NPs [104]. Therefore, the oxygen moieties on GNMs have an important role in the heterogeneous nucleation of NPs. GO, which has a higher oxygen content compared to rGO, will form composites with more abundant, smaller, and better distributed NPs [99].

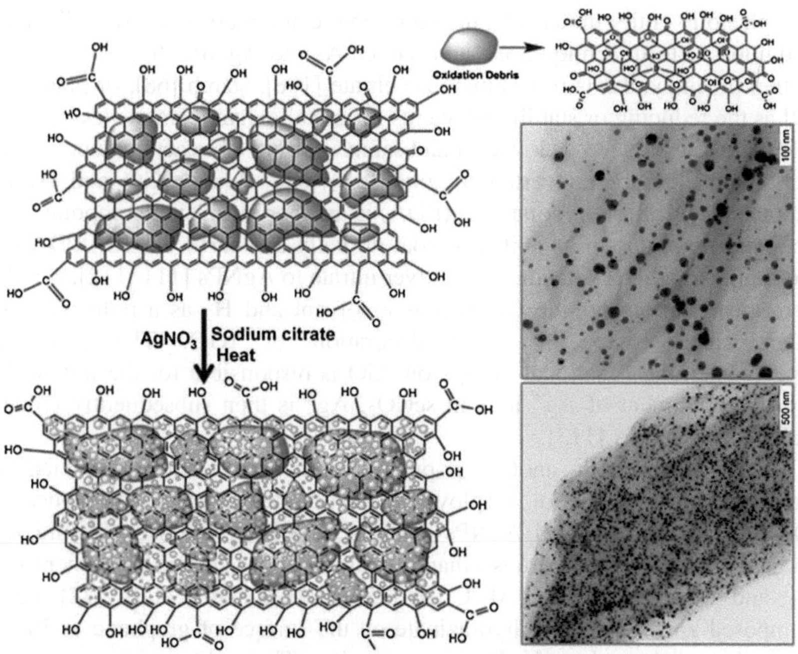

Fig. 4 In situ reduction of AgNPs on GO and the role of oxidation debris in the nucleation process. Adapted from Faria et al. [99] with permission of Elsevier Ltd., © 2014

Morphology of in situ formed NPs also can be guided by controlling the temperature of nucleation, the reaction time, or by using capping agents such as sodium citrate and poly(*N*-vinyl-2-pyrrolidone) [81, 82, 104–107]. The ratio of precursor to GNMs is also used to control the final NP size and density on the surface of GNM. This ratio should be optimized because a low ratio will result in low NPs density, while a high ratio can lead to NPs aggregation and the formation of large NPs, which also decrease the antimicrobial activity [105].

While all these approaches were found to generate nanocomposites with good performance, they also involve potentially hazardous chemicals. Therefore, several attempts have been done to find environment-friendly synthesis route. The easiest way to achieve green NP synthesis is to use GO as the substrate, the reducing agent, and the stabilizer, thus reducing the amount of chemicals needed for synthesis. Vijay et al. developed a one-step method to prepare GO/Ag nanocomposites using GO sheets first modified by reaction with hydrophilic *N*-(trimethoxysilylpropyl) ethylenediamine triacetic acid sodium salt (TETA). The functionalized GO benefits from the TETA's large number of carboxylic groups, along with its own unreacted hydroxyl groups attached to the basal planes. The Ag^+ ion reacts with these oxygen moieties through electrostatic bonding. Then, the unreacted hydroxyl groups, which act as phenolic entity under alkaline condition, can undergo electrophilic aromatic substitution to form semiquinone and reduce Ag^+ to AgNPs [69].

Other green synthesis routes may use citrate [108], starch [68], or plant extracts [109] as the reducing or stabilizing agents for metallic nanocomposites. Starch is of particular interest due to its low cost and safety. During the reaction, a portion of the starch molecules are converted to glucose, which reduces GO and $Ag(NH_3)_2^+$ simultaneously [68]. Polydopamine (PDA) is another promising compound for GO functionalization due to its ability to coat virtually any surface [110]. PDA is used to both functionalize GO and reduce silver nitrate to AgNPs [111, 112]. Employing supercritical carbon dioxide ($scCO_2$) as a solvent and H_2 as a reducing agent is another green route for silver decoration of rGO. The presence of oxygen-containing functional groups on rGO is responsible for the initial electrostatically attachment of Ag^+ ions in $scCO_2$; Ag^+ is then subsequently reduced to metallic silver by H_2 [113].

Thermal reduction is another popular approach to reduce both GNMs and metallic precursors. He et al. employed a solvothermal approach to reduce GO to rGO followed by TiO_2 and AuNPs decoration. This reaction used a mixture of trisodium citrate and anhydrous ethanol solution and a thermal treatment of GO, TiO_2, and Au precursor at 120 °C for 6.5 h [84]. Kavitha et al. [72] thermally decomposed zinc benzoate dihydrazinate on the surface of graphene at 200 °C to form antibacterial graphene/ZnO nanocomposites. Thermal treatment are especially desirable in applications where the reduced form of GO is needed to enhance the thermal and electronic conductivity of the materials, such as in photocatalysis or photothermal materials.

1.3.1 Ex Situ Formation of Graphene-Based Nanocomposites

In situ synthesis often lacks good control over the reaction process, which results in a lack of control over NPs size and distribution. It is also difficult to fabricate metal-decorated GO by reductive deposition, since GO will be reduced as well [114]. In ex situ synthesis, where as-prepared NPs are bonded to GNMs, these limitations can be overcome. The size, shape, and morphology of NPs can be precisely controlled in a first synthesis step, and then NPs are linked to GNMs. Because of the abundance of functional groups available, GO is often more amenable to functionalization than other GNMs.

GO can be functionalized by coupling reactions using its carboxylic acid groups. These reactions require the activation of the acid group using compounds like thionyl chloride ($SOCl_2$) or 1-ethyl-3-(3-dimethylaminopropyl)-carbodiimide (EDC). Subsequent addition of nucleophilic species, such as amines or hydroxyls, produce covalently attached functional groups to GO via the formation of amides or esters. The combination of EDC and *N*-hydroxy succinimide (NHS) is used to form amine-reactive esters, which makes GO ready to react with other types of functional materials [115]. In addition to small molecules, polymers have also been attached to the surface of GO. These attachments are typically made by either grafting-onto or grafting-from approaches. Lee et al. [116] modified the surface of GO and rGO through the reaction of carboxylic groups of GO with cysteamine and EDC to form thiol (–SH) groups on the edges of the sheet. As-prepared AuNPs then were added to the functionalized GO suspension to form a covalent bond with its thiol groups, thus resulting in a precise and controllable bonding of AuNPs on the edges of the functionalized GO sheet [116].

Zhang et al. studied ex situ approach to prepare GO/Ag nanocomposite using gelatin. Due to its oxygen- and nitrogen-rich structures in carboxyl amine groups, the use of gelatin led to a strong electrostatic bond between metal clusters and NPs. When GO is added to a colloid suspension of Ag nanoprisms, the carboxy—on the edges of GO interact with amino—of gelatin to attach Ag nanoprisms [117]. Faria et al. [118] also attached AgNPs, synthesized by the fungi *Fusarium oxysporum*, on GO through electrostatic adsorption. Wu et al. synthesized antibacterial graphene decorated stannous dioxide SnO_2@G through hydrothermal ex situ approach. SnO_2 was produced using a hydrothermal method in which $SnCl_2 \cdot 5H_2O$ was dissolved in water followed by the addition of sulfuric acid. The final mixture was heated at 200 °C for 24 h. As-prepared SnO_2NPs were then added to a GO suspension to be electrostatically adsorbed on GO [85].

1.4 Antimicrobial Surface Design Using Graphene-Based Materials

The antimicrobial potential of the aforementioned GNMs have multiple applications for the control of microbial growth. Due to their contact-mediated effect, the most viable approach to use GNMs for antimicrobial applications is through their application as a surface coating for textiles and polymers to be used for biomedical or environmental applications.

1.5 Antimicrobial Textiles

Textile fibers, when in contact with human skin, provide very rich environments for microbial growth. They accumulate nutrients and humidity, in the form of sweat and dead skin cells, and microorganisms from the body and the environment [121]. In medical settings, this may represent sources of infection and pathogens that can be avoided by imparting antimicrobial properties to textiles. In this regard, GNMs are useful materials for antimicrobial textile design.

Fibrous textile materials can be modified with GNMs by a simple dip-coating approach, where electrostatic interactions capture GNMs in the fibrous structure of the fabric. However, dip-coating do not provide strong binding of GNMs to the fabric and potential leaching of GNMs can be of concern. To alleviate this, GNMs

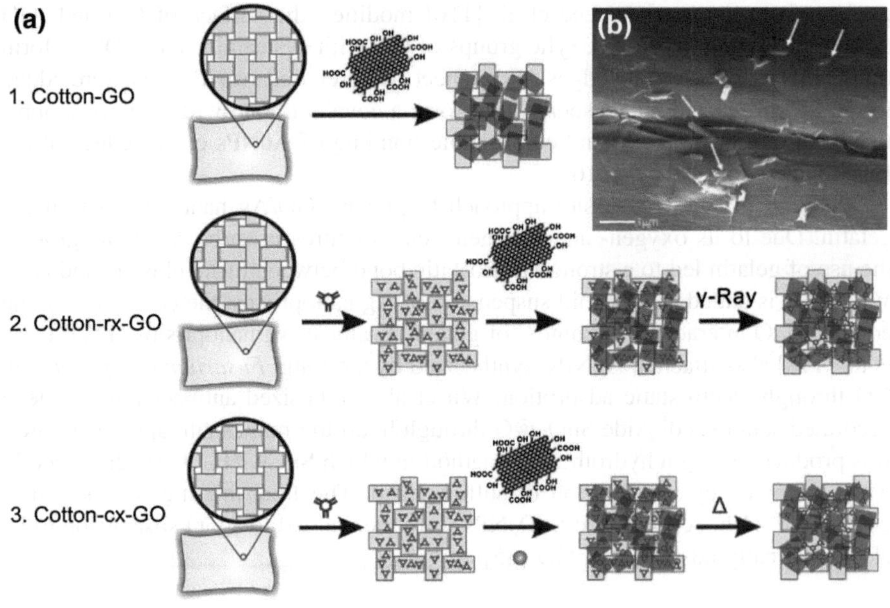

Fig. 5 Pathways for the functionalization of cotton with GNMs. **a** Cotton is functionalized by dip-coating (*1*), γ-ray cross-linking (*2*) or chemical cross-linking (*3*); **b** electron micrograph of bacteria deposited on GO-functionalized cotton. Dead cells are indicated by an *arrow*. Adapted from Zhao et al. [122] with permissions, © 2013, John Wiley and Sons

have been cross-linked to the cotton material by γ-ray irradiation or chemical reaction (Fig. 5). Cross-linking the material was found to reduce the risk of GNMs leaching without reducing their antimicrobial potential [122].

Using graphene-based composite materials, the antimicrobial activity of the textile materials can be further enhanced. Karimi et al. [82] produced a rGO–TiO_2 cotton fabric by first coating the cotton with GO sheets by dip-coating, followed by the addition of titanium trichloride and thermal treatment for the simultaneous reduction of GO to rGO and TiO_2 NPs formation. Loading TiO_2 NPs on rGO significantly increased the antimicrobial properties of the cotton fabric compared to pristine and GO-loaded cotton.

In a biomedical setting, quick bacterial inactivation is needed and graphene-based composite materials with strong antimicrobial potential can offer the most promising results. Combining GNMs with TiO_2, AgNPs, or even conventional antibiotics can offer potential benefits due to the synergistic interactions between the two components of the composite materials. The higher antimicrobial potential of graphene-based composites compared to a single biocide may allow us to reduce the amount of biocide used in a biomedical settings and, by doing so, reduce the risk of bacterial resistance development [8].

1.5.1 Antimicrobial and Antifouling Membranes

Membranes are used to purify water for industrial uses, the treatment of drinking water and wastewater, and for desalination [123–126]. While membranes have proven to be very performant separation processes, over time their efficacy drops due to fouling by organic, inorganic, or microbial foulants. Membrane fouling reduces permeate flux and membrane selectivity [127]. To compensate, operators must increase the applied pressure, wash membranes more frequently, and replace membranes more often. These practices add to the economic and environmental costs of membrane processes and, as such, represent one of the main technical challenges in membrane-based water treatment [6, 128].

Several factors are known to impact the ability of foulants to adsorb to membranes. High surface roughness increases the propensity of the membrane to foul. A rough surface provides a larger surface area for foulants to adhere to and forms stagnant pockets protected from the shear forces of the flowing water [129]. Hydrophilicity is also important, as hydrophilic surfaces have a tight layer of bound water that foulants must displace to adsorb to the surface [130]. Finally, electrostatic interactions also play an important role in adhesion. Membranes with a negative charge are more likely to repel foulants, as many proteins, microbes, and organic material have a negative surface charge in water [129].

The first step to membrane fouling is the adhesion of foulants to the surface. However, in the specific case of biofouling, microorganism will also multiply on the surface and secrete an adhesive matrix full of nutrients, termed extracellular polymeric substances (EPS), to change the surface properties of the membrane and strengthen cell adhesion to the surface. In mature biofilms, EPS, rather than the cells, accounts for the majority of organic carbon (50–90 %) in biofilms [129, 130] and, consequently, will be a major component of the reduced system performance. Preventing microbial activity after cell adhesion can thus have an important role in reducing the deleterious impacts of biofouling.

This interest in GO for fouling control in membranes is due to its two-dimensional structure, hydrophilic nature, and intrinsic antimicrobial activity [42, 131]. When applied to membranes, GO reduces surface roughness, increases hydrophilicity, and lowers the surface zeta potential [132–134]. All of these traits can reduce the rate of membrane fouling. In addition, the biocidal effects of GO can inactivate bacteria depositing on the membrane and reduce the rate of biofilm development (Fig. 6). Biofouling mitigation of up to 98 % was found by Chae et al. [132] when GO sheets were embedded in the polyamide layer of reverse osmosis (RO) membranes. Finally, GO provides membranes with a layer of protection from chlorine, allowing membranes to be chemically cleaned without degrading the membrane.

Several methods have been used to add GNMs to membranes. The mixed matrix technique blends GNMs into the membrane by adding it as an additional ingredient during membrane synthesis. GO can be mixed into polysulfone, polyester, and polyamide, meaning it can be applied to microfiltration, ultrafiltration, nanofiltration, and RO membranes. When considering the factors that measure a membrane's

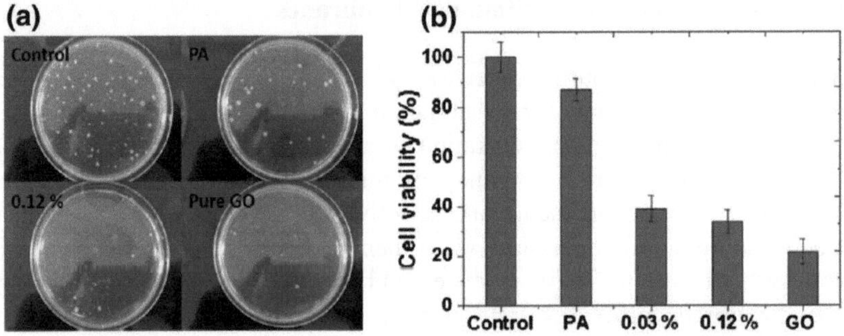

Fig. 6 Viable bacterial count on a RO membrane functionalized with GO sheets using a MMM approach. Control is the bacterial suspension, PA the pristine polyamide TFC membranes, 0.12 % is a GO-functionalized TFC membranes with 0.12 % GO content, and Pure GO a pure GO surface. **a** Pictures of the colony counts on agar media; **b** change in cell viability of different membranes and surfaces exposed to an *E. coli* suspension. Adapted from He et al. [135] with permission of Elsevier Ltd., © 2015

performance (permeability, selectivity, fouling resistance), the optimal GO concentration typically falls between 0.5 and 1.5 % wt% for mixed matrix membranes [136–140], and between 0.01 and 0.1 % for active layer mixed matrix membranes [133, 141]. Beyond these concentrations, the polymer matrix is weakened and the performance of the membranes is compromised. Antimicrobial properties can usually be obtained before any significant decline in membrane performance [132, 133].

While part of the GO stays trapped within the membrane, some of the GO is exposed at the surface. This exposed GO interacts with the fluid and its foulants by increasing the membrane's flux, hydrophilicity, and antifouling properties. Conversely, GO can also be applied in a more targeted approach called layer-by-layer (LbL) [142]. In this method, GO is added solely to the surface of the membrane, resulting in a thin coating of GO. As with the mixed matrix approach, the GO-LbL membranes exhibit enhanced transport properties, hydrophilicity, and resistance to fouling [139]. However, by focusing the application of GO to the membrane surface, LbL concentrates GNMs at the surface of the membrane and, as such, uses less GNMs to obtain similar anti-biofouling properties [132].

One of the determining factors of foulant adhesion is the size of the stagnant layer that separates the bulk water flow from the membrane surface [143]. When foulants enter this domain, they are shielded from the sheer forces of the bulk flow and can attach to the membrane surface. Due to polyamide's roughness, this fouling process is accelerated. Polyamide's rough surface creates pockets of stagnant water protected from the bulk fluid's sheer forces and accelerate fouling. Roughness also increases the membrane's surface area, leading to more locations for foulants to adhere to [129]. When applied to membranes, GO's relatively flat and homogenous form can cover the polyamide's rough surface, leading to higher sheer forces along the membrane and reduced fouling [132, 139].

Table 3 Overview of the different methodologies used to impact antimicrobial and antifouling properties to membranes using GNMs

GNMs	Application	Membrane modification	Main findings	References
GO	Microfiltration	Mixed matrix membrane	–Taguchi experimental design identifies optimal membrane as 12 wt% PVDF, 5 wt% PVP and 3 wt% GO in DMAC –GO reduces flux decline due to microbial fouling	[149]
GO	Ultrafiltration	Mixed matrix membrane	–GO blended in polysulfone membrane –Increased Young modulus, pore size, water flux and hydrophilicity –Lower biofouling propensity	[150]
GO	Ultrafiltration	Mixed matrix membrane	–Improved water flux and due removal –Higher hydrophilicity and reduced surface roughness –Reduced protein fouling –Optimal performance at 0.5 wt%	[151]
GO	Ultrafiltration	Mixed matrix membrane	–GO blended in polyethersulfone membranes –Reduced protein adsorption –Improved hydrophilicity and water flux	[152]
GO	Ultrafiltration	Mixed matrix membrane	–PVDF–GO blends –Increased permeability until a loading of 2 % GO –Reduced protein adhesion due to higher hydrophilicity and smoother surface	[138]
GO GO-APTS	Ultrafiltration	Mixed matrix membrane	–Blend of PVDF and GO or GO-APTS –PVDF/GO-APTS membranes have superior hydrophilicity, water flux, and rejection rate –APTS-GO reduces the adhesion of proteins	[153]
GO–SiO_2	Ultrafiltration	Mixed matrix membrane	–GO–SiO_2 in polysulfone up to 0.8 wt% –Increased permeability up to 0.3 % –Reduced protein adhesion due to higher hydrophilicity –Psf–GO–SiO_2 performs better than Psf–GO or Psf–SiO_2	[154]

(continued)

Table 3 (continued)

GNMs	Application	Membrane modification	Main findings	References
GO-HPEI	Ultrafiltration	Mixed matrix membrane	–GO-HPEI composite blended with polyethersulfone –Improved mechanical properties –Improved hydrophilicity and fouling propensity –Reduced biofilm formation –HPEI increases affinity between GO and PES	[155]
GO–CNTs	Ultrafiltration	Mixed matrix membranes	–Optimal performance at a 5:5 CNT–GO ratio in PVDF –Improved hydrophilicity and higher flux –Reduced protein adhesion –Blending GO and CNTs has a synergistic impact	[156]
GO–lysozyme rGO–lysozyme	Ultrafiltration	Mixed matrix membrane	–GO-Ly MMM is more hydrophilic than pristine or and rGO–Ly membranes –Permeability increases from Pristine < GO–Ly < rGO–Ly –GO-Ly and rGO-Ly impart bacteriostasic properties	[144]
GO–AgNPs	Ultrafiltration	Mixed Matrix Membrane	–GO–AgNPs improve hydrophilicity, flux, and rejection –Optimal improvement at 0.5 % wt. GO–AgNPs –No *E. coli* colony growths on 0.5 % wt GO–AgNPs	[157]
GO	Reverse osmosis	Mixed matric polyamide layer	–GO cross-linked into the polyamide active layer –GO decreases membrane surface energy –GO increases bacterial inactivation by fourfolds	[135]
GO	Reverse osmosis	Mixed matrix polyamide layer	–GO cross-linked into the polyamide active layer –Increasing GO level improves hydrophilicity –Increasing GO level decreased biofouling –Increased chlorine resistance	[158]

(continued)

Table 3 (continued)

GNMs	Application	Membrane modification	Main findings	References
Graphene GO	Microfiltration	Surface adsorption	–Cellulose nitrate filters modified with a composite of poly(*N*-vinylcarbazole) and GO or graphene –Increased antimicrobial activity and bacteria removal	[43]
Crumpled GO with Ag or TiO_2 NPs	Ultrafiltration	Surface deposition and cross-linking	–Crumpled GO resists compression –Crumpled GO–TiO_2 provides photocatalytic activity –Crumpled GO–TiO_2 have three logs higher antimicrobial effect than crumpled GO alone	[147]
GO	Reverse osmosis	Surface functionalization by amide coupling	–EDC and NHS mediated covalent binding of GO to the polyamide layer –Antimicrobial properties and increased hydrophilicity –No effect on membrane transport properties	[44]
GO	Forward osmosis	Surface functionalization by amide coupling	–EDC and NHS mediated covalent binding of GO to the polyamide layer –Antimicrobial and anti-adhesive properties –Reduced biofilm formation by GO functionalization –No effect on membrane transport properties	[134]
GO–Ag	Forward osmosis	Surface functionalization by cysteamine click chemistry	–AgNPs made in situ via wet chemical reduction on GO –GO–Ag covalently bonded to TFC membranes –GO–Ag improves membrane hydrophilicity –GO–Ag imparts high antimicrobial activity	[76]
GO–Ag	Forward osmosis	Surface functionalization by amide coupling	–AgNPs made in situ via wet chemical reduction on GO –GO–Ag covalently bonded to TFC membranes –Improved membrane hydrophilicity and roughness –Higher antimicrobial effect of GO–Ag compared to AgNPs alone	[70]

(continued)

Table 3 (continued)

GNMs	Application	Membrane modification	Main findings	References
GO	Forward osmosis	Surface functionalization by layer-by-layer or amide coupling	–Surface functionalization of polyamide TFC membranes –GO–TFC by amide coupling binding improved antimicrobial activity and antifouling properties	[159]
GO aminated GO	Reverse osmosis	Layer-by-layer	–Electrostatic assembly of GO/aminated GO on polyamide TFC membranes –Go coating increased protein fouling resistance –Go coating increased chlorine resistance. –NO effect on the membrane transport properties	[142]

Moreover, GO can be used to efficiently and effectively load membranes with other antifouling materials, including biocidal metals and enzymes. Compared to pristine membranes, GO surfaces encourage a higher and more dispersed loading of metallic antimicrobial nanocomposites [70, 144], both of which contribute to more sustained, comprehensive antifouling effect from the metals. GO better retains AgNPs and CuNPs than polyamide does, thereby increasing their biocidal benefits [70, 145]. Additionally, when membranes are functionalized with GO that has been loaded with additional biocidal particles, the membrane benefits from both compounds, leading to enhanced antimicrobial properties. Finally, it is worth noting that the flux and solute rejection rates of GO membranes and NPs-loaded GO membranes are quite similar, meaning the additional nanoparticle loading does not significantly impact membrane performance [76, 137].

TiO_2 is one particularly promising NPs that can be loading onto membranes with GO. When combined with UV light, TiO_2 is capable of destroying foulants by oxidizing them, which converts organic foulants to harmless byproducts [146]. This results in less fouling on the TiO_2-loaded membrane. The interaction of TiO_2 with GNMs, by increasing the electrons' excitability, increases the efficiency of photocatalytic efficiency of TiO_2, leading to more oxidation reactions [146, 147]. GO–TiO_2 membranes are found to have higher flux and solute rejection rates than pristine membranes [148], while decreasing fouling rates and the total number of organic contaminants in the system [146, 147].

Due to the hydrophilicity and antimicrobial effects of GNMs, they have been investigated extensively for membrane applications. Their capacity to decrease fouling without compromising the transport properties makes them particularly attractive for membrane development. Below is an overview of the different approaches used to integrate GNMs into membrane design and their impact on the membrane performance and fouling resistance (Table 3).

1.6 Conclusion and Outlook

In the design of antimicrobial materials and surfaces, GNMs offer multiple advantages that can result in improved performances as well as reduced costs and environmental impacts. This is made possible by the unique properties of GNMs, which include high degree of functionalization, excellent thermal and electron conductivity, and hydrophilicity.

By serving as a platform for the design of advanced nanocomposites with antimicrobial properties, GNMs can enhance the performance of conventional biocides, which allows for reduced biocide loadings or improved inactivation rates. These synergistic properties of GNMs make them very interesting for biomedical applications, where drug-resistance and pathogen survival are major concerns.

The ease of functionalization of GNMs also enables their integration into a wide variety of materials for the design of antimicrobial surfaces. In that regard, GNMs most commonly serve as scaffolds to obtain a more homogeneous coverage, higher surface loading, and overall better antimicrobial performances. In the field of membrane-based water treatment, the hydrophilicity of GO-based materials also enables their application on membranes without compromising the membrane permeability. In addition to antimicrobial properties, this hydrophilic character was also found to reduce the fouling propensity of membranes.

In the synthesis of these advanced nanocomposites with antimicrobial properties, GNMs have also allowed for greener synthesis pathways, more complex materials with multiple functionalities, and enhanced antimicrobial effect. This area of research may represent promising field for antimicrobial GNMs, as they offer antimicrobial performances higher than both pure GNMs and conventional biocides. More complex and efficient materials are continuously being made. However, detailed life-cycle analysis still remains to be done in order to evaluate if this improvement in performances is worth the additional synthesis complexity.

However, for an efficient application of GNMs in antimicrobial applications, a better understanding of the interactions of GNMs with bacterial systems is needed. Despite the recent progress in this area, there are still multiple unanswered questions on what are the main mechanisms of bacterial inactivation, what properties govern these interactions, and how the specific properties of the cellular system may affect these interactions. Understanding these relationships is essential when designing novel antimicrobial composites that will benefit for the synergy between GNMs and other types of biocides. Through this fundamental understanding of the mechanisms of bacterial interactions of GNMs, optimal antimicrobial performance may then be achieved, facilitating their eventual transition to commercial applications.

Acknowledgments F.P. acknowledges the financial support from the NSF Nanosystems Engineering Research Center for Nanotechnology-Enabled Water Treatment (ERC-1449500) and the Ira A. Fulton Schools of Engineering at Arizona State University. D.Z.R. acknowledges the support from the Arizona State University Fulton Schools of Engineering Deans Fellowship program. M.S.R. thanks the Natural Sciences and Engineering Research Council (NSERC) of Canada for providing financial support for this project. A.S. acknowledges the support of a graduate entrance scholarship from Concordia University.

References

1. Aminov RI (2010) A brief history of the antibiotic era: lessons learned and challenges for the future. Front Microbiol 1:1–7. doi:10.3389/fmicb.2010.00134
2. Cohen ML (2000) Changing patterns of infectious disease. Nature 406:762–767. doi:10.1038/35021206
3. Melo LF, Bott TR (1997) Biofouling in water systems. Exp Therm Fluid Sci 14:375–381. doi:10.1016/S0894-1777(96)00139-2
4. Schultz MP, Bendick JA, Holm ER, Hertel WM (2011) Economic impact of biofouling on a naval surface ship. Biofouling 27:87–98. doi:10.1080/08927014.2010.542809
5. Little BJ, Lee JS, Ray RI (2008) The influence of marine biofilms on corrosion: a concise review. Electrochim Acta 54:2–7. doi:10.1016/j.electacta.2008.02.071
6. Flemming HC, Schaule G, Griebe T et al (1997) Biofouling—the Achilles heel of membrane processes. Desalination 113:215–225. doi:10.1016/S0011-9164(97)00132-X
7. Azis PKA, Ai-tisan I, Sasikumar N (2001) Biofouling potential and environmental factors of seawater at a desalination plant intake. Water 135:69–82. doi:10.1016/S0011-9164(01)00140-0
8. Huh AJ, Kwon YJ (2011) "Nanoantibiotics": a new paradigm for treating infectious diseases using nanomaterials in the antibiotics resistant era. J Control Release 156:128–145. doi:10.1016/j.jconrel.2011.07.002
9. Xiu Z, Zhang Q, Puppala HL et al (2012) Negligible particle-specific antibacterial activity of silver nanoparticles. Nano Lett 12:4271–4275. doi:10.1021/nl301934w
10. Liu J, Sonshine DA, Shervani S, Hurt RH (2010) Controlled release of biologically active silver from nanosilver surfaces. ACS Nano 4:6903–6913. doi:10.1021/nn102272n
11. Bondarenko O, Ivask A, Käkinen A et al (2013) Particle-cell contact enhances antibacterial activity of silver nanoparticles. PLoS One 8:e64060. doi:10.1371/journal.pone.0064060
12. Bondarenko O, Ivask A, Käkinen A, Kahru A (2012) Sub-toxic effects of CuO nanoparticles on bacteria: kinetics, role of Cu ions and possible mechanisms of action. Environ Pollut 169:81–89. doi:10.1016/j.envpol.2012.05.009
13. Perreault F, Oukarroum A, Melegari SP et al (2012) Polymer coating of copper oxide nanoparticles increases nanoparticles uptake and toxicity in the green alga *Chlamydomonas reinhardtii*. Chemosphere 87:1388–1394. doi:10.1016/j.chemosphere.2012.02.046
14. Gilbertson LM, Albalghiti EM, Fishman ZS et al (2016) Shape-dependent surface reactivity and antimicrobial activity of nano-cupric oxide. Environ Sci Technol 50:3975–3984. doi:10.1021/acs.est.5b05734
15. Chen CZ, Cooper SL (2000) Recent advances in antimicrobial dendrimers. Adv Mater 12:843–846. doi:10.1002/(SICI)1521-4095(200006)12:11<843::AID-ADMA843>3.0.CO;2-Tw
16. Wang B, Navath RS, Menjoge AR et al (2010) Inhibition of bacterial growth and intramniotic infection in a guinea pig model of chorioamnionitis using PAMAM dendrimers. Int J Pharm 395:298–308. doi:10.1016/j.ijpharm.2010.05.030
17. Perreault F, Melegari SP, Fuzinatto CF et al (2014) Toxicity of pamam-coated gold nanoparticles in different unicellular models. Environ Toxicol 29:328–336. doi:10.1002/tox.21761

18. Tong T, Wilke CM, Wu J et al (2015) Combined toxicity of nano-zno and nano-TiO_2: From single- to multinanomaterial systems. Environ Sci Technol 49:8113–8123. doi:10.1021/acs.est.5b02148
19. Wang Y-W, Cao A, Jiang Y et al (2014) Superior antibacterial activity of zinc oxide/graphene oxide composites originating from high zinc concentration localized around bacteria. ACS Appl Mater Interfaces 6:2791–2798. doi:10.1021/am4053317
20. Tong T, Shereef A, Wu J et al (2013) Effects of material morphology on the phototoxicity of nano-TiO_2 to bacteria. Environ Sci Technol 47:12487–12495. doi:10.1021/es403079h
21. Kubacka A, Diez MS, Rojo D et al (2014) Understanding the antimicrobial mechanism of TiO_2-based nanocomposite films in a pathogenic bacterium. Sci Rep 4:4134. doi:10.1038/srep04134
22. Lyon DY, Alvarez PJJ (2008) Fullerene water suspension (nC60) exerts antibacterial effects via ROS-independent protein oxidation. Environ Sci Technol 42:8127–8132. doi:10.1021/es801869m
23. Lyon DY, Brunet L, Hinkal GW et al (2008) Antibacterial activity of fullerene water suspensions (nC 60) is not due to ROS-mediated damage. Nano Lett 8:1539–1543. doi:10.1021/nl0726398
24. Natalio F, André R, Hartog AF et al (2012) Vanadium pentoxide nanoparticles mimic vanadium haloperoxidases and thwart biofilm formation. Nat Nanotechnol 7:530–535. doi:10.1038/nnano.2012.91
25. Kang S, Herzberg M, Rodrigues DF, Elimelech M (2008) Antibacterial effects of carbon nanotubes: size does matter! Langmuir 24:6409–6413. doi:10.1021/la800951v
26. Pasquini LM, Sekol RC, Taylor AD et al (2013) Realizing comparable oxidative and cytotoxic potential of single- and multiwalled carbon nanotubes through annealing. Environ Sci Technol 47:8775–8783. doi:10.1021/es401786s
27. Perreault F, de Faria AF, Nejati S, Elimelech M (2015) Antimicrobial properties of graphene oxide nanosheets: why size matters. ACS Nano 9:7226–7236. doi:10.1021/ascnano.5b02067
28. Liu S, Hu M, Zeng TH et al (2012) Lateral dimension-dependent antibacterial activity of graphene oxide sheets. Langmuir 28:12364–12372. doi:10.1021/la3023908
29. Liu S, Zeng TH, Hofmann M et al (2011) Antibacterial activity of graphite, graphite oxide, graphene oxide, and reduced graphene oxide: membrane and oxidative stress. ACS Nano 5:6971–6980. doi:10.1021/nn202451x
30. Mangadlao JD, Santos CM, Felipe MJL et al (2015) On the antibacterial mechanism of graphene oxide (GO) Langmuir–Blodgett films. Chem Commun 1:1–4. doi:10.1039/C4CC07836E
31. Mao J, Guo R, Yan L-T (2014) Simulation and analysis of cellular internalization pathways and membrane perturbation for graphene nanosheets. Biomaterials 35:6069–6077. doi:10.1016/j.biomaterials.2014.03.087
32. Wang J, Wei Y, Shi X, Gao H (2013) Cellular entry of graphene nanosheets: the role of thickness, oxidation and surface adsorption. RSC Adv 3:15776–15782. doi:10.1039/c3ra40392k
33. Dallavalle M, Calvaresi M, Bottoni A et al (2015) Graphene can wreak havoc with cell membranes. ACS Appl Mater Interfaces 7:4406–4414. doi:10.1021/am508938u
34. Guo R, Mao J, Yan L-T (2013) Computer simulation of cell entry of graphene nanosheet. Biomaterials 34:4296–4301. doi:10.1016/j.biomaterials.2013.02.047
35. Li Y, Yuan H, von dem Bussche A et al (2013) Graphene microsheets enter cells through spontaneous membrane penetration at edge asperities and corner sites. Proc Natl Acad Sci USA 110:12295–12300. doi:10.1073/pnas.1222276110
36. Tu Y, Lv M, Xiu P et al (2013) Destructive extraction of phospholipids from Escherichia coli membranes by graphene nanosheets. Nat Nanotechnol 8:594–601. doi:10.1038/nnano.2013.125
37. Lei H, Zhou X, Wu H et al (2014) Morphology change and detachment of lipid bilayers from the mica substrate driven by graphene oxide sheets. Langmuir 30:4678–4683. doi:10.1021/la500788z

38. Frost R, Jönsson GE, Chakarov D et al (2012) Graphene oxide and lipid membranes: interactions and nanocomposite structures. Nano Lett 12:3356–3362. doi:10.1021/nl203107k
39. Liu X, Chen KL (2015) Interactions of graphene oxide with model cell membranes: probing nanoparticle attachment and lipid bilayer disruption. Langmuir 31:12076–12086. doi:10.1021/acs.langmuir.5b02414
40. Chen KL, Bothun GD (2014) Nanoparticles meet cell membranes: probing nonspecific interactions using model membranes. Environ Sci Technol 48:873–880. doi:10.1021/es403864v
41. Castrillón SR-V, Perreault F, de Faria AF, Elimelech M (2015) Interaction of graphene oxide with bacterial cell membranes: insights from force spectroscopy. Environ Sci Technol Lett 2:112–117. doi:10.1021/acs.estlett.5b00066
42. Perreault F, Faria AF, Elimelech M et al (2015) Environmental applications of graphene-based nanomaterials. Chem Soc Rev 44:5861–5896. doi:10.1039/C5CS00021A
43. Musico YLF, Santos CM, Dalida MLP, Rodrigues DF (2014) Surface modification of membrane filters using graphene and graphene oxide-based nanomaterials for bacterial inactivation and removal. ACS Sustain Chem Eng 2:1559–1565. doi:10.1021/sc500044p
44. Perreault F, Tousley ME, Elimelech M (2014) Thin-film composite polyamide membranes functionalized with biocidal graphene oxide nanosheets. Environ Sci Technol Lett 71–76. doi:10.1021/ez4001356
45. Chen J, Peng H, Wang X et al (2014) Graphene oxide exhibits broad-spectrum antimicrobial activity against bacterial phytopathogens and fungal conidia by intertwining and membrane perturbation. Nanoscale 6:1879–1889. doi:10.1039/c3nr04941h
46. Pham VTH, Truong VK, Quinn MDJ et al (2015) Graphene induces formation of pores that kill spherical and rod-shaped bacteria. ACS Nano 9:8458–8467. doi:10.1021/acsnano.5b03368
47. Gurunathan S, Han JW, Dayem AA et al (2012) Oxidative stress-mediated antibacterial activity of graphene oxide and reduced graphene oxide in *Pseudomonas aeruginosa*. Int J Nanomed 7:5901–5914. doi:10.2147/IJN.S37397
48. Fuentes AM, Amábile-Cuevas CF (1998) Antioxidant vitamins C and E affect the superoxide-mediated induction of the soxRS regulon of *Escherichia coli*. Microbiology 144:1731–1736. doi:10.1099/00221287-144-7-1731
49. Melegari SP, Perreault F, Moukha S et al (2012) Induction to oxidative stress by saxitoxin investigated through lipid peroxidation in neuro 2A cells and *Chlamydomonas reinhardtii* alga. Chemosphere 89:38–43. doi:10.1016/j.chemosphere.2012.04.009
50. Krishnamoorthy K, Veerapandian M, Zhang L, Yun K (2012) Antibacterial efficiency of graphene nanosheets against pathogenic bacteria via lipid peroxidation. J Phys Chem C 116:17280−17287. doi:10.1021/jp3047054
51. Liu X, Sen S, Liu J et al (2011) Antioxidant deactivation on graphenic nanocarbon surfaces. Small 7:2775–2785. doi:10.1002/smll.201100651
52. Yahraus T, Chandra S, Legendre L, Low PS (1995) Evidence for a mechanically induced oxidative burst. Plant Physiol 109:1259–1266. doi:10.1104/pp.109.4.1259
53. Akhavan O, Ghaderi E, Esfandiar A (2011) Wrapping bacteria by graphene nanosheets for isolation from environment, reactivation by sonication, and inactivation by near-infrared irradiation. J Phys Chem B 115:6279–6288. doi:10.1021/jp200686k
54. Johnston HJ, Hutchison GR, Christensen FM et al (2009) The biological mechanisms and physicochemical characteristics responsible for driving fullerene toxicity. Toxicol Sci 114:162–182. doi:10.1093/toxsci/kfp265
55. Nel AE, Mädler L, Velegol D et al (2009) Understanding biophysicochemical interactions at the nano-bio interface. Nat Mater 8:543–557. doi:10.1038/nmat2442
56. Krug HF, Wick P (2011) Nanotoxicology: an interdisciplinary challenge. Angew Chem Int Ed Engl 50:1260–1278. doi:10.1002/anie.201001037

57. Hurum DC, Agrios AF, Gray KA et al (2003) Explaining the enhanced photocatalytic activity of Degussa P25 mixed-phase TiO_2 using EPR. J Phys Chem B 107:4545–4549. doi:10.1021/Jp0273934
58. Pasquini LM, Hashmi SM, Sommer TJ et al (2012) Impact of surface functionalization on bacterial cytotoxicity of single-walled carbon nanotubes. Environ Sci Technol 46:6297–6305. doi:10.1021/es300514s
59. Dreyer DR, Park S, Bielawski CW, Ruoff RS (2010) The chemistry of graphene oxide. Chem Soc Rev 39:228–240. doi:10.1039/b917103g
60. Akhavan O, Ghaderi E (2010) Toxicity of graphene and graphene oxide nanowalls against bacteria. ACS Nano 4:5731–5736. doi:10.1021/nn101390x
61. Gurunathan S, Han JW, Dayem AA et al (2013) Antibacterial activity of dithiothreitol reduced graphene oxide. J Ind Eng Chem 19:1280–1288. doi:10.1016/j.jiec.2012.12.029
62. Das S, Singh S, Singh V et al (2013) Oxygenated functional group density on graphene oxide: its effect on cell toxicity. Part Part Syst Charact 30:148–157. doi:10.1002/ppsc.201200066
63. Chng ELK, Pumera M (2013) The toxicity of graphene oxides: dependence on the oxidative methods used. Chemistry 19:8227–8235. doi:10.1002/chem.201300824
64. Sharma R, Baik JH, Perera CJ, Strano MS (2010) Anomalously large reactivity of single graphene layers and edges toward electron transfer chemistries. Nano Lett 10:398–405. doi:10.1021/nl902741x
65. Upadhyayula VKK, Gadhamshetty V (2010) Appreciating the role of carbon nanotube composites in preventing biofouling and promoting biofilms on material surfaces in environmental engineering: a review. Biotechnol Adv 28:802–816. doi:10.1016/j.biotechadv.2010.06.006
66. Lin Y, Watson KA, Fallbach MJ et al (2009) Rapid, solventless, bulk preparation of metal nanoparticle-decorated carbon nanotubes. ACS Nano 3:871–884. doi:10.1021/nn8009097
67. Bao Q, Zhang D, Qi P (2011) Synthesis and characterization of silver nanoparticle and graphene oxide nanosheet composites as a bactericidal agent for water disinfection. J Colloid Interface Sci 360:463–470. doi:10.1016/j.jcis.2011.05.009
68. Han Y, Luo Z, Yuwen L et al (2013) Synthesis of silver nanoparticles on reduced graphene oxide under microwave irradiation with starch as an ideal reductant and stabilizer. Appl Surf Sci 266:188–193. doi:10.1016/j.apsusc.2012.11.132
69. Vijay Kumar S, Huang NM, Lim HN et al (2013) One-step size-controlled synthesis of functional graphene oxide/silver nanocomposites at room temperature. Chem Eng J 219:217–224. doi:10.1016/j.cej.2012.09.063
70. Soroush A, Ma W, Cyr M et al (2016) In situ silver decoration on graphene oxide-treated thin film composite forward osmosis membranes: biocidal properties and regeneration potential. Environ Sci Technol Lett 3:13–18. doi:10.1021/acs.estlett.5b00304
71. Cai X, Lin M, Tan S et al (2012) The use of polyethyleneimine-modified reduced graphene oxide as a substrate for silver nanoparticles to produce a material with lower cytotoxicity and long-term antibacterial activity. Carbon N Y 50:3407–3415. doi:10.1016/j.carbon.2012.02.002
72. Kavitha T, Gopalan AI, Lee K-P, Park S-Y (2012) Glucose sensing, photocatalytic and antibacterial properties of graphene–ZnO nanoparticle hybrids. Carbon N Y 50:2994–3000. doi:10.1016/j.carbon.2012.02.082
73. He T, Liu H, Zhou Y et al (2014) Antibacterial effect and proteomic analysis of graphene-based silver nanoparticles on a pathogenic bacterium *Pseudomonas aeruginosa*. Biometals 27:673–682. doi:10.1007/s10534-014-9756-1
74. Sharma VK, Yngard RA, Lin Y (2009) Silver nanoparticles: green synthesis and their antimicrobial activities. Adv Colloid Interface Sci 145:83–96. doi:10.1016/j.cis.2008.09.002
75. Marambio-Jones C, Hoek EMV (2010) A review of the antibacterial effects of silver nanomaterials and potential implications for human health and the environment. J Nanopart Res 12:1531–1551. doi:10.1007/s11051-010-9900-y

76. Soroush A, Ma W, Silvino Y, Rahaman MS (2015) Surface modification of thin film composite forward osmosis membrane by silver-decorated graphene-oxide nanosheets. Environ Sci Nano 2:395–405. doi:10.1039/C5EN00086F
77. Faria AF, Perreault F, Shaulsky E et al (2015) Antimicrobial electrospun biopolymer nanofiber mats functionalized with graphene oxide-silver nanocomposites. ACS Appl Mater Interfaces 7:12751–12759. doi:10.1021/acsami.5b01639
78. Li C, Wang X, Chen F et al (2013) The antifungal activity of graphene oxide-silver nanocomposites. Biomaterials 34:3882–3890. doi:10.1016/j.biomaterials.2013.02.001
79. Ahamed M, Alhadlaq HA, Khan MAM et al (2014) Synthesis, characterization, and antimicrobial activity of copper oxide nanoparticles. J Nanomater 2014:1–4. doi:10.1155/2014/637858
80. Ben-Sasson M, Zodrow KR, Genggeng Q et al (2014) Surface functionalization of thin-film composite membranes with copper nanoparticles for antimicrobial surface properties. Environ Sci Technol 48:384–393. doi:10.1021/es404232s
81. Ouyang Y, Cai X, Shi Q et al (2013) Poly-L-lysine-modified reduced graphene oxide stabilizes the copper nanoparticles with higher water-solubility and long-term additively antibacterial activity. Colloids Surf B Biointerfaces 107:107–114. doi:10.1016/j.colsurfb.2013.01.073
82. Karimi L, Yazdanshenas ME, Khajavi R et al (2014) Using graphene/TiO_2 nanocomposite as a new route for preparation of electroconductive, self-cleaning, antibacterial and antifungal cotton fabric without toxicity. Cellulose 21:3813–3827. doi:10.1007/s10570-014-0385-1
83. Xu C, Cui A, Xu Y, Fu X (2013) Graphene oxide–TiO_2 composite filtration membranes and their potential application for water purification. Carbon N Y 62:465–471. doi:10.1016/j.carbon.2013.06.035
84. He W, Huang H, Yan J, Zhu J (2013) Photocatalytic and antibacterial properties of Au–TiO_2 nanocomposite on monolayer graphene: from experiment to theory. J Appl Phys 114:204701. doi:10.1063/1.4836875
85. Wu B-S, Abdelhamid HN, Wu H-F (2014) Synthesis and antibacterial activities of graphene decorated with stannous dioxide. RSC Adv 4:3722. doi:10.1039/c3ra43992e
86. Ullah K, Kim Y-H, Lee B-E et al (2014) Visible light induced catalytic properties of CdSe–graphene nanocomposites and study of its bactericidal effect. Chin Chem Lett 25:941–946. doi:10.1016/j.cclet.2014.03.050
87. Ristic BZ, Milenkovic MM, Dakic IR et al (2014) Photodynamic antibacterial effect of graphene quantum dots. Biomaterials 35:4428–4435. doi:10.1016/j.biomaterials.2014.02.014
88. Sreeprasad TS, Maliyekkal SM, Lisha KP, Pradeep T (2011) Reduced graphene oxide-metal/metal oxide composites: facile synthesis and application in water purification. J Hazard Mater 186:921–931. doi:10.1016/j.jhazmat.2010.11.100
89. Tian T, Shi X, Cheng L et al (2014) Graphene-based nanocomposite as an effective, multifunctional, and recyclable antibacterial agent. ACS Appl Mater Interfaces 6:8542–8548. doi:10.1021/am5022914
90. Wu M-C, Deokar AR, Liao J-H et al (2013) Graphene-based photothermal agent for rapid and effective killing of bacteria. ACS Nano 7:1281–1290. doi:10.1021/nn304782d
91. Santhosh C, Kollu P, Doshi S et al (2014) Adsorption, photodegradation and antibacterial study of graphene–Fe_3O_4 nanocomposite for multipurpose water purification application. RSC Adv 4:28300. doi:10.1039/c4ra02913e
92. Deng C-H, Gong J-L, Zeng G-M et al (2014) Inactivation performance and mechanism of *Escherichia coli* in aqueous system exposed to iron oxide loaded graphene nanocomposites. J Hazard Mater 276:66–76. doi:10.1016/j.jhazmat.2014.05.011
93. Cai X, Tan S, Lin M et al (2011) Synergistic antibacterial brilliant blue/reduced graphene oxide/quaternary phosphonium salt composite with excellent water solubility and specific targeting capability. Langmuir 27:7828–7835. doi:10.1021/la201499s

94. Mejias Carpio IE, Mangadlao JD, Nguyen HN et al (2014) Graphene oxide functionalized with ethylenediamine triacetic acid for heavy metal adsorption and anti-microbial applications. Carbon N Y 77:289–301. doi:10.1016/j.carbon.2014.05.032
95. Maktedar SS, Mehetre SS, Singh M, Kale RK (2014) Ultrasound irradiation: a robust approach for direct functionalization of graphene oxide with thermal and antimicrobial aspects. Ultrason Sonochem 21:1407–1416. doi:10.1016/j.ultsonch.2014.02.022
96. Some S, Ho SM, Dua P et al (2012) Dual functions of highly potent graphene derivative-poly-l-lysine composites to inhibit bacteria and support human cells. ACS Nano 6:7151–7161. doi:10.1021/nn302215y
97. Yuan B, Zhu T, Zhang Z et al (2011) Self-assembly of multilayered functional films based on graphene oxide sheets for controlled release. J Mater Chem 21:3471. doi:10.1039/c0jm03643a
98. Zhu Z, Su M, Ma L et al (2013) Preparation of graphene oxide-silver nanoparticle nanohybrids with highly antibacterial capability. Talanta 117:449–455. doi:10.1016/j.talanta.2013.09.017
99. de Faria AF, Martinez DST, Meira SMM et al (2014) Anti-adhesion and antibacterial activity of silver nanoparticles supported on graphene oxide sheets. Colloids Surf B Biointerfaces 113:115–124. doi:10.1016/j.colsurfb.2013.08.006
100. Das MR, Sarma RK, Saikia R et al (2011) Synthesis of silver nanoparticles in an aqueous suspension of graphene oxide sheets and its antimicrobial activity. Colloids Surf B Biointerfaces 83:16–22. doi:10.1016/j.colsurfb.2010.10.033
101. Polte J, Tuaev X, Wuithschick M et al (2012) Formation mechanism of colloidal silver nanoparticles: analogies and differences to the growth of gold nanoparticles. ACS Nano 6:5791–5802. doi:10.1021/nn301724z
102. Thanh NTK, Maclean N, Mahiddine S (2014) Mechanisms of nucleation and growth of nanoparticles in solution. Chem Rev 114:7610–7630. doi:10.1021/cr400544s
103. Liao H-G, Niu K, Zheng H (2013) Observation of growth of metal nanoparticles. Chem Commun 49:11720. doi:10.1039/c3cc47473a
104. Tang X-Z, Li X, Cao Z et al (2013) Synthesis of graphene decorated with silver nanoparticles by simultaneous reduction of graphene oxide and silver ions with glucose. Carbon N Y 59:93–99. doi:10.1016/j.carbon.2013.02.058
105. Das MR, Sarma RK, Borah SC et al (2013) The synthesis of citrate-modified silver nanoparticles in an aqueous suspension of graphene oxide nanosheets and their antibacterial activity. Colloids Surf B Biointerfaces 105:128–136. doi:10.1016/j.colsurfb.2012.12.033
106. Zhou Y, Yang J, He T et al (2013) Highly stable and dispersive silver nanoparticle-graphene composites by a simple and low-energy-consuming approach and their antimicrobial activity. Small 9:3445–3454. doi:10.1002/smll.201202455
107. Zhang Z, Xu F, Yang W et al (2011) A facile one-pot method to high-quality Ag–graphene composite nanosheets for efficient surface-enhanced Raman scattering. Chem Commun (Camb) 47:6440–6442. doi:10.1039/c1cc11125f
108. Yuan W, Gu Y, Li L (2012) Green synthesis of graphene/Ag nanocomposites. Appl Surf Sci 261:753–758. doi:10.1016/j.apsusc.2012.08.094
109. Barua S, Thakur S, Aidew L et al (2014) One step preparation of a biocompatible, antimicrobial reduced graphene oxide–silver nanohybrid as a topical antimicrobial agent. RSC Adv 4:9777. doi:10.1039/c3ra46835f
110. Liu Y, Ai K, Lu L (2014) Polydopamine and its derivative materials: synthesis and promising applications in energy, environmental, and biomedical fields. Chem Rev 114:5057–5115. doi:10.1021/cr400407a
111. Zhang Z, Zhang J, Zhang B, Tang J (2013) Mussel-inspired functionalization of graphene for synthesizing Ag-polydopamine-graphene nanosheets as antibacterial materials. Nanoscale 5:118–123. doi:10.1039/c2nr32092d
112. Tang L, Livi KJT, Chen KL (2015) Polysulfone membranes modified with bioinspired polydopamine and silver nanoparticles formed in situ to mitigate biofouling. Environ Sci Technol Lett 2:59–65. doi:10.1021/acs.estlett.5b00008

113. Nguyen VH, Kim B-K, Jo Y-L, Shim J-J (2012) Preparation and antibacterial activity of silver nanoparticles-decorated graphene composites. J Supercrit Fluids 72:28–35. doi:10.1016/j.supflu.2012.08.005
114. Liu J, Fu S, Yuan B et al (2010) Toward a universal "adhesive nanosheet" for the assembly of multiple nanoparticles based on a protein-induced reduction/decoration of graphene oxide. J Am Chem Soc 132:7279–7281. doi:10.1021/ja100938r
115. Sehgal D, Vijay IK (1994) A method for the high efficiency of water-soluble carbodiimide-mediated amidation. Anal Biochem 218:87–91. doi:10.1006/abio.1994.1144
116. Lee JU, Lee W, Yoon SS et al (2014) Site-selective immobilization of gold nanoparticles on graphene sheets and its electrochemical properties. Appl Surf Sci 315:73–80. doi:10.1016/j.apsusc.2014.07.099
117. Zhang D, Liu X, Wang X (2011) Green synthesis of graphene oxide sheets decorated by silver nanoprisms and their anti-bacterial properties. J Inorg Biochem 105:1181–1186. doi:10.1016/j.jinorgbio.2011.05.014
118. de Faria AF, de Moraes ACM, Marcato PD et al (2014) Eco-friendly decoration of graphene oxide with biogenic silver nanoparticles: antibacterial and antibiofilm activity. J Nanopart Res 16:2110. doi:10.1007/s11051-013-2110-7
119. Mondal T, Bhowmick AK, Krishnamoorti R (2012) Chlorophenyl pendant decorated graphene sheet as a potential antimicrobial agent: synthesis and characterization. J Mater Chem 22:22481. doi:10.1039/c2jm33398h
120. Wahid MH, Stroeher UH, Eroglu E et al (2015) Aqueous based synthesis of antimicrobial-decorated graphene. J Colloid Interface Sci 443:88–96. doi:10.1016/j.jcis.2014.11.043
121. Dastjerdi R, Montazer M (2010) A review on the application of inorganic nano-structured materials in the modification of textiles: focus on anti-microbial properties. Colloids Surfaces B Biointerfaces 79:5–18. doi:10.1016/j.colsurfb.2010.03.029
122. Zhao J, Deng B, Lv M et al (2013) Graphene oxide-based antibacterial cotton fabrics. Adv Healthc Mater 2:1259–1266. doi:10.1002/adhm.201200437
123. Yee KWK, Bao J, Wiley DE (2012) Dynamic operability analysis of an industrial membrane separation process. Chem Eng Sci 71:85–96. doi:10.1016/j.ces.2011.11.046
124. Shamsuddin N, Das DB, Starov VM (2015) Filtration of natural organic matter using ultrafiltration membranes for drinking water purposes: circular cross-flow compared with stirred dead end flow. Chem Eng J 276:331–339. doi:10.1016/j.cej.2015.04.075
125. Rautenbach R, Vossenkaul K, Linn T, Katz T (1997) Waste water treatment by membrane processes—new development in ultrafiltration, nanofiltration and reverse osmosis. Desalination 108:247–253. doi:10.1016/S0011-9164(97)00032-5
126. Hegab HM, Zou L (2015) Graphene oxide-assisted membranes: fabrication and potential applications in desalination and water purification. J Memb Sci 484:95–106. doi:10.1016/j.memsci.2015.03.011
127. Banerjee I, Pangule RC, Kane RS (2011) Antifouling coatings: recent developments in the design of surfaces that prevent fouling by proteins, bacteria, and marine organisms. Adv Mater 23:690–718. doi:10.1002/adma.201001215
128. Elimelech M, Phillip WA (2011) The future of seawater and the environment: energy, technology, and the environment. Science 333:712–718. doi:10.1126/science.1200488
129. Kochkodan V, Hilal N (2015) A comprehensive review on surface modified polymer membranes for biofouling mitigation. Desalination 356:187–207. doi:10.1016/j.desal.2014.09.015
130. Rana D, Matsuura T (2010) Surface modifications for antifouling membranes. Chem Rev 110:2448–2471. doi:10.1021/cr800208y
131. Sanchez VC, Jachak A, Hurt RH, Kane AB (2012) Biological interactions of graphene-family nanomaterials—an interdisciplinary review. Chem Res Toxicol 15–34. doi:10.1021/tx200339h

132. Chae H-RR, Lee J, Lee C-HH et al (2015) Graphene oxide-embedded thin-film composite reverse osmosis membrane with high flux, anti-biofouling, and chlorine resistance. J Memb Sci 483:128–135. doi:10.1016/j.memsci.2015.02.045
133. He L, Dumée LF, Feng C et al (2015) Promoted water transport across graphene oxide–poly (amide) thin film composite membranes and their antibacterial activity. Desalination 365:126–135. doi:10.1016/j.desal.2015.02.032
134. Perreault F, Jaramillo H, Xie M et al (2016) Biofouling mitigation in forward osmosis using graphene oxide functionalized thin-film composite membranes. Environ Sci Technol. doi:10.1021/acs.est.5b06364
135. He L, Dumee LF, Feng C et al (2015) Promoted water transport across graphene oxide-poly (amide) thin film composite membranes and their antibacterial activity. Desalination 365:126–135. doi:10.1016/j.desal.2015.02.032
136. Lee J, Chae H, June Y et al (2013) Graphene oxide nanoplatelets composite membrane with hydrophilic and antifouling properties for wastewater treatment. J Memb Sci 448:223–230. doi:10.1016/j.memsci.2013.08.017
137. Mahmoudi E, Yong L, Ba-abbad MM, Mohammad AW (2015) Novel nanohybrid polysulfone membrane embedded with silver nanoparticles on graphene oxide nanoplates. Chem Eng J 277:1–10. doi:10.1016/j.cej.2015.04.107
138. Zhao C, Xu X, Chen J, Yang F (2013) Journal of Environmental Chemical Engineering Effect of graphene oxide concentration on the morphologies and antifouling properties of PVDF ultrafiltration membranes. Biochem Pharmacol 1:349–354. doi:10.1016/j.jece.2013.05.014
139. Zinadini S, Akbar A, Rahimi M, Vatanpour V (2014) Preparation of a novel antifouling mixed matrix PES membrane by embedding graphene oxide nanoplates. J Memb Sci 453:292–301. doi:10.1016/j.memsci.2013.10.070
140. Ionita M, Pandele AM, Crica L, Pilan L (2014) Composites: part B Improving the thermal and mechanical properties of polysulfone by incorporation of graphene oxide. Compos Part B 59:133–139. doi:10.1016/j.compositesb.2013.11.018
141. Yin J, Zhu G, Deng B (2016) Graphene oxide (GO) enhanced polyamide (PA) thin-film nanocomposite (TFN) membrane for water purification. Desalination 379:93–101. doi:10.1016/j.desal.2015.11.001
142. Choi W, Choi J, Bang J, Lee J (2013) Layer-by-layer assembly of graphene oxide nanosheets on polyamide membranes for durable reverse-osmosis applications. ACS Appl Mater Interfaces 5:12510–12519. doi:10.1021/am403790s
143. Marselina Y, Le-clech P, Stuetz RM, Chen V (2009) Characterisation of membrane fouling deposition and removal by direct observation technique. J Memb Sci 341:163–171. doi:10.1016/j.memsci.2009.06.001
144. Duan L, Wang Y, Zhang Y, Liu J (2015) Applied surface science graphene immobilized enzyme/ polyethersulfone mixed matrix membrane: enhanced antibacterial, permeable and mechanical properties. Appl Surf Sci 355:436–445. doi:10.1016/j.apsusc.2015.07.127
145. Sun X, Qin J, Xia P et al (2015) Graphene oxide–silver nanoparticle membrane for biofouling control and water purification. Chem Eng J 281:53–59. doi:10.1016/j.cej.2015.06.059
146. Gao Y, Hu M, Mi B (2014) Membrane surface modification with TiO_2–graphene oxide for enhanced photocatalytic performance. J Memb Sci 455:349–356. doi:10.1016/j.memsci.2014.01.011
147. Jiang Y, Wang WN, Liu D et al (2015) Engineered crumpled graphene oxide nanocomposite membrane assemblies for advanced water treatment processes. Environ Sci Technol 49:6846–6854. doi:10.1021/acs.est.5b00904
148. Safarpour M, Vatanpour V, Khataee A, Esmaeili M (2015) Development of a novel high flux and fouling-resistant thin film composite nanofiltration membrane by embedding reduced graphene. Sep Purif Technol 154:96–107. doi:10.1016/j.seppur.2015.09.039

149. Zhao C, Xu X, Chen J, Yang F (2014) Optimization of preparation conditions of poly (vinylidene fluoride)/graphene oxide microfiltration membranes by the Taguchi experimental design. Desalination 334:17–22. doi:10.1016/j.desal.2013.07.011
150. Lee JJ, Chae H-R, Won YJ et al (2013) Graphene oxide nanoplatelets composite membrane with hydrophilic and antifouling properties for wastewater treatment. J Memb Sci 448:223–230. doi:10.1016/j.memsci.2013.08.017
151. Zinadini S, Zinatizadeh AA, Rahimi M et al (2014) Preparation of a novel antifouling mixed matrix PES membrane by embedding graphene oxide nanoplates. J Memb Sci 453:292–301. doi:10.1016/j.memsci.2013.10.070
152. Jin F, Lv W, Zhang C et al (2013) High-performance ultrafiltration membranes based on polyethersulfone–graphene oxide composites. RSC Adv 3:21394. doi:10.1039/c3ra42908c
153. Xu Z, Zhang J, Shan M et al (2014) Organosilane-functionalized graphene oxide for enhanced antifouling and mechanical properties of polyvinylidene fluoride ultrafiltration membranes. J Memb Sci 458:1–13. doi:10.1016/j.memsci.2014.01.050
154. Wu H, Tang B, Wu P (2014) Development of novel SiO_2–GO nanohybrid/polysulfone membrane with enhanced performance. J Memb Sci 451:94–102. doi:10.1016/j.memsci.2013.09.018
155. Yu L, Zhang Y, Zhang B et al (2013) Preparation and characterization of HPEI-GO/PES ultrafiltration membrane with antifouling and antibacterial properties. J Memb Sci 447:452–462. doi:10.1016/j.memsci.2013.07.042
156. Zhang J, Xu Z, Shan M et al (2013) Synergetic effects of oxidized carbon nanotubes and graphene oxide on fouling control and anti-fouling mechanism of polyvinylidene fluoride ultrafiltration membranes. J Memb Sci 448:81–92. doi:10.1016/j.memsci.2013.07.064
157. Mahmoudi E, Ng LY, Ba-Abbad MM, Mohammad AW (2015) Novel nanohybrid polysulfone membrane embedded with silver nanoparticles on graphene oxide nanoplates. Chem Eng J 277:1–10. doi:10.1016/j.cej.2015.04.107
158. Chae HR, Lee J, Lee CH et al (2015) Graphene oxide-embedded thin-film composite reverse osmosis membrane with high flux, anti-biofouling, and chlorine resistance. J Memb Sci 483:128–135. doi:10.1016/j.memsci.2015.02.045
159. Hegab HM, ElMekawy A, Barclay TG et al (2015) Fine-tuning the surface of forward osmosis membranes via grafting graphene oxide: performance patterns and biofouling propensity. ACS Appl Mater Interfaces 7:18004–18016. doi:10.1021/acsami.5b04818

Toxicity and Environmental Applications of Graphene-Based Nanomaterials

Enrico Tapire Nadres, Jingjing Fan and Debora Frigi Rodrigues

Abstract Graphene can be found in pure form or as derivatives of graphene; both forms are known as graphene-based nanoparticles (GNPs). These derivatives of graphene include graphene oxide (GO), reduced GO, GNP–polymer nanocomposites, and GNP–metal hybrids. These modifications of graphene nanoparticles can lead to nanomaterials or nanocomposites with different and novel properties, such as antimicrobial, adsorbent, and catalytic properties. As antimicrobials, GNPs can be used in environmental and medical applications. In environmental application, as an antimicrobial, the particles of GNPs have shown to inactivate both pure cultures and wastewater microbial communities. When using the GNPs as coatings in medical devices or water treatment membranes, the surface inhibits microbial survival and biofilm growth. Aside from antimicrobial applications, GNPs have also been used as adsorbent; owing to their large surface area and presence of functional groups. These GNPs have the ability to remove both heavy metals and organic contaminants from water. In addition, GNPs can serve as semiconductors to increase the efficiencies of photocatalytic and electrocatalytic systems, which can be used to inactivate microorganisms and degrade organic chemicals in water. The many uses and applications of GNPs will inevitably lead to their way to the environment through manufacturing byproducts and wastes, as well as weathering of commercial products containing GNP-based nanomaterials. GNPs are bioactive and they can impact the environment. While GNPs might be extremely useful, we should find a middle ground between toxicity and applications to minimize risks to the ecosystem.

E.T. Nadres · J. Fan · D.F. Rodrigues (✉)
Department of Civil and Environmental Engineering,
University of Houston, Houston, TX, USA
e-mail: dfrigirodrigues@uh.edu

G. Gonçalves et al. (eds.), *Graphene-based Materials in Health and Environment*,
Carbon Nanostructures, DOI 10.1007/978-3-319-45639-3_11

1 Introduction to Graphene-Based Nanoparticles (GNPs)

Graphene is a carbon allotrope with two-dimensional structure [1]. The carbon atoms in graphene are arranged in hexagonal lattice, one-atom thick, with all carbons sp^2 hybridized [2]. Lauded as the material of the future, graphene has superlative properties. For instance, pristine graphene has strong mechanical properties [3], excellent electrical and thermal conductivities [4, 5] and an astounding surface area of 2630 m^2/g [6]. These exceptional electronic properties are the reasons for graphene being initially tapped as a material with great potential for electronic applications [2].

Besides pure graphene, there are graphene-derived nanoparticles that are not all-carbon, and have defects on their honeycomb structure. These graphene-derived nanoparticles might contain other atoms such as oxygen, nitrogen, halogens, metals, and other elements. The electronic properties of graphene with imperfections are, however, inferior to pristine graphene, but they can lead to nanomaterials with different and novel properties. Furthermore, many types of graphene-derived nanoparticles are cheaper to obtain than pristine graphene, making them more available for future scale up operations. Applications of graphene and its derivatives are distributed in many fields, including semiconductor [7], medical [8], environmental [9], antimicrobial [10], and catalysis [11]. In this chapter, we will use the term graphene-based nanoparticles (GNPs) for both graphene and its derivatives.

This chapter will focus on toxicity and antimicrobial properties of GNPs, as well as their environmental applications and implications. The mechanisms of toxicity of these GNPs will be also discussed in detail in this chapter to provide fundamental understanding of the behavior of GNP's interactions with biological systems. Finally, we will present the current knowledge of the risks associated to the release of GNPs in different environments, such as aquatic and terrestrial systems. Based on the unique and valuable properties of GNPs, it is expected their application in diverse industries and consumer products. The expected extensive use of GNPS would lead to their release to the environment. For instance, production facilities could discharge GNP byproduct in the sewage system; or devices containing GNPs could potentially be disposed in landfills. The life cycle of GNPs will be important for future applications as it will determine where the GNPs could potentially exert the most environmental impact [12].

2 Antimicrobial Properties of GNPs

GNPs have been described extensively in the literature to have antimicrobial properties. Their antimicrobial properties depend largely on the type of graphene and its modifications. GNPs typically originate from bulk carbon materials, such as carbon black. These bulk materials are biologically inert and nontoxic; but when unraveled to nanosizes, they become extremely active and are capable to inactivate many types of cells, including microbial cells [13]. In this section, we will focus on the antimicrobial properties of graphene, graphene oxide and graphene-based composites.

2.1 Graphene

The property of graphene as an antibacterial agent against bacterial proliferation has recently received significant attention. For instance, pure graphene in water at 1000 μg/mL is toxic to both *Escherichia coli* and *Bacillus subtilis* [14]. Graphene also inhibits *Salmonella enterica* and *Listeria monocytogenes* growth at 250 μg/mL [15]. Several studies have shown that some microorganisms are more resistant than others [16]. Typically microorganisms with outer membranes (Gram-negative), such as *E. coli*, tend to be more resistant than microorganisms lacking the outer membrane (Gram-positive), such as *Staphylococcus aureus* [17]. However, under aqueous conditions, such as growth cultures, graphene tend to aggregate overtime, exerting less toxicity [18]. On the other hand, smaller graphene, termed as graphene quantum dots (GQD), whose lateral size is 20–67 nm and less than 3 nm thickness, can be dispersed really well and exhibit good antibacterial capacity [19]. When photoexcited at 470 nm (1 W), this GNP generates reactive oxygen species (ROS), including singlet oxygen, hence becomes lethal to different types of cells [19]. The antimicrobial activities of graphene for different microorganisms are summarized in Table 1.

Besides planktonic (free-swimming) bacteria, biofilm toxicity of graphene has also been evaluated. Studies of toxicity of graphene on biofilms have been very conflicting. For instance, comparison of the biofilm thickness, on graphene-coated surfaces and control surfaces without graphene, showed significant biofilm inhibition on the coated surfaces [14]. The inhibition is attributed to the direct contact of the cells with the sharp edges of graphene, which would inactivate them by acting like a blade [14]. It is also possible that the bacteria could sense the presence of graphene on the coated surface and by chemotaxis would avoid attaching to it [14].

On the other hand, other studies showed increasing biofilm thickness on increasing concentrations of graphene coated on anodes of microbial fuel cells [20]. The enhanced biofilm thickness could be attributed to the excellent electrochemical activity of graphene, which allows sufficient electron transfer via c-type cytochromes associated with the bacteria outer membranes. These electron transfers could have promoted bacterial metabolism and biofilm growth [21]. Another explanation for biofilm resistance to graphene is that mature biofilms can be formed on top of an initial layer of dead cells on graphene-coated surfaces and the large amounts of extracellular polymeric substance (EPS) produced by the biofilms could protect the cells from the antibacterial activity of graphene. These conflicting findings on the graphene toxicity toward biofilms suggest that more investigations with different types of microorganisms are necessary to allow a comprehensive understanding of graphene toxicity on biofilms.

Table 1 Toxicity of GNPs toward different bacteria

Type	Bacteria strains	Toxicity	References
Graphene	*Shewanella oneidensis*	1000 μg/mL 86.1 % biofilm inactivation	[20]
Graphene	*Listeria monocytogenes*	250 μg/mL 100 % inactivation	[15]
Graphene	*Salmonella enterica*	250 μg/mL 100 % inactivation	[15]
Graphene quantum dots	*Staphylococcus aureus*	MIC 200 μg/mL	[17]
GO nanosheets	*Escherichia coli*	97.3 % inactivation at 10 μg/mL	[51]
GO nanosheets	*Streptococcus mutans*	83.7 % inactivation at 80 μg/mL	[152]
GO nanosheets	*Fusobacterium nucleatum*	78.5 % inactivation at 80 μg/mL	[152]
GO nanosheets	*Porphyromonas gingivalis*	85.3 % inactivation at 80 μg/mL	[152]
GO nanosheets	*Pseudomonas aeruginosa*	No growth higher than 175 μg/mL	[22]
GO Langmuir–Blogdett film	*Escherichia coli*	89 % inactivation	[68]
rGO (reduced graphene oxide via the hydrothermal flow method)	*Escherichia coli*	MIC 500 μg/mL	[153]
rGO	*Pseudomonas aeruginosa*	MIC 175 μg/mL	[22]
rGO deposited on stainless steel	*Escherichia coli*	1 mg/mL 16 % survival	[30]
Reduced graphene oxide	*Escherichia coli*	MIC 1 μg/mL	[139]
Reduced graphene oxide	*Salmonella typhimurium*	MIC 1 μg/mL	[139]
Reduced graphene oxide	*Enterococcus faecalis*	MIC 8 μg/mL	[139]
Reduced graphene oxide	*Bacillus subtilis*	MIC 4 μg/mL	[139]

2.2 *Graphene Oxide*

Among all GNPs, graphene oxide (GO) has been one of the most widely investigated antimicrobial agents. GO is easily dispersed in water and has been reported to inactive 50 % of *E. coli* with 40 μg/mL GO within 1 h [10]. The plate counting method revealed that GO is able to inactivate 91 % of *E. coli* with a concentration of 80 μg/mL [10]. Other microorganisms such as *S. enterica* and *L. monocytogenes* are inhibited by 100 % at a concentration of 25 μg/mL [15]. Additionally, *Pseudomonas aeruginosa* showed 92 % growth inhibition after incubation for 4 h with GO [22]. These results showed the GNPs are potent antimicrobials on pure cultures.

Besides pure cultures, GO has also a detrimental effect in wastewater borne microbial communities. In wastewater, where there are different types of microbial communities present, the toxicity is attenuated, but is still present. Significant reduction ($\sim$20–70 %) of metabolic activity in wastewater is observed in the presence of 10 mg/L GO [23]. The presence of organic matter and ions are responsible for reducing the toxicity of GNPs in wastewater. Hence, higher concentrations are necessary to significantly affect microbial communities in wastewater. For instance, concentrations at 100–300 mg/L GO results in $\sim$35 % bacterial growth inhibition, which affects degradation of organic carbon and nutrient removal in the wastewater treatment [23].

Similar to graphene, the toxicity of graphene oxide towards biofilms was also evaluated in several studies. A GO concentration of 1000 μg/mL was shown to inhibit 67 % biofilm growth of *E. coli*. For *B. subtilis* and *Rhodococcus opacus*, GO showed biofilm inhibition of up to 60.9 and 42.8 %, respectively [24]. Furthermore, GO modified membrane filters also showed high antibacterial and antifouling properties. Through confocal laser scanning microscopy, it was observed that the amount of *P. aeruginosa* attached on the membrane was reduced by more than 80 %, as the GO content increased from 1 to 3.5 % [25]. The anti-biofouling property of GO could be attributed to the permanent inactivation effects on planktonic microorganisms, and therefore bacterial growth and subsequent biofilm formation on the surface.

In conclusion, most studies based on the antimicrobial properties of GO were done using model microorganisms, like *E. coli* and *B. subtilis*, in pure cultures under controlled laboratory conditions. However, natural and engineered aquatic systems are more complex than simplified systems in terms of microbial community. Therefore, more thorough studies on possible mechanisms behind different kinds of bacteria on complex environments are needed to fully understand the toxicity of GO.

2.3 Reduced Graphene Oxide

In the case of rGO, traces of functional groups are still present after reduction of GO. This property is sufficient to allow rGO to stay suspended long enough in solution to exert antimicrobial action [26]. Through the classic plate counting method, a study showed that rGO almost completely suppressed the growth of *E. coli*, leading to a viability loss of up to 95 % [27]. Furthermore, it was also demonstrated that rGO could serve as a good antibacterial material. In another study, rGO showed a minimum inhibitory concentration (MIC) against *E. coli*, *S. typhimurium*, *P. aeruginosa* and *A. calcoaceticus* of 0.4, 0.35, 0.05 and 0.15 mg/mL, respectively [28].

Since rGO is prepared from GO, most of the studies to date compare the antibacterial capacity of rGO with GO. In some of these studies, rGO exhibits superior antibacterial activity than GO toward *E. coli* and *S. aureus* [29]. The authors attributed the stronger bacterial toxicity of the rGO to its shaper edges, which provide a damaging interaction with the cell membrane [30]. Another study, demonstrated that rGO is found to have a lower antibacterial activity than GO at low concentrations, whereas at high concentrations, the growth inhibition of *S. enterica* and *L. monocytogenes* is similar in the presence of either GO or rGO [14]. Similarly, results of lower antibacterial activity of rGO compared to GO against *P. aeruginosa* was observed [31]. Interestingly, some studies also revealed that GO dispersion have much higher antibacterial activities than rGO dispersion toward *E. coli* [31]. One study even compared the antibacterial capacity among pure graphene, GO, and rGO. According to their report, GO is the most efficient antibacterial material, followed by rGO and pure graphene [14]. These findings suggest that the toxicity of GO and rGO towards microorganisms is different. In some cases, certain microorganisms are more sensitive to GO than rGO and vice versa. These findings suggest that toxicity of GNPs may be species dependent.

The biofilm inhibition capacity of rGO and GO have also been compared in few studies. Significantly fewer cells adhere to rGO-coated surfaces as observed through SEM images; whereas a dense and uniform bacterial biofilm can be observed on control surfaces [28]. Compared to GO, rGO appear to play a less effective role on biofilm inhibition since rGO has the tendency to aggregate during the long incubation period required for biofilm formation. This tendency of aggregation is due to the hydrophobic nature of rGO.

In summary, rGO presents toxicity toward planktonic cells and biofilms. However, the toxicity comparison studies among different GNPs show controversial results, especially between GO and rGO on planktonic cells. These different observations of antimicrobial properties of rGO compared with GO may be explained by the lack of a standard procedure to synthesize rGO. Different studies employ distinct chemical reducing agents, which may have caused the observed contradictory results [32]. In addition, differences in the methodologies employed to assess antibacterial activity may have also influenced the results obtained by the different studies. Another potential reason for the discrepancy in the results is that different graphene-based nanomaterial may have different toxicity mechanisms towards different species of bacteria.

2.4 Graphene-Based Composites

Nanocomposites are advanced functional materials composed of nanomaterials dispersed inside polymer matrices. Due to excellent mechanical, thermal and optical properties, graphene, and its derivatives (e.g., GO and rGO) are excellent candidates as filler materials [21]. Graphene can be modified in several ways with different molecules and atoms to form nanocomposites or nanohybrid materials. In this chapter, we will focus on polymer and transition metals and metal oxides modified graphene nanomaterials, since they have been used extensively for environmental applications and as antimicrobials. Polymer composites can be produced by noncovalent and covalent modifications of graphene, GO or rGO; while carbon–metal hybrid nanoparticles are typically synthesized by covalent reactions with transition metals or metal oxides. The main reason for modifying graphene and its derivatives is because most of these GNPs tend to aggregate in physiological solutions due to electrostatic charges and nonspecific binding to other macromolecules [33]. GNP-based nanocomposites can improve their solubility and biocompatibility, as well as increase toxicity toward microorganisms.

2.4.1 Noncovalent and Covalent Polymer Modifications

Noncovalent methods of functionalization employ hydrophobic interactions, π–π interactions, van der Waals forces and electrostatic binding [21]. Reports of antibacterial capacity for noncovalent modification of GNPs include (*N*-vinyl-carbazole) PVK, chitosan, sulphonated polyaniline, poly(ethylene oxide) (PEO) poly(propylene oxide) (PPO), and others. Taking PVK as an example, this polymer has exceptional electronic and mechanical properties, as well as anticorrosion capability. Dispersing graphene nanoplatelets into the PVK matrix can create a PVK–graphene solution for fabrication of PVK–graphene thin films though electrodeposition [14]. Studies with this polymer showed that the presence of PVK enhances more than 10 % of the bactericidal capacity of graphene toward *E. coli* and *B. subtilis* than pristine graphene [14]. The authors demonstrated that better dispersion of graphene in the presence of PVK could be the reason for the higher toxicity of the PVK–graphene nanocomposite. In addition, synergistic effect caused by electronic interactions or morphological interactions between PVK and graphene could have also contributed to the antibacterial effect. Comparing with pure graphene, PVK–graphene shows similar biofilm inhibition capacity, but lower cytotoxicity to human cells. This property makes PVK–graphene a better candidate for biomedical and environmental applications involving human contact.

Another toxicity study with PVK and GO demonstrated that PVK–GO with only 3 % GO in the nanocomposite presented 30 and 57 % higher antimicrobial effects toward planktonic cells and biofilms, respectively, than GO alone. This property of PVK–GO, suggests that this nanocomposite can be cost effective against biofouling, due to the low cost and high manufacturing yield of this nanocomposite [14]. The toxicity of rGO nanocomposite has not been extensively investigated like GO and graphene nanocomposites. Therefore, rGO will not be discussed in this section.

In addition to the noncovalent modifications of GNPs, it is possible to modify GNPs through covalent modifications. Covalent modification of GNPs can include many types of aliphatic and aromatic amines, amino acids, amine terminated biomolecules, and enzymes. In this chapter, we will focus on the polymers, since many antimicrobial investigations and environmental applications employ polymer nanocomposites. Examples of graphene-based nanomaterial modified by covalent reactions include polyethylenimine (PEI) [34], polyvinyl alcohol (PVA), plasticized poly(lactic acid) (PLA), and others. In the study of Cao and collaborators, graphene was embedded in a PVA matrix with a polymer biocide (quaternary ammonium modified polyvinyl benzyl chloride (qPvB/Cl–)) to improve the mechanical strength of the polymer and enhance the antibacterial property of graphene. As a result, PVA–graphene with 10 % loading of polymer biocide was able to inactive 97.1 % of *E. coli* and 99.7 % of *S. aureus* without exhibiting human cell toxicity [27]. They also demonstrated that PVA–graphene–biocide nanocomposite is able to improve the antibacterial toxicity by 92 %, when comparing with pure graphene. The enhanced toxicity was explained by the increasing number of functional groups in the nanocomposite [27]. The efficient antibacterial capacity and mechanical reinforcement of PVA–graphene–biocide nanocomposite produced an advanced functional material suitable for hygiene and food packaging.

Biocompatible and biodegradable polymers, such as polylactic acid (PLA) and polyurethane (PU), are commonly used to prepare PLA/PU–GO nanocomposites through electrospinning [35]. The addition of 5 % GO to the polymer composite PLA/PU showed remarkable improvement (more than 90 % enhancement) of antibacterial capacity without inhibiting normal human cell proliferation and differentiation. Another study also showed that introduction of PLA increases the microbial inactivation by more than 1 log toward four different types of bacteria [36]. Therefore, the good antibacterial activity and biocompatibility of PLA–GO nanocomposite make it attractive for environmental and clinical applications, as well as for tissue engineering. Similarly, rGO-based polyester (PE) nanocomposites present strong antibacterial activity toward five different kinds of bacterial strains [32]. Hence, PE/rGO nanocomposite with high mechanical strength, thermal stability, electrical property, and antibacterial capacity could offer new opportunities for environmentally friendly applications.

Alternatively, incorporation of the GNPs into polymers could also assist in the development of membrane systems with less biofouling due to higher antimicrobial properties. For example, the ability to embed GO in polysulfone (PSF) membranes to mitigate biofouling was evaluated with *P. aeruginosa* biofilms. The results indicated that 1 % of GO added to the Psf membrane showed 47 % biofilm reduction compared to pure Psf membranes. These results were explained by the increasing hydrophilicity and electrostatic repulsion characteristics of the membranes containing GO [25]. The anti-biofouling property of the membranes suggests that they can potentially be used in wastewater treatment and water filtration systems.

Overall, different types of GNP nanocomposites have been investigated to achieve high antibacterial toxicity and low human cell cytotoxicity. These properties make them very attractive for water treatment applications, especially in the development of anti-biofouling coatings or membrane systems for water filtration. So far, most of the studies exposing GNPs to microorganisms were done for a short period and in controlled environments. However, the effectiveness of these nanocomposites after long exposure time to bacteria needs to be investigated with various environmental conditions and water chemistries. More importantly, the chronic exposure effects of these nanocomposites need to be further investigated before such material can be widely used for water treatment or other applications.

2.4.2 Metal Modification

In environmental applications, transition metals and metal oxides coupled to graphene-based nanomaterials have also been explored as antimicrobial agents, mostly for water disinfection. In these studies, GNPs were shown to serve as scaffold for other nanoparticles. The most common transition metals and metal oxides used with GNPs are silver, magnetite, zinc oxide and titanium dioxide, copper and others. In this section, we will focus on the most popular ones for different GNPs.

Among metal oxides used to modify graphene, zinc oxide has been widely investigated. The graphene–ZnO nanohybrid can be fabricated by depositing ZnO nanoparticles or Zn(II) ions on graphene nanosheet surfaces [37]. Studies with graphene/ZnO nanoparticles presented excellent activity against *E. coli*. The authors explained that the controlled dissolution of ZnO into Zn(II) ions in solution and synergistic antimicrobial effects of GO and ZnO are responsible for its superior antimicrobial properties [38]. Another study, suggested that the antimicrobial property of graphene/ZnO is due to its photocatalytic activity [39].

TiO_2 nanoparticle used in combination with GO also showed excellent antimicrobial properties. A study with GO/TiO_2 nanocomposite annealed at 400 °C showed that 0.2 μg/mL of GO/TiO_2 is able to inactivate *E. coli* 25 % more than bare TiO_2 [40]. The better antibacterial activity of GO/TiO_2 is attributed to higher production rate of hydroxyl radicals through photocatalytic reaction. The latter reaction happened since GO could be used as an electron acceptor and transporter, which promoted the efficient separation of photogenerated electron hole pairs, and led to production of more hydroxyl radicals involved in the antibacterial activity [41]. The low cost of GO/TiO_2 manufacturing procedure and its effective antibacterial properties indicate that GO/TiO_2 is a promising candidate for water disinfection.

In the case of silver nanoparticles, they can be deposited onto rGO surfaces to form nanohybrid materials with superior antibacterial activity against *E. coli* (MIC 0.16 μg/mL) due to the synergistic effect of GO and silver ions [42]. Ag/rGO nanocomposites can also be synthesized by a facile synthesis approach, where silver nitrate is reduced by rGO without additional reductants [43]. This study demonstrated that the MIC of *E. coli* for Ag/rGO is 25 times higher than rGO. This excellent antibacterial property is attributed to the synergistic effects of rGO nanosheets adsorption properties, and silver nanoparticles' bactericidal property. Hence, this nanohybrid material is attractive to fields requiring materials with antimicrobial and anti-biofouling properties.

In addition to studies with planktonic cells, other studies also investigated metal GNPs nanocomposites as an anti-biofouling agent. For instance, GO/Ag at 2.5 μg/mL was shown to inhibit *P. aeruginosa* biofilm formation by 100 % after 1 h of contact. Similarly, it was observed that 2 μg/mL of GO/Ag nanoparticles can inhibit more than 95 % of *P. aeruginosa* biofilm formation [44]. Other toxicity studies among different graphene-based nanocomposites are listed in Table 2.

Table 2 Toxicity of different graphene based nanocomposites against bacteria

Type	Bacteria strains	Toxicity	References
TiO_2/graphene sheets nanocomposites	*Escherichia coli*	TiO_2/4.2 wt.% GSs showed 9.5 % viability	[154]
Magnetic graphene oxide–TiO_2 composites	*Escherichia coli*	2.83 % viability	[41]
Graphene and layered titanate nanosheet	*Escherichia coli*	Almost 100 %	[155]
Ag–rGO–Fe_3O_4–PEI composite	*Escherichia coli*	0.1 μg mL 99.9 % inactivation	[156]
Silver nanoparticles decorate graphene composites	*Escherichia coli*	MIC 0.16 μg/mL	[157]
Silver nanoparticles decorate graphene composites	*Listonella anguillarum*	MIC 0.08 μg/mL	[157]
Silver nanoparticles decorate graphene composites	*Bacillus cereus*	MIC 0.04 μg/mL	[157]
Silver nanoparticles decorate graphene composites	*Staphylococcus aureus*	MIC 0.16 μg/mL	[157]
PDDA-protected Ag–RGO	*Escherichia coli*	100 % disinfection	[158]
Graphene oxide–chitosan layers	*Staphylococcus aureus*	77 % bacterial reduction	[72]
Poly(vinyl alcohol) (PVA) and graphene nanosheets composite	*Escherichia coli*	99.7 % inactivation	[159]
Co_3O_4/RGO nanocomposites	*Escherichia coli*	inhibition with a diameter 12 mm	[160]
Graphene–Fe_3O_4 composite	*Escherichia coli*	99.56 % inhibition	[161]
GO immobilized lysozyme (GO-Ly)	*Escherichia coli*	MIC 512 μg/mL	[162]
Reduced graphene oxide (CRGO) immobilized lysozyme (CRGO-Ly)	*Escherichia coli*	MIC 256 μg/mL	[162]
Polyethylene glycol (PEG) and polyhexamethylene guanidine hydrochloride (PHGC) dual-polymer-functionalized graphene oxide (GO) (GO–PEG–PHGC)	*Escherichia coli*	100 % inactivation	[163]
(GO–PEG–PHGC)	*Staphylococcus aureus*	100 % inactivation	[163]

In summary, various types of metal and metal oxide GNP modification have been investigated and are still under investigation to enhance the antibacterial activities of GNPs for different environmental applications, like wastewater treatment and drinking water disinfection. As of right now, most of these materials have been investigated with pure cultures, it is, however, unclear how water chemistry and complexity of real water samples would affect the antimicrobial activity of these new composites. Future studies with more complex water chemistries should be investigated to determine the effectiveness of these composites and GNPs for real environmental applications.

3 Mechanisms of Antimicrobial Activity

The previous section described the ability of GNPs to inactivate microorganisms. In this section, we will discuss the mechanisms of microbial inactivation by GNPs. In general, the mechanisms are divided into two major categories. The first category is physical interaction, which involves membrane damage or cell growth inhibition caused by interactions of the cell or cell membrane with the GNPs. The second category involves chemical reactions leading to the production of reactive oxygen species (ROS).

3.1 Physical Mechanisms of Antimicrobial Activity

According to recent studies, the physical interaction of GNPs with microorganisms is the primary antimicrobial activity of these nanostructures [30]. Researchers describe that bacterial cell membrane damage caused by direct contact of the bacteria with extremely sharp edges of the nanomaterials is the most effective mechanism in bacterial inactivation [24]. By taking graphene as an example, researchers demonstrated that the corners of graphene sheets are very sharp and able to penetrate the membrane lipid bilayer. The membrane damage leads to leakage of intracellular materials (e.g. cytoplasm, ribosomes, and nucleic acids), which will eventually lead to cellular death [45]. Computer model has suggested that graphene buries itself in the cell membrane in three stages: (a) initial penetration; (b) rotation of graphene sheet and partial membrane encapsulation; followed by (c) reverse rotation to get the graphene sheet fully inserted in the membrane [46]. From this model, graphene can interact with the cellular contents and become trapped in the lipid bilayer causing cell leakage and inactivation [47]. Independent of the method for synthesis, rGO has similar surface chemistry to graphene [48]. Therefore, the physical interaction of rGO is suggested to be similar to graphene for cell inactivation.

In the case of GO, which contains carboxyl, hydroxyl, and epoxy groups, it can form more stable dispersions in aqueous solutions than graphene [30]. Hence, the stable dispersion of GO offers more opportunities for cellular interactions. Studies have suggested that GO disturbs primarily the cell membrane and enhance cell permeability [49]. Through TEM analyses, three stages of cell damage caused by GO were observed for *E. coli* (Fig. 1). In stage I, the cells tolerated GO; after a certain contact time, the cell membranes partially lose integrity (Stage II); and in stage III, the cell membranes are severely damaged and some lose completely the

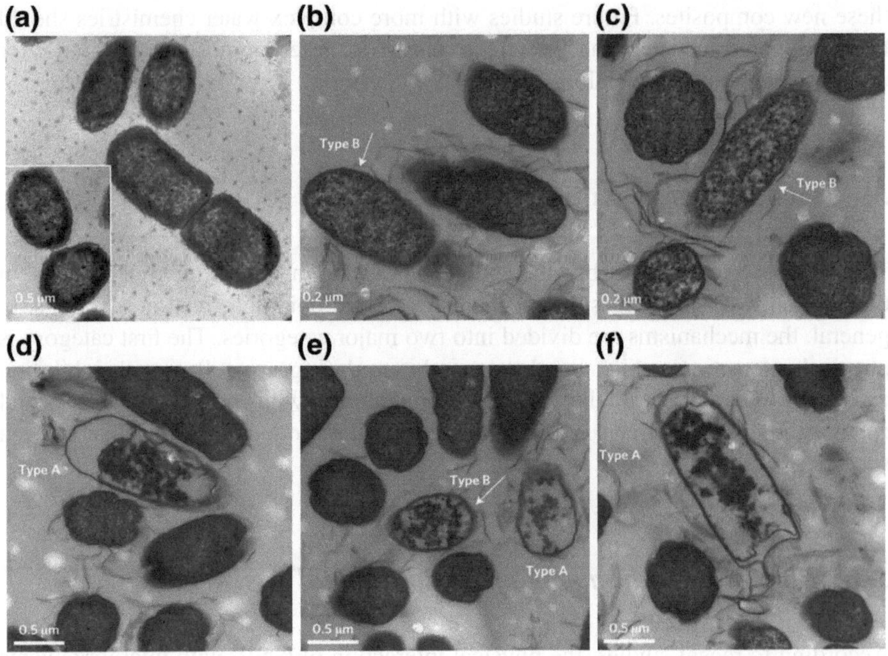

Fig. 1 Morphology of *E. coli* exposed to graphene oxide nanosheets. TEM images showing *E. coli* undergoing changes in morphology after incubation with 100 µg/mL graphene oxide nanosheets at 37 °C for 2.5 h. Three stages of destruction can be seen. **a** Initial morphology of *E. coli* (control or Stage 1). **b**, **c** Partial damage of cell membranes, with some bacteria showing a lower density of surface phospholipids (Stage II). *Arrows* indicate a Type B mechanism, where graphene nanosheets extract phospholipids from the cell membrane. **d–f** Three representative images showing the complete loss of membrane integrity, with some showing 'empty nests' and missing cytoplasm (Stage III). **d** and **f** are representative images showing a Type A mechanism, where graphene nanosheets cut off large areas of membrane surfaces. In e, both Type A and Type B mechanisms are shown [50]. (Image reproduced with the publisher's permission)

intracellular material [50]. However, the direct contact between the sharp edges of GO and microorganisms is not the only mechanism responsible for cell death. Through SEM observation, it was determined that large GO sheets can wrap bacterial cells, isolating them from the growth medium, where the nutrients are found (Fig. 2) [48]. Hence, cells can neither consume the nutrients nor proliferate because GO forms a physical barrier [51].

Overall, it is evident that physical membrane damage is one of the main mechanisms behind toxicity of all the graphene-based nanomaterials. This mechanism will be affected greatly by the specific properties of the nanomaterial, such as size, contact time, concentration, functionalization, and others.

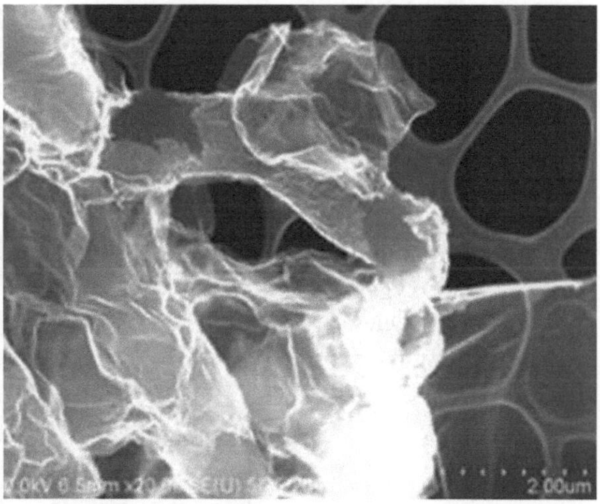

Fig. 2 Illustration of graphene oxide wrapping a bacterial cell [24]

3.2 *Chemical Mechanisms of Antimicrobial Toxicity*

The majority of the studies to date has suggested that ROS generation and subsequent oxidative stress are frequently associated to GNPs toxicity. This section will focus on reactive oxygen species production by GNPs and oxidative stress as mechanisms for GNP-induced toxicity.

3.2.1 Reactive Oxygen Species

GNPs have been frequently described to produce reactive oxygen species (ROS), which can ultimately lead to cell death. By definition, superoxide anion ($^{\cdot}O_2^-$), hydrogen peroxide (H_2O_2), and hydroxyl radical ($^{\cdot}OH$) collectively are called ROS [52]. These ROS are natural byproducts of the metabolism of respiring organisms. The generation, interconversion, and elimination of ROS are presented in Fig. 3 [52]. Briefly, $^{\cdot}O_2^-$ is generated by losing one electron from oxygen. After $^{\cdot}O_2^-$ is generated, it can accept two protons to yield hydrogen peroxide. Hydrogen peroxide is not a free radical, but is more active chemically than molecular oxygen; hence it is included in the ROS group. Hydrogen peroxide molecule can be split up to produce hydroxyl radical and hydroxyl anion by accepting an electron. Finally,

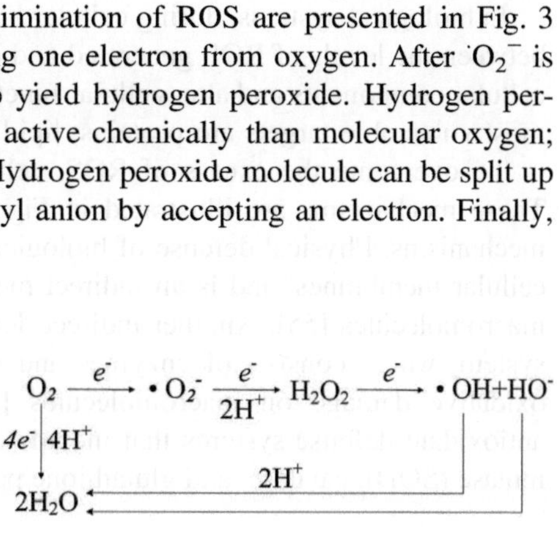

Fig. 3 Reactive oxygen species (ROS) generation, interconversion and elimination

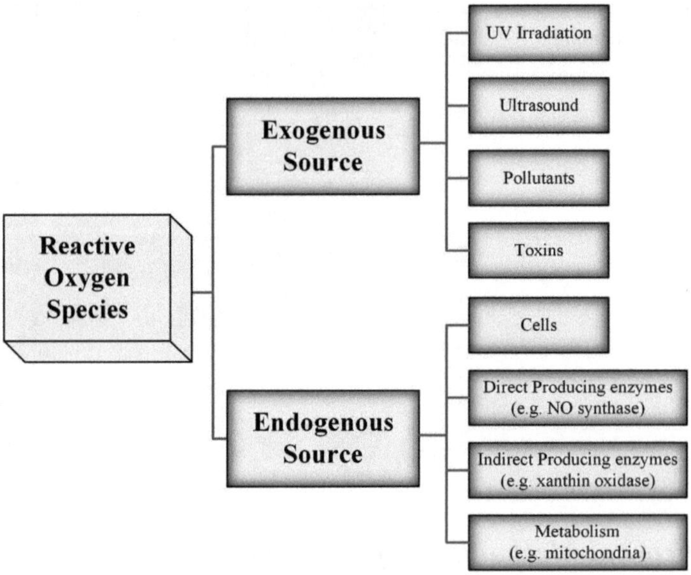

Fig. 4 Sources of reactive oxygen species (ROS)

hydroxyl radical can interact with an electron and proton resulting in formation of a water molecule.

ROS are generated intrinsically or extrinsically within the cell. Inside the cells, nicotinamide adenine dinucleotide phosphate (NADPH) oxidase will reduce the molecular oxygen into $^{\cdot}O_2^{-}$. Further reduction of oxygen may either lead to H_2O_2 and $^{\cdot}OH$ via dismutation and metal-catalyzed Fenton reaction, respectively [53, 54]. Some of the endogenous sources of ROS include mitochondrial respiration, inflammatory response, and peroxisome, while engineered nanomaterials and environmental pollutants act as exogenous ROS inducers. The sources of ROS are shown in Fig. 4 [55].

In biological systems, during cellular homeostasis, the cells maintain the balance between the levels of ROS generated and eliminated. Excess ROS will damage the cellular components and alter cellular functions. Among the biological damages, the most vulnerable targets are proteins, lipid membranes, and nucleic acids [54].

To overcome the excess of ROS, cells have developed defense mechanisms. These mechanisms are illustrated in Fig. 5 and include both indirect and direct mechanisms. Physical defense of biological systems could enhance the stability of cellular membranes, and is an indirect method to prevent ROS to approach target macromolecules [55]. Another indirect defense mechanism includes the cell repair system, which consists of enzymes and molecules that can efficiently repair the oxidative damage on macromolecules [55]. Aerobic cells also possess direct antioxidant defense systems that include the enzymatic scavenger superoxide dismutase (SOD), catalase, and glutathione peroxidase. SOD speeds up the conversion

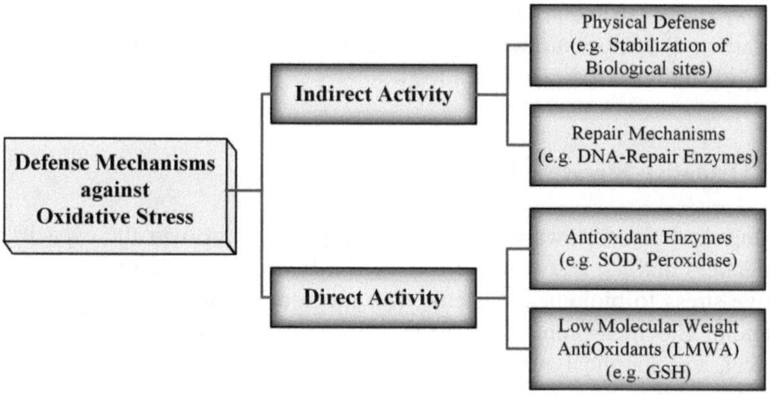

Fig. 5 Defense mechanisms against oxidative stress

of superoxide to hydrogen peroxide, whereas catalase and glutathione peroxidase convert hydrogen peroxide to water. At the same time, low-molecular-weight antioxidants (LMSA) work as another group of direct defense mechanism, which include ascorbate, pyruvate, flavonoids and the most important one, glutathione (γ-L-glutamyl-L-cysteinyl-glycine, GSH) [56]. Overall, through these defense mechanisms, cells are able to prevent oxidative damage when interacting with ROS.

3.2.2 ROS Generated by GNPs

GNPs generate ROS through different pathways. Graphene acts as a semimetal or a zero-band gap semiconductor with remarkably high electron mobility at room temperature. This property provides good electron transfer capacity. Therefore, conductive graphene nanosheets can act as a conductive bridge over the insulating lipid bilayer to mediate electron transfer from bacterial intracellular components to the external environmental to generate ROS [30]. Similar reactions also occur for rGO, which has high conductivity. In the case of rGO, ROS is produced during light excitation that makes rGO to become excited (rGO* form). After that, the excited rGO* reduces O_2 to O_2^-. The specific reaction is shown below:

$$\mathrm{rGO} \xrightarrow{h\nu} \mathrm{rGO}*$$

$$O_2 + \mathrm{rGO}^* \rightarrow {}^{\cdot}O_2^- + \mathrm{rGO}^+$$

In case of GO, it is believed that GO can generate ROS from aerial oxygen. The plausible mechanism is that the unpaired electrons of GO reduce molecular oxygen to form $^{\cdot}O_2^-$. The $^{\cdot}O_2^-$ can interact with a hydrogen atom to produce H_2O_2 [57]. Since hydroxyl groups are abundant in GO, hydroxyl groups can be directly detached from GO to form $^{\cdot}OH$. Furthermore, GO can catalyze the formation of

$^{\cdot}OH$ from H_2O_2 [58]. The large numbers of oxygen-containing functional groups (e.g., carboxyl) in the surface of GO can facilitate the production of ROS and increase its antibacterial properties [59].

3.2.3 Oxidative Stress and ROS Measurement

Changing the balance towards an increase in pro-oxidant over the capacity of the antioxidant is defined as oxidative stress [55]. In order to determine conditions of oxidative stress to biological systems, few methods are used to quantify the ROS generation. The measurement of ROS can be done with instrumental analysis, biological, and chemical analyses.

In the case of instrumental analysis, the unpaired electron of free radicals can be directly observed using electron spin resonance (ESR) spectroscopy. For free radicals, whose lifetime is very short, a technique called spin trapping is employed. In spin trapping, the reactive free radical is reacted with a diamagnetic molecule to form a more stable free radical, which can now be detected by ESR [60].

The high reactivity and relative instability of ROS make them extremely difficult to be detected and measured in biological systems. Thus, the assessment of ROS and free radical generation has been extensively investigated by indirect measurements of biological cellular components, such as lipids, protein, and DNA. Damage of lipid will alter and modify cellular membranes. Therefore, lipid peroxidation could be a signature of ROS damage. To determine ROS production, it is possible to quantify malondiadehyde (MDA) production, an oxidized product of polyunsaturated fatty acids. In such ROS assay, MDA forms an adduct with thiobarbituric acid (TBA) resulting in a pink product with increasing absorbance at 535 nm [61].

Additionally, oxidative stress can be also evaluated by measuring cellular SOD levels. SODs are the first biological mechanism of defense against oxidative stress. SOD catalyzes the disproportionation reaction of superoxide radicals to form oxygen and hydrogen peroxide. The reaction is shown below. More than 50 % reduction in the cellular SOD level induced by GO can be observed with 1 mg/L GO-exposed cells compared to control samples [62].

$$^{\cdot}O_2^- + SOD{-}M^{2+} \rightarrow O_2 + M^+ \qquad \text{(Reaction1)}$$

$$^{\cdot}O_2^- + 2H^{+} + SOD{-}M^+ \rightarrow H_2O_2 + M^{2+} \qquad \text{(Reaction2)}$$

Sum:

$$2\,^{\cdot}O_2^- + 2H^+ \rightarrow H_2O_2 + O_2 \qquad \text{(Reaction3)}$$

Chemical analysis methods to identify specific ROS are based on reactions with various molecules that are modified to elicit luminescent or fluorescent signals. GSH is a key endogenous antioxidant, whose depletion is proportional to the generation of ROS, such as hydrogen peroxide. Therefore, the consumption of GSH is widely used as a measure of oxidative stress in biological systems [22]. Briefly, glutathione peroxidase will reduced H_2O_2 into H_2O, while the GSH will be oxidized into glutathione (GSSG) at the same time. Total loss of glutathione can be determined colorimetrically by reaction of GSH with 5,5′-dithiobis-(2-nitrobenzoic acid) DTNB (Ellman's reagent). This reaction will yield a yellow colored 5-thion-2-nitrobenzoic acid (TNB), which can be absorbed at 412 nm [63].

Besides hydrogen peroxide, superoxide can also be quantified based on the interaction of superoxide with tetrazolium dye (2,3-*bis*-(2-methoxy-4-nitro-5-sulfophenyl)-2*H*-tetrazolium-5-carboxanilide (XTT). XTT can be reduced by $^{\cdot}O_2^{-}$ to a soluble formazan, which is detected at 470 nm [64]. The GO and rGO in a concentration of 100 μg/mL induce two and onefold production of superoxide radical anions, respectively, when compared to the control. This remarkable amounts of ROS generated in the presence of GNPs are one of the reasons for cell death [31].

In conclusion, both physical and chemical mechanisms contribute greatly to the toxicity of GNPs. Understanding the mechanisms of GNP toxicity is important for the design and fabrication of nanomaterials and their related applications.

4 Environmental Applications of GNPs

The versatility of GNPs has triggered researchers to investigate their potential biomedical and environmental engineering applications, such as water treatment and anti-biofouling properties. Applications that require exposure to environmental conditions usually call for a more robust material that could withstand extreme temperatures, wind and water current. GNPs, as free nanoparticles, would easily scatter in the environment. For this reason, most of the environmental applications of GNPs require them to be used as coatings or immobilized as polymer nanocomposites or nanometal–GNP hybrids.

GNPs have been shown to effectively inactivate microorganisms through different mechanisms. GNPs as coatings can use their inherent antimicrobial properties to protect surfaces from microbial colonization. In addition to the antimicrobial properties of GNPs, these nanomaterials have also been shown to have promising properties to remove chemicals contaminants from water. The unique antimicrobial and chemical removal properties of GNPs make them valuable materials for removal of water contaminants. Hence, in this section, we will discuss the current trend of applications of GNPs for removal of biological and chemical contaminants.

4.1 Removal of Microorganisms by GNPs

GNPs can be utilized in different ways as antimicrobial agents in environmental applications. Pure GNPs, as well as polymer nanocomposites, can be used directly as coatings on different types of surfaces, such as metals and plastics. GNPs can also serve as scaffolds for nanometals for more efficient antimicrobial action. In addition, GNPs are also utilized as part of photocatalytic and electrocatalytic systems for water treatment to inactivate microorganisms.

On surfaces, the antimicrobial properties of GNPs make them very attractive to prevent microbial colonization. For instance, biofilm formation studies showed that graphene can prevent microbial growth and subsequent biocorrosion [65]. Similar antimicrobial action against *S. aureus* and *E. coli* is seen for copper and germanium surfaces when a graphene monolayer is deposited on their surfaces [66]. Studies have also shown that rGO can be deposited on surfaces of poly-L-lysine and hyaluronic acid composites using the layer-by-layer method to render the polymer resistant to bacterial colonization [67]. Pure GO deposited on poly(ethylene terephthalate) also showed antimicrobial activities, even on single layered GO Langmuir–Blodgett films [68]. These properties make GNPs very attractive to prevent corrosion and for coating of medical devices in biomedical applications.

Alternatively, GNPs can also be used to change the surface to favorably enhance bacterial adhesion, as investigated in microbial fuel cells applications [69]. The rGO formed from electrodeposition of GO on the surface of graphite blocks are shown to enhance the formation of *Shewanella oneidensis* biofilms on the anode of microbial fuel cells [20]. The thick biofilm on the anode could be attributed to the excellent electrochemical activity of GNP, which allows sufficient electron transfer from cells to the electrode to promote bacterial metabolism and biofilm growth [70]. *E. coli* was also shown to proliferate on GO deposited on filters [69]. For these microorganisms, the matured biofilm contained a large amount of extracellular polymeric substance (EPS), which may have protected the cells from the antibacterial activity of GNPs by preventing direct contact [70].

Another important application of GNPs is in water treatment. For instance, graphene can effectively coat filters without affecting the porosity and resulting in a product with superior antimicrobial activity [59]. Alternatively, GNPs can produce composites, containing lower concentrations of these nanoparticles, which can be very effective in preventing biofilm formation. For instance, nanocomposite coatings made by blending GO with poly-*N*-vinylcarbazole (PVK) showed higher antimicrobial activity than pure GO films, owing to a better dispersion of GO in the polymer matrix because of π–π interaction with the side chains of PVK [71]. This nanocomposite also showed strong antimicrobial activity against *E. coli*, *B. subtilis*, *R. opacus*, and *Cupriavidus metallidurans* but low toxicity toward mammalian fibroblast cells [24]. The PVK–GO nanocomposite was also successfully applied as coating on membrane filters and was able to remove *E. coli* and *B. subtilis* by 4 and 3 logs, respectively [59]. Other blends, such as chitosan–GO composite, also produce surfaces with antibacterial activity [72].

Even though the GNPs have their own antimicrobial properties; addition of metallic nanoparticles, such as Au and Ag, provides a more potent antimicrobial material. The composite Au–GO can be prepared by deposition of Au(III) ions on the surface of GO, followed by reduction with sodium citrate to form a covalently linked Au nanoparticle to rGO [73]. Later, the GNP–metal hybrid Au–rGO was shown to exhibit 100 % inactivation of Gram-positive (e.g., *S. aureus* and *B. subtilis*) and Gram-negative (e.g., *E. coli* and *P. aeruginosa*) bacteria [74]. Another example of application of GNPs for water treatment is a porous hydrogel made from Ag–GO. This hydrogel was able to inactivate 99 % of coliforms from natural water samples from a lake and a creek [75]. The inactivation of the microorganisms is thought to be due to the slow release of Au or Ag ions, which complexes with the thiol group in the active site of bacterial enzymes, while GNPs accelerates the ions' action by disrupting the cell membranes [42, 76].

Another strategy by which GNPs can participate in antimicrobial applications is when GNPs are part of a redox system, as an electron conductor. The interface of GNPs with photocatalysts creates a region of increased charge separation, which leads to a more efficient ROS generation and microbial inactivation [77]. Photocatalyst made from Au–TiO_2–graphene composite has been shown to inactivate *E. coli*, *Rhodopseudomonas palustris* and *Candida* sp. after exposure to visible light. The light assists in the production of ROS [78]. A 100 % inactivation is observed for the Au–TiO_2–graphene composite and fourfold more activity is observed for this nanohybrid material when compared to TiO_2-only photocatalyst [78]. Not only microorganisms, but also the bacteriophage MS2 can be inactivated using graphene–tungsten oxide composites under visible light [79]. Aside from the use of GNPs as semiconductor in photocatalytic systems, GNPs is also useful in electrochemical redox systems. The boron-doped diamond and rGO electrochemical oxidation system was found to be a superior tandem for generating hydroxyl radicals that inactivates 4.5 log of *E. coli* after only 5 min exposure [80]. These two examples show the potential use of GNPs in water disinfection.

In summary, the antimicrobial properties of GNPs can be utilized in three ways—the first is by directly using GNPs as antimicrobial agents, the second is by using GNPs as a nanometal scaffold, and the third is by using GNPs as electron conductors in photo- and electrocatalytic systems. The variety of these applications show the versatility of GNPs in environmental antimicrobial applications.

4.2 Removal of Organic Contaminants by Adsorption and Photodegradation

In addition to the unique antimicrobial properties of GNPs, these nanomaterials can also be used for the removal of organic contaminants. GNPs have been described to remove organic contaminants by adsorption and by photocatalytic activity. In the case of adsorption, the large surface areas of GNPs make them effective adsorbents for aromatic organic contaminants [81]. Aromatic hydrophobic compounds have extensive sp^2 structure, which allow π–π interactions with the hydrophobic graphene. For instance, aromatic organic contaminants, such as trichlorophenol, trichlorobenzene, and naphthalene can be captured by graphene from water [82]. Other hydrophobic polyaromatic hydrocarbon compounds, such as phenanthrene and biphenyl, have also been described to adsorb more efficiently to graphene than to multiwall carbon nanotubes [83]. Graphene exhibits three times higher adsorption capacity for these organic contaminants from water compared to multiwall carbon nanotubes [83]. Even when present as a coating, graphene is still effective in extracting hydrophobic organics from water. For instance, a silica gel coated with graphene is very effective in adsorbing organic pollutants, such as phenanthrene [84]. In a more extreme case of environmental chemical contamination, such as oil spill, a foam made of nickel coated with graphene has also been described to effectively separate oil from water [85].

Dyes are another class of aromatic organics that usually contain several fused aromatic rings and polar groups in their structure, which increase their water solubility. Given GO hydrophilicity, GO has been demonstrated to adsorb cationic dyes (methylene blue, methyl violet, and Rhodamine B) and anionic dye (Orange G) from water through electrostatic and hydrogen bonds, in addition to π–π interactions [86]. Acrydine orange, a powerful DNA intercalating agent [87], can be adsorbed by GO at a maximum adsorption capacity of 3300 mg/g [88]. Another highly hydrophilic environmental contaminant, the tetracycline antibiotic, is adsorbed effectively from water (adsorption capacity 313 mg/g) using GO [89]. GO functionalized with magnetic particles are also effective in the removal of other organic pollutants, such as phthalate esters, estriol, and the aromatic compound fluorene [90].

In the case of rGO, incomplete reduction of GO will provide residual hydroxyl groups that are important in the adsorption of the carcinogen Bisphenol A [91, 92], and hydrophobic pesticides, such as chlorpyrifos, endosulfan, and malathion. These pesticides are more effectively removed using the less functionalized rGO (adsorption capacities of 1200, 1100, and 800 mg/g, respectively) than GO. This best removal by rGO is due to a better balance between electrostatic and hydrophobic interactions, which can cause the rGO and the aromatic contaminants to precipitate out together in water [93].

In general, the effective removal of organic materials by GNPs is directly related to the surface properties of the particular GNP. The more hydrophobic GNPs, such as graphene, are more suitable for removing more hydrophobic organic compounds, i.e., polyaromatic hydrocarbons. The GO, with the presence of carbons in different oxidation states, is more effective in removing charged and hydrophilic organic compounds, which includes different antibiotics and dyes. GO offers H-bonding sites to these groups of organic compounds in addition to π–π interactions. The rGO, with its sparse functionality and being mostly hydrophobic, is effective in adsorption of pesticides with similar hydrophobicity to rGO. However, for all of these examples, most of the mechanisms of chemical removal are due to approximate interactions based on hydrophobicity/hydrophilicity of GNP and solutes. In the future, we hope to see development of methods to tailor GNP derivatives with precise functional group arrangements on the surface. These groups could tune the adsorption sites to remove specific contaminants. The summary of adsorption of organic contaminants from water is presented in Table 3.

Table 3 Removal of organic contaminants from water using GNPs

Organic compound	GNPs	Capacity (mg/g)	References
Aromatic compounds			
m-Dinitrobenzene	Reduced GO	265.7	[164]
Nitrobenzene	Reduced GO	260.9	[164]
p-Nitrotoluene	Reduced GO	238.8	[164]
Aniline	GO	115.1	[165]
Chlorobenzene	GO	67.2	[165]
Polyaromatic hydrocarbons			
Biphenyl	Graphene	54.6–61.4	[83]
Biphenyl	GO	44.8	[83]
Naphthalene	Graphene	127.7	[166]
Phenanthrene	Graphene	143.7–156.6	[83]
Phenanthrene	GO	163.9	[83]
Phenanthrene	Graphene-coated silica	0.775	[84]
Pyrene	Graphene	170.2	[166]
Pesticide residues			
Chlorpyrifos	Reduced GO	~1200	[93]
Diethylphthalate	GO–magnetic	8.71	[90]
Endosulfan	Reduced GO	1100	[93]
Malathion	Reduced GO	800	[93]
Endocrine disrupting chemicals			
Bisphenol A	Reduced GO	36.6	[91]
Diethylphthalate	GO–magnetic	8.71	[90]
Dyes			
Acridine orange	GO	3300	[87]
Methyl violet	GO	2.47	[86]
Methylene blue	GO	17.3	[86]
Orange G	Reduced GO	5.98	[86]
Rhodamine B	GO	1.24	[86]
Antibiotics			
Doxytetracycline	GO	398.4	[89]
Oxyteracycline	GO	212.3	[89]
Tetracycline	GO	313.5	[89]

In addition to adsorption, some GNP composites can also be used for photodegradation of organic pollutants. Recent studies have shown that the use of GNPs together with a semiconductor improve their photocatalytic activity [94]. For example, the photocatalytic activity of titanium dioxide (TiO_2) is enhanced when GNPs are added to the catalyst system. The activity of the photocatalyst is enhanced due to several factors [9]. The first one is the adsorption of the organic contaminant to the GNP, making the organic contaminant readily accessible for catalytic action [95]. The light absorption wavelength is also shifted by 30–40 nm, making the absorption of light more efficient [96, 97]. In addition, the presence of GNP in the system provides an additional electron acceptor that allows the excited electron to move easier through the conductive sp^2-hybridized carbon network [95]. The effectiveness of this photocatalytic activity is, however, directly related to the method of preparation of GNPs. For instance, graphene prepared using the catalyst-assisted CVD exhibits higher activity than graphene from direct exfoliation of graphite and solution processing of GO or rGO [98]. This could be due to higher electrical conductivity and larger interfacial area (higher surface roughness) of CVD graphene [98].

Across most of the systems studied, the photodegradation of organic compounds by addition of GNPs has been shown to present an enhanced performance. For instance, a graphene–TiO_2 system is four times more active than TiO_2 alone for the degradation of Rhodamine B [99]. Ag–TiO_2 photocatalyst wrapped in rGO degrades Bisphenol A with an efficiency of 62 %; a much higher value compared to 25 % efficiency of Ag–TiO_2 alone [100]. Other photocatalysts that have shown increased photocatalytic activity when composited with GNPs are tin oxide [96], cadmium sulfide [101], bismuth vanadate [102], vanadium oxide [103], silver halides [104], and silver phosphate [105]. These photocatalytic systems have also been shown to be effective in the degradation of organic chemicals present in water.

In summary, there are two main categories by which GNPs are employed in organic chemical removal in water. The first category is composed of methods based on the adsorptive efficiency of GNPs. The high surface area combined with presence of various functional groups on the surface of GNPs allows them to adsorb various classes of organic chemicals. The second category is based on actual degradation of organic contaminants due to photocatalytic action. The use of GNPs in photocatalytic systems increases their efficiency.

4.3 Removal of Cationic Heavy Metal Contaminants

Besides the removal of organic compounds, GNPs are extremely effective in the removal of cationic heavy metal contaminants. The removal of heavy metals can be achieved by GO and rGO nanoparticles, since the presence of functional groups on GO and rGO surfaces, such as carboxyl and hydroxyl groups, enhances their affinity for many types of metal ions [106]. On the other hand, the absence of functional groups in graphene nanoparticles makes graphene less attractive for heavy metal removal. Among all GNPs, GO is typically the most effective. For instance, divalent ions of copper, zinc, cadmium, and lead can be removed by GO with adsorption capacities as high as 300, 350, 530, and 1100 mg/g, respectively [107]. In the case of rGO, synthesized using acetyl acetone, the adsorption capacities of cadmium and cobalt are 50 and 28 mg, respectively. These results show the superior performance of GO compared to rGO.

The excellent adsorbent property of GO can be further enhanced by surface modifications. For instance, composites of β-cyclodextrin and GO can remove chromium(VI) from water, and at the same time, convert chromium(VI) into less toxic chromium(III) [108]. A chitosan–GO composite can recover gold(II) from aqueous solutions with adsorption capacity exceeding 1000 mg/g [109]. The attachment of ethylenediaminetetraacetic acid (EDTA) chelator on GO provides another mechanism for heavy metal removal, which now includes superior lead(II) and mercury(II) removal in its capabilities [110]. To recover these nanoparticles from water, it is possible to attach magnetite to GO and obtain a reusable EDTA–GO adsorbent [111]. Recoverable magnetite–GO composites can also be applied for the removal of cadmium(II) [112], cobalt(II) [112] and arsenic(III) and (V) [113]. Addition of thiol groups on the surface of GO increases the adsorption capacity of mercury(II) from 30 to 200 mg/g [114].

Besides using the nanoparticles for contaminant removal, other studies have conjugated GNPs to membranes or filters for water filtration and simultaneous removal of heavy metal contaminants, such as arsenic(III), arsenic(V) and lead(II), and pathogenic *E. coli* [115]. For instance, PVK–GO nanocomposites membrane filters (10 % GO content) can remove lead (II) from aqueous solution [116] and show antimicrobial activities against *E. coli* and *B. subtilis* with 3 and 4 log removals, respectively [59]. Another filter that was successfully modified with GO is a polysulfone (PSF) membrane geared for arsenic removal. Addition of 1 % GO to PSF while casting resulted into a PSF–GO membrane that is capable of rejecting twice the amount of arsenic compared to PSF-only membrane during filtration. In both cases, only a relatively small amount of GO was required in order to see enhancement of metal removal.

For some metals, the conversion of GO to rGO could also provide better adsorbents as their new derivative can now provide a new mechanism for removal, such as cation–π interaction and metal reduction. For example, polyvinylpyrrolidone–rGO (PVP–rGO) has an adsorption capacity for copper(II) of 1690 mg/g, which is more than ten times the adsorption capacity of copper by GO alone [117]. The polymer, in this case, serves as both reductant and dispersant. The better dispersion of rGO in the polymer matrix enhances the surface area of rGO, which improves cation–π interaction of the metal with the aromatic rings of rGO [117]. In another example, the GO was reduced by *S. oneidensis* to form rGO hydrogel. The rGO hydrogel reduced gold(III) and served as nucleation points to the recovered gold (0) nanoparticles [118]. The rGO hydrogel recovered gold with 99.98 % efficiency [118]. These examples show that not only electrostatic interaction and chelation are the mechanism by which GNPs remove heavy metals from water; specific interactions such as cation–π and redox between the metal contaminant and the GNPs are present as well.

Hence, these studies have shown that both GO and rGO by themselves or associated with polymers can effectively remove cationic heavy metals. The effective removal of heavy metals is linked to the presence of functional groups in GNPs, which serves as sites for metal adsorption. In addition, the same functional groups allow GNPs to be easily modifiable to enhance chemical removal and, in some cases, produce a more selective metal adsorbent. Representative studies on the removal of different metals by GNPs and composites are listed in Table 4.

Table 4 Removal of heavy contaminants from water using GNPs

Metals	GNP	Adsorption capacity (mg/g)	References
Ag(I)	Graphene–carbon nanotube	64	[167]
As(III)	rGO–magnetite	13.10	[168]
As(V)	rGO–magnetite	5.83	[168]
Au(II)	GO–chitosan	1000	[109]
Cd(II)	GO	530	[107]
Co(II)	GO	68.2	[112]
Co(II)	GO–magnetite	12.98	[169]
Cr(VI)	GO–cyclodextrin	120	[108]
Cu(II)	GO	300	[107]
Cu(II)	PVP–rGO	1690	[117]
Eu(II)	GO	175	[170]
Fe(III)	GO foam	587.6	[171]
Hg(II)	GO–thiol	200	[114]
Ni(II)	Reduced GO–MnO_2	42.9	[172]
Pb(II)	GO	1100	[107]
Pb(II)	GO–EDTA	479	[110]
Sb(II)	Reduced GO	10.92	[173]
U(VI)	GO	300	[170]
Zn(II)	GO	350	[107]

4.4 Graphene-Based Nanomaterials as Antioxidant and Coatings

In addition to the unique antimicrobial and chemical removal properties of GNPs, this group of nanomaterials also has antioxidant activity. This activity can be harnessed for environmental applications by encapsulation and coating surfaces with GNPs, or incorporating GNPs into polymer mixtures.

For instance, GO was found to be particularly effective against hydroxyl radicals when used as an encapsulating agent [119]. The GO acts as a physical barrier that traps the hydroxyl radicals produced by TiO_2 and prevents hydroxyl from diffusing to nearby organic targets [119]. In addition of acting as a barrier, the phenolic groups of GO can also react with hydroxyl radicals [120], however this pathway is limited due to relatively small amounts of phenolic groups on GO [119]. The rGO, on the other hand, is efficient in quenching superoxide radicals [121]. The excellent antioxidant activity of rGO is retained even when dispersed in a polypropylene (PP) composite matrix, thereby helping in the improvement of the composite properties with regard to thermo-oxidative stability [122]. The stabilization of the PP is due to free radical scavenging property of rGO during thermal oxidation. More importantly, rGO sheets act as a localized barrier that delays oxygen permeation [123]. Another composite of rGO, phosphomolybic acid–rGO composite, improves the flame retardant property of PP by enhancing radical trapping [124].

Besides encapsulation, coated metal surfaces with GNPs have been shown to serve as a barrier material and to enhance corrosion resistance. For example, oxidation was not observed on graphene-coated nickel after being subjected to a pressurized water reactor, suggesting that graphene can be an effective antioxidant coating [125]. Similarly, a copper surface coated with graphene (via CVD) prevented corrosion upon exposure to salt solution [126]. Electrochemical corrosion of copper and nickel surfaces were also prevented when graphene was deposited on their surfaces via CVD [127]. Another study, also showed that few layers of graphene deposited on a nickel surface, by mechanical transfer, reduced the rate of corrosion by 20 times [128]; while a carbon steel coated with graphene using laser presented 89 % less corrosion than its non-coated counterpart exposed to 3.5 % NaCl [129]. Other GNP-coated metal surfaces that inhibited corrosion include magnesium alloys [130] and copper/nickel alloy [131]. Besides pure GNPs, composites of GNPs were also found to be effective coating materials. Well-dispersed graphene in polyaniline was shown to effectively coat steel and protect against water and oxygen gas [132]. In addition to metal surfaces, GNPs can also be used to coat plastic substrates. For instance, spray coating rGO on polyethylene terephthalate (PET) films protected the polymer from degradation under exposure to corrosive acid, such as hydrofluoric acid [133].

The present section presented some of the most current environmental applications of GNPs. One of the major applications of GNP and GNP-based materials is environmental remediation, such as removal of heavy metals, organic compounds, and microorganisms. Another application of GNPs is protection against corrosion and degradation caused by chemicals or microorganisms. Even though the mechanism of action for some of these applications is not yet fully elucidated; we cannot ignore the contribution of GNPs in the formulation of the newly developed materials. In the future, we hope to better understand the properties of GNPs to be able to maximize their benefits.

5 Potential Environmental Impacts of GNPs

GNPs are promising materials for many industrial applications, ranging from medical, optics, electronics, and environmental applications. Like for many new materials and chemicals, caution on their utilization is necessary, since over the years, we have developed very important chemicals and materials that turned out to be extremely hazardous. For instance, the discovery of radioactive elements has led many creative and enterprising individuals to sell them to the general public as wonder substances capable of healing most maladies. As a result, elements like radium-226 were added from toothpaste to chocolates [134]. Soon enough, we learned that the use of radioactive materials could have lethal consequences. This was not an isolated incident in history. Another example is the pesticide DDT. This pesticide was the wonder chemical in the 1940s, until researchers figured out that this pesticide was responsible for making birds to lay soft egg shells, eventually thinning their population [135].

The burgeoning applications of GNPs and the unknown environmental and health effects of GNPs could lead to serious environmental impacts [136]. In the last few years, industries across the globe have revved up their engines in producing GNPs in massive scales [137]. Eventually, GNPs will be leashed into the environment as by-products and wastes from the manufacture of devices and materials containing them. The slow release from polymeric matrices or weathering of materials containing GNPs will also be a valid concern. Ultimately, the final disposal and resting place of GNPs will be landfills, wastewater treatment plants, and the environment. Contamination of the environment will be easily spread through water systems and runoff, as for example, GO has been shown to be very mobile in water when in the presence of naturally occurring ions and organic matter [138].

Once in the environment, GNPs might wreak havoc natural microbial populations. The GNPs have a broad-spectrum antimicrobial activity that is sometimes even broader than traditional antibiotics [139]. Experiments on the effects of GNPs on environmental microbial population are slowly beginning to appear and indicate that the natural microbial populations are affected. Studies have also shown that microorganisms are eventually able to cope with environmental stressors, however, it is unknown their tolerance and long term exposure effects of GNPs. Nanomaterials in general have negative impact on microbial communities on activated sludge [140]. Carbon-based nanoparticles is no exception. For instance, structurally similar carbon nanotubes and GO were proven to be toxic to microbes in activated sludge [23, 141].

Cytotoxicity of GNPs against animal and plants cells is evident and should be investigated carefully [24, 142, 143]. There are many other variables that are affecting the toxicity of GNPs in the environment. For instance, humic acids and potential biodegradation of GNPs have been shown to play major roles on the toxic effects of GNPs. The presence of humic acid could neutralize the effect of oxidative stress. This toxicity neutralization was observed with zebrafish [144]. Biodegradation is also potentially another mechanism to reduce the toxicity of GNPs. Although biodegradation studies of GNPs is still in its infancy, preliminary studies showed that biodegradation of more soluble and highly oxidized GNPs seems to occur [145]. More hydrophobic, less oxidized graphene, on the other hand, would probably persist in the environment, and settle at the bottom of bodies of water. In this case, benthic organisms could consume and lead to accumulation of GNPs in the food chain [146].

In soil, the effects of GNP are controversial. When graphene was mixed with soil-containing coriander and garlic, it enhanced their growth [147]. Different results were observed for cabbage, tomato, spinach and lettuce. In the latter plants, soil contaminated with GO decreased plant growth and damaged plant tissues due to oxidative stress [148].

Some studies, however, showed that GNPs could have positive effects to certain microbial populations in wastewater. The presence of GNPs has shown to help enhance the biodegradation capability of pollutants by microbial populations. For example, when rGO was incorporated into anaerobic sludge, the digestion of nitrobenzene was accelerated due rGO's electron conducting properties [149]. This and other evidences point to GNPs as useful and benign to the environment. However, little is known on the long-term fate of GNPs.

We are on the dawn of GNP devices era. It will not take longer until GNPs show up in water and soil ecosystems. Therefore, further research needs to be done to guide and help lawmakers and governmental agencies developing fair laws and guidelines in GNP utilization, transport, treatment, and disposal.

6 Conclusions

GNPs can be prepared in various ways depending on the desired characteristic of the final material required. They can be modified to suit a particular application. Because of the diversity of GNPs, some are more toxic than others. The toxicity typically depends on the synthesis and types of nanomaterials or nanocomposites, environmental chemistry, as well as the type of interactions GNPs have with biological systems [150]. The source of toxicity is usually a combination of at least two mechanisms, mechanical and chemical. The first can cause physical damage to cells, leading to cytoplasmic content leakage. The other involves the formation of ROS by GNPs, which reacts with cell components rendering them inactive [151]. The toxicity of GNPs is the major reason leading us to believe that certain GNPs can potentially impact the environment. It is not, however, our place to discredit GNPs due to their potential environmental impact. The main challenge is to find a middle ground between toxicity and application, where GNPs are useful but have minimum risk to the ecosystem.

References

1. Wu J, Pisula W, Müllen K (2007) Graphenes as potential material for electronics. Chem Rev 107(3):718–747
2. Miller RD, Chandross EA (2010) Introduction: materials for electronics. Chem Rev 110 (1):1–2
3. Lee C et al (2008) Measurement of the elastic properties and intrinsic strength of monolayer graphene. Science 321(5887):385–388
4. Balandin AA et al (2008) Superior thermal conductivity of single-layer graphene. Nano Lett 8(3):902–907
5. Hwang EH, Das Sarma S (2008) Acoustic phonon scattering limited carrier mobility in two-dimensional extrinsic graphene. Phys Rev B 77(11):115449
6. Stoller MD et al (2008) Graphene-based ultracapacitors. Nano Lett 8(10):3498–3502
7. Gao N, Fang X (2015) Synthesis and development of graphene–inorganic semiconductor nanocomposites. Chem Rev 115(16):8294–8343
8. Mao HY et al (2013) Graphene: promises, facts, opportunities, and challenges in nanomedicine. Chem Rev 113(5):3407–3424

9. Perreault F, Fonseca de Faria A, Elimelech M (2015) Environmental applications of graphene-based nanomaterials. Chem Soc Rev 44(16):5861–5896
10. Liu S et al (2011) Antibacterial activity of graphite, graphite oxide, graphene oxide, and reduced graphene oxide: membrane and oxidative stress. ACS Nano 5(9):6971–6980
11. Navalon S et al (2014) Carbocatalysis by graphene-based materials. Chem Rev 114 (12):6179–6212
12. Dale AL et al (2015) Modeling nanomaterial environmental fate in aquatic systems. Environ Sci Technol 49(5):2587–2593
13. Jung S-K et al (2015) Multi-endpoint, high-throughput study of nanomaterial toxicity in *Caenorhabditis elegans*. Environ Sci Technol 49(4):2477–2485
14. Catherine MS et al (2012) Graphene nanocomposite for biomedical applications: fabrication, antimicrobial and cytotoxic investigations. Nanotechnology 23(39):395101
15. Kurantowicz N et al (2015) Interaction of graphene family materials with *Listeria monocytogenes* and *Salmonella enterica*. Nanoscale Res Lett 10:23
16. Bansal P et al (2015) Exoelectrogens leading to precise reduction of graphene oxide by flexibly switching their environment during respiration. ACS Appl Mater Interfaces 7 (37):20576–20584
17. Ristic BZ et al (2014) Photodynamic antibacterial effect of graphene quantum dots. Biomaterials 35(15):4428–4435
18. Efremova LV et al (2015) Toxicity of graphene shells, graphene oxide, and graphene oxide paper evaluated with *Escherichia coli* biotests. BioMed Res Int 2015:10
19. Markovic ZM et al (2012) Graphene quantum dots as autophagy-inducing photodynamic agents. Biomaterials 33(29):7084–7092
20. Chen J et al (2015) Antibacterial activity of graphene-modified anode on *Shewanella oneidensis* MR-1 biofilm in microbial fuel cell. J Power Sources 290:80–86
21. Kim H, Abdala AA, Macosko CW (2010) Graphene/polymer nanocomposites. Macromolecules 43(16):6515–6530
22. Gurunathan S et al (2012) Oxidative stress-mediated antibacterial activity of graphene oxide and reduced graphene oxide in *Pseudomonas aeruginosa*. Int J Nanomed 7:5901–5914
23. Ahmed F, Rodrigues DF (2013) Investigation of acute effects of graphene oxide on wastewater microbial community: a case study. J Hazard Mater 256–257:33–39
24. Mejias Carpio IE et al (2012) Toxicity of a polymer–graphene oxide composite against bacterial planktonic cells, biofilms, and mammalian cells. Nanoscale 4(15):4746–4756
25. Lee J et al (2013) Graphene oxide nanoplatelets composite membrane with hydrophilic and antifouling properties for wastewater treatment. J Membr Sci 448:223–230
26. Camden AN, Barr SA, Berry RJ (2013) Simulations of peptide–graphene interactions in explicit water. J Phys Chem B 117(37):10691–10697
27. Cao Y-C et al (2015) The preparation of graphene reinforced poly(vinyl alcohol) antibacterial nanocomposite thin film. Int J Polym Sci 2015(407043):1–7
28. Kumar D et al (2015) Microwave-assisted synthesis, characterization of reduced graphene oxide, and its antibacterial activity. Bull Korean Chem Soc 36(8):2034–2038
29. Akhavan O, Ghaderi E (2010) Toxicity of graphene and graphene oxide nanowalls against bacteria. ACS Nano 4(10):5731–5736
30. Akhavan O, Ghaderi E (2010) Toxicity of graphene and graphene oxide nanowalls against bacteria. ACS Nano 4(10):5731–5736
31. Gurunathan S et al (2013) Antibacterial activity of dithiothreitol reduced graphene oxide. J Ind Eng Chem 19(4):1280–1288
32. Bora C et al (2013) Strong and conductive reduced graphene oxide/polyester resin composite films with improved mechanical strength, thermal stability and its antibacterial activity. Compos Sci Technol 87:1–7
33. Pan Y, Sahoo NG, Li L (2012) The application of graphene oxide in drug delivery. Expert Opin Drug Deliv 9(11):1365–1376

34. Cai X et al (2012) The use of polyethyleneimine-modified reduced graphene oxide as a substrate for silver nanoparticles to produce a material with lower cytotoxicity and long-term antibacterial activity. Carbon 50(10):3407–3415
35. An X et al (2013) Graphene oxide reinforced polylactic acid/polyurethane antibacterial composites. J Nanomater 2013:18
36. Chieng BW et al (2015) Reinforcement of graphene nanoplatelets on plasticized poly(lactic acid) nanocomposites: mechanical, thermal, morphology, and antibacterial properties. J Appl Polym Sci 132(11): n/a–n/a
37. Zhou X, Shi T, Zhou H (2012) Hydrothermal preparation of ZnO-reduced graphene oxide hybrid with high performance in photocatalytic degradation. Appl Surf Sci 258(17): 6204–6211
38. Wang Y-W et al (2014) Superior antibacterial activity of zinc oxide/graphene oxide composites originating from high zinc concentration localized around bacteria. ACS Appl Mater Interfaces 6(4):2791–2798
39. Kavitha T et al (2012) Glucose sensing, photocatalytic and antibacterial properties of graphene–ZnO nanoparticle hybrids. Carbon 50(8):2994–3000
40. Akhavan O, Ghaderi E (2009) Photocatalytic reduction of graphene oxide nanosheets on TiO_2 thin film for photoinactivation of bacteria in solar light irradiation. J Phys Chem C 113 (47):20214–20220
41. Chang Y-N et al (2015) Synthesis of magnetic graphene oxide–TiO_2 and their antibacterial properties under solar irradiation. Appl Surf Sci 343:1–10
42. Shao W et al (2015) Preparation, characterization, and antibacterial activity of silver nanoparticle-decorated graphene oxide nanocomposite. ACS Appl Mater Interfaces 7 (12):6966–6973
43. Xu W-P et al (2011) Facile synthesis of silver@ graphene oxide nanocomposites and their enhanced antibacterial properties. J Mater Chem 21(12):4593–4597
44. Kalishwaralal K et al (2010) Silver nanoparticles impede the biofilm formation by *Pseudomonas aeruginosa* and *Staphylococcus epidermidis*. Colloids Surf B 79(2):340–344
45. Jastrzębska AM, Kurtycz P, Olszyna AR (2012) Recent advances in graphene family materials toxicity investigations. J Nanopart Res 14(12):1–21
46. Guo R, Mao J, Yan L-T (2013) Computer simulation of cell entry of graphene nanosheet. Biomaterials 34(17):4296–4301
47. Titov AV, Král P, Pearson R (2010) Sandwiched graphene—membrane superstructures. ACS Nano 4(1):229–234
48. Upadhyay RK, Soin N, Roy SS (2014) Role of graphene/metal oxide composites as photocatalysts, adsorbents and disinfectants in water treatment: a review. RSC Adv 4 (8):3823–3851
49. Wu C et al (2015) Vacuolization in cytoplasm and cell membrane permeability enhancement triggered by micrometer-sized graphene oxide. ACS Nano 9(8):7913–7924
50. Tu Y et al (2013) Destructive extraction of phospholipids from *Escherichia coli* membranes by graphene nanosheets. Nat Nano 8(8):594–601
51. Liu S et al (2012) Lateral dimension-dependent antibacterial activity of graphene oxide sheets. Langmuir 28(33):12364–12372
52. Lushchak VI (2014) Free radicals, reactive oxygen species, oxidative stress and its classification. Chem Biol Interact 224:164–175
53. Vallyathan V, Shi X (1997) The role of oxygen free radicals in occupational and environmental lung diseases. Environ Health Perspect 105(Suppl 1):165
54. Thannickal VJ, Fanburg BL (2000) Reactive oxygen species in cell signaling. Am J Physiol Lung Cell Mol Physiol 279(6):L1005–L1028
55. Kohen R, Nyska A (2002) Invited review: oxidation of biological systems: oxidative stress phenomena, antioxidants, redox reactions, and methods for their quantification. Toxicol Pathol 30(6):620–650
56. Finkel T, Holbrook NJ (2000) Oxidants, oxidative stress and the biology of ageing. Nature 408(6809):239–247

57. Su C et al (2012) Probing the catalytic activity of porous graphene oxide and the origin of this behaviour. Nat Commun 3:1298
58. Zhang W et al (2012) Unraveling stress-induced toxicity properties of graphene oxide and the underlying mechanism. Adv Mater 24(39):5391–5397
59. Musico YLF et al (2014) Surface modification of membrane filters using graphene and graphene oxide-based nanomaterials for bacterial inactivation and removal. ACS Sustain Chem Eng 2(7):1559–1565
60. Rana S et al (2010) Electron paramagnetic resonance spectroscopy in radiation research: current status and perspectives. J Pharm Bioallied Sci 2(2):80
61. Applerot G et al (2012) Understanding the antibacterial mechanism of CuO nanoparticles: revealing the route of induced oxidative stress. Small 8(21):3326–3337
62. Hu X et al (2015) Effects of graphene oxide and oxidized carbon nanotubes on the cellular division, microstructure, uptake, oxidative stress, and metabolic profiles. Environ Sci Technol 49(18):10825–10833
63. Apel K, Hirt H (2004) Reactive oxygen species: metabolism, oxidative stress, and signal transduction. Annu Rev Plant Biol 55:373–399
64. Amanda JA, David IG, Mark WS (1998) Use of a new tetrazolium-based assay to study the production of superoxide radicals by tobacco cell cultures challenged with avirulent zoospores of *Phytophthora parasitica* var *nicotianae*. Plant Physiol 117(2):491–499
65. Krishnamurthy A et al (2013) Passivation of microbial corrosion using a graphene coating. Carbon 56:45–49
66. Li J et al (2014) Antibacterial activity of large-area monolayer graphene film manipulated by charge transfer. Sci Rep 4:4359
67. Major R et al (2014) Graphene based porous coatings with antibacterial and antithrombogenous function—materials and design. Arch Civil Mech Eng 14(4):540–549
68. Mangadlao JD et al (2015) On the antibacterial mechanism of graphene oxide (GO) Langmuir–Blodgett films. Chem Commun 51(14):2886–2889
69. Yuan H, He Z (2015) Graphene-modified electrodes for enhancing the performance of microbial fuel cells. Nanoscale 7(16):7022–7029
70. Rabaey K, Verstraete W (2005) Microbial fuel cells: novel biotechnology for energy generation. Trends Biotechnol 23(6):291–298
71. Santos CM et al (2011) Antimicrobial graphene polymer (PVK–GO) nanocomposite films. Chem Commun 47(31):8892–8894
72. Mazaheri M, Akhavan O, Simchi A (2014) Flexible bactericidal graphene oxide–chitosan layers for stem cell proliferation. Appl Surf Sci 301:456–462
73. Goncalves G et al (2009) Surface modification of graphene nanosheets with gold nanoparticles: the role of oxygen moieties at graphene surface on gold nucleation and growth. Chem Mater 21(20):4796–4802
74. Hussain N et al (2014) Reduced graphene oxide nanosheets decorated with au nanoparticles as an effective bactericide: investigation of biocompatibility and leakage of sugars and proteins. ChemPlusChem 79(12):1774–1784
75. Zeng X et al (2015) Silver/reduced graphene oxide hydrogel as novel bactericidal filter for point-of-use water disinfection. Adv Funct Mater 25(27):4344–4351
76. Tedesco S et al (2010) Oxidative stress and toxicity of gold nanoparticles in *Mytilus edulis*. Aquat Toxicol 100(2):178–186
77. Akhavan O, Ghaderi E, Rahimi K (2012) Adverse effects of graphene incorporated in TiO_2 photocatalyst on minuscule animals under solar light irradiation. J Mater Chem 22 (43):23260–23266
78. He W et al (2013) Photocatalytic and antibacterial properties of Au–TiO_2 nanocomposite on monolayer graphene: from experiment to theory. J Appl Phys 114(20)
79. Akhavan O, Choobtashani M, Ghaderi E (2012) Protein degradation and RNA efflux of viruses photocatalyzed by graphene–tungsten oxide composite under visible light irradiation. J Phys Chem C 116(17):9653–9659

80. Qi X et al (2015) Synergetic antibacterial activity of reduced graphene oxide and boron doped diamond anode in three dimensional electrochemical oxidation system. Sci Rep 5:10388
81. Smith SC, Rodrigues DF (2015) Carbon-based nanomaterials for removal of chemical and biological contaminants from water: a review of mechanisms and applications. Carbon 91:122–143
82. Pei Z et al (2013) Adsorption characteristics of 1,2,4-trichlorobenzene, 2,4,6-trichlorophenol, 2-naphthol and naphthalene on graphene and graphene oxide. Carbon 51:156–163
83. Apul OG et al (2013) Adsorption of aromatic organic contaminants by graphene nanosheets: comparison with carbon nanotubes and activated carbon. Water Res 47(4):1648–1654
84. Yang K, Chen B, Zhu L (2015) Graphene-coated materials using silica particles as a framework for highly efficient removal of aromatic pollutants in water. Sci Rep 5:11641
85. Bong J et al (2015) Dynamic graphene filters for selective gas–water–oil separation. Sci Rep 5:14321
86. Ramesha GK et al (2011) Graphene and graphene oxide as effective adsorbents toward anionic and cationic dyes. J Colloid Interface Sci 361(1):270–277
87. Lenglet G et al (2010) DNA-destabilizing agents as an alternative approach for targeting DNA: mechanisms of action and cellular consequences. J Nucleic Acids
88. Sun L, Yu H, Fugetsu B (2012) Graphene oxide adsorption enhanced by in situ reduction with sodium hydrosulfite to remove acridine orange from aqueous solution. J Hazard Mater 203–204:101–110
89. Gao Y et al (2012) Adsorption and removal of tetracycline antibiotics from aqueous solution by graphene oxide. J Colloid Interface Sci 368(1):540–546
90. Yin L, Lin Y, Jia L (2014) Graphene oxide functionalized magnetic nanoparticles as adsorbents for removal of phthalate esters. Microchim Acta 181(9–10):957–965
91. Xu J, Wang L, Zhu Y (2012) Decontamination of bisphenol A from aqueous solution by graphene adsorption. Langmuir 28(22):8418–8425
92. Michałowicz J (2014) Bisphenol A—sources, toxicity and biotransformation. Environ Toxicol Pharmacol 37(2):738–758
93. Maliyekkal SM et al (2013) Graphene: a reusable substrate for unprecedented adsorption of pesticides. Small 9(2):273–283
94. Zhang N et al (2015) Waltzing with the versatile platform of graphene to synthesize composite photocatalysts. Chem Rev 115(18):10307–10377
95. Perera SD et al (2012) Hydrothermal synthesis of graphene–TiO_2 nanotube composites with enhanced photocatalytic activity. ACS Catal 2(6):949–956
96. Zhang J, Xiong Z, Zhao XS (2011) Graphene–metal-oxide composites for the degradation of dyes under visible light irradiation. J Mater Chem 21(11):3634–3640
97. Liu J et al (2010) Self-assembling TiO_2 nanorods on large graphene oxide sheets at a two-phase interface and their anti-recombination in photocatalytic applications. Adv Funct Mater 20(23):4175–4181
98. Sellappan R et al (2013) Influence of graphene synthesizing techniques on the photocatalytic performance of graphene–TiO_2 nanocomposites. Phys Chem Chem Phys 15(37):15528–15537
99. Liang Y et al (2010) TiO_2 nanocrystals grown on graphene as advanced photocatalytic hybrid materials. Nano Res 3(10):701–705
100. Leong KH et al (2015) Reduced graphene oxide and Ag wrapped TiO_2 photocatalyst for enhanced visible light photocatalysis. APL Mater 3(10):104503–104503
101. Khan Z et al (2012) Visible light assisted photocatalytic hydrogen generation and organic dye degradation by CdS–metal oxide hybrids in presence of graphene oxide. RSC Adv 2 (32):12122–12128
102. Wang Y et al (2014) Electrostatic self-assembly of $BiVO_4$–reduced graphene oxide nanocomposites for highly efficient visible light photocatalytic activities. ACS Appl Mater Interfaces 6(15):12698–12706

103. Shanmugam M et al (2015) Enhanced photocatalytic performance of the graphene–V_2O_5 nanocomposite in the degradation of methylene blue dye under direct sunlight. ACS Appl Mater Interfaces 7(27):14905–14911
104. Zhu M, Chen P, Liu M (2011) Graphene oxide enwrapped Ag/AgX (X = Br, Cl) nanocomposite as a highly efficient visible-light plasmonic photocatalyst. ACS Nano 5 (6):4529–4536
105. Yang X et al (2013) Fabrication of Ag_3PO_4–graphene composites with highly efficient and stable visible light photocatalytic performance. ACS Catal 3(3):363–369
106. Yusuf M et al (2015) Applications of graphene and its derivatives as an adsorbent for heavy metal and dye removal: a systematic and comprehensive overview. RSC Adv 5(62):50392–50420
107. Sitko R et al (2013) Adsorption of divalent metal ions from aqueous solutions using graphene oxide. Dalton Trans 42(16):5682–5689
108. Fan L et al (2012) Synthesis of graphene oxide decorated with magnetic cyclodextrin for fast chromium removal. J Mater Chem 22(47):24577–24583
109. Liu L et al (2012) Preparation and characterization of chitosan/graphene oxide composites for the adsorption of Au(III) and Pd(II). Talanta 93:350–357
110. Madadrang CJ et al (2012) Adsorption behavior of EDTA-graphene oxide for Pb(II) removal. ACS Appl Mater Interfaces 4(3):1186–1193
111. Cui L et al (2015) EDTA functionalized magnetic graphene oxide for removal of Pb(II), Hg (II) and Cu(II) in water treatment: adsorption mechanism and separation property. Chem Eng J 281:1–10
112. Zhao G et al (2011) Few-layered graphene oxide nanosheets as superior sorbents for heavy metal ion pollution management. Environ Sci Technol 45(24):10454–10462
113. Luo X et al (2012) Adsorption of As(III) and As(V) from water using magnetite Fe_3O_4-reduced graphite oxide–MnO_2 nanocomposites. Chem Eng J 187:45–52
114. Gao W et al (2011) Engineered graphite oxide materials for application in water purification. ACS Appl Mater Interfaces 3(6):1821–1826
115. Viraka Nellore BP et al (2015) Bio-conjugated CNT-bridged 3D porous graphene oxide membrane for highly efficient disinfection of pathogenic bacteria and removal of toxic metals from water. ACS Appl Mater Interfaces 7(34):19210–19218
116. Musico YLF et al (2013) Improved removal of lead(II) from water using a polymer-based graphene oxide nanocomposite. J Mater Chem A 1(11):3789–3796
117. Zhang Y et al (2014) Highly efficient adsorption of copper ions by a PVP-reduced graphene oxide based on a new adsorptions mechanism. Nano-Micro Lett 6(1):80–87
118. He Y-R et al (2015) A green approach to recover Au(III) in aqueous solution using biologically assembled rGO hydrogels. Chem Eng J 270:476–484
119. Qiu Y et al (2014) Antioxidant chemistry of graphene-based materials and its role in oxidation protection technology. Nanoscale 6(20):11744–11755
120. Lipinski B (2011) Hydroxyl radical and its scavengers in health and disease. Oxid Med Cell Longev
121. Zhang C et al (2015) Reduced graphene oxide enhances horseradish peroxidase stability by serving as radical scavenger and redox mediator. Carbon 94:531–538
122. Yang J et al (2015) The synergistic mechanism of thermally reduced graphene oxide and antioxidant in improving the thermo-oxidative stability of polypropylene. Carbon 89:340–349
123. Yang J et al (2013) The intrinsic thermal-oxidative stabilization effect of chemically reduced graphene oxide on polypropylene. J Mater Chem A 1(37):11184–11191
124. Yuan B et al (2015) Solid acid-reduced graphene oxide nanohybrid for enhancing thermal stability, mechanical property and flame retardancy of polypropylene. RSC Adv 5 (51):41307–41316
125. Ming H et al (2014) Multilayer graphene: a potential anti-oxidation barrier in simulated primary water. J Mater Sci Technol 30(11):1084–1087
126. Singh Raman RK et al (2012) Protecting copper from electrochemical degradation by graphene coating. Carbon 50(11):4040–4045

127. Kirkland NT et al (2012) Exploring graphene as a corrosion protection barrier. Corros Sci 56:1–4
128. Prasai D et al (2012) Graphene: corrosion-inhibiting coating. ACS Nano 6(2):1102–1108
129. Ye X et al (2015) Protecting carbon steel from corrosion by laser in situ grown graphene films. Carbon 94:326–334
130. Qiu Z et al (2015) Graphene oxide as a corrosion-inhibitive coating on magnesium alloys. RSC Adv 5(55):44149–44159
131. Chen S et al (2011) Oxidation resistance of graphene-coated Cu and Cu/Ni Alloy. ACS Nano 5(2):1321–1327
132. Chang C-H et al (2012) Novel anticorrosion coatings prepared from polyaniline/graphene composites. Carbon 50(14):5044–5051
133. Su Y et al (2014) Impermeable barrier films and protective coatings based on reduced graphene oxide. Nat Commun 5
134. Harvie DI (1999) The radium century. Endeavour 23(3):100–105
135. Bouwman H et al (2013) DDT: fifty years since. Silent Spring
136. Smith SC, Rodrigues DFR (2013) The fate of carbon-based nanomaterials in the environment. J Bioremed Biodeg 3(1):2155–6199
137. Zurutuza A, Marinelli C (2014) Challenges and opportunities in graphene commercialization. Nat Nano 9(10):730–734
138. Lanphere JD et al (2014) Stability and transport of graphene oxide nanoparticles in groundwater and surface water. Environ Eng Sci 31(7):350–359
139. Krishnamoorthy K et al (2012) Antibacterial efficiency of graphene nanosheets against pathogenic bacteria via lipid peroxidation. J Phys Chem C 116(32):17280–17287
140. Brar SK et al (2010) Engineered nanoparticles in wastewater and wastewater sludge—evidence and impacts. Waste Manag 30(3):504–520
141. Rodrigues DF, Elimelech M (2010) Toxic effects of single-walled carbon nanotubes in the development of *E. coli* biofilm. Environ Sci Technol 44(12):4583–4589
142. Oleszczuk P, Jośko I, Xing B (2011) The toxicity to plants of the sewage sludges containing multiwalled carbon nanotubes. J Hazard Mater 186(1):436–442
143. Hu X et al (2014) Interactions between graphene oxide and plant cells: regulation of cell morphology, uptake, organelle damage, oxidative effects and metabolic disorders. Carbon 80 (2014):665–676
144. Chen Y et al (2015) Mitigation in multiple effects of graphene oxide toxicity in zebrafish embryogenesis driven by humic acid. Environ Sci Technol 49(16):10147–10154
145. Kurapati R et al (2015) Dispersibility-dependent biodegradation of graphene oxide by myeloperoxidase. Small 11(32):3985–3994
146. Jackson P et al (2013) Bioaccumulation and ecotoxicity of carbon nanotubes. Chem Cent J 7:154
147. Chakravarty D, Erande MB, Late DJ (2015) Graphene quantum dots as enhanced plant growth regulators: effects on coriander and garlic plants. J Sci Food Agric 95(13):2772–2778
148. Begum P, Ikhtiari R, Fugetsu B (2011) Graphene phytotoxicity in the seedling stage of cabbage, tomato, red spinach, and lettuce. Carbon 49(12):3907–3919
149. Wang J et al (2016) Effect of bioreduced graphene oxide on anaerobic biotransformation of nitrobenzene in an anaerobic reactor. Environ Technol 37(1):39–45
150. Hu X, Zhou Q (2013) Health and ecosystem risks of graphene. Chem Rev 113(5):3815–3835
151. Dreyer DR et al (2010) The chemistry of graphene oxide. Chem Soc Rev 39(1):228–240
152. He J et al (2015) Killing dental pathogens using antibacterial graphene oxide. ACS Appl Mater Interfaces 7(9):5605–5611
153. Kellici S et al (2014) A single rapid route for the synthesis of reduced graphene oxide with antibacterial activities. RSC Adv 4(29):14858–14861
154. Cao B et al (2013) High antibacterial activity of ultrafine TiO_2/graphene sheets nanocomposites under visible light irradiation. Mater Lett 93:349–352

155. Kim IY et al (2014) Strongly-coupled freestanding hybrid films of graphene and layered titanate nanosheets: an effective way to tailor the physicochemical and antibacterial properties of graphene film. Adv Funct Mater 24(16):2288–2294
156. Ning W et al (2015) Polyethylenimine mediated silver nanoparticle-decorated magnetic graphene as a promising photothermal antibacterial agent. Nanotechnology 26(19):195703–195703
157. Nguyen VH et al (2012) Preparation and antibacterial activity of silver nanoparticles-decorated graphene composites. J Supercrit Fluids 72:28–35
158. Shen J et al (2012) Polyelectrolyte-assisted one-step hydrothermal synthesis of Ag-reduced graphene oxide composite and its antibacterial properties. Mater Sci Eng C 32(2012):2042–2047
159. Cao Y-C et al (2015) The preparation of graphene reinforced poly(vinyl alcohol) antibacterial nanocomposite thin film. Int J Polym Sci 2015:1–7
160. Alsharaeh E et al (2016) Novel route for the preparation of cobalt oxide nanoparticles/reduced graphene oxide nanocomposites and their antibacterial activities. Ceram Int 42(2, Part B):3407–3410
161. Santhosh C et al (2014) Adsorption, photodegradation and antibacterial study of graphene–Fe_3O_4 nanocomposite for multipurpose water purification application. RSC Adv 4 (54):28300–28308
162. Duan L et al (2015) Graphene immobilized enzyme/polyethersulfone mixed matrix membrane: enhanced antibacterial, permeable and mechanical properties. Appl Surf Sci 355:436–445
163. Li P et al (2015) Developing of a novel antibacterial agent by functionalization of graphene oxide with guanidine polymer with enhanced antibacterial activity. Appl Surf Sci 355:446–452
164. Chen X, Chen B (2015) Macroscopic and spectroscopic investigations of the adsorption of nitroaromatic compounds on graphene oxide, reduced graphene oxide, and graphene nanosheets. Environ Sci Technol 49(10):6181–6189
165. Yan H et al (2015) Influence of the surface structure of graphene oxide on the adsorption of aromatic organic compounds from water. ACS Appl Mater Interfaces 7(12):6690–6697
166. Wang J, Chen Z, Chen B (2014) Adsorption of polycyclic aromatic hydrocarbons by graphene and graphene oxide nanosheets. Environ Sci Technol 48(9):4817–4825
167. Sui Z et al (2012) Green synthesis of carbon nanotube-graphene hybrid aerogels and their use as versatile agents for water purification. J Mater Chem 22(18):8767–8771
168. Chandra V et al (2010) Water-dispersible magnetite-reduced graphene oxide composites for arsenic removal. ACS Nano 4(7):3979–3986
169. Liu M et al (2011) Synthesis of magnetite/graphene oxide composite and application for Cobalt(II) removal. The Journal of Physical Chemistry C 115(51):25234–25240
170. Sun Y et al (2012) Interaction between Eu(III) and graphene oxide nanosheets investigated by batch and extended X-ray absorption fine structure spectroscopy and by modeling techniques. Environ Sci Technol 46(11):6020–6027
171. Lei Y et al (2014) Synthesis of three-dimensional graphene oxide foam for the removal of heavy metal ions. Chem Phys Lett 593:122–127
172. Ren Y et al (2011) Graphene/δ-MnO_2 composite as adsorbent for the removal of nickel ions from wastewater. Chem Eng J 175:1–7
173. Leng Y et al (2012) Removal of antimony(III) from aqueous solution by graphene as an adsorbent. Chem Eng J 211–212:406–411

MIX
Papier aus verantwortungsvollen Quellen
Paper from responsible sources
FSC® C105338

If you have any concerns about our products,
you can contact us on
ProductSafety@springernature.com

In case Publisher is established outside the EU,
the EU authorized representative is:
Springer Nature Customer Service Center GmbH
Europaplatz 3, 69115 Heidelberg, Germany

Printed by Libri Plureos GmbH
in Hamburg, Germany